HANDBOOK OF DEVELOPMENTAL SYSTEMS THEORY AND METHODOLOGY

Handbook of Developmental Systems Theory and Methodology

Edited by

Peter C. M. Molenaar
Richard M. Lerner
Karl M. Newell

THE GUILFORD PRESS
New York London

A Division of Guilford Publications, Inc.
72 Spring Street, New York, NY 10012
www.guilford.com

Printed in the United States of America

This book is printed on acid-free paper.

Last digit is print number: 9 8 7 6 5 4 3 2 1

Library of Congress Cataloging-in-Publication Data

Handbook of developmental systems theory and methodology / edited by Peter C. M. Molenaar, Richard M. Lerner, Karl M. Newell.
pages cm
Includes bibliographical references and index.
ISBN 978-1-60918-509-1 (hardcover)
1. Developmental psychology. 2. Developmental psychobiology. 3. Systems theory. I. Molenaar, Peter C. M. II. Lerner, Richard M. III. Newell, Karl M., 1945–
BF713.H3645 2014
155—dc23

2013019533

Preface

The chapters in this volume were derived from the Conference on Inductive Developmental Systems Theory that took place at The Pennsylvania State University on October 31–November 2, 2011. The conference brought together leading scientists in the fields of developmental psychology, biological systems theory, econometrics, systems science, and psychometrics to frame the theoretical, experimental, and methodological integrations we seek to elucidate. The contributions made at the conference have evolved into the chapters in this handbook. Together, these chapters address the integration of diverse strands of specialized systems theory and simulation methodology with innovative statistical modeling techniques into an effective research agenda for the study of development.

Consequently, this handbook can be conceived as providing a roadmap for integrating key theoretical concepts of developmental systems theory (e.g., self-organization, reciprocal dynamic interaction, probabilistic epigenesis) and simulation models (e.g., connectionistic and agent-based models) into advanced statistical dynamic modeling approaches—approaches that are deployed in empirical tests using innovative design and data acquisition techniques. In addition, the chapters lay the groundwork for further extensions of integrative work that are aimed at the creation of optimal guidance for developmental and disease processes, involving subject- and patient-specific dynamic systems approaches (adaptive guidance in real time). A general goal of this handbook is, then, to advance the innovative integration of systems approaches to development with state-of-the-art statistical dynamic modeling tools.

In short, we believe that this handbook makes a timely and important contribution pertinent to the need to fully integrate developmental systems theory and innovative inductive statistical dynamic systems methodology. We hope that this contribution will advance significantly the theoretical and methodological tools that developmental science can use effectively to better describe, explain, and optimize the individual developmental trajectories of diverse people across the life span.

There are numerous people to thank in regard to the preparation of this handbook. First and foremost, we are indebted to our contributing authors. Their scholarship and dedication to excellence enabled this work to be produced. We express our appreciation also to The Guilford Press for the generous funding of the Penn State conference and for the support of our work that was unflaggingly provided by our editor at Guilford, C. Deborah Laughton. We thank Susan LeWay, Director of Outreach at the College of Health and Human Development of Penn State, for her many contributions to the organization of the conference. We are appreciative as well for the support of several programs within Penn State. Building on the longstanding strong history of statistical-methodological research at Penn State, the following programs contributed substantial funding to the conference: the Social Science Research Institute, the Sol Kramer Foundation, and the Department of Human Development and Family Studies.

Peter C. M. Molenaar was supported by Grant No. 1157220 from the National Science Foundation; Richard M. Lerner by the John Templeton Foundation, the National 4-H Council, and the Thrive Foundation for Youth; and Karl M. Newell by Grant No. 0848339 from the National Science Foundation.

We are especially grateful to Jarrett M. Lerner, Managing Editor at the Institute for Applied Research at Tufts University, for his superb editorial work. His commitment to quality and productivity, his efficiency and judgment, and his resilience in the face of the tribulations of manuscript production are greatly admired and deeply appreciated. The excellence of his editorial skills contributed mightily to the cohesiveness and clarity of this handbook.

Finally, we thank our respective families for their love and support during the development and completion of this project. We dedicate this work to them.

P. C. M. M.
R. M. L.
K. M. N.

Contents

PART VII. NONLINEAR DYNAMICAL MODELS OF DEVELOPMENT

PART VIII. NONERGODIC DEVELOPMENTAL SYSTEMS

PART IX. COMPLEX SYSTEMS MODELS IN HUMAN DEVELOPMENT: REEVALUATION AND FUTURE DIRECTIONS

PART I

INTRODUCTION

CHAPTER 1

Developmental Systems Theory and Methodology

A View of the Issues

Peter C. M. Molenaar, Richard M. Lerner, and Karl M. Newell

There exists a long tradition in theoretical psychology and theoretical biology in which developmental processes are explained as the result of self-organizing processes with emergent properties that have complex, dynamic interactions with environmental influences. The general denotation for this tradition is *developmental systems theory*. Important contributions to developmental systems theory include Ford and Lerner's (1992) integrative approach, based on the interplay between intraindividual variation and interindividual variation and change; Gottlieb's (1992, 2003) theoretical work on probabilistic epigenetic development; and Overton's discussions of relational metatheory and relational developmental systems theories (RDST; e.g., Overton, 2010, 2012; Overton & Müller, 2012).

Oyama, Griffiths, and Gray (2001) present a compilation of contributions of developmental systems theory to theoretical biology and, in turn, both the 1998 and the 2006 editions of the *Handbook of Child Psychology* (Damon, 1998; Damon & Lerner, 2006) devote the first volume of the four-volume work to theoretical models of human development; most of the theories represent variants of developmental systems theories. Indeed, in his introduction to the 2006 edition of this volume, Lerner (2006) notes that developmental systems models are at the cutting edge of theory in developmental science. This genre of theory frames as well the presentations in other major compendiums in developmental theories, for example, the second and third editions of the *Handbook of Adolescent Psychology* (Lerner & Steinberg, 2004, 2009), the two volumes of the *Handbook of Life-Span Development* (Lamb & Freund, 2010; Lerner & Overton, 2010), and the *Handbook of Developmental Science, Behavior and Genetics* (Hood, Halpern, Greenberg, & Lerner, 2010).

Accordingly, the tenets of developmental systems theory are well established as the superordinate developmental frame in contemporary developmental science. In addition, there are strong conceptual links between these theories and other contemporary theoretical models, such as dynamical systems, biological systems theory, and artificial neural networks (e.g., connectionism). However, researchers' interest in such theories has been limited by inadequate methods of measuring and analyzing change in continuous time-dependent streams of multivariate data. Such methods are necessary in decisive empirical tests of theoretical predictions derived from developmental systems theory. Impressive progress has been made in implementing developmental systems in sophisticated simulation programs (cf. Spencer, Thomas, & McClelland, 2009); yet these simulation models can only indirectly address the empirical validity of developmental process models. The ultimate test of theoretical models in empirical science is to fit models directly to appropriate empirical data.

Only recently have the required mathematical–statistical tools become available to fit developmental systems models directly to intensive measurements of developmental processes (for overviews, see Molenaar & Newell, 2010; Newell & Molenaar, 1998). These new modeling tools are predominantly based on dynamic systems modeling, having roots in applied nonlinear dynamics, computational engineering, econometrics, and statistical signal analysis. Initial applications of these tools have, for instance, resulted in major restructuring of (connectionistic) simulation programs of sudden qualitative transitions in development (cf. van der Maas & Raijmakers, 2009). The combination of advanced developmental systems theoretical modeling and sophisticated statistical–methodological approaches of empirical validation holds the potential to yield a powerful new paradigm for social science and medicine (cf. Schwartz & Collins, 2007).

In its most elementary form this new paradigm is characterized by the use of intensive measurements of single subjects (or patients) in real time and in their natural environments, often by means of advanced sensoring techniques. The data streams thus obtained are analyzed by means of recursive dynamic modeling techniques in order to (semi-)continuously assess and optimally guide the psychological, biobehavioral, and/or disease processes of interest. It has been demonstrated recently (Molenaar, 2004, 2007) that this intensive subject-specific data acquisition and dynamic modeling, which constitutes the first level of the new paradigm prior to pooling across subjects, yields powerful explanatory process models.

In order to propel forward the innovative integration of developmental systems theory with state-of-the-art statistical dynamic modeling tools, we organized the Conference on Inductive Developmental Systems Theory. During the conference, lively interactions took place that inspired some of the contents of the present *Handbook* and initiated further collaboration among the participants and colleagues beyond those at the conference. At the theoretical level, several chapters present important extensions of developmental systems theory, captured under the denotation *relational developmental systems*. At the level of inductive dynamic systems theory, many chapters present innovative applications of powerful statistical techniques to real and simulated time-dependent data streams in order to test theoretical predictions derived from developmental systems theory.

The Plan of the *Handbook*

We conclude this brief introductory chapter with short characterizations of each chapter in the *Handbook*. This section presents chapters that discuss key features and conceptual implications of relational developmental models of human development.

Willis F. Overton, in a chapter that constitutes a keynote presentation for the *Handbook*, builds on his prior discussions of relational developmental systems and focuses on methodological issues. He notes that if one's approach to methodology assumes that development is the systematic study of changes in intraindividual variability, intensive assessment, multivariate, within-subject methods and designs emerge and become critical. He points to the ideas of Nesselroade and Molenaar (2010, p. 31) that clearly articulate this point:

> Attention to intraindividual variability leads to favoring some kinds of research designs over others, how and what one measures, and the data analyses one performs. Even more fundamentally, intraindividual variability concerns help to delimit the very way one formulates his or her research questions and the manner in which one conceptualizes and deals with fundamental scientific matters such as prediction and generalizability. These latter concerns, in turn, rightfully have strong "trickle-down" effects on the design, measurement, and modeling efforts of students of development.

Accordingly, Overton argues that the "trickle-down" effects Nesselroade and Molenaar describe are real and important, but the sources of these effects begin much further upstream. He traces these effects to their headwaters, located in the highly abstract conceptual space of metatheory, and from there through their tributaries of various developmental conceptual models and theories, and to specific methods.

In the next chapter, G. John Geldhof and colleagues provide an illustration of the use of relational developmental systems ideas in regard to a specific portion of the life span—that is, adolescence—and note the methodological issues and implications of using relational developmental systems ideas for the study of this age period. The authors note that interest in the strengths of youth, the plasticity of human development, and the concept of resilience coalesced in the 1990s to foster the evolution of the positive youth development (PYD) concept. As discussed by Hamilton (1999), the concept of PYD can be understood in three interrelated but distinct ways: as a developmental process, as a philosophy or approach to youth programming, and as instances of youth programs and organizations focused on fostering healthy or positive development among youth. Geldhof and colleagues focus on the idea of PYD as a developmental process by using a model framed by relational developmental systems ideas; they describe the nature and implications of the considerable research across the adolescent period that has been inspired by this model.

In the next chapter, Jennifer Brown Urban and her colleagues note how developmental systems science extends the conceptual and methodological boundaries of developmental science. They note that the term *developmental systems science* refers to the application of systems science methodologies (e.g., social network analysis, system dynamics, and agent-based modeling) to developmental science questions, particularly those derived from a developmental systems theoretical perspective. The phrase *developmental systems science* deliberately combines *developmental science* with *systems science* and is meant to reflect the joining together of these two fields. *Developmental science* is an approach to the study of human development that emphasizes multidisciplinary and systemic thinking and includes the spectrum from basic to applied forms of inquiry (Lerner, 2006). *Systems science* refers to a family of methodologies that enables the study of complex problems and typically involves modeling and simulation.

This chapter aims to present a compelling rationale for the application of systems science methods in developmental science as well as an introduction to three specific systems science methodologies. This chapter builds on the authors' previous work in this area (Urban, Osgood, & Mabry, 2011) by giving a much richer and more detailed description of the selected systems science methods and by including a brief review of systems science applications in developmental science.

The next section of the *Handbook* presents two chapters that discuss epigenetic development and evolution. Mae-Wan Ho links epigenetics and generative dynamics to explain how development directs evolution. She notes that, whereas the epigenetic approach fully reaffirms the fundamental holistic nature of life and discredits any theory ascribing putative group differences in human attributes to genes (Ho, 2010), it also gives no justification to *simplistic mechanistic* ideas of arbitrary effects arising from use and disuse or the inheritance of acquired characters. It does not lead to any kind of determinism, environmental or genetic. Organisms are, above all, complex, nonlinear dynamical systems (Saunders, 1993), and as such, they have regions of stability and instability that enable them to maintain homeostasis or to adapt to change, or not, as the case may be. The appearance of novelties and of mass extinctions alike in evolutionary history are but two sides of the same coin; we cannot be complacent about the capacity of organisms to adapt to any and all environmental insults that are perpetrated, the most pressing of which is anthropogenic global warming. The dynamics of the developmental process ultimately holds the key to heredity and evolution by determining the sorts of changes that can occur and in its resilience to certain perturbations and susceptibility to others. Our knowledge in this crucial area is urgently required.

What implications are there for evolution? Just as interaction and selection cannot be separated, nor can variation (or mutation) and selection, for the "selective" regimen may itself cause specific epigenetic variations or "directed" mutations. The organism experiences its environment in one continuous nested *process*, adjusting and changing, leaving imprints in its epigenetic system, its genome, as well as on the environment—all of which are passed on to subsequent generations. Thus, *there is no separation between development and evolution*. In that way, the organism actively participates in shaping its own development as well as the evolution of its ecological and social community. We do hold the future in our hands—it is precious; be careful.

In the succeeding chapter Peter T. Saunders discusses dynamical systems, the epigenetic landscape, and punctuated equilibria. He explains that complex nonlinear dynamical systems can have many properties that are not found in linear systems. These properties include multiple steady states, abrupt changes, chaotic behavior, and self-organization. Simply realizing that systems may have these properties can enable us to understand many phenomena without the need to postulate special forces or complicated and implausible scenarios. This chapter illustrates the idea by applying it to biological development and evolution. In particular, a simple explanation is provided for punctuated equilibria.

The next section of the *Handbook* presents two chapters elucidating the links between neural networks and development. In the first of these chapters, Maartje E. J. Raijmakers and colleagues discuss nonlinear epigenetic variance in developmental processes. They point out that the assumption in twin studies—that genotypic and environmental factors reflect the underlying mechanisms causing phenotypic individual differences (e.g., Plomin, DeFries, McClearn, & McGuffin, 2008)—has important problems. First, a limiting feature of behavior–genetic methodology is that the causal interpretation pertains to phenotypic individual differences and not to phenotypes themselves (Dolan & Molenaar, 1995; Lewontin, 1974; Oyama, 1985). Second, the validity of the methodology may be undermined by the limitations of the statistical model employed to carry out the decomposition of phenotypic variance. For instance, genotype–environment interaction, genotypic–environmental covariance, and assortative mating are not taken into account in the standard linear model (but see Plomin et al., 2008). Third, the search for specific genetic and environmental variables may be limited in that an important part of phenotypic variance may be due to nonlinear (epigenetic) processes (Molenaar, Boomsma, & Dolan, 1993; Zuk, Hechter, Sunyaev, & Lander, 2012). Because these effects will appear unsystematic, they are difficult to distinguish from specific environmental effects and measurement error.

Accordingly, the aim of the Raijmakers and colleagues chapter is to study the third limitation: the role of nonlinear epigenetic processes as a source of phenotypic variance. They review the relevant literature and present ample evidence in support of these processes. In addition, they present the results of computer simulations. The implications and consequences of the presence of nonlinear epigenetic variance are particularly interesting in the light of the largely unsuccessful attempts to identify specific, nonshared, environmental influences (e.g., see Turkheimer & Waldron, 2000).

In turn, in the second chapter in this section, Gregor Schöner presents a penetrating discussion of the self-organization metaphor, emphasizing that mere mathematization in the form of, for instance, reaction–diffusion models explaining biological pattern formation is not sufficient to transform this metaphor into productive theory.

He then moves to an equally penetrating discussion of the dynamic systems metaphor, taking Waddington's epigenetic landscape as example. Schöner lists the strengths, but also several important weaknesses, of the dynamic systems metaphor. He then shows that dynamic field theory resolves the weaknesses of the dynamic systems metaphor, thus providing a strong theoretical framework for understanding development. This is illustrated by means of two worked examples, the "*A*-not-*B*" paradigm and the habituation paradigm, both providing impressive examples of how dynamic field theory is applied in concrete situations.

The next section of the *Handbook* focuses on the dynamics of development. Flavio Cunha and James Heckman discuss methods useful for estimating the technology of cognitive and noncognitive skill formation. They treat the linear case and therefore include estimates from linear models of the evolution of cognitive and noncognitive skills. They explore the role of family environments in shaping these skills at different stages of the life cycle. Central to this analysis is identification of the technology of skill formation.

The authors estimate a dynamic factor model to solve the problem of endogeneity of inputs and multiplicity of inputs relative to instruments. They identify the scale of the factors by estimating their effects on adult outcomes. In this fashion, they avoid reliance on test scores and changes in test scores that have no natural metric. Parental investments are generally more effective in raising noncognitive skills, and noncognitive skills promote the formation of cognitive skills—but, in most specifications of their model, cognitive skills do not promote the formation of noncognitive skills. Parental inputs have different effects at different stages of the child's life cycle, with cognitive skills affected more at early ages and noncognitive skills affected more at later ages.

In turn, in the next chapter in this section, Han L. J. van der Maas and colleagues use a complex systems approach to discuss the dynamics of development. They provide three examples of modeling and investigating complex systems. First, they present a new model for general intelligence based on a mathematical model for ecological networks. Second, they discuss ways to study phase transitions in psychological systems. Third, they introduce a completely new approach to collect high-frequency data on children's development, which is a necessity for studying complex systems. Through these three examples, the authors aim is to demonstrate the viability of the complex system approach to the study of human development.

The next chapter in this section, by Kurt W. Fischer and Paul van Geert, discusses the dynamic development of brain and behavior. The authors ask how the growth of neural systems in the brain relates to children's psychological development. They note that the study of neuroscience is moving beyond speculation to discovering how brain and behavior connect—how development of brain functioning relates to actions, thoughts, and emotions. This knowledge provides possibilities for moving beyond global correlations to real breakthroughs in the understanding of developmental processes. Analysis of patterns of growth can illuminate how brain functions and behaviors develop through common developmental mechanisms and produce similar growth curves.

The authors propose that two characteristics are especially important for analyzing and explaining the developmental mechanisms underlying brain–behavior relations. First, many growers are connected, with important variations in the types of connections; growers powerfully influence each other's growth. Second, growers commonly move through periods of rapid change or developmental discontinuity, in which new capacities or forms of activity emerge, creating the transformations of childhood and adolescence. Advances in methods for studying development make it possible to analyze the processes of development of brain and behavior.

In addition, Fischer and van Geert propose a framework for analyzing brain–behavior relations in development based on the hypothesis that major developmental changes involve coordination of brain–behavior components into higher-order control systems, which they call *dynamic skills* (Bullock & Grossberg, 1988; Fischer, 1980; Fischer & Bidell, 2006; Grossberg, 1988; Mascolo & Fischer, 2010). Skills comprise multiple elements, following the principles of dynamic systems (Smith & Thelen, 1993; Stein, Dawson, & Fischer, 2010; van Geert, 1991, 1998). Before coordination, connections of these elements are mostly weak. With development of coordination, connections become strong and shape growth functions. The authors conclude that dynamic systems models portray a wide range of phenomena in approximate terms and thus sketch the kinds of phenomena that these models can explain. This important function lays out the sorts of phenomena that dynamic systems models can elucidate.

Karl M. Newell and Yeou-Teh Liu, in the next chapter, discuss the dynamics of motor learning and development across the life span, emphasizing that the scientific subdomains of motor learning and motor development have largely evolved independently of each other. In contrast, Newell and Liu indicate that the existing theoretical approaches to motor development (e.g., the emergence of new movement forms) and motor learning (e.g., task-relevant changes of existing movement forms) share substantial common ground and therefore have to be integrated. To accomplish this integration, they present the epigenetic landscape model as a viable low-dimensional approximation of the high-dimensional dynamic systems underlying motor learning and development. In particular, they show how continuity and discontinuity, as well as the existence of different timescales associated with motor learning and development, are captured by the epigenetic landscape model. These ideas are illustrated by applications to experimental movement learning data involving two timescales and to the occurrence of phase transitions in movement learning.

The next section of the *Handbook* presents two chapters that treat the dynamics of social interaction. Emilio Ferrer and Joel Steele present differential equations for evaluating theoretical models of dyadic interactions. They illustrate differential equation models (DEMs) as suited to examine the interdependence of dyadic members over time. First, they describe theoretical models developed for dyadic interactions. They then apply differential equations representing such models to daily data on affect from individuals in couples. Third, they evaluate the parameter estimates from each model and compare across the different specifications. Finally, they examine the predictive validity of the models by using their corresponding estimates to predict the couples' relationship quality and status 1 and 2 years later.

In turn, Steven M. Boker and colleagues present a differential equations model for the ovarian hormone cycle. They note that dynamical systems models of behavior and regulation have become increasingly popular due to the promise that within-person mechanisms can be modeled and explained. However, it can be difficult to construct differential equation models of regulatory dynamics that test specific theoretically interesting mechanisms.

Accordingly, the authors use the example of ovarian hormone regulation and develop a model to capture features of observed hormone levels and, as well, link parameters of the model to biological mechanisms. Ovarian hormones regulate the monthly female reproductive cycle and have been implicated to have effects on affective states and eating behavior. The three major hormones in this system are estrogen, progesterone, and lutenizing hormone. These hormones are coupled together as a regulatory system. Estrogen level is associated with the release of lutenizing hormone by the hypothalamus. Lutenizing hormone triggers ovulation and the transformation of the dominant follicle into the corpus luteus, which, in turn, produces progesterone. A differential equations model is developed that is biologically plausible and produces nonlinear cycling similar to that seen in a large, ongoing daily-measure study of ovarian hormones and eating behavior.

The next section presents a nonlinear dynamical model of development. Sy-Miin Chow and colleagues discuss a regime-switching longitudinal model of alcohol lapse–relapse. The authors note that general contemporary linear models assume that continuous changes in the predictor variables result in proportionate amounts of (linear) change in the outcome variable. Empirical evidence from the alcohol treatment literature, however, favors the application of nonlinear dynamics models over the general linear model because of their ability to capture sudden, discontinuous jumps in individuals' drinking tendency.

One example of such nonlinear models is the cusp catastrophe model used by Witkiewitz and Marlatt (2004) to represent the complex interplay between different risk factors in triggering sudden shifts in individuals' tendency to drink. Although the cusp catastrophe model has been promising in capturing some aspects of alcohol use dynamics, current approaches of fitting variations of this model do not address several practical data-analytic problems commonly seen in empirical data, including the presence of incomplete data, measurement and/or process noise, the lagged effects of previous drinking on current alcohol use, heterogeneous timing of lapse–relapse within and across subjects, and the large number of abstainers at any given time—commonly referred to as the "zero inflation" phenomenon.

Accordingly, the authors propose a mixture structural equation model with regime-switching (MSEM-RS) as an alternative approach to account for these data-analytic issues, while retaining some of the key features of the cusp catastrophe model. The proposed model is illustrated using longitudinal drinking data from the COMBINE study (COMBINE Study Research Group, 2003).

The next section of the *Handbook* presents two chapters that discuss nonergodic developmental systems. Wayne F. Velicer and colleagues discuss idiographic applications involving issues of ergodicity and generalizability. The authors note that idiographic methods focus on the time-dependent variation within a single individual or unit (intrasubject variability) in contrast to methods that focus on group-level relationships (intersubject variability). Idiographic methods are widely used in other disciplines, such as engineering, business, and economics, but only recently employed in the behavioral sciences. This method is an alternative to the dominant scientific approach in the behavioral sciences, the nomothetic approach, which focuses on group-level analysis.

Accordingly, the authors present three different examples from behavioral medicine to illustrate both the challenges and rewards of using idiographic methods. These include a nicotine harm reduction study, a study of the patterns of adherence in sleep apnea, and an intervention study that identifies different patterns of arousal in children with autism. The studies illustrate how idiographic methods can address unique and important research questions.

In the second chapter in this section, Peter C. M. Molenaar and John R. Nesselroade discuss new trends in inductive developmental systems theory: ergodicity, idiographic filtering, and alternative specifications of measurement equivalence. The authors note that methodological implications of some aspects of development emphasized by RDST have been considered before. For instance, Sidman (1960) presented an overview of early methodological work addressing the (lack of) relationship between individual and average learning curves, starting with Merrill (1931). Wohlwill (1973) criticized the use of interindividual variation in studying developmental processes and, in its stead, recommended a focus on individual developmental functions.

Accordingly, the authors seek to strengthen and generalize these early approaches. That is, they first consider a general mathematical theory, ergodic theory, which specifies necessary conditions for a valid inductive relational developmental systems model based on fitting subject-specific stochastic dynamic systems models with time-varying parameters to appropriate empirical data. Although ergodic theory was founded more than a century ago, its surprisingly direct relevance to inductive relational developmental systems models was made explicit only recently (Molenaar, 2004). Next, a number of important implications of ergodic theory for inductive relational developmental systems models is elaborated, in particular, how to deal with pervasive subject specificity and change in fitting appropriate dynamic systems models to the data. In this context, the authors consider a new approach inspired by these models to estimate subject-specific heritabilities.

The final section of the *Handbook*, by Phillip K. Wood, is a one-chapter integrative summary of the preceding chapters. He presents the interesting point of view that science in general can be characterized as an inductive developmental system in which different scientific models constitute competing webs of beliefs. Wood emphasizes the central theme of this handbook, namely that empirical tests of developmental systems theory are possible, but require appropriate data and models. He presents an insightful discussion of the epigenetic landscape metaphor that figures in several previous chapters, adding important considerations about its manifold theoretical implications and its use in data analysis. The chapter closes with some noteworthy observations and comments on research inspired by developmental systems theory.

Conclusions

The chapters in this handbook are all aimed at fulfilling the scientific promise of developmental systems theories by coupling such models within inductive methods. Across the chapters in this handbook, authors capitalize on the important ingredients of an inductive methodological approach to developmental systems theories that have become available recently, as documented by Molenaar and Newell (2010). This handbook takes an additional and crucial next step. Across the chapters in this handbook (and as summarized by Wood, in the concluding chapter in this volume), the authors integrate these statistical modeling approaches within developmental systems theory. Together, the contributions of the authors culminate in a new powerful set of methodologies and accompanying statistical modeling approaches to fit developmental systems theoretical models to empirical data.

Given this contribution, our hope is that this handbook provides to current and future developmental scientists an understanding of the use of developmental systems theory and methods in the description, explanation, and optimization of intraindividual changes across life and of interindividual differences in such life-span change. We believe that the chapters offer integrative and authoritative discussions of the theory–method synergies in this cutting-edge framework for understanding human behavior and development. As such, we believe as well that this handbook provides to current doctoral-level researchers and professors and to their students a rich and detailed depiction of the nature of past, contemporary, and likely future scholarship pertinent to theory and method in this area of developmental science.

To the extent that the chapters in this handbook enhance understanding of how to employ theory-predicated methods to enhance understanding of the mutually influential relations between individuals and the multiple levels of their context that constitute the developmental system, we believe that the scholarship in this book will contribute to a new era in the conduct of developmental science—one that captures the complexity of the developmental system and enhances the means to not only describe and explain intraindividual change and interindividual differences in intraindividual change but, as well, provides new means to generate evidence-based actions that optimize the course of health and positive functioning across the life span.

ACKNOWLEDGMENTS

The preparation of this chapter was supported in part by grants to Peter C. M. Molenaar from the National Science Foundation (Grant No. 1157220); to Richard M. Lerner from the John Templeton Foundation, the National 4-H Council, and the Thrive Foundation for Youth; and to Karl M. Newell from the National Science Foundation (Grant No. 0848339). We thank Jarrett M. Lerner for his editorial assistance.

REFERENCES

Bullock, D., & Grossberg, S. (1988). Neural dynamics of planned arm movements: Emergent invariants and speed-accuracy properties during trajectory formation. *Psychological Review, 95*, 49–90.

COMBINE Study Research Group. (2003). Testing pharmacotherapies and behavioral interventions for alcohol dependence (the COMBINE Study): A pilot feasibility study. *Alcohol: Clinical and Experimental Research, 27*(7), 1123–1131.

Damon, W. (Ed.). (1998). *Handbook of child psychology* (5th ed.). New York: Wiley.

Damon, W., & Lerner, R. M. (Eds.). (2006). *Handbook of child psychology* (6th ed.). Hoboken, NJ: Wiley.

Dolan, C. V., & Molenaar, P. C. M. (1995). A note on the scope of developmental behaviour genetics. *International Journal of Behavioural Development, 18*, 749–760.

Fischer, K. W. (1980). A theory of cognitive development: The control and construction of hierarchies of skills. *Psychological Review, 87*, 477–531.

Fischer, K. W., & Bidell, T. R. (2006). Dynamic development of action, thought, and emotion. In R. M. Lerner (Ed.), *Handbook of child psychology: Vol. 1. Theoretical models of human development* (6th ed., pp. 313–399). New York: Wiley.

Ford, D. H., & Lerner, R. M. (1992). *Developmental systems theory.* Newbury Park, CA: Sage.

Gottlieb, G. (1992). *Individual development and evolution: The genesis of novel behavior.* New York: Oxford University Press.

Gottlieb, G. (2003). On making behavioral genetics truly developmental. *Human Development, 46*, 337–355.

Grossberg, S. (1988). *Neural networks and natural intelligence.* Cambridge, MA: MIT Press.

Hamilton, S. F. (1999). *A three-part definition of positive youth development.* Unpublished manuscript, Cornell University, Ithaca, NY.

Ho, M.-W. (2010). Development and evolution revisited. In K. E. Hood, C. Tucker Halpern, G. Greenberg, & R. M. Lerner (Eds.), *Handbook of developmental science, behavior, and genetics* (pp. 61–109). Malden, MA: Wiley-Blackwell.

Hood, K. E., Halpern, C. T., Greenberg, G., & Lerner, R. M. (Eds.). (2010). *The handbook of developmental science, behavior and genetics.* Malden, MA: Wiley-Blackwell.

Lamb, M. E., & Freund, A. M. (Vol. Eds.), Lerner, R. M. (Ed.-in-Chief). (2010). *The handbook of life-span development: Vol. 2. Social and emotional development.* Hoboken, NJ: Wiley.

Lerner, R. M. (2006). Developmental science, developmental systems, and contemporary theories of human development. In R. M. Lerner (Ed.), *Handbook of child psychology: Vol. 1. Theoretical models of human development* (6th ed., pp. 1–17). Hoboken, NJ: Wiley.

Lerner, R. M. (Ed.-in-Chief) & Overton, W. F. (Vol. Ed.). (2010). *Handbook of life-span development: Vol. 1. Cognition, biology, and methods.* Hoboken, NJ: Wiley.

Lerner, R. M., & Steinberg, L. (Eds.). (2004). *Handbook of adolescent psychology* (2nd ed.). New York: Wiley.

Lerner, R. M., & Steinberg, L. (Eds.). (2009). *Handbook of adolescent psychology* (3rd ed.). Hoboken, NJ: Wiley.

Lewontin, R. C. (1974). *The genetic basis of evolutionary change.* New York: Columbia University Press.

Mascolo, M. F., & Fischer, K. W. (2010). The dynamic development of thinking, feeling, and acting over the lifespan. In R. M. Lerner (Ed.-in-Chief) & W. F. Overton (Vol. Ed.), *Handbook of life-span development: Vol. 1. Biology, cognition, and methods* (pp. 149–194). Hoboken NJ: Wiley.

Merrill, M. (1931). The relationship of individual growth to average growth. *Human Biology, 3*, 37–70.

Molenaar, P. C. M. (2004). A manifesto on psychology as idiographic science: Bringing the person back into scientific psychology, this time forever. *Measurement, 2*, 201–218.

Molenaar, P. C. M. (2007). On the implications of the classical ergodic theorems: Analysis of developmental processes has to focus on intra-individual variation. *Developmental Psychobiology, 50*, 60–69.

Molenaar, P. C. M., Boomsma, D. I., & Dolan, C. V. (1993). A third source of developmental differences. *Behavior Genetics, 23*, 519–524.

Molenaar, P. C. M., & Newell, K. M. (Eds.). (2010). *Individual pathways of change: Statistical models for analyzing learning and development*. Washington, DC: American Psychological Association.

Nesselroade, J. R., & Molenaar, P. C. M. (2010). Emphasizing intraindividual variability in the study of development over the life span. In W. F. Overton & R. M. Lerner (Eds.), *The handbook of life-span development: Vol. 1. Cognition, biology, and methods* (pp. 30–54). Hoboken, NJ: Wiley.

Newell, K. M., & Molenaar, P. C. M. (Eds.). (1998). *Applications of nonlinear dynamics to developmental process modeling*. Hillsdale, NJ: Erlbaum.

Overton, W. F. (2010). Life-span development: Concepts and issues. In R. M. Lerner (Ed.-in-Chief) & W. F. Overton (Vol. Ed). *Handbook of life-span development: Vol. 1. Cognition, biology, and methods* (pp. 1–29). Hoboken, NJ: Wiley.

Overton, W. F. (2012). Evolving scientific paradigms: Retrospective and prospective. In L. L'Abate (Ed.), *The role of paradigms in theory construction* (pp. 31–65). New York: Springer.

Overton, W. F., & Müller, U. (2012). Development across the life span: Philosophy, concepts, theory. In R. M. Lerner, M. A. Easterbrooks, & J. Mistry (Eds.), *Handbook of psychology: Vol. 6. Developmental psychology* (pp. 19–58). New York: Wiley.

Oyama, S. (1985). *The ontogeny of information: Developmental systems and evolution*. Cambridge, UK: Cambridge University Press.

Oyama, S., Griffiths, P. E., & Gray, R. D. (Eds.). (2001). *Cycles of contingency: Developmental systems and evolution*. Cambridge, MA: MIT Press.

Plomin, R., DeFries, J. C., McClearn, G. E., & McGuffin, P. (2008). *Behavior genetics* (5th ed.). New York: Worth.

Saunders, P. T. (1993). The organism as a dynamical system. In F. Varela & W. Stein (Eds.), *Thinking about biology* (pp. 41–63). Reading, MA: Addison-Wesley.

Schwartz, D., & Collins, F. (2007). Environmental biology and human disease. *Science, 316*, 695–696.

Sidman, M. (1960). *Tactics of scientific research*. New York: Basic Books.

Smith, L. B., & Thelen, E. (1993). *A dynamic systems approach to development*. Cambridge, MA: MIT Press.

Spencer, J. P., Thomas, M. S. C., & McClelland, J. L. (2009). *Toward a unified theory of development: Connectionism and dynamic systems theory reconsidered*. New York: Oxford University Press.

Stein, Z., Dawson, T., & Fischer, K. W. (2010). Redesigning testing: Operationalizing the new science of learning. In M. S. Khine & I. M. Saleh (Eds.), *New science of learning: Cognition, computers and collaboration in education* (pp. 207–224). New York: Springer.

Turkheimer, E., & Waldron, M. (2000). Statistical analysis, experimental method, and causal inference in developmental behavior genetics. *Human Development, 43*, 51–52.

Urban, J. B., Osgood, N., & Mabry, P. (2011). Developmental systems science: Exploring the application of non-linear methods to developmental science questions. *Research in Human Development, 8*, 1–25.

van der Maas, H. L. J., & Raijmakers, M. E. J. (2009). Transitions in cognitive development: Prospects and limitations of a neural dynamic approach. In J. P. Spencer, M. S. C. Thomas, & J. L. McClelland (Eds.), *Toward a unified theory of development: Connectionism and dynamic systems theory reconsidered* (pp. 299–312). New York: Oxford University Press.

van Geert, P. (1991). A dynamic systems model of cognitive and language growth. *Psychological Review, 98*, 3–53.

van Geert, P. (1998). A dynamic systems model of basic developmental mechanisms: Piaget, Vygotsky and beyond. *Psychological Review, 105*, 634–677.

Witkiewitz, K., & Marlatt, G. A. (2004). Relapse prevention for alcohol and drug problems. *American Psychologist, 59*, 224–235.

Wohlwill, J. F. (1973). *The study of behavioral development.* New York: Academic Press.

Zuk, O., Hechter, E., Sunyaev, S. R., & Lander, E. S. (2012). The mystery of missing heritability: Genetic interactions create phantom heritability. *Proceedings of the National Academy of Sciences, 109*, 1193–1198.

PART II

RELATIONAL DEVELOPMENTAL SYSTEMS THEORY

CHAPTER 2

Relational Developmental Systems and Developmental Science

A Focus on Methodology

Willis F. Overton

In a recent paper (Overton, 2013) I described relational developmental systems as the core of a scientific paradigm that provides an advantageous conceptual framework for developmental science. As the inadequacies of the standard Cartesian scientific paradigm become increasingly evident in developmental science's subfields, including inheritance, evolution, and organismic—prenatal, cognitive, emotional, motivational, sociocultural—development across the life span (i.e., from conception until the late adult years), relational developmental systems serves as an alternative scientific paradigm, providing an alternative conceptual framework that adequately grounds many recent discoveries and profitably points to the generation of many new discoveries.

In the earlier paper I focused on how the relational developmental systems paradigm supports and offers guidelines for theoretical and empirical progress in the various subfields of developmental science. In the present chapter I repeat several of my arguments, but I focus on methodological issues. By *methodology*, I do not mean methods. *Methodology* refers to a set of guiding principles for the development of specific methods. By way of illustration, if your methodology assumes that the living organism is a linear input–output machine with strictly decomposable parts, then you will be prone to exclusively develop and use linear additive methods such as the analysis of variance (ANOVA) and regression models. If, on the other hand, your methodology assumes that the living organism is a nonlinear self-organizing dynamic system, you will be prone to employ consistent methods, such as, for example, regime-switching models (Kim & Nelson, 1999), nonlinear dynamics systems models with bifurcation (van der Maas & Molenaar, 1992), and state–space models (Chow & Filteau, 2010), among others (see Molenaar & Nesselroade, Chapter 17, this volume; Molenaar & Newell, 2010; Urban, Osgood, & Mabry, 2011, for multiple examples). Similarly, if your methodology assumes that individual differences (interindividual variability) stand at the center of understanding development processes, then various single-assessment, single-variable, between-subject designs will be favored. If, on the other hand, your methodology assumes that development is the systematic study of changes in intraindividual variability, intensive assessment, multivariate, within-subject methods and designs emerge and become critical. Nesselroade and Molenaar (2010) clearly articulate this point:

> Attention to intraindividual variability leads to favoring some kinds of research designs over others, how and what one measures, and the data analyses one performs. Even more fundamentally, intraindividual variability concerns help to delimit the very way one formulates his or her research questions and the manner in which one conceptualizes and deals with fundamental scientific matters such as prediction and generalizability. These latter concerns, in turn, rightfully have strong "trickle-down" effects on the design, measurement, and modeling efforts of students of development. (p. 31)

In the present chapter I argue that the "trickle-down" effects Nesselroade and Molenaar describe are real and important, but the sources of these effects begin much further upstream than the quote suggests. I trace these effects to their headwaters, located in the highly abstract conceptual space of metatheory, and from there through their tributaries of various developmental conceptual models and theories, and to specific methods.

Scientific Research Programs and Paradigms: Metatheories, Theories, and Observations

To understand relational developmental systems as a scientific paradigm, it is necessary to recognize that any contemporary scientific research program operates within a layered nested conceptual framework. One layer of this framework is composed of background ideas or what are usually termed *metatheoretical* concepts. The metatheoretical concepts transcend theories in the sense that they define the context in which theoretical concepts are constructed. Metatheoretical concepts ground, constrain, and sustain theoretical concepts, and they serve the same functions for empirical methods of investigation. Generally, when referring to the former function the concept of *metatheory* is used, whereas references to the latter function are termed *methodology*. The primary function of the *metatheory–methodology* framework is to provide a rich source of concepts out of which theories and methods emerge. Metatheory–methodology also provides guidelines that help avoid conceptual confusions and inappropriate methods.

The Structure of Scientific Discourse: From Observations to the Metatheoretical–Methodological and Back

An elaboration of the nature and role of specific metatheories–methodologies in science requires that we keep in mind that the structure of scientific discourse maintains a distinct demarcation between the various levels of this structure (see Figure 2.1). Theories and methods refer directly to the empirical world, whereas metatheories–methodologies refer to the theories and methods themselves. The most concrete and circumscribed level of scientific discourse is the *observational level*. This level is one's current commonsense means of conceptualizing the nature of objects and events in the world. There is uniform agreement across a wide spectrum of philosophical systems that commonsense observations constitute the base of the structure of scientific discourse (Nagel, 1967, 1979; Pepper, 1942; Wartofsky, 1968). These commonsense observations are not the pristine observations of neopositivist methodology; rather they are folk beliefs and distinctions that surround us in everyday life. That people take risks is common sense and that some take more risks than others is common sense. That some people are smart and that some are smarter than others is common sense. In the domain of interest to developmental science, we begin from the common sense observation that living organisms change over time in their actions, their

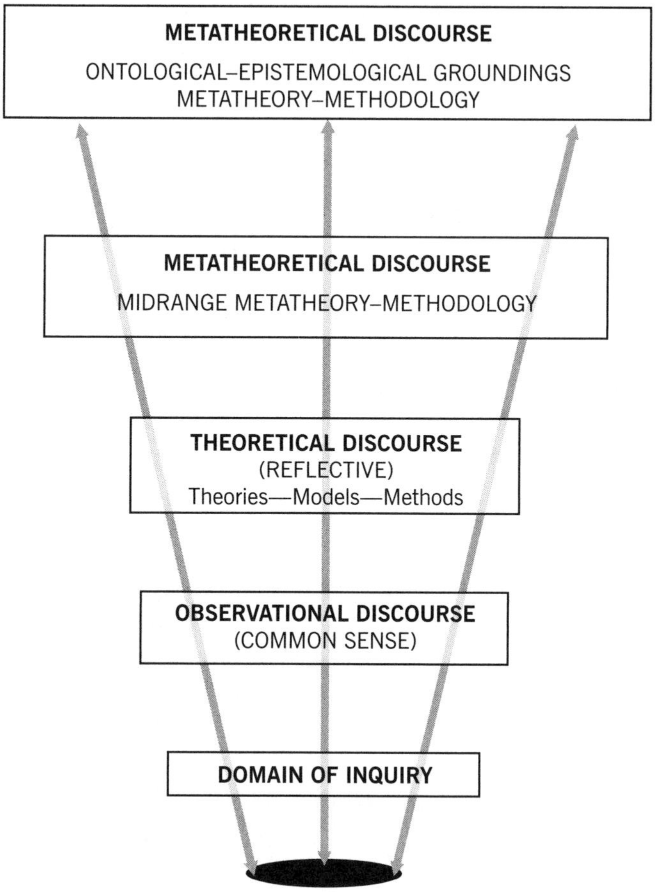

FIGURE 2.1. Levels of discourse in a scientific research program.

perceiving, thinking, feeling, relating, remembering, valuing, intending, playing, creating, languaging, comparing, reasoning, wishing, willing, judging, and so on.

Although the observational, commonsense, or folk level of discourse has a sense of immediacy and concreteness, science—often defined as the reflective criticism of common sense (Wartofsky, 1968)—requires movement to a level of critical reflection on this commonsense understanding. This *reflective level* of discourse itself entails several sublevels. The first sublevel is *theoretical* discourse. Here, thought is about organizing and reformulating observational understandings in a broader and more coherent fashion. At this level concepts are *about* the observational level. These concepts include theories, models, and methods, and they range from informal hunches to highly refined concepts concerning the nature of, and relations among, objects and events, including human behavior and change. Here, for example, classical developmental theories such as Piaget's (1970) and Werner's (1948, 1957) contain theoretical principles (e.g., "stages") that hypothesize that ultimately a combination of continuous and discontinuous changes best defines human development, and this conception implicates methods for assessing both forms of change. Skinnerian and social learning theories, on the other hand, hypothesize that all change is best represented as strictly continuous and additive, thus implicating methods with similar characteristics.

The next level of critical reflective thought above the level of theoretical discourse is the metatheoretical level. Here thought is about basic concepts (metatheories and methodologies) that impact both the theoretical and observational levels. A *metatheory* itself is a coherent set of rules or principles that both describes and prescribes what is acceptable and unacceptable as theory—the means of conceptual exploration of any scientific domain. For example, in a metatheory often termed *atomism*, which entails the assumption that all complex events can be reduced to foundational discrete elements, change itself must be reducible to the additive combination of these elements. As a consequence, true novelty is not possible within this metatheory and, hence, all apparent change must be considered purely quantitative or continuous. Thus, theories constructed upon this metatheoretical assumption are committed to strictly quantitative change. Qualitative change is excluded a priori. A *methodology* is also a set of rules or principles, but this set describes and prescribes the nature of acceptable methods—the means of observational exploration—in a scientific discipline. For example, a methodology closely related to the atomistic metatheory asserts the principle that all change is additive in nature; thus, additive methods are created.

Discourse at the metatheoretical level is itself divided along dimensions of abstractness and breadth. At the pinnacle are metatheories–methodologies that operate at the broadest level of generality and abstractness. This level is one of *worldviews*, and like other metatheories–methodologies these worldviews are composed of coherent sets of rules or principles. Specifically, the principles of worldviews are *epistemological* (i.e., pertaining to sources and justification of knowledge) and *ontological* (i.e., pertaining to fundamental categories of reality) in nature. In essence, a worldview is a framework that presents a vision of the nature of the world (e.g., Is stasis or change, substance or process, asserted as the fundamental categories of the world?) and the nature of how we know that world (e.g., Is knowledge the reflection of a mind-independent reality, or do minds actively participate in the constitution of the world as known?).

Nested within worldviews are *metatheories–methodologies of a middle range.* These concepts are less abstract, less broad, and entail principles that are identifiably more specific to the observational domains of interest. For example, nested within one particular worldview, a midrange metatheory–methodology conceptualizes the human organism as an *active agent* or as a *dynamic system*, whereas nested within another worldview a midrange metatheory conceptualizes the organism as an *input–output device.*

This described structure of scientific discourse, from commonsense observations to worldviews and back again, also represents the structure of what Imere Lakatos (1978a, 1978b) referred to as a "scientific research program" (see also Laudan, 1977) and the metatheoretical–methodological features of such a program (i.e., what Lakatos referred to as the program's "hard core" and "positive heuristic") identifies Thomas Kuhn's (1962, 1970, 1977) concept of a *scientific paradigm* (see Figure 2.2 for levels with specific examples).

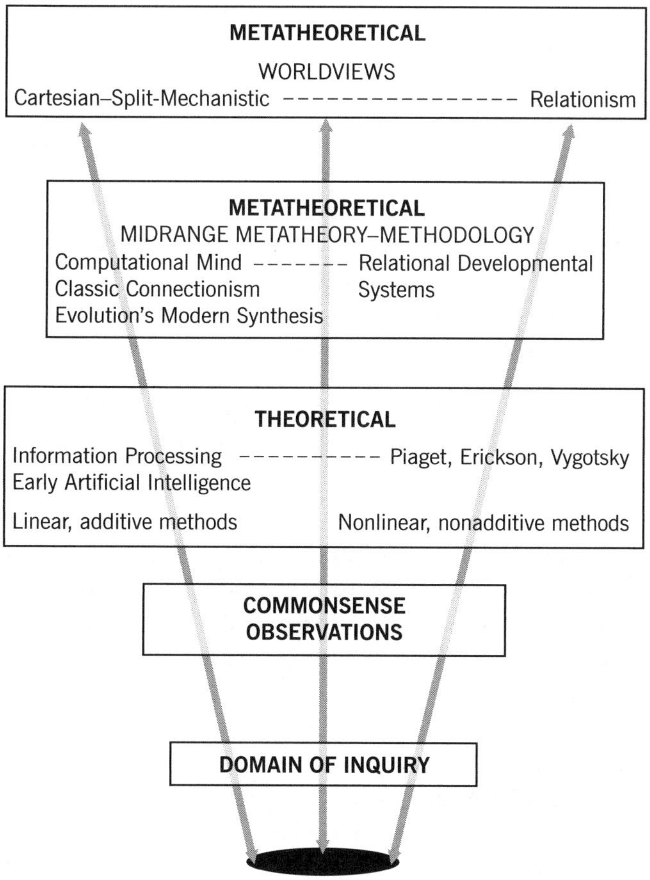

FIGURE 2.2. Levels of discourse and examples of alternative Cartesian and relational scientific research programs.

The Cartesian–Split-Mechanistic Worldview and Split-Mechanistic Midrange Metatheories: Methodologies as Scientific Paradigm

As suggested earlier, until recently the hard core of the worldview as paradigm that constituted the broad abstract framework of many scientific paradigms has been the Cartesian worldview, or, more broadly, the Cartesian–split-mechanistic worldview. Lakatos (1978b) has referred to this as the "Cartesian metaphysics, that is, the mechanistic theory of the universe—according to which the universe is a huge clockwork [and system of vortices] with push as the only cause of motion" (p. 47) and in which fundamental features of this world are split into dichotomous pure forms. A worldview is rarely, if ever, developed by one individual, and this point holds for the Cartesian–split-mechanistic worldview.

Cartesian–Split-Mechanistic Metatheory–Methodology

The early protagonists who developed the basic tenets of this broad metatheory were Galileo Galilei and his physics of a natural world disconnected from mind; René Descartes, whose epistemology elevated disconnection or splitting to a first principle and whose ontology began the path to viewing the world in terms of the categories of the machine, a view that was later elaborated by Newton's admirers such as John Locke; and Thomas Hobbes, who envisioned both mind and nature in terms of an ontology of mechanically operating atomistic materialism. Of the three main figures, Descartes was, perhaps, to have the greatest and most lasting impact on the text and subtexts of this particular metatheoretical–methodological story.

Here I focus primarily on Descartes's major epistemological contributions, although the ontological contribution of Hobbs's and later Newton's static, fixed, materialism cannot be ignored. Descartes's epistemological contributions consisted of the introduction of *splitting*, *foundationalism*, and *atomism* as key interrelated themes in the story of scientific knowing. *Splitting* is the separation of components of a whole into mutually exclusive pure forms or elements. In splitting these ostensibly pure forms are cast into an exclusive "either–or" framework that forces them to be understood as contradictions in the sense that one category *absolutely* excludes the other (i.e., follows the logical law of contradiction: It is never the case that A = not A). But, in order to split, one must accept the twin principles of *foundationalism* and *atomism*. These are the metatheoretical axioms that there is ultimately a rock bottom, unchanging nature to Reality (i.e., with a capital R, distinguishing this ultimate reality from the commonsense reality of everyday objects [see Putnam, 1987]). This conception is Descartes's foundationalism, describing a final fixed secure base. It constitutes absolute, fixed, unchanging bedrock; a final Archimedean point (Descartes, 1969). Furthermore, this rock bottom is composed of elements—pure forms—(the atoms of atomism) that preserve their identity regardless of context. A corollary principle here is the assumption that all complexity is *simple complexity* or *simply complicated* in the sense that any whole is taken to be a purely additive combination of its elements.

Splitting, foundationalism, and atomism are all principles of decomposition; that is, of breaking the aggregate down to its smallest pieces, to its bedrock (Overton, 2006). This process also goes by other names, including *reductionism* and the *analytic attitude.* Split metatheory–methodology, however, requires another principle to reassemble or recompose the whole. This is the principle of *unidirectional, linear, and additive associative or causal sequences.* The elements must be related either according to their contiguous co-occurrence in space and time, or according to simple efficient or material mechanical cause–effect sequences[1] that proceed in a single direction (Bunge, 1962; Overton & Reese, 1973). In fact, split metatheory–methodology admits no determination other than individual efficient and material causes, or these individual causes operating in a conjunctive (i.e., additive) plurality. Truly reciprocal bidirectional causality (↔),[2] circular causality, and downward causality (Witherington, 2011) are *not permitted* in this system (Bunge, 1962; Overton & Reese, 1973).

Midrange Cartesian Metatheories–Methodologies

As suggested above, a scientific paradigm entails both the ontological and epistemological assumptions of a worldview as well as midrange metatheory–methodologies nested within this worldview. The midrange metatheory–methodologies vary across the several subfields of developmental science. Thus, for example, in the field of population genetics, a midrange metatheory–methodology has utilized the Fisher–Wright ANOVA model; in traditional evolutionary biology a midrange metatheory–methodology is the modern synthesis; in cognition the computational model of mind is a midrange metatheory–methodology; and in traditional cultural approaches it has been the person–culture dichotomy.

There are numerous ways in which the Cartesian–split-mechanistic worldview and its midrange metatheory–methodologies have impacted issues in developmental science. Here, two are briefly mentioned. The first example entails the nature–nurture issue. In classic genetics the gene was introduced as an analogue to the chemical element as the foundational biological element (Keller, 2010); the gene was conceptualized as the "master molecule" that "causes" the production of proteins; a linear additive, unidirectional causal path was asserted to operate from DNA to RNA to protein as defined by the "central dogma of molecular biology" (Gottlieb, 2000). With respect to population (quantitative) genetics, the relation of genes to environment has been conceptualized within a completely additive (Lewontin, 1974; Overton & Reese, 1973) model, and statistical gene × environment interactions are themselves completely decomposable into strictly additive elements. As Turkheimer (2011) points out, this assumption of additivity is "the foundation of modern quantitative genetics" (p. 228).

The second example concerns evolution. In the evolutionary modern synthesis model, development and evolution are split (e.g., see Lickliter & Honeycut, 2010). Furthermore, internal is split off from external, yielding an isolated internalism of gene centrism (i.e, gene as the sole unit of variation), and an isolated externalism of change (i.e., emphasis on natural selection as the virtually sole vehicle of change) (e.g., see Pigliucci & Mueller 2010a). The modern synthesis model also entails commitment to evolutionary "gradualism" (i.e., additive continuity) that derives from the same mathematical formalism identified with the ANOVA model that Pigliucci and Mueller (2010a) refer to as the "backbone" of population genetics, and, hence the backbone of the modern synthesis.

Relationism and Relational Developmental Systems: A Paradigm for Developmental Science

While the Cartesian paradigm undoubtedly continues to exert a significant influence, it has come under increased criticism as an adequate scientific paradigm for developmental science. These criticisms are cross-disciplinary, coming from many fields, including biology and neuroscience (e.g., Damasio, 1994; Edelman, 1992, 2006; Gallese & Lakoff, 2005; Gilbert & Epel, 2009; Mayr, 1996), philosophy (Gallagher, 2005; Taylor, 1995), anthropology (Ingold, 2000; Sheets-Johnstone, 1990), and psychology (Barsalou, Simmons, Barbey, & Wilson, 2003; Colombetti & Thompson, 2008; Goode, 2007; Hobson, 2002; Müller & Newman, 2008; Rowlands, 2010; Smith, 2005). As a consequence, many in the developmental sciences (e.g., Blair, 2010; Bundick, Yeager, King, & Damon, 2010; Carpendale & Lewis, 2010; Geldhof, Little, & Colombo, 2010; Kuczynski & Parkin, 2009; Kurtines et al., 2008; Lerner, 2006; Lerner et al., 2012; Lewis, 2010; MacWhinney, 2010; Marshall, 2009, in press; Mascolo & Fischer, 2010; McClelland, Ponitz, Messersmith, & Tominey, 2010; Müller & Racine, 2010; Ricco & Overton, 2011; Santostefano, 2010; Stewart, Gapenne, & Di Paolo, 2010; Turiel, 2010; Witherington, 2011; Witherington & Heying, 2013) have begun to shape their science within a conceptual framework that replaces the ontology of fixed and static with active and changing, and contingent organization with necessary organization. This framework further replaces the epistemology of atomism and reductionism with holism and system analyses; splitting with relations; foundational elements with networks; and unidirectional, linear, additive associative or causal sequences with emergent directional sequences entailing nonlinear, nonadditive reciprocal and circular feedback systems. This framework is the worldview of relationism and the midrange metatheory–methodology of relational developmental systems.

Relationism

As with any worldview, relationism has a generative history. It owes much to Aristotle's treatment of form and matter, as well as being (fixed) and becoming (change) as relational, not isolated elements. Similarly important was Aristotle's insistence that complete explanation must involve the coordination of several types of causality (Witherington, 2011), including formal and final, and not limited to split-off efficient or material cause. Kant's insistence that mind is a system, an active system, constitutes a major contribution, as does Hegel's introduction of the dialectic as an active holistic developing process. Brentano's (1973) act psychology, with its focus on the centrality of intentional act, along with James and Dewey's development of pragmatism, were also critical contributions. And, more recently, Bruno Latour (1993, 2004) proposed a move away from the extremes of Cartesian splits to a center or *middle kingdom* position, wherein entities and ideas are represented, not as pure forms, but as forms that flow across fuzzy boundaries. This movement is one toward what Latour terms *relationism*, a metatheoretical space where foundations are groundings, not bedrocks of certainty, and analysis is about creating categories, not about cutting nature at its joints. The present version of relationism builds on Latour's proposal.

Relationism as a worldview formed as a principled synthesis of Pepper's (1942) organicism and contextualism (for details, see Overton, 2007; Overton & Ennis, 2006a, 2006b). As a worldview it is composed of a coherent set of intertwined ontological and epistemological principles. The ontology of relationism offers a Real based on *process–substance* rather than a split-off substance (Bickhard, 2008). This ontology is what Gadamer (1989) argues to be the movement of *to and fro* and what has been sometimes defined as an ontology of *Becoming* (Allport, 1955; Overton, 1991). It includes process, activity, dialectical change, emergence, and necessary organization as fundamental defining categories, but it does not exclude categories of substance, stability, fixity, additive, and contingent organization.

The epistemology of relationism is, first and foremost, a *relatively inclusive* epistemology, involving both knowing and known as equal and indissociable complementary processes in the construction, acquisition, and growth of knowledge. It is *relatively* inclusive because *inclusion* itself—much like Hegel's master–slave dialectic—can be grasped only in relation to its complement, *exclusion*. Thus, just as *freedom* must be identified in the context of *constraint, inclusion* must be identified in the context of *exclusion*. Relational epistemology specifically excludes Cartesian dualistic ways of knowing because Cartesian epistemology trades on *absolute* exclusivity; it constitutes a *nothing-but* epistemology. For the same reason, relationalism rejects both the mechanistic worldview and a strict contextualist interpretation of the contextualist worldview (Overton, 2007; Witherington, 2007, 2011). Epistemologically, relationism begins by clearing the "nothing-but" *splitting, foundationalism, atomism*, and *objectivism* from the field of play, and in so doing, it moves toward transforming antinomies into coequal, indissociable complementarities. In the relational frame, fixed absolute elements are replaced by contextually defined parts.

In place of the rejected splitting, foundationalism, and atomism, relationism installs *holism* as the overarching epistemological first principle. Although holism is central to relationism, holism does not, in itself, offer a detailed program for resolving the many dualisms that have framed scientific knowing and knowledge under a Cartesian paradigm. A complete relational program requires principles according to which the individual identity of each concept of a formerly dichotomous pair is maintained, while simultaneously it is affirmed that each concept constitutes, and is constituted by, the other. This understanding is accomplished by considering identity and differences as two *moments of analysis*. The first moment is based on the principle of the *identity* of opposites; the second moment is based on the principle of the *opposites* of identity.

The Identity of Opposites

The principle of the identity of opposites establishes the *identity among parts* of a whole by casting them, not as exclusive contradictions, as in the split epistemology, but as differentiated polarities (i.e., coequals) of a unified (i.e., indissociable) inclusive matrix—as a *relation*. As differentiations, each pole is defined recursively; each pole defines and is defined by its opposite. In this identity moment of analysis, the law of contradiction is suspended and each category contains and, in fact, *is* its opposite. Further—and centrally—as a differentiation, this moment pertains to character, origin, and outcomes. The character of any contemporary behavior, for example, is 100% nature because it is 100% nurture; 100% biology because it is 100% culture. There is no origin to this behavior that was some other percentage—regardless of whether we climb back into the womb, back into the cell, or back into the DNA—nor can there be a later behavior that will be a different percentage.

There is a number of ways to illustrate this principle; one particularly clear illustration is found in the famous ink sketch by M. C. Escher, titled *Drawing Hands* (Overton, 2006). In this sketch a left and a right hand assume a relational posture according to which each is simultaneously drawing and being drawn by the other (see Figure 2.3a as a schematic illustration). In this matrix there is a sense in which each hand is different (opposite left and right hands) and a sense in which the hands are identical (each is drawing and being drawn). In the latter analytic identity of opposites, the hands are identical (i.e., *A* and not *A*)—thus coequal and indissociable. This moment of analysis is one in which the law of contradiction (i.e., not the case that *A* = not *A*) is relaxed, and identity (i.e., *A* = not *A*) reigns. In this identity moment of analysis, pure forms or the notion of "natural kinds" collapse, and categories flow into each other. Here each category contains, and is, its opposite. As a consequence, there is a broad inclusivity established among categories.

Within the identity moment of analysis, it is often a useful exercise to write on each hand (or the arrows of the schematic) one of the bipolar terms of often-split dualisms (e.g., biology and culture, genotype and phenotype, development and evolution, encapsulated and extended mental processes, person and culture) and to explore the resulting effect (e.g., see Figures 2.3b and 2.3c). This exercise is quite different from an illustration of a familiar, simple bidirectionality of mechanical cause and effect. This exercise makes tangible a central tenet of the relational metatheory–methodology: Seemingly dichotomous ideas often viewed as competing alternatives (see Table 2.1) can, in fact, enter into inquiry as coequal and indissociable. This exercise also concretizes the meaning of "causality" within relationism. In this framework, the concepts of *reciprocal determination* (Overton & Reese, 1973), *coaction* (Gottlieb, Wahlsten, & Lickliter, 2006), *fusion* (Greenberg, 2011; Partridge, 2011), as well as *reciprocal* and *relational bidirectional* (↔) *causality* (Lerner, 2006), *relational causality* (Gottlieb, 2003), and *circular causality* (Witherington, 2011) are synonyms used to differentiate the positive and negative feedback loops of relationism from additive (even bidirectionally additive mechanistic) causality of the Cartesian–split-mechanistic worldview. It would, in fact, significantly reduce conceptual confusions if the terms *cause* and

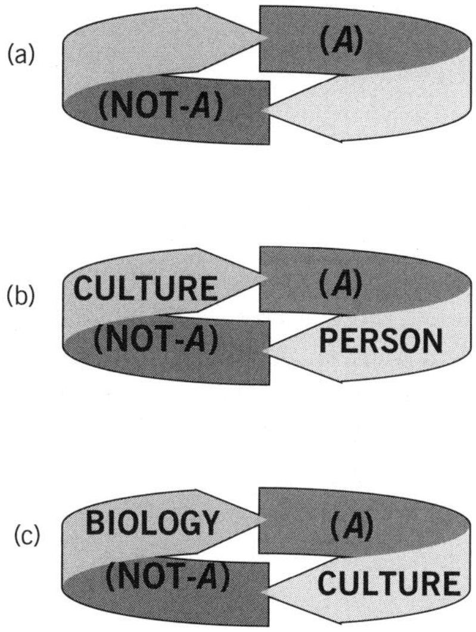

FIGURE 2.3. Relationism. (a) The identity of opposites; (b) culture–biology as an identity of opposites; (c) biology–culture as an identity of opposites.

causation were reserved for mechanical cause–effect and the causation of constant conjunction, whereas the several relational determinants were referred to simply as *explanations*. Miller (2010) suggests generally that in place of *causality* we use "perhaps with less baggage, *explanation*, or even just discoveries of regularities in nature" (p. 725), and Redhead (1990) goes further and argues against the need for the concept of causality at all in science.

The principle of the identity of opposites imposes theoretical and methodological constraints on any field of inquiry—biological, evolutionary, person-centered, cultural—just as other metatheories impose constraints on any field of inquiry. The primary constraints within relationism are that (1) splits are not permitted (e.g., the split of genotype and phenotype in genetics; the split of internalism and externalism in

TABLE 2.1. Fundamental Antimonies/False Dichotomies

Brain	Body	Continuous	Discontinuous
Mind	Body	Quantitative	Qualitative
Biology	Culture	Internal	External
Person	Biology	Universal	Particular
Culture	Person	Transcendent	Immanent
Nature	Nurture	Analysis	Synthesis
Subject	Object	Unity	Diversity
Form	Matter	Interpretation	Observation
Stability	Change		

the modern synthesis; the split of brain, body, and culture in cognitivism; the cultural split of person and culture) and (2) phenomena cannot be conceived as being decomposable into independent and additive pure forms (e.g., the Fisher–Wright AVOVA model in genetics and in the modern synthesis).

If the principle of the identity of opposites introduces constraints, it also opens possibilities. One of these is the recognition that, to paraphrase Searle (1992), the fact that a behavior implicates activity of the biological system does not imply that it does not implicate activity of the cultural system, and the fact that the behavior implicates activity of the cultural system does not imply that it does not implicate activity of the biological system. In other words, the identity of opposites establishes the metatheoretical rationale for the theoretical position that biology, person, and culture operate in a truly *interpenetrating, relational* manner.

The Opposites of Identity

Although the identity of opposites sets constraints and opens possibilities, it does not, in itself, set a positive agenda for empirical scientific inquiry. The limitation of the identity moment of analysis is that, in establishing a flow of categories of one into the other, a stable base for inquiry that was provided by bedrock material *atoms* of the split metatheory is eliminated. In the split approach no relativity entered the picture; all was absolute. Reestablishing a *stable base*—not an absolute fixity, nor an absolute relativity, but a *relative relativity* (Latour, 1993)—within relational metatheory requires moving to a second moment of analysis. This is the oppositional moment, wherein the figure of identity and ground of opposites reverses and opposites become figure. This moment becomes dominated by a *relational exclusivity*. Thus, in this opposite moment of analysis, it becomes clear that, despite the earlier identity, the schematic of Escher's sketch does illustrate both a *right* hand and a *left* hand (see Figure 2.4 for *culture* and *biology* example). In this moment of opposition, the law of contradiction is reasserted and categories again exclude each other. As a consequence of this exclusion, parts exhibit *unique* identities that differentiate each from the other. These unique differential qualities are stable within any holistic system and thus may form relatively stable platforms for empirical inquiry. The platforms created according to the principle of the opposites of identity become *standpoints, points of view,*

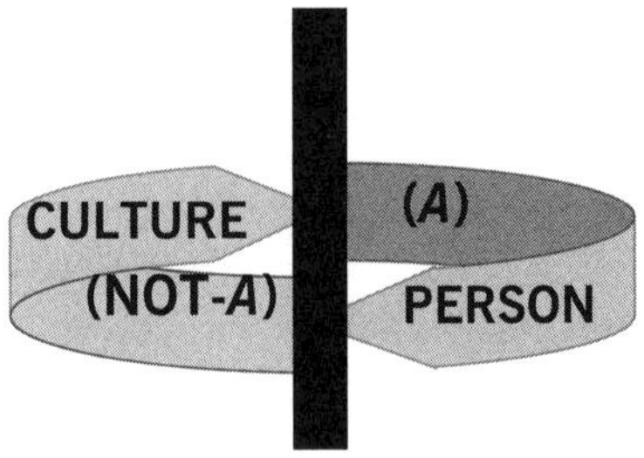

FIGURE 2.4. Relationism. The opposites of identity.

or *lines of sight*, in recognition that they do not reflect absolute foundations (Latour, 1993, 2004) but perspectives in a multiperspective world. They may also be considered under the common rubric *levels of analysis*, when these are not understood as bedrock foundations.

Again, thinking of the Escher sketch (or the schematic of Figure 2.3), when left hand as left hand (*A*) and right as right (not *A*) are separately the focus of attention, it then becomes quite clear that, were they large enough, one could stand at either hand and examine the structures and functions of that location, as well as its relation to the other location (i.e., the *coactions* of parts). Thus, to return to the example of nature–nurture, although explicitly recognizing that any behavior is both 100% biology and 100% culture, alternative points of view permit the scientist to analyze the acts of the person from a *biological* or from a *cultural standpoint. Biology and culture no longer constitute competing alternative explanations; rather, they are two points of view on an object of inquiry that has been created by, and will be fully understood only through, multiple viewpoints.* More generally, *the unity that constitutes the organism and its development becomes discovered only in the diversity of multiple interrelated lines of sight.*[3]

Relational Developmental Systems

Taken as a whole—including both its ontological and epistemological assumptions—relationism operates as the contextual frame for the construction of midrange metatheories. These latter metatheory–methodologies are less broad in scope, more specific to particular domains of inquiry, and together with relationism, constitute a *conceptual framework* for a scientific paradigm. Relational developmental systems is itself the broadest of these midrange metatheory–methodologies—all of which incorporate systems concepts—including developmental, dynamic, dialectical, transactional systems, and enaction. It is important to note that there is no relational developmental systems theory—relational developmental systems is a *metatheory*. There are, however, theories that are informed by relational developmental systems and, hence, one might say that there are relational developmental systems theories. Relational developmental systems represents an extension (Lerner, 2006, 2012; Lerner & Overton, 2008; Overton, 2006, 2010, 2012, 2013; Overton & Lerner, 2012) of the original developmental systems "theory" described by Ford and Lerner (1992) and Gottlieb (1996). This extension was motivated by an increasing recognition of relationism as a central feature of the conceptual framework of an alternative scientific paradigm to that formulated within the Cartesian–split-mechanistic worldview.

Relational developmental systems is a perspective on developmental science (i.e., development [behavioral, cognitive, motivational, emotional, sociocultural], inheritance, and evolution). The relational nature of the system emphasizes causality as reciprocal bi- or multidirectional (↔) or circular (positive and negative feedback loops). All facets of the person and the context exist in mutually influential relations (Elder, 1998; Molenaar, 2007). Accordingly, the potential for *plasticity* (Batson & Gluckman, 2011; Charney, 2012; Lerner, 1984; West-Eberhard, 2003) of *intraindividual change* (Nesselroade & Molenaar, 2010; Nesselroade & Ram, 2004) is a hallmark of relational developmental systems.

This metatheory–methodology conceptualizes living organisms as *active agents*—that is, as relational, spontaneously active, complex adaptive systems, that are self-creating (i.e., enactive, autopoetic), self-organizing (i.e., a process according to which higher-level-system organization arises solely from the coaction of lower-level components of the system), and self-regulating. Furthermore, the developmental process—including embryogenesis, ontogenesis, and phylogenesis—is conceptualized as entailing five defining features: (1) *nonlinearity* (i.e., inputs are not proportional to outputs), (2) *order* and *sequence*, (3) *direction*, (4) *relative permanence* and *relative irreversibility*, and (5) *epigenesis* and *emergence*. *Epigenesis* is conceptualized as "probabilistic epigenesis" (Gottlieb, 1992), which designates a *holistic* approach to understanding developmental complexity. *Probabilistic epigenesis* is the principle that the role played by any part of a relational developmental system—DNA, cell, tissue, organ, organism, physical environment, culture—is a function of all of the interpenetrating and coacting parts of the system. It is through complex *reciprocal bidirectional* and *circular reciprocal interpenetrating* actions among the coacting parts that the system moves to levels of increasingly organized complexity. Thus, epigenesis identifies the system as being completely *contextualized* and *situated; time and place matter* (Elder, 1998).

Epigenesis entails the closely related feature of *emergence of system novelty*. As systems change, they become increasingly complex. This increased complexity is a complexity of form rather than an additive complexity of elements. The butterfly emerges from the caterpillar through the differentiation and reintegration of organization, the frog from the tadpole, the plant from the seed, and the organism from the zygote. In an identical fashion, higher-order psychological structures emerge from lower-order structures; also in an identical fashion, new forms of organization exhibit novel features that cannot be reduced to (i.e., completely explained by) or predicted from earlier forms. This impossibility of reduction asserts what Witherington refers to as ontological or "strong" emergence (Witherington, 2011; see also Blachowicz, 2012). The novel features are termed *systemic*, indicating that they are properties of the whole system and not properties of any individual part. This emergence of novelty is commonly referred to as *qualitative* change in the sense that it is change that cannot be represented as purely additive. Similarly, reference to a "discontinuity" in development is simply the recognition of emergent novelty and qualitative change in a system.

System constitutes the core concept of this metatheory, and this concept has been defined in various ways. For example, van Geert (2003) offers "any collection of phenomena, components, variables" (p. 655). However, this conception and other "collection" or aggregate-like definitions are inconsistent with holism and, consequently, inconsistent with relational developmental systems. A more adequate relational definition of system is "a whole which functions as a whole by virtue of the interdependence of its parts" (Overton, 1975, p. 73). Thus, a system is, by its nature, organized and organized holistically. Furthermore, the relational system is an adaptive system. Here *adaptation* refers to how the system responds to changing environments—to *perturbations* in systems language—so as to increase its probability of survival, not in the sense of *adjusting* to an environment. Adaptive systems are defined in contrast with "determined" systems. In determined systems the relation between inputs and outputs are exactly and reproducibly connected. For example, an automobile is a determined system. When the driver presses the accelerator or turns the steering wheel, both driver and passenger expect the auto to speed up or turn. All components of the auto must be fully determined to achieve this collective response. And determined systems are linear in the sense that small inputs result in small outputs; large inputs in large outputs—thus, outputs are predictable. In adaptive systems, the parts follow simple rules, whereas the behavior of the whole system is not determined; it is probabilistic because it is a function of all of the timing of the relations among the interpenetrating events, including non-normative events.

A second core concept of relational developmental systems is *action*. The relational developmental system is the source of action. At *subpersonal levels*, where it is not necessary to limit a definition to organismic development or even to living systems, action is defined as the *characteristic functioning of any complex adaptive system-organizing system*. For example, weather systems form high- and low-pressure areas and move from west to east. Living systems, on the other hand, organize and adapt to their biological, sociocultural, and environmental worlds. At the *person level*, organismic developmental action is defined as *intentional activity* (i.e., meaning-giving activity). Intentionality, however, is not to be identified solely with consciousness. Although all acts are intentional, only some intentions are conscious or self-conscious. In a similar fashion, intention is not to be identified solely with a symbolic or reflective level of knowing. Following Brentano (1973), all acts, even those occurring at early sensorimotor levels of functioning, intend some object.

The primary function of action is that at the microscopic level, it represents the *general mechanism for all development*. It is through the co-constituting actions of any target system of interest (e.g., genetic, epigenetic, cell, zygote, embryo, fetus, infant . . . species) with its environments, as well as the resistances (perturbations) the target system encounters, that the system changes and thereby becomes differentiated and reintegrated at increasingly complex and novel levels of organization.

A final core concept is that of *embodiment* (Overton, 1994, 2008). All acts are embodied acts and consequently the general case is that *embodied action is the general mechanism for all development*. Embodiment represents the interpenetrating relations between person, biology, and culture. It is the claim that perception, thinking, feelings, desires—*the way we behave, experience, and live the world*—are contextualized by our being *active agents* with this *particular kind of body* (Taylor, 1995). The kind of body we have is a constitutive precondition for having the kind of behaviors, experiences, and meanings that we have. Embodiment includes not merely the physical structures of the body but *the body as a form of lived experience, actively engaged with the world of sociocultural and physical objects*. The *body as form* references the biological point of view; the *body as lived experience* references the psychological subject standpoint; and the *body actively engaged with the world* represents the sociocultural point of view. Within a relational context, embodiment is a concept that bridges and joins these several research points of synthesis in a unified whole without any appeal to splits, foundationalism, elements, atomism, or reductionism.

A Historical Account of Changes in Methodological Strategies: From Cartesian to Relational (Relationism–Relational Developmental Systems)

To this point I have described two broad scientific paradigms—Cartesian and relational—that function as alternative conceptual frameworks for any broad scientific research program. Each paradigm functions at metatheoretical–methodological levels of scientific discourse. With respect to the specifically methodological, several features of each paradigm have been discussed (e.g., reductionism vs. holism, additive vs. nonadditive, linear vs. nonlinear). However, across history both the Cartesian and relational paradigms have also generated sets of alternative methodological strategies that set the standards for what is and what is not considered legitimate science. An understanding of the nature of these methodological strategies is critical to the continuing development of productive, fruitful methods in developmental science, and it further demonstrates the continuity of concepts from the broad headwaters of worldviews to specific theoretical concepts and down to issues of design, measurement, and modeling efforts. Figure 2.5 illustrates the history, background, and main features of Cartesian methodological strategies, including neopositivism, conventionalism/instrumentalism, and relational methodology. Also included is the transitional methodological strategy of hermeneutics.

Cartesian Methodology: The Newtonian Tradition

Cartesian methodology begins by splitting inference from observation, observation from explanation, and explanation from understanding. Thus, since the 18th century, science within a Cartesian framework has been dichotomized into two airtight compartments consisting of *pristine observations* and *explanations*. The pristine observations compartment is the bedrock of the epistemology of empiricism—the idea asserted by British empiricists John Locke and David Hume—that ultimately knowledge must rest on *sensory* experience—untouched by inference or background ideas (i.e., metatheoretical)—and only sensory experience. The explanation compartment has entailed either linear causal sequences of, or empirical correlations among, observations. Each of these was generated by Newton's methodology—called *mechanical explanation* (Prosch, 1964)—which became the Cartesian paradigm's route from common sense to critically refined scientific knowledge. This Cartesian methodology has taken several forms over the centuries, from the 19th-century "positivism" of Comte and Mach, to the early 20th-century "neopositivism" of the logical positivists, and the mid-20th-century "instrumentalism" or "conventionalism" (Popper, 1959, 1963, 1970). And, most importantly, the legacy of this Cartesian methodology often continues to shape the thought of some contemporary developmental scientists. Sterba and

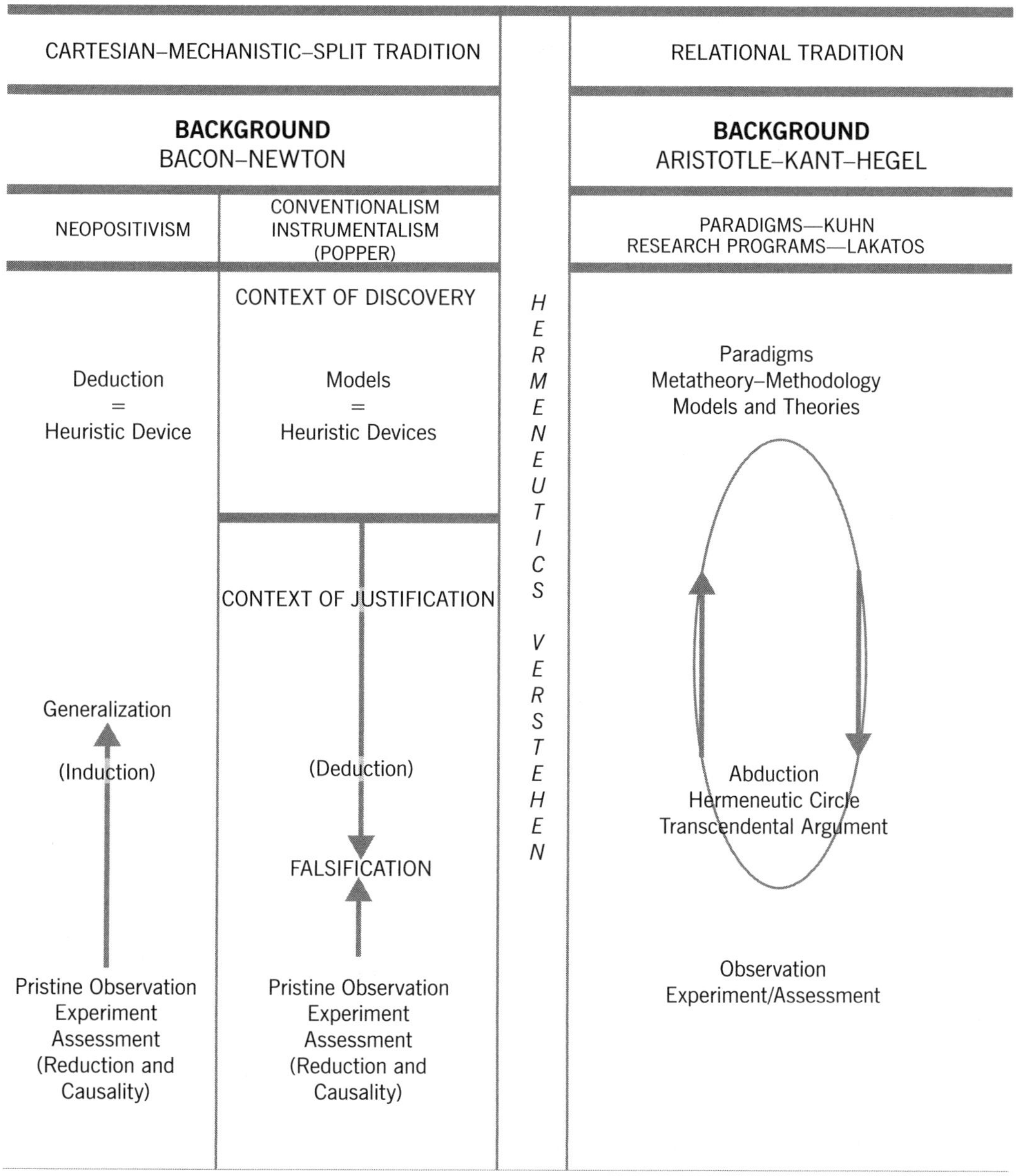

FIGURE 2.5. The history and critical features of scientific methodologies.

Bauer (2010) capture this impact as it affects methods when they point out that despite the rise of several influential non-Cartesian system perspectives on development that have emerged over the past 25 years,

> the most commonly used longitudinal methods of the last quarter century for examining stability (e.g., panel and cross-lag models) and change (repeated measures analysis of variance, analysis of differences scores, analysis of residual change) largely preclude . . . [these] perspectives. These highly restrictive *variable-oriented methods* (Block, 1971) assume that, given prior behavior scores, genetic markers, contextual risk and protective factors, and so forth individuals are interchangeable units who, apart from random error, differ neither quantitatively nor qualitatively in behavioral course. (p. 239)

Cartesian methodology can most clearly be viewed through the lens of Newton's mechanical explanation. There are three steps in this process. The first is considered descriptive, and the second and third are considered explanatory.

Step 1: Reduction–Description

The first step of mechanical explanation entails addressing the commonsense object of inquiry and *reducing* it to the absolute material, objective, fixed, unchanging, foundational elements or atoms. As stated earlier, terms such as *reductionism, atomism, elementarism*, and *analytic attitude* all identify this step. In psychology for many years the atoms were *stimuli* and *responses*. Today they tend to be *neurons* and *genes* and *contextual variables*. Within this paradigm the storyline changes over time, but the theme remains the same. In keeping to the epistemology of empiricism, the broad stricture here is to reduce all scientific concepts to the visible.

Briefly consider one impact of this first step on developmental inquiry. Immediately, "transformational change," "stages" of development, and the "mental organizations" and "systems" that emerge during development become suspect as being somehow derivative because they are not directly observable. Similarly, efforts at constructing statistical models for discontinuous change (e.g., Chow & Filteau, 2010; Kaplan, 2008; van der Maas & Molenaar, 1992) become vacuous for the same reason. At best under this storyline, transformations, stages, and systems can only function as summary statements for an underlying, more molecular-based, continuous ontological reality. In fact, the drive throughout this step is toward the ever more molecular in the belief that it is in the realm of the molecular that the ultimate Real is directly observed. This is particularly well illustrated in the enthusiasm for a *microgenetic* method (e.g., Kuhn, Garcia-Mila, Zohar, & Andersen, 1995; Siegler, 1996; Siegler & Crowley, 1991, 1992) as a method that has been presented as offering "a *direct* means for studying cognitive development" (Siegler & Crowley, 1991, p. 606, emphasis added). This use of the microgenetic method as an intensive "trial-by-trial analysis" reduces the very notion of development to a molecular bedrock of visible behavioral *differences* as they appear across learning trials. This approach is very different from the relational work of Nesselroade and Molenaar (2010), which, although also arguing for extensive repeated observations of single cases, does so in the service of modeling intraindividual changes in latent processes.

It is important to recognize that the aim of Step 1 is to drive out *interpretations or anything but inductive inferences* from the commonsense phenomena under investigation. Under the empiricist theme, commonsense observation is error laden, and it is only through ever more careful *neutral* observation that science can eliminate this error and ultimately arrive at the elementary bedrock that constitutes the level of "facts" or "data" (i.e., invariable observations).

Step 2: Explanation

Step 2, mechanical explanation, begins to move inquiry into the second compartment—*explanation.* Step 2 consists of the instruction to find the relations among the elements described in Step 1. More specifically, given our objects of study in developmental science—behavior and behavior change—this step directs inquiry to locate antecedents. These antecedents may be correlational in nature, in which case the explanation is not causal. However, when the antecedents meet certain criteria of necessity and sufficiency, they are termed *causes*, and *the discovery of cause often defines explanation* within this methodology. The antecedents are also often referred to as *mechanisms*, but the meaning here is identical. This point is another place at which to pause and notice an important impact of this paradigm methodology. Here, because of the particular methodological principles involved, the word *explanation* is defined exclusively as an antecedent–consequent relation, or the efficient/material proximal cause of the object of inquiry. Furthermore, within this framework science itself often is defined as the (causal) explanation of natural phenomena.

It is critically important to remember here that Aristotle had earlier produced a very different, more relational, methodological story of scientific explanation. Aristotle's schema entailed *complementary relations* among four types of explanation, rather than a splitting. Aristotle (Randall, 1960; Ross, 1959) proposed that the complete explanation of any phenomenon requires four interconnected types of explanation: (1) material, (2) efficient, (3) formal, and (4) final. Aristotle referred to each of these as "causes," but as we shall see, the first two differ from the second two to an extent that the application of *cause* to these is questionable.

Material cause refers to the substance that constitutes the object of inquiry. Contemporary examples in the human sciences include the various hereditary, physiological, and neurological features of the organism. *Efficient* cause describes that which "moves" the object of inquiry or the force applied to the object. In the human sciences efficient and material causes have fuzzy boundaries, especially in the case of a biological explanation as a material cause (e.g., neuron), which can also often be thought of as an efficient (e.g., the firing of the neuron) cause. Efficient cause also has come to include the split-off stimulus, the split-off situation, the split-off environment, the split-off social–cultural context, and so on.

Formal cause is the articulation of the form, structure, pattern, order, morphology, or organization of the action or activity of the object of inquiry. This type of determinant is particularly critical to any type of action system approach. *Final cause* is the articulation of the directional order of the action or activity of the object of inquiry; it is a necessary determinant for any developmental action system. A critical distinction here is that Aristotle did not present formal and final as causes in the sense that a cause implies something productive, as in "Cause *A* produces effect *B*." Formal and final were presented as *principles of intelligibility* or *regulative principles* (Randall, 1960; Taylor, 1995). They give meaning to the activity of interest, both in terms of its present organization and in terms of the directed sequence of emergent changes that the active system undergoes. As suggested earlier, it would add significantly to conceptual clarity if the term *cause* were limited to efficient and material factors, while referring to other determinants, including formal and final, as simply *explanations*. Here, this will be done whenever possible. However, because *reciprocal, circular*, and *downward* determinants entail issues of formal and final explanation, but have traditionally employed the qualifier *cause*, conceptual clarity is sometimes sacrificed to tradition.

Today, the structure of the atom, the structure of DNA, the structure of the solar system, and the structure of the universe are all familiar examples of formal explanation drawn from the natural sciences. Kinship structures, mental structures, mental organization, dynamic systems, the attachment behavior system, structures of language, ego and superego, dynamisms, schemes, operations, and cognitive structures are familiar examples of formal explanation drawn from the human sciences. Witherington (2011) presents an insightful and detailed discussion of the central role of formal explanation in the understanding of circular causality in dynamic systems. With respect to contemporary uses of the notion of final explanation, reference to the sequence and directionality found in the Second Law of Thermodynamics, self-organizing systems, and the psychological developmental concepts of "equilibration process," "reflective abstraction," the "orthogenetic principle," and "probabilistic epigenesis" are all relevant examples (Overton, 1994).

It is of note that material and efficient causes are potentially observable, whereas formal and final must be known by interpretive inference. This means that in the Cartesian *split* story of mechanical explanation—as guided by reductionism—formal and final principles completely lose any explanatory status; this is because in the Cartesian scheme mechanical explanation is limited to *nothing but observable* efficient and material causes. At best, within the mechanical story formal and final principles may reappear in the descriptive compartment as mere summary statements of the underlying molecular descriptive Real discussed in Step 1. In this way, relational developmental systems and the directed sequence of emergent changes they engender become eliminated or marginalized with respect to developmental inquiry.

Step 3: Induction of Interpretation-Free Hypotheses, Theories, and Laws

Step 3 of mechanical explanation installs *induction* as the foundational logic of science. Step 3 instructs the investigator that ultimate explanations in science must be found in fixed, unchanging laws, and these must be inductively derived as *empirical generalizations* from the repeated observation of associative or cause–effect relations found in Step 2. Weak generalizations from Step 2 regularities constitute interpretation-free "hypotheses." Stronger generalizations constitute interpretation-free theoretical propositions. Theoretical propositions joined as logical conjunctions (*and* connectives) constitute interpretation-free theories. Laws represent the strongest and final inductions.

Deduction enters the methodological picture here as a split-off heuristic method of moving from inductively derived hypotheses and theoretical propositions to further empirical observations. When in the 1940s a variation on the Step 3 theme was introduced with the development of the so-called *hypothetico-deductive method*, it was simply more variation on the same theme. In this variant the hypothesis of "hypothetico" has nothing to do with an interpretive inference or conjecture, but is simply an empirical generalization driven by pristine data that then serve as a major premise in a formal deductive argument. Similarly, at a later time when *instrumentalism* or *conventionalism*—best illustrated in the work of Karl Popper (1959, 1963, 1970)—moved away from the hypothetico-deductive[4] stance to allow the introduction of theoretical models into scientific discourse, models themselves functioned merely as the same type of heuristic devices as that served by deduction in neopositivism. That is, models were split off and marginalized into a box called the "context of discovery," as the empiricist epistemological real work of science functioned in another box, called the "context of justification."

Another important variation—but a variation nevertheless—on this same theme was the so-called *covering law model* of scientific explanation. This concept was introduced by Carl Hempel (1942; Hemple & Oppenheim, 1948) and became an important prototype of all later explanations formulated within this methodology. Understanding the covering law model is particularly important for developmental science because it treats historical events as analogous to physical events in the sense that earlier events are considered the causal antecedents of later events (Ricoeur, 1984). This treatment of historical events leads directly to the "highly restrictive variable-oriented methods" discussed earlier by Sterba and Bauer (2010).

According to the covering law model (also termed the *subsumptive model* and the *deductive-nomological model*), scientific explanation takes a deductive logical form; that is, particular events are explained when they are logically subsumed under a universal law or law-like statement (i.e., *a highly confirmed inductive empirical generalization*) (Ayer, 1970; Hempel, 1942).

As an example of the covering law model, assume that we have available an inductively drawn universal law of socialization that states:

> For any *x*, if *x* is a human child, then *x* increases in social competence when parents act responsively to the child's needs.

This law becomes the major premise of the deductive argument. Next, the initial conditions of an experiment present antecedent conditions (causal conditions) of a minor premise:

This is a human child (or group of human children), and the parents of this child (these children) have acted responsively.

Finally, assume that the experiment yields the observed event or conclusion:

This child (group of children) increase(s) in social competence.

Major premise, minor premise, and conclusion in this example are consistent, and the observed event (group of children increasing in social competence) is said to be explained by the law that covers it. It should be noted, as Popper (1959, 1963, 1970) later did, that the actual logic of this proposal is faulty in that it entails an invalid argument form called the *fallacy of the affirmed consequence.* Stated without the logical formality, this simply means that the observed event could have occurred for many other reasons having nothing to do with the *law* itself. It should also be noted that if the observed event had been "This child (group of children) does *not* increase in social competence," then that observation would have, indeed, constituted a valid logical refutation or falsification of the law itself (or of a theoretical proposition, or hypothesis, if these had been the major premise). In this latter example, major premise, minor premise, and conclusion comprise the valid deductive argument form *modus tollens* (If *p*, then *q*; not *q*; therefore, not *p*). It was this distinction between the invalid *affirmed consequence* argument of the covering law model and the valid argument of *modus tollens* that later became the base of Popper's (1959) contention that science proceeds by falsification rather than confirmation.

When the covering law model includes the antecedent as cause, then *cause, law, explanation*, and *prediction* are tightly interwoven. An event is explained when it is "covered" by a law or law-like statement (i.e., a universal conditional) and when its antecedents are legitimately called its causes. Furthermore, prediction is simply the inversion of the universal conditional (i.e., "If parents act responsively to the child's needs, then there is an increase in the child's social competence"). As a result of this close relation of explanation and prediction within this scheme, the predictive value of a hypothesis becomes an important criterion of the validity of an explanation. Contemporary Cartesian proponents often explicitly split cause from the covering law model and argue over whether scientific explanation is better understood as mechanistic causal explanation *or* covering law explanations (e.g., Bechtel & Abrahamsen, 2005), and at least one mechanist argues that the reciprocal causality of dynamic systems can be reduced to mechanistic terms (Zednik, 2011). However, all such arguments continue to be framed by the Cartesian–split-mechanistic methodology.

Here, then, is the basic outline of the Cartesian empiricist story of scientific methodology:

Step 1: Reduce to the objective (interpretation-free) observable foundation.

Step 2: Find the correlations or causes.

Step 3: Induce the hypothesis, theory, law.

As noted, variations have appeared throughout history. In fact, it would be misleading not to acknowledge that "probability" has replaced "certainty" as the favored *lexical item* in the story as it is told today. Indeed, induction is itself statistical and probabilistic in nature. However, this change from a search for certainty to one of probability represents much more style than substance, as the aim remains to move toward 100% probability—that is, probability is invoked *only* because of error of measurement, and *not* because laws are probabilistic—thereby arriving at certainty or its closest approximation to an absolute objective mind-independent reality. This type of fallibilistic stance continues to pit doubt against certainty as competing alternatives rather than understanding doubt and certainty as a dialectical relation, framed by the concept of *plausibility*. More generally, all of the variations that have been introduced since the origin of Newtonian explanation—including those formulated under the methodological banners of neopositivism, instrumentalism, and conventionalism—have changed the basic epistemological and ontological features *not at all*.

The neopositivist form of the Cartesian methodology reached its zenith in the 1940s and 1950s, but ultimately both the friends and the foes of neopositivism recognized its inadequacy as a broad scientific strategy. These inadequacies included the increasing recognition that:

1. Rich theories are not reducible to a neutral observational language, as demonstrated in the work of Quine (1953) and others (e.g., Lakatos, 1978b; Popper, 1959; Putnam, 1983).
2. Induction is a deeply flawed method for arriving at theoretical propositions (Hanson, 1958, 1970; Lakatos, 1978b; Popper, 1959).[5]
3. The covering law model is highly restricted in its application (Ricoeur, 1984) and faulty in its logic (Popper, 1959).
4. There are theories that warrant the attribution *scientific* despite the fact that they lead to no testable predictions (Putnam, 1983; Toulmin, 1961).

Both neopositivism and its successor Cartesian methodology—Popper's (1959, 1963, 1970) conventionalism/instrumentalism—suffered further devastating wounds as the idea began to enter science that "all data are theory laden" (Hanson, 1958), or to state this slightly differently, background ideas, including ontological and epistemological concepts, enter into the warp and woof of the scientific enterprise. The growth of this idea owes much to numerous historians and philosophers of science and other scholars, including Elizabeth Anscombe (1957), Richard Bernstein (1983), William Dray (1957), Charles Frankel (1957), Hans Georg Gadamer (1989, 1993), N. R. Hanson (1958), Thomas Kuhn (1962, 1970, 1977), Hilary Putnam (1983, 1987), Paul Ricoeur (1984, 1991), Charles Taylor (1964), Steven Toulmin (1953, 1961), Georg Henrik von Wright (1971), Peter Winch, (1958), and Ludwig Wittgenstein (1958).

Although the many critiques have been sufficient to discredit Cartesian methodology for all but the most fervid Cartesian advocates, the ghosts of the Cartesian scheme continue to haunt the halls of science. One methodologically relevant example is found in a text that discusses multilevel modeling (Luke, 2004). The author begins the work by raising the question that, given the availability of sophisticated multilevel modeling approaches, why do investigators continue to use more simplistic single-level techniques. He answers:

> A number of epistemological traditions have shaped this behavior. First is the long reach of the [neo]positivist tradition. Even many years after philosophers of science have established the inadequacy of logical positivism as a framework for the biological, health, and social sciences, we still tend to emphasize research designs and analytic tools that trace their roots back to a [neo]positive science. For example, the emphasis on control over experimental and observational conditions, the reliance on control and comparison groups, and the use of modeling techniques that statistically "remove" or control for the effects of covariates all combine to provide a lot of precision over inference. However, at the same time, they severely restrict the ability to measure or evaluate extra-individual contextual effects. (p. 2)

There are also still those who argue that emergence of genuine novel behavior is not possible, and that any apparent novelty must be completely explained by antecedent efficient/material causal mechanisms (see Blachowicz, 2012; Witherington, 2011, for extended discussions of problems associated with this argument). Indeed, the claim is still put forth that if an efficient/material causal mechanism is not identified, then there is no "real" explanation, only "mysticism" (Elman et al., 1996) or "miracles" (Siegler & Munakata, 1993). This assertion reflects the Cartesian mechanistically defined argument that there can be no nonreducible emergence in development. All change, according to this mechanistic argument, is (i.e., must be) totally additive in nature; all apparent qualitative change must be reduced to nothing but quantitative change. And there are also those who still argue that developmental science has no room for Aristotlean final explanations, calling for a movement away from the positing of developmental endpoints and toward "a more neutral, person-time-and-situation-geared conception of development" (Demetriou & Raftopoulos, 2004, p. 91).

There are probably several reasons for the failure to recognize and accept the demise of Cartesian empiricist methodology, particularly in the biological, behavioral, and social sciences. One of these has to do with socialization. For biological, behavioral, and social scientists who were reared in the environment of the epistemological strictures of mechanical explanation, these strictures are difficult to abandon, and these values tend to be passed from generation to generation without deep reflection. Indeed, as this methodology is virtually inscribed with the motto "Don't think, find out" (Cohen, 1931, p. 76), it is not surprising that fledgling investigators are often discouraged from taking the very notion of metatheory–methodology seriously and, hence, seldom do they evaluate the merits and flaws of alternative background assumptions and their impact on methodology and methods. Another perhaps more important reason, however, was the apparent lack of viable empirical scientific alternatives and the seeming abyss of uncertainty that is faced when one abandons a secure, rock-solid base. The failures of neopositivism and conventionalism/instrumentalism did nothing to allay this uncertainty. Clearly what was needed was a new non-Cartesian scientific methodology.

Methodology: The Hermeneutic Alternative

Emerging from the hermeneutic and phenomenological philosophical tradition (Latour, 1993), a scientific methodology was developed that has come to operate parallel to, and as a reaction against, the Cartesian quest for reductionist mechanical explanation. This alternative picture champions *understanding*, in contrast to *explanation*, as the base of scientific knowledge, at least as this scientific knowledge pertains to the behavioral and social sciences, including the humanities.

Broadly, hermeneutics is the theory or philosophy of the interpretation of meaning. Hermeneutics elevates to a heroic role the very concept that Cartesian methodology casts as demon error—*interpretive inference*. For present purposes, we may pass by the periods of classical, biblical, and romantic hermeneutics, as well as Vico's historical hermeneutics. Our brief focus here is on the effort that Dilthey (1972) promoted at the end of the 19th century to construct a methodology for the social sciences. This was *Verstehen* or *understanding*. Within this methodology, *understanding* operates as an epistemological rather than a psychological concept. And, most importantly, *interpretive inference* operates as the procedure that results in understanding.

As a methodology of the social and behavioral sciences, understanding is closely related to the relational developmental systems' *embodied action perspective* mentioned earlier. This scheme takes the person and intraindividual change as its focus in inquiring into the origin and nature of processes and operations of the meaning-generating, living, embodied agent (Brandtstädter & Lerner, 1999; Dreyfus & Dreyfus, 1999; Overton, 1997a, 1997b; Overton, Müller, & Newman, 2008). The person-oriented embodied action perspective stands in contrast to split-off externalist (i.e., exclusively biological or exclusively environmental) and split-off event-oriented approaches. Ricoeur has outlined—in the context of Wittgenstein's (1958) *language games*, which are themselves metatheoretical background principles—the distinction between externalist events and person actions, and in this outline he suggests the distinction between mechanical explanation and hermeneutic understanding:

> It is not the same language game that we speak of events occurring in nature or of actions performed by people. For, to speak of events, we enter a language game including notions like cause, law, fact, explanation and so on. . . . It is . . . in another language game and in another conceptual network that we can speak of human action [i.e., a person-oriented frame]. For, if we have begun to speak in terms of action, we shall continue to speak in terms of projects, intentions, motives, reasons for acting, agents, [interpretation, understanding] and so forth. (1991, pp. 132–133)

Unfortunately, the creation of a *separate* methodology for the social sciences is yet another example of proceeding within a split background frame. Within the hermeneutic tradition, *Verstehen* is presented as a *competing* account of human functioning to that found in the natural sciences. However, from the perspective of relationism, the articulation of this dichotomy also suggests a means of resolution.

Verstehen as a methodology and a relational developmental systems' embodied action perspective as an approach to human functioning are closely related through the intentional quality of action. Intention is never directly observable by a third party. To intend is to do something for the sake of it. Intention involves direction and order. There is a goal toward which embodied action moves, and a sequence of acts lead to that goal. To explain (understand) an act, it is necessary to make interpretive inferences about patterns of embodied acts, as these make the specific behavioral movement of an act intelligible and give a reason for the movement. For example, the act we term *reaching* in the young infant is only that if the inference is made that the infant intends a particular goal object. Under another inference, the observed movements might be termed *stretching*. Making interpretive inferences about action patterns is, in fact, identical to Aristotle's "formal" and "final" explanations that, as stated earlier, were designed to make the object of inquiry *intelligible* and give *reasons* for the nature and functioning of the object. Thus, a rapprochement in developmental science between adherents of a natural science perspective and adherents of a relational developmental systems' embodied action perspective can be found in a relational methodology that integrates mechanical and action explanations.

Relational Methodology: A Rapprochement among the Sciences

A relational methodological rapprochement among the sciences emerges by transforming the historically traditional dichotomies of pristine observation versus interpretive observation, and theory versus data, into relational bipolarities. Given this movement in grounding, mechanical explanation and hermeneutic understanding become an integrated methodology in the following manner:

Step 1: Relational Analysis—Synthesis Replaces Split Reductionism

Clearly, the reductionism and atomism of mechanical explanation are split principles and need to be replaced. Anointing holism as the guiding principle without further comment is not workable because *holism* in a split conceptual world is frequently identified with *synthesis* as *reductionism* is identified with *analysis*. A rapprochement requires that *analysis* and *synthesis* operate as a relational polarity. Analysis must occur in the context of some integrated whole, and the integrated whole operates in the context of its analytic parts. Because relational methodology is sometimes, however incorrectly, viewed as less rigorous than Cartesian methodology, a major feature of this first step is the affirmation of the importance of analysis and the analytic tools of any empirical science. The provisos here are that it simultaneously be recognized that the analytic moment always occurs in the context of a moment of synthesis, and that the analysis of parts can neither eliminate nor marginalize the synthetic whole. Thus, no observation can ever be pristinely context free. As discussed earlier, science begins with commonsense observations, and a relational methodology recognizes that context is an inseparable feature of any observation.

Step 2: Relational Action Patterns—Explanation of Conditions Replaces Split Causes

As noted earlier, the defining marks of Cartesian mechanical explanation and hermeneutic understanding have been the "nothing-but" reliance of the former on efficient/material causes and inductive correlations, and the latter on action patterns. By entering into a relational context, these different types of determinants become integrated. In a relational context, when considering living organisms, material and efficient causes are transformed from interpretation-free observed objects or events that produce changes in other objects or events, into *conditions* that are associated with resources for, or developmental assets for, change (see Benson, Scales, & Syvertsen, 2011, for an extended discussion of developmental assets). A cause is interpretation free only when analysis is split off from synthesis; in a relational model, conditions—as an analytic moment of inquiry—are understood as functioning under some interpretation, some synthesis (Hanson, 1958). A cause can be a "force" that "produces" or "influences" or "affects" the status or change of an object in a model that splits system and activity. In a relational model, system and activity are joined as a dynamic structure–function relation and *conditions* are identified as necessary and/or sufficient to the occurrence of the phenomenon under investigation (von Wright, 1971). Thus, rather than inquiry into the *causes* of behavior or development, inquiry from a relational perspective examines conditions that are associated with behavior or that can serve as resources or assets for intraindividual change in the relational developmental system. For example, if inquiry concerned the development of a plant, food and water would represent necessary conditions for the plant to grow, but they would not be offered as the "cause" of the plant's development in the sense of producing that development. Similarly, neither nature factors nor nurture factors can be considered the causes of human development; they represent conditions that are associated with that development—resources or assets that serve the relational developmental system (e.g., community resource programs or adult mentors are the ecological assets associated with positive youth development [Lerner et al., 2012]).

The relational methodological alternative to the claims of efficient/material causality is a reminder that (1) this causality has not been uniformly accepted as the sole type of scientific explanation. Action pattern (developing action systems) explanation—Aristotle's formal and final explanations—has come to be an equally viable and scientifically acceptable type of explanation (e.g., see Bunge, 1962; Cartwright, 1995; Hanson, 1958; Kitcher, 1981; Laudan, 1996; Overton & Reese, 1981; Putnam, 1983; Toulmin, 1953). (2) It has become recognized that efficient/material causes are themselves never free of the interpretive frames that give them meaning as causes (e.g., see Hanson, 1958; Holton, 1973; Kuhn, 1962, 1977; Lakatos, 1978b; Laudan, 1977; Putnam, 1983). What Hanson (1958) said about Galileo's approach to the law of falling bodies, and about cause broadly, illustrates these facets of the relational alternative:

> He [Galileo] seeks not a descriptive formula; nor does he seek to predict observations of freely falling bodies. He already has a formula. . . . He seeks more: an *explanation* of these data. They must be intelligibly systematized. . . . He has no confidence in observaions which cannot be explained theoretically. Galileo was not seeking the cause of the acceleration; that was Descartes' programme. Galileo wished only to understand. His law of constant acceleration (1632) is not a causal law. (p. 37)

> What we refer to as "causes" are theory-loaded from the beginning to end. They are not simple, tangible links in the chain of sense experience, but rather details in an intricate pattern of concepts. (p. 54)

> Causes certainly are connected with effects; but this is because our theories connect them, not because the world is held together by cosmic glue. (p. 64)

The assertion that *causes*, when applied to living organisms, are best understood as *conditions* leaves open the question of what, in fact, does determine behavior and change. The issue here is that of *mechanism*. As is the case with other key terms, *mechanism* has several, often incompatible, definitions. In the present case, the meaning is closer to "a process, physical or mental, by which something is done or comes into being" than to "the doctrine that all natural phenomena are explicable by material causes and mechanical principles" (*American Heritage Dictionary of the English Language*, 2000). Hence, for present purposes, *mechanism* is defined as an active method or process rather than as a cause or set of causes.[6] Process, in turn, is reasonably defined by Nesselroade and Molenaar (2010) as "highly structured temporal [active] organizations" (p. 32). Thus, mechanisms are found in the structure–function relations that identify patterns of action.

Any relational developmental system is constituted by a set of structure–function relations. The system is not a random aggregate of elements; it has a specific dynamic organization or architecture (i.e., a structure). Furthermore, this structure is not randomly active; it has a characteristic activity (i.e., a function). Even computers (structure)—when they are turned on—compute (function). However, computers do not change—at least, they do not demonstrate a directed sequence of emergent organizational changes—and for this reason they are limited as models of the human mind (Fodor, 2000) and its development. The input and output of a computer may change—and this is the basis for traditional and contemporary split "functionalist" approaches to explanation (Overton, 1994). However, the organization/activity of the computer itself does not undergo a directed sequence of emergent change. In contrast, living organisms are relational developmental systems; they are organizations (structures) of processes that are inherently active (function) and undergo a directed sequence of emergent organizational change involving novel structure–function relations.

When a relational developmental system is viewed from the *standpoint of function*, it is the function itself (i.e., the *characteristic* action of the developmental system) that constitutes the mechanism and explanation of behavior and change. Systems change through their characteristic action in the context of external conditions. Thus, the explanation of behavior and change is given by the functioning of the system

(Witherington, 2007). However, because of the inseparable relation of structure and function, when a system is viewed from the *standpoint of structure*, then structure explains function. Consequently, both structure and function enter centrally into the explanatory process.

Structure and function are central to explanation, but they are also fundamentally interpretive in nature; they are not directly observable. Structure–function relations are systems or patterns of action, but patterns are never directly observable; they must be inferred. When examined from the structural standpoint, the patterns constitute Aristotle's formal and final explanations. From the *structural standpoint*, the *pattern or system of actions* makes the object of inquiry *intelligible* and gives *reasons* for the nature and functioning of the object. From the *functional standpoint*, the *action patterns explain by presenting the mechanism of behavior and development*. Action patterns, however, necessarily operate within the context of material conditions, both internal to the system and external to it. Thus, the introduction of structure–function relations serves to integrate the hermeneutic explanation in terms of understanding and the natural science explanation in terms of conditions. Both types of explanation are necessary, but each operates from a different standpoint.

Developmental science offers several illustrations of this explanatory integration. For example, Witherington (2011; see also Witherington & Heying, 2013) employs the relational Aristotelian scheme in differentiating dynamic systems approaches based on strictly contextualist accounts, which are reductionistic, from those, including his own, that take an integrated organicism–contextual (i.e., relational) approach. Furthermore, Whitherington demonstrates how formal explanation contributes to our understanding of what has been traditionally termed *downward causality* and *circular causality*, each being an exemplar of *reciprocal* (↔) *causality*. Basically, downward "causality" entails a higher level of organization (pattern of action) operating as the active context, which provides constraints to lower-level patterns of action.[7] Consider, for example, riding a bicycle. Much of the synchronized action involved reflects a sensorimotor level of functioning. However, that action occurs and is constrained by feedback loops from a higher symbolic level of functioning. Circular "causality" in turn joins downward "causality" with the fact that it is out of the functioning at the lower levels (e.g., sensorimotor) that higher levels (e.g., the symbolic) emerge (Blachowicz, 2012).

With respect to broader examples, Bowlby's (1958) theory of infant–caregiver attachment posits a behavioral attachment *system* (structure) in relation to actions that serve the adaptive function of keeping the caregiver in close proximity. Furthermore, although the emphasis is often on individual differences in attachment styles, Bowlby offers a developmental theory that describes emergent levels of organization from infancy through adulthood.

Piaget's (1952, 1985) theory offers a more complex and more complete example. This theory represents an attempt to make sense of (i.e., explain) the development of knowing. Like Bowlby's model, Piaget's is a relational developmental systems approach that takes seriously the background ideas of dynamic structure–function relations and conditions. Because Piaget's theoretical goal is to explain the person and the development of the knowing person, Piaget takes a *person (and epistemic) standpoint* rather than a biological or a cultural standpoint. The theory conceptualizes the person as a complex adaptive system of self-organizing action systems *operating in a world of biological and environmental resources* (conditions). Structure and function constitute thesis and antithesis, and the resulting synthesis is an emergent system, or *stage*, of novel structure–function relations. Structures are the mental organizations that are expressed as patterns of action. On the *structural* side of the equation, Piaget introduces the theoretical concepts of *schemes, coordination of schemes, operations, groupings*, and *group*. Each explains (i.e., *formal explanation*)—at successive novel levels of emergent systems—the cognitive equipment that the infant, toddler, child, and adolescent come to have available for constructing their known world.

Theoretical concepts of *adaptation, assimilation–accommodation, equilibrium, equilibration*, and *reflective abstraction* constitute the *functional* side of Piaget's equation. Schemes, coordinated schemes, operations, and the like *function*; they are active and it is through their embodied action in a world of biological, physical, and sociocultural conditions that they change. Piaget's is an embodied action theory, and embodied action is the general mechanism of development. Through the organized actions of the person in the world, the person's mode of knowing the world becomes transformed, and these novel systems are adaptive. As the mechanism of development embodied action becomes more specific through recognition of its biphasic nature. Assimilation is the phase of action that expresses the mental organization. This expression gives meaning to the world; it constitutes the world as known. However, these meanings—including meanings at a presymbolic, preconceptual level of organization—have an instrumental function as well as the expressive function. When the instrumental function of the action is not completely successful in securing an adaptive goal, *variation* occurs in the action. For example, an infant may intend (assimilate) the side of the breast as a nipple by sucking it, but when the satisfaction of feeding does not occur, variations arise in the action, and this is exemplified by the infant sucking in various new locations. Variations open up new possibilities that both secure a goal and provide feedback that result in the transformation (differentiations and novel coordinations) of the system itself. This action phase of variation and organizational modification is the accommodation phase of any action.

Organization or system explains in the sense of establishing the *form* (structure), and action of the system yields the *explanatory mechanism* (function). This relational polarity operates in the context of *conditions*, such as parents who do or do not provide appropriate opportunities for the adequate exercise of functioning. It is also the case that at the beginning of any stage of novel structure–function relations, the capacity for successful adaptation at that level is limited. This situation is theoretically expressed in the idea that there is more assimilation than accommodation at the beginning of a stage, and hence, there is a lack of balance or *equilibrium* between assimilation and accommodation. Through action, this imbalance changes, and the two phases of action eventually move into equilibrium within a given stage. Of course, given the relational nature of the theory, equilibrium of assimilation and accommodation also means that the underlying structures have reached a stable state (equilibrium) of differentiation and intercoordination.

The movement toward equilibrium of the action phases of assimilation and accommodation describes the development mechanism *within* a stage. To explain development *across* stages, Piaget introduces a principle that also has both a structural and a functional face. *Structurally*, this is the *equilibration principle* (Piaget, 1985), and it asserts that developmental change is directed toward "improved" states or patterns of the just-described equilibrium. *Improved* here is defined in terms of the adaptive value of one stage of cognitive structures relative to the adaptive value of other stages of cognitive structures. For example, the formal operational structures associated with adolescence represent an improved equilibrium over the sensorimotor structures associated with infancy, in that the formal operational structures are more stable, more flexible, and describe a much broader range of potential cognitive experiences than sensorimotor structures. The equilibration principle introduces hierarchical organization into the theory and explains sequence, order, and direction in the emergence of novel cognitive abilities, just as the Second Law of Thermodynamics explains sequence, order, and direction with respect to the physical world. It reflects Aristotle's metatheoretical *final explanation*, and it is consistent with the final structural explanations offered in other developmental theories, including Heinz Werner's (1948, 1957) "orthogenetic principle" and Erik Erikson's (1968) "epigenetic principle."

The *functional* face of the mechanism of development *across* stages is termed *reflective abstraction* (Piaget, 1985; see Campbell, 2009). Reflective abstraction is action, but it is action that has its own biphasic character consisting of "reflecting," in the sense of projecting something from a lower to a higher level, and "reflexion," which is the reorganization of what has been projected. The alternation of the reflection–reflexion phases produces each new stage of cognitive reorganization. Reflection is similar to the act of generalizing; reflexion is acting from the generalized position to consolidate the gains made through generalizing. What is abstracted in this process is the coordination of the differentiated structures of the lower level of organization.

Step 3: Abductive/Retroductive Logic Replaces Split Induction and Deduction

The third step toward a relational developmental system—a relational methodology that integrates mechanical explanation and hermeneutic understanding—addresses the nature of scientific *logic*. To the early positivist of the 19th century, the neopositivists of the 20th century, and the contemporary adherents of these methodologies, *induction* has constituted the *fundamental* logic of science, with *deduction* introduced as an ancillary heuristic tool (see Figure 2.5). In fact, John Stuart Mill's five canons of induction (methods of agreement, difference, joint agreement and difference, residues, and concomitant variation [Copi & Cohen, 2006]) represented the prototype for later experimental designs. For those who follow Popper's conventionalism/instrumentalism, inductive inference is seen as lacking rational justification and "must lead to logical inconsistencies" (1977, p. 29). In its place *deduction* becomes the fundamental and sole logic of science, as inferences are deductively drawn from conjectured hypotheses to be tested under the methodological rule of falsification.

A relational methodology introduces the logic of *abduction* as the synthesis of the opposite identities of theory (broadly considered, including background ideas) and data. Abduction (also called *retroduction*) was originally described by pragmatist philosopher Charles Sanders Pierce (1992) and historian of science N. R. Hanson (1958, 1970) who argued that it has long been the fundamental, if often invisible, logic of scientific activity. Earlier, a form of abduction, termed the *transcendental argument*, was introduced by Kant and has been elaborated by Charles Taylor (1995; see also Grayling, 1992). In a contemporary version this logic is also termed "inference to the best explanation" (Fumerton, 1992; Harman, 1965; Lipton, 1991; Putnam, 1994).

Abductive inference operates by arranging the observation under consideration and all background ideas (here including specific theoretical and metatheoretical concepts) as two Escherian hands. The question is then asked as to what must necessarily be assumed in order to have that observation (see Figure 2.6). The inference to, or interpretation of, what must necessarily be assumed in the context of background ideas then comes to constitute the explanation of the phenomenon.

The basic logic of abduction operates as follows:

1. Step 1 entails the description of some highly reliable phenomenological observation (O is the case).
2. For Step 2, with O as the explanandum, an inference or interpretation is made to an action-pattern (system) explanation (*E*). This results in the conditional proposition "If *E* is the case, then O is expected."
3. Step 3 entails the conclusion that *E* is indeed the case. The logical form of the argument is:

   O (phenomenological observation) is the case.

   If *E* (action-pattern [system] explanation) is the case, then O is expected.

   Therefore, E is the case.

As Fumerton (1992) points out, it is obvious that if the conditional in Step 2 is read as material implication, the argument would be hopeless because it would then describe the fallacy of the affirmed consequent (i.e., would be viciously circular). Quite correctly, Fumerton recognizes that the "if . . . then" relation asserts some other sort of connection. Specifically, the connection is one of what logicians refer to as *meaning relevance* between *E* and O, where *relevance* is defined in terms of the intelligibility of the relation between *E* and O (Overton, 1990).

Abductive/retroductive inference is illustrated in virtually any psychological work that assumes a person-oriented methodology (e.g., see Molenaar & Campbell, 2009; Sterba & Bauer, 2010; von Eye, Bergman, & Hsieh, in press) and a focus on emotional, motivational, or cognitive mental systems. Russell (1996), for example, has discussed the significance of abduction to the area of cognition. Chomsky's (1957,

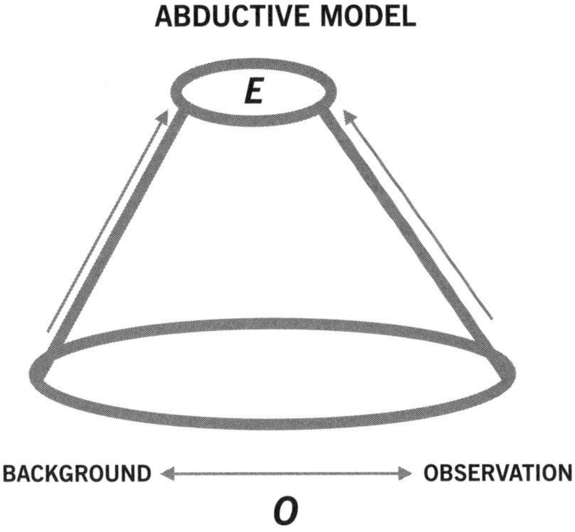

FIGURE 2.6. Constructing the abductive/retroductive model.

2000) work in language and Piaget's work in cognitive development are particularly rich in abductive inference. Consider the following example as an illustration of the process, drawn from Piaget: (1) There is the phenomenal observation (O) that it is the case that a certain group of people (children around 6–7 years of age) understands that physical substances maintain the same quantity despite changes in qualitative appearances (i.e., conservation). (2) Given the relational background ideas discussed in this chapter, Piaget forms the abductive/retroductive inference that the explanation of this observation (*E*) is that a certain type of action system, having specified features including reversibility (i.e., concrete operational groupings model), must be available to these people. This inference forms the conditional statement "If (*E*) concrete operational model, then (O) conservation is expected." (3) Given (O), the conclusion is, "Therefore, concrete operational model explains the understanding of conservation." To avoid vicious circularity, the *E* must then be tested in domains beyond those constituting the observational base for the generation of *E*. For example, Piaget's concrete operational structures were tested in several quantitative domains, including "amount," "volume," "number," "area," as well as other tasks that involved the addition and subtraction of classes. Or, as another example, Chomsky's (1957) original model was formulated for the English language and the task then became to assess the model in other languages. And finally, as some have pointed out (e.g., Patokorpi, 2007; Thagard, 1988), the logic of Sherlock Holmes—who amazed his loyal friend Dr. Watson by drawing a correct conclusion from an array of seemingly disparate and unconnected facts and observations—was abduction.

More broadly, abductive inference is the foundation for the construction and assessment of models in systems research. Here the focus is on articulating models of systems and their development in propositional or mathematical terms, and then testing these models to determine whether they lead to the developmental outcomes claimed by the models. When formulated in mathematical terms, such model-driven research (e.g., Chow & Filteau, 2010; Fischer & Kennedy, 1997; Kan, Ploeger, Raijmakers, Dolan, & van der Maas, 2010; Kaplan, 2008; Molenaar, Sinclair, Rovine, Ram & Corneal, 2009; Montpetit, Bergeman, Deboeck, Tiberio, & Boker, 2010; Nesselroade & Molenaar, 2010; van der Maas & Molenaar, 1992; van Geert, 1991) has explicated developmental processes with mathematical rigor, giving new life and power to many concepts, including *equilibration, critical periods, resilience*, and *stage* (Fischer & Rose, 1999, p. 200).

It is in the testing of the model—the *E* (explanation)—that decisions are made as to whether the chosen model is, in fact, the "best" model. This step entails not merely the support or nonsupport of predictions from the model, but also all of the traditional criteria for theory *selection* that have been accepted within the scientific community. These criteria include the model's depth; coherence; logical consistency; extent to which it reduces the proportion of unsolved to solved conceptual and/or empirical problems in a domain (Laudan, 1977); and last but not least, scope, empirical support, and empirical fruitfulness.

Scope, fruitfulness, and empirical support as partial criteria for choice of a best model explanation all demand a return to the observational grounds for empirical assessment. Some of the statistical and research strategies associated with this return are described by Rozeboom (1997). *Scope* is assessed through testing the abductive explanation in observational contexts that go beyond the context that generated the explanation. For example, as described above, conservation can be assessed in the contexts of number, weight, area, and volume, and it can be assessed in relation to other skills that should, in the context of the explanation, be associated with it. The assessment of scope also serves, as described above, the function of establishing that the abductive explanation–observation relation, although circular, is not viciously circular.

The *fruitfulness* of an explanation is measured in terms of the extent to which the model generates (predicts) new observations, which lead to a further elaboration of the model, then leading to novel observations, which lead to an elaboration of the model, and so on. As the model becomes elaborated, it creates a newly elaborated background (see Figure 2.7). The new background generates novel observations, but these too—because they constitute a background–observation relation—yield opposite identities that require further abductive inferences. Piaget's (1985) equilibration model is again illustrative (developmental change being directed toward "improved" states or patterns of action). Here, for example, the concrete operational model becomes background for the formulation of the "improved" formal operational model. The issue of fruitfulness generally points to an important relational progressive feature of abductive inference: It assumes the form of the familiar hermeneutic circle—projection and correction—by moving from the phenomenological level (the commonsense object) to explanation and back in an ever-widening cycle.

Empirical support for an abductive explanation is the outcome of any assessment of scope—including various forms of replication. Here, another central feature of a relational methodology needs to be differentiated from the traditional Cartesian split

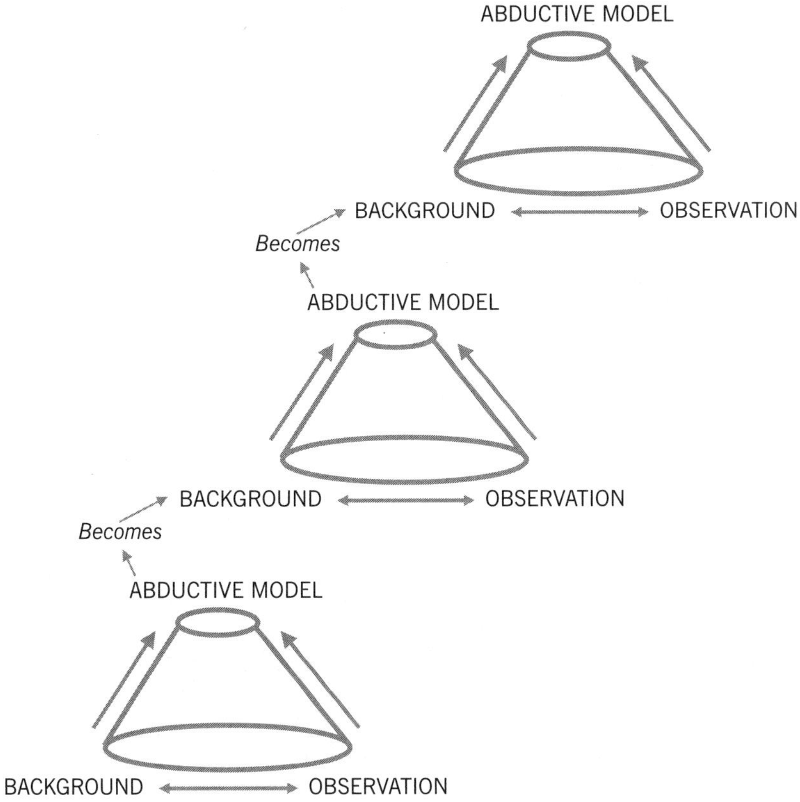

FIGURE 2.7. The progression of abductive/retroductive models.

methodology. Under the rule of split-off induction and deduction, it was assumed that scientific progress moved forward through the *deductive falsification* of theories (Popper, 1959). The criterion of theory falsification, however, fell into disrepute through demonstrations by several historians and philosophers of science (e.g., Hanson, 1958; Kuhn, 1970; Lakatos, 1978a, 1978b; Laudan, 1977; Putnam, 1983; Quine, 1953) that although deductive logic, and hence falsification, may be logically applicable to a specific empirical hypothesis, this form of falsification does not reach to the level of a rich model or theory. Furthermore, it should be noted that conventional significance testing plays, at best, a "minor and misleading" (Meehl, 1990a, p. 115) role in issues of failed or supported empirical hypotheses (e.g., see Lambdin, 2012; Meehl, 1990a, 1990b). As Lambdin (2012) points out with respect to significance testing, "a *p* value is the probability of obtaining the results in hand assuming the statistical null hypothesis is true in the population" (p. 74), and this has little to do with whether a specific theoretically derived empirical hypothesis is correct or incorrect. Within a relational metatheory, empirical hypotheses that are not supported by relevant empirical data are important in that they constitute failures of support for the broader abductive explanation, but they are not important in the sense of constituting a refutation of the explanation itself. Kuhn, Lakatos, and Laudan describe these failures as *anomalous instances*, and as such they require evaluation, especially in terms of the model's or theory's fruitfulness. Following from Serlin and Lapsley's (1985) notion that it is more realistic to think of theories as being "good enough" rather than establishing a rigid true–false dichotomy, Meehl (1990a) argues that "good enough" be identified with fruitfulness or what "might be called the 'track record' or 'money in the bank' principle" (p. 115). Furthermore, he argues that "the main way a theory gets money in the bank is by predicting facts that, absent the theory, would be antecedently improbable" (p. 115). Thus, anomalies do not, in and of themselves, require abandonment of the abductive explanation (see Overton, 1984, 1994). However, when an abductively generated model does fail in the face of the aforementioned criteria, it then becomes clear that the abductive process is itself *recursive* in nature (Romesburg, 2009). That is, under the aforementioned criteria, when a model fails, the abductive process continues through the construction of another abductive model (e.g., see the change in Chomsky's original [1957] syntactical model to his more recent [2002] "minimalist program").

Conclusions

Relationism as metatheoretical–methodological worldview and relational developmental systems as a midrange metatheory–methodology offer a broad scientific paradigm for developmental science. The ontological base of relationism presents a world that is complex (vs. complicated), spontaneously active, and constantly undergoing dialectical change; its epistemology offers understanding that is holistic, where analysis proceeds in the context of part–whole relations, and constructivist, where the activity of human agency participates in the construction of the world as known.

Operating within relationism's broad context, relational developmental systems offers a vision of organisms as active, complex, open, and adaptive relational systems. At a psychological level, persons are organisms that act, perceive, think, reason, speak, feel, wish, will, and desire. At the subpersonal level persons are *complex, open, adaptive, relational systems* operating far from entropic equilibrium. These systems are *self-creating* (autopoetic, enactive); *autonomously self-organizing* (i.e., proceed according to which higher-level system organization arises from the endogenous coactions of lower-level components of the system, functioning without explicit instructions or guidance from outside forces); *self-regulating* (i.e., maintaining structure and order without explicit instructions or guidance from outside forces); and *nonlinear* (i.e., input is not proportional to output). They develop through complex *reciprocal bidirectional* (↔) and *circular* interpenetrating *embodied action* in a world of sociocultural and physical objects. This development is characterized by *probabilistic epigenetic processes* (i.e., holistically, the role played by any part of a relational developmental system—DNA, cell, tissue, organ, organism, physical environment, culture—is a function of all the timing of the relations among the interpenetrating and coacting parts of the system) leading to *emergent* (i.e., novel, nonadditive) directional sequences.

The reciprocal coaction with the world and holistic epigenesis identifies the person as being completely *contextualized* and *situated*. Accordingly, the potential for *plasticity* of *intraindividual change* is a hallmark of the relational developmental system.

Methodologically, these features of relationism and relational developmental systems represent the headwaters from which issues of method, including design, measurement, and modeling, ultimately flow. These broad metatheoretical–methodological principles point first and foremost to a holistic, and hence contextualized, analysis of intraindividual change across the life span of the person. These principles suggest the need for specific methods that are intensive and multivariate and can accommodate change that is nonlinear (i.e., not the result of classical antecedent causes) and that is nonadditive (i.e., discontinuous) as well as additive (for some currently available methods, see Boker & Graham, 1998; Molenaar & Newell, 2010; Partridge, 2011; Urban, Osgood, & Mabry, 2011). And they suggest the need for the continued development of methods that are process- and person-oriented (e.g., see Molenaar & Campbell, 2009; Sterba & Bauer, 2010; von Eye et al., in press).

Finally, these principles establish a broad set of strategies as context for the further development of methods for the developmental sciences. The first of these is a strategy of approaching developmental issues with recognition that description is never pristine, and that traditional mechanical cause–effect is a highly limited form of explanation. Explanation can take many forms, and those most relevant to the developmental sciences entail positive and negative feedback loops involving complex reciprocal bidirectionality (↔) and circularity. A final, equally important strategy emerges from an understanding that abduction/retroduction is a fundamental logic of any science. This yields the central fact that the construction of explanatory models based on abductive/retroductive inference is a central focus of the developmental sciences.

NOTES

1. Material and efficient causes entail understanding "cause–effect" from an ontological perspective, whereas "cause–effect" as constant conjunction refers to an epistemological understanding and is often referred to as *Humean causation*, following David Hume (see Lorkowski, 2010).

2. In the philosophical literature this idea is referred to as *continuous reciprocal causation*. It is defined as the relation between two systems, such that each system's behavior is at all times determining, as well as being determined by, the other's—a strongly *relational* definition (Clark, 1997).

3. A final principle of relationism, titled the *synthesis of wholes*, establishes a third relational standpoint. See Overton (2013) for details.

4. Popper's own deductive system is sometimes confused with this method (e.g., Pidwirny, 2006).

5. It should be emphasized that the flaws of induction do not reach Molenaar and Nesselroade's (Chapter 17, this volume) use of induction as "the fit of dynamic systems models inspired by relational developmental systems to appropriate empirical data."

6. This definition is similar—*when it is slightly modified to reflect a holistic perspective*—to a definition given by a leading member of the "new mechanistic" group in philosophy, William Bechtel: "A mechanism is a structure performing a function in virtue of its component parts, component operations, and their organization. The orchestrated functioning of the mechanism, manifested in patterns of change over time in properties of its [organization,] parts and operations, is responsible for one or more phenomena" (Bechtel & Abrahamsen, 2010, p. 323).

7. As stated earlier, it would be conceptually clearer to refer to reciprocal, circular, and downward causality as *explanations*, given that they entail formal and final features. An illustration of the confusion that can arise by using the term *causality* for all types of determinants appears in the writings of Pigliucci and Mueller (2010b) and Mueller (2010). In discussions of new trends in evolution, these authors acknowledge and support the centrality of systems concepts—which are clearly nonmechanistic. However, they simultaneously describe this as a "shift towards a causal-mechanistic approach," a "shift . . . to a causal-mechanistic theory" (Pigliucci & Mueller, 2010b, p. 12), and a "turn towards the mechanistic explanation of phenotypic change" (Mueller, 2010, p. 309). There is a profound difference between the claim that there has been a trend away from correlational approaches to other explanatory approaches, and the claim that there has been a trend away from correlational approaches to *mechanistic causal* approaches. The latter claim entails a significant conceptual confusion.

REFERENCES

Allport, G. (1955). *Becoming.* New Haven, CT: Yale University Press.

American heritage dictionary of the English language (4th ed.). (2000). Boston: Houghton Mifflin Harcourt.

Anscombe, G. E. M. (1957). *Intention.* Oxford, UK: Basil Blackwell.

Ayer, A. J. (1970). What is a law of nature? In B. A. Brody (Ed.), *Readings in the philosophy of science* (pp. 39–54). Englewood Cliffs, NJ: Prentice-Hall.

Barsalou, L. W., Simmons, W. K., Barbey, A. K., & Wilson, C. D. (2003). Grounding conceptual knowledge in modality-specific systems. *Trends in Cognitive Sciences, 7,* 84–91.

Batson, P., & Gluckman, P. (2011). *Plasticity, robustness, development and evolution.* New York: Cambridge University Press.

Bechtel, W., & Abrahamsen, A. (2002). *Connectionism and the mind: Parallel processing, dynamics, and evolution in networks* (2nd ed.). Oxford, UK: Blackwell.

Bechtel, W., & Abrahamsen, A. (2005). Explanation: A mechanistic alternative. *Studies in History and Philosophy of Biological and Biomedical Sciences, 36,* 421–441.

Bechtel, W., & Abrahamsen, A. (2010). Dynamic mechanistic explanation: Computational modeling of circadian rhythms as an exemplar for cognitive science. *Studies in History and Philosophy of Science, 41,* 321–333.

Benson, P. L., Scales, P. C., & Syvertsen, A. K. (2011). The contribution of the developmental assets framework to positive youth development theory and practice. *Advances in Child Development and Behavior, 41,* 197–230.

Bernstein, R. J. (1983). *Beyond objectivism and relativism: Science, hermeneutics, and praxis.* Philadelphia: University of Pennsylvania Press.

Bickhard, M. H. (2008). Are you social?: The ontological and developmental emergence of the person. In U. Müeller, J. I. M. Carpendale, N. Budwig, & B. Sokol (Eds.), *Social life and social knowledge: Toward a process account of development* (pp. 17–42). New York: Taylor & Francis.

Blachowicz, J. (2012). *Essential difference: Toward a metaphysics of emergence.* Buffalo: State University of New York.

Blair, C. (2010). Fluid cognitive abilities and general intelligence: A life-span neuroscience perspective. In R. M. Lerner (Ed.-in-Chief) & W. F. Overton (Vol. Ed.), *Handbook of life-span development: Vol. 1. Cognition, biology, and methods* (pp. 226–258). Hoboken, NJ: Wiley.

Block, J. (1971). *Lives through time.* Berkeley, CA: Bancroft.

Boker, S. M., & Graham, J. (1998). A dynamical systems analysis of adolescent substance abuse. *Multivariate Behavioral Research, 33*(4), 479–507.

Bowlby, J. (1958). The nature of the child's tie to his mother. *International Journal of Psychoanalysis, 39,* 350–373.

Brandtstädter, J., & Lerner, R. M. (Eds.). (1999). *Action and self development: Theory and research through the life span.* London: Sage.

Brentano, E. (1973). *Psychology from an empirical standpoint* (A. C. Rancurello, D. B. Terrell, & L. McAlister, Trans.). London: Routledge. (Original work published 1874)

Bundick, M. J., Yeager, D. S., King, P. E., & Damon, W. (2010). Thriving across the life span. In R. M. Lerner (Ed.-in-Chief) & W. F. Overton (Vol. Ed.), *Handbook of life-span development: Vol. 1. Cognition, biology, and methods* (pp. 882–926). Hoboken, NJ: Wiley.

Bunge, M. (1962). *Causality: The place of the causal principle in modern science.* New York: World.

Campbell, R. (2009). Constructive processes: Abstraction, generalization and dialectics. In U. Müeller, J. I. M. Carpendale, & L. Smith (Eds.), *Cambridge companion to Piaget* (pp. 150–170). New York: Cambridge University Press.

Carpendale, J. I. M., & Lewis, C. (2010). The development of social understanding: A relational perspective. In R. M. Lerner (Ed.-in-Chief) & W. F. Overton (Vol. Ed.), *Handbook of life-span development: Vol. 1. Cognition, biology, and methods* (pp. 584–627). Hoboken, NJ: Wiley.

Cartwright, N. (1995). The reality of causes in a world of instrumental laws. In R. Boyd, P. Gasper, & J. D. Trout (Eds.), *The philosophy of science* (pp. 379–386). Cambridge, MA: MIT Press.

Charney, E. (2012). Behavior genetics and postgenomics. *Behavioral and Brain Sciences, 35,* 331–358.

Chomsky, N. (1957). *Syntactic structures.* The Hague, Netherlands: Mouton.

Chomsky, N. (2000). *New horizons in the study of language and mind.* Cambridge, UK: Cambridge University Press.

Chow, S., & Filteau, G. (2010). A state space approach to representing discontinuous shifts in change processes. In P. C. M. Molenaar & K. M. Newell (Eds.), *Individual pathways of change: Statistical models for analyzing learning and development* (pp. 87–108). Washington, DC: American Psychological Association.

Clark, A. (1997). *Being there: Putting brain, body, and world back together again.* Cambridge, MA: MIT Press.

Cohen, M. R. (1931). *Reason and nature: An essay on the meaning of scientific method.* New York: Harcourt, Brace.

Colombetti, G., & Thompson, E. (2008). The feeling body: Towards an enactive approach to emotion. In W. F. Overton, U. Müeller, & J. L. Newman (Eds.), *Developmental perspectives on embodiment and consciousness* (pp. 313–342). New York: Taylor & Francis.

Copi, M., & Cohen, C. (2006). *Essentials of logic* (2nd ed.). Upper Saddle River, NJ: Prentice-Hall.

Damasio, A. (1994). *Descartes' error: Emotion, reason, and the human brain.* New York: Grosset/Putnam.

Demetriou, A., & Raftopoulos, A. (2004). The shape and direction of development: Teleologically but erratically lifted up or timely harmonious? *Journal of Cognition and Development, 5*, 89–95.

Descartes, R. (1969). *The philosophical works of Descartes* (2 vols.; E. S. Haldane & G. R. T. Ross, Trans.). Cambridge, UK: Cambridge University Press.

Dilthey, W. (1972). The rise of hermeneutics (F. Jameson, Trans.). *New Literary History, 3*, 229–244.

Dray, W. H. (1957). *Laws and explanation in history.* Oxford, UK: Oxford University Press.

Dreyfus, H. L., & Dreyfus, S. E. (1999). The challenge of Merleau-Ponty's phenomenology of embodiment for cognitive science. In G. Weiss & H. F. Haber (Eds.), *Perspectives on embodiment: The intersections of nature and culture* (pp. 217–246). New York: Routledge.

Edelman, G. M. (1992). *Bright air, brilliant fire: On the matter of the mind.* New York: Basic Books.

Edelman, G. M. (2006). *Second nature: Brain science and human knowledge.* New Haven, CT: Yale University Press.

Elder, G. H., Jr. (1998). The life course and human development. In W. Damon (Editor-in-Chief) & R. M. Lerner (Ed.), *Handbook of child psychology: Vol. 1. Theoretical models of human development* (5th ed., pp. 939–991). New York: Wiley.

Elman, J. L., Bates, E. A., Johnson, M. H., Karmiloff-Smith, A., Parisi, D., & Plunkett, K. (1996). *Rethinking innateness: A connectionist perspective on development.* Cambridge, MA: MIT Press.

Erikson, E. H. (1968). *Identity youth and crisis.* New York: Norton.

Fischer, K. W., & Kennedy, B. (1997). Tools for analyzing the many shapes of development: The case of self-in-relationships in Korea. In E. Amsel & K. A. Renninger (Eds.), *Change and development: Issues of theory, method, and application* (pp. 117–152). Mahwah, NJ: Erlbaum.

Fischer, K. W., & Rose, S. P. (1999). Rulers, clocks, and nonlinear dynamics: Measurement and method in developmental research. In G. Savelsbergh, H. van der Maas, & P. van Geert (Eds.), *Nonlinear developmental processes* (pp. 197–212). Amsterdam: Royal Netherlands Academy of Arts and Sciences.

Fodor, J. (2000). *The mind doesn't work that way: The scope and limits of computational psychology.* Cambridge, MA: MIT Press.

Ford, D. H., & Lerner, R. M. (1992). *Developmental systems theory: An integrative approach.* Newbury Park, CA: Sage.

Frankel, C. (1957). Explanation and interpretation in history. *Philosophy of Science, 24*, 137–155.

Fumerton, R. (1992). Inference to the best explanation. In J. Dancy & E. Sosa (Eds.), *A companion to epistemolgy* (pp. 207–209). Oxford, UK: Basil Blackwell.

Gadamer, H. G. (1989). *Truth and method* (2nd ed., rev.; J. Weinsheimer & D. Marshall, Trans.). New York: Crossroad.

Gadamer, H. G. (1993). *Reason in the age of science.* Cambridge, MA: MIT Press.

Gallagher, S. (2005). *How the body shapes the mind.* Oxford, UK: Clarendon Press.
Gallese, V., & Lakoff, G. F. (2005). The brain's concepts: The role of the sensory–motor system in conceptual knowledge. *Cognitive Neuropsychology, 22,* 455–479.
Geldhof, G. J., Little, T. D., & Colombo, J. (2010). Self-regulation across the lifespan. In R. M. Lerner (Editor-in-Chief) & M. E. Lamb & A. M. Freund (Vol. Eds.), *The handbook of lifespan development: Vol. 2. Social and emotional development* (pp. 78–102). Hoboken, NJ: Wiley.
Gilbert, S. F., & Epel, D. (2009). *Ecological developmental biology: Integrating epigenetics, medicine, and evolution.* Sunderland, MA: Sinauer Associates.
Goode, J. M. M. (2007). The affordances for social psychology of the ecological approach to social knowing. *Theory and Psychology, 17*(2), 265–295.
Gottlieb, G. (1992). *Individual development and evolution: The genesis of novel behavior.* New York: Oxford University Press.
Gottlieb, G. (1996). A systems view of psychobiological development. In D. Magnusson (Ed.), *Individual development over the lifespan: Biological and psychosocial perspectives* (pp. 76–103). New York: Cambridge University Press.
Gottlieb, G. (2000). Environmental and behavioral influences on gene activity. *Current Directions in Psychological Science, 9*(3), 93–97.
Gottlieb, G. (2003). On making behavioral genetics truly developmental. *Human Development, 46,* 337–355.
Gottlieb, G., Wahlsten, D., & Lickliter, R. (2006). The significance of biology for human development: A developmental psychobiological systems view. In R. M. Lerner (Vol. Ed.) & W. Damon & R. M. Lerner (Series Eds.), *Handbook of child psychology: Vol. 1. Theoretical models of human development* (6th ed., pp. 210–257). New York: Wiley.
Grayling, A. C. (1992). Transcendental arguments. In J. Dancy & E. Sosa (Eds.), *A companion to epistemology* (pp. 506–509). Oxford, UK: Basil Blackwell.
Greenberg, G. (2011). The failure of biogenetic analysis in psychology: Why psychology is not a biological science. *Research in Human Development, 8*(3–4), 173–191.
Hanson, N. R. (1958). *Patterns of discovery.* London: Cambridge University Press.
Hanson, N. R. (1970). Is there a logic of scientific discovery? In B. A. Brody (Ed.), *Readings in the philosophy of science* (pp. 620–633). Englewood Cliffs, NJ: Prentice-Hall.
Harman, G. (1965). The inference to the best explanation. *Philosophical Review, 74,* 88–95.
Hempel, C. G. (1942). The function of general laws in history. *Journal of Philosophy, 39,* 35–48.
Hempel, C. G., & Oppenheim, P. (1948). Studies in the logic of explanation. *Philosophy of Science, 15*(2), 135–175.
Hobson, R. P. (2002). *The cradle of thought.* London: Macmillan.
Holton, G. (1973). *Thematic origins of scientific thought: Kepler to Einstein.* Cambridge, MA: Harvard University Press.
Ingold, T. (2000) Evolving skills. In H. Rose & S. Rose (Eds.), *Alas, poor Darwin: Arguments against evolutionary psychology* (pp. 273–297). New York: Harmony Books.
Kan, K. J., Ploeger, A., Raijmakers, M. E. J., Dolan, C. V., & van der Maas, H. L. J. (2010). Nonlinear epigenetic variance: Review and simulations. *Developmental Science, 13*(1), 11–27.
Kaplan, D. (2008). An overview of Markov chain methods for the study of stage-sequential developmental processes. *Developmental Psychology, 44*(2), 457–467.
Keller, E. F. (2010). *The mirage of a space between nature and nurture.* Durham, NC: Duke University Press.

Kim, C. J., & Nelson, C. R. (1999). *State–space models with regime switching: Classical and Gibbs-sampling approaches with applications.* Cambridge, MA: MIT Press.

Kitcher, P. (1981). Explanatory unification. *Philosophy of Science, 48*, 507–531.

Kuczynski, L., & Parkin, C. M. (2009). Pursuing a dialectical perspective on transaction: A social relational theory of micro family processes. In A. Sameroff (Ed.), *The transactional model of development: How children and contexts shape each other* (pp. 247–260). Washington, DC: American Psychological Association.

Kuhn, D., Garcia-Mila, M., Zohar, A., & Andersen, C. (1995). Strategies of knowledge acquisition. *Monographs of the Society for Research in Child Development, 60*(4, Serial No. 245).

Kuhn, T. S. (1962). *The structure of scientific revolutions.* Chicago: University of Chicago Press.

Kuhn, T. S. (1970). *The structure of scientific revolutions* (2nd ed.). Chicago: University of Chicago Press.

Kuhn, T. S. (1977). *The essential tension: Selected studies in the scientific tradition and change.* Chicago: University of Chicago Press.

Kurtines, W. M., Ferrer-Wreder, L., Berman, S. L., Lorente, C. C., Silverman, W. K., & Montgomery, M. J. (2008). Promoting positive youth development: New directions in developmental theory, methods, and research. *Journal of Adolescent Research, 23*(Whole Issue), 233–378.

Lakatos, I. (1978a). *Mathematics, science and epistemology: Philosophical papers* (Vol. 2). New York: Cambridge University Press.

Lakatos, I. (1978b). *The methodology of scientific research programmes: Philosophical papers* (Vol. 1). New York: Cambridge University Press.

Lambdin, C. (2012). Significance tests as sorcery: Science is empirical–significance tests are not. *Theory and Psychology, 22*(1), 67–90.

Latour, B. (1993). *We have never been modern.* Cambridge, MA: Harvard University Press.

Latour, B. (2004). *Politics of nature.* Cambridge, MA: Harvard University Press.

Laudan, L. (1977). *Progress and its problems: Towards a theory of scientific growth.* Berkeley: University of California Press.

Laudan, L. (1996). *Beyond positivism and relativism: Theory, method, and evidence.* Boulder, CO: Westview Press.

Lerner, J., Bowers, E., Minor, K., Boyd, M. J., Mueller, M. K., Schmid, K. L., et al. (2012). Positive youth development: Processes, philosophies, and programs. In I. B. Weiner (Editor-in-Chief) & R. M. Lerner, M. A. Easterbrooks, & J. Mistry (Eds.), *Comprehensive handbook of psychology: Developmental psychology* (Vol. 6, pp. 365–392). New York: Wiley.

Lerner, R. M. (1984). *On the nature of human plasticity.* New York: Cambridge University Press.

Lerner, R. M. (2006). Developmental science, developmental systems, and contemporary theories of human development. In R. M. Lerner (Vol. Ed.) & W. Damon & R. M. Lerner (Series Eds.), *Handbook of child psychology: Vol. 1. Theoretical models of human development* (6th ed., pp. 1–17). New York: Wiley.

Lerner, R. M. (2012). Developmental science: Past, present and future. *International Journal of Developmental Science, 15*(6), 29–36.

Lerner, R. M., & Overton, W. F. (2008). Exemplifying the integrations of the relational developmental system: Synthesizing theory, research, and application to promote positive development and social justice. *Journal of Adolescent Research, 23*, 245–255.

Lewis, M. (2010). The emergence of consciousness and its role in human development. In R. M. Lerner (Ed.-in-Chief) & W. F. Overton (Vol. Ed.), *Handbook of life-span development: Vol. 1. Cognition, biology, and methods* (pp. 628–670). Hoboken, NJ: Wiley.

Lewontin, R. C. (1974). The analysis of variance and the analysis of causes. *American Journal of Human Genetics, 26*, 400–411.

Lickliter, R., & Honeycutt, H. (2010). Rethinking epigenesis and evolution in light of developmental science. In M. S. Blumberg, J. H. Freeman, & S. R. Robinson (Eds.), *Oxford handbook of developmental behavioral neuroscience* (pp. 30–47). New York: Oxford University Press.

Lipton, P. (1991). *Inference to the best explanation*. New York: Routledge.

Lorkowski, C. M. (2010). David Hume: Causation. *Internet Encyclopedia of Philosophy*. Retrieved from *www.iep.utm.edu/hume-cau/#h6*.

Luke, D. A. (2004). *Multilevel modeling*. London: Sage.

Mayr, E. (1996). The autonomy of biology: The position of biology among the sciences. *Quarterly Review of Biology, 71*(1), 97–106.

MacWhinney, B. (2010). Language development. In R. M. Lerner (Ed.-in-Chief) & W. F. Overton (Vol. Ed.), *Handbook of life-span development: Vol. 1. Cognition, biology, and methods* (pp. 467–508). Hoboken, NJ: Wiley.

Marshall, P. J. (2009). Relating psychology and neuroscience. *Perspectives on Psychological Science, 4*, 113–125.

Marshall, P. J. (in press). Coping with complexity: Developmental systems and multilevel analyses in developmental psychopathology. *Development and Psychopathology*,

Mascolo, M. F., & Fischer, K. W. (2010). The dynamic development of thinking, feeling, and acting over the life span. In R. M. Lerner (Ed.-in-Chief) & W. F. Overton (Vol. Ed.), *Handbook of life-span development: Vol. 1. Cognition, biology, and methods* (pp. 149–194). Hoboken, NJ: Wiley.

McClelland, M. M., Ponitz, C. C., Messersmith, E. E., & Tominey, S. (2010). Self-regulation: Integration of cognition and emotion. In R. M. Lerner (Ed.-in-Chief) & W. F. Overton (Vol. Ed.), *Handbook of life-span development: Vol. 1. Cognition, biology, and methods* (pp. 509–553). Hoboken, NJ: Wiley.

Meehl, P. E. (1990a). Apprising and amending theories: The strategy of Lakatosian defense and two principles that warrant it. *Psychological Inquiry, 1*, 108–141.

Meehl, P. E. (1990b). Why summaries of research on psychological theories are often uninterruptable. *Psychological Reports, 66*, 195–244.

Miller, G. (2010). Mistreating psychology in the decades of the brain. *Perspectives on Psychological Science, 5*(6), 716–743.

Molenaar, P. C. M. (2007). On the implications of the classical ergodic theorems: Analysis of developmental processes has to focus on intra-individual variation. *Developmental Psychobiology, 50*, 60–69.

Molenaar, P. C. M., & Campbell, C. G. (2009). The new person-specific paradigm in psychology. *Current Directions in Psychology, 18*, 112–117.

Molenaar, P. C. M., & Newell, K. M. (Eds). (2010). *Individual pathways of change: Statistical models for analyzing learning and development*. Washington, DC: American Psychological Association.

Molenaar, P. C. M., Sinclair, K. O., Rovine, M. J., Ram, N., & Corneal, S. E. (2009). Analyzing developmental processes on an individual level using non-stationary time series modeling. *Developmental Psychology, 45*, 260–271.

Montpetit, M. A., Bergeman, C. S., Deboeck, P. R., Tiberio, S. S., & Boker, S. M. (2010). Resilience-as-process: Negative affect, stress, and coupled dynamical systems. *Psychology and Aging, 25*(3), 631–640.

Mueller, G. B. (2010). Epigenetic innovation. In M. Pigliucci & G. B. Mueller (Eds.), *Evolution: The extended synthesis* (pp. 307–332). Cambridge, MA: MIT Press.

Müller, U., & Newman, J. L. (2008). The body in action: Perspectives on embodiment and development. In W. F. Overton, U. Müller, & J. L. Newman (Eds.), *Developmental perspectives on embodiment and consciousness* (pp. 313–342). New York: Taylor & Francis.

Müller, U., & Racine, P. (2010). The development of representation and concepts. In R. M. Lerner (Ed.-in-Chief) & W. F. Overton (Vol. Ed.), *Handbook of life-span development: Vol. 1. Cognition, biology, and methods* (pp. 346–390). Hoboken, NJ: Wiley.

Nagel, E. (1967). The nature and aim of science. In S. Morgenbesser (Ed.), *Philosophy of science today* (pp. 5–13). New York: Basic Books.

Nagel, E. (1979). *The structure of science* (2nd ed.). Cambridge, MA: Hackett.

Nesselroade, J. R., & Molenaar, P. C. M. (2010). Emphasizing intraindividual variability in the study of development over the life span: Concepts and issues. In R. M. Lerner (Ed.-in-Chief) & W. F. Overton (Vol. Ed.), *Handbook of life-span development: Vol. 1. Cognition, biology, and methods* (pp. 30–54). Hoboken, NJ: Wiley.

Nesselroade, J. R., & Ram, N. (2004). Studying intraindividual variability: What we have learned that will help us understand lives in context. *Research in Human Development, 1*(1 & 2), 9–29.

Overton, W. F. (1975). General systems, structure and development. In K. Riegel & G. Rosenwald (Eds.), *Structure and transformation: Developmental aspects* (pp. 61–81). New York: Wiley InterScience.

Overton, W. F. (1984). World views and their influence on psychological theory and research: Kuhn–Lakatos–Laudan. In H. W. Reese (Ed.), *Advances in child development and behavior* (Vol. 18, pp. 191–226). New York: Academic Press.

Overton, W. F. (1990). Competence and procedures: Constraints on the development of logical reasoning. In W. F. Overton (Ed.), *Reasoning, necessity, and logic: Developmental perspectives* (pp. 1–32). Hillsdale, NJ: Erlbaum

Overton, W. F. (1991). Historical and contemporary perspectives on developmental theory and research strategies. In R. Downs, L. Liben, & D. Palermo (Eds.), *Visions of aesthetics, the environment, and development: The legacy of Joachim Wohlwill* (pp. 263–311). Hillsdale, NJ: Erlbaum.

Overton, W. F. (1994). The arrow of time and cycles of time: Concepts of change, cognition, and embodiment. *Psychological Inquiry, 5*, 215–237.

Overton, W. F. (1997a). Beyond dichotomy: An embodied active agent for cultural psychology. *Culture and Psychology, 3*, 315–334.

Overton, W. F. (1997b). Relational-developmental theory: A psychology perspective. In D. Görlitz, H. J. Harloff, J. Valsiner, & G. Mey (Eds.), *Children, cities and psychological theories: Developing relationships* (pp. 315–335). Berlin & New York: de Gruyter.

Overton, W. F. (2006). Developmental psychology: Philosophy, concepts, and methodology. In W. Damon (Editor-in-Chief) & R. M. Lerner (Ed.), *The handbook of child psychology: Vol. 1. Theoretical models of human development* (pp. 18–88). New York: Wiley.

Overton, W. F. (2007). A coherent metatheory for dynamic systems: Relational organicism-contextualism. *Human Development, 50*, 154–159.

Overton, W. F. (2008). Embodiment from a relational perspective. In W. F. Overton, U. Müller, & J. L. Newman (Eds.), *Developmental perspective on embodiment and consciousness* (pp. 1–18). Hillsdale, NJ: Erlbaum.

Overton, W. F. (2010). Life-span development: Concepts and issues. In R. M. Lerner (Ed.-in-Chief) & W. F. Overton (Vol. Ed.), *Handbook of life-span development: Vol. 1. Cognition, biology, and methods* (pp. 1–29). Hoboken, NJ: Wiley.

Overton, W. F. (2012). Evolving scientific paradigms: Retrospective and prospective. In L. L'Abate (Ed.), *The role of paradigms in theory construction* (pp. 31–65). New York: Springer.

Overton, W. F. (2013). Relationism and relational-developmental-systems: A paradigm for developmental science in the post-Cartesian era. *Advances in Child Development and Behavior, 44*, 21–94.

Overton, W. F., & Ennis, M. (2006a). Cognitive-developmental and behavior-analytic theories: Evolving into complementarity. *Human Development, 49*, 143–172.

Overton, W. F., & Ennis, M. (2006b). Relationism, ontology, and other concerns. *Human Development, 49*, 180–183.

Overton, W. F., & Lerner, R. M. (2012). Relational-developmental-systems: Paradigm for developmental science in the postgenomic era. *Brain and Behavioral Science, 35*, 375–376.

Overton, W. F., Müller, U., & Newman, J. L. (Eds.). (2008). *Developmental perspectives on embodiment and consciousness*. Hillsdale, NJ: Erlbaum.

Overton, W. F., & Reese, H. W. (1973). Models of development: Methodological implications. In J. R. Nesselroade & H. W. Reese (Eds.), *Life-span developmental psychology: Methodological issues* (pp. 65–86). New York: Academic Press.

Overton, W. F., & Reese, H. W. (1981). Conceptual prerequisites for an understanding of stability-change and continuity-discontinuity. *International Journal of Behavioral Development, 4*, 99–123.

Partridge, T. (2011). Methodological advances toward a dynamic developmental behavioral genetics: Bridging the gap. *Research in Human Development, 8*, 242–257.

Patokorpi, E. (2007). Logic of Sherlock Holmes in technology enhanced learning. *Educational Technology and Society, 10*(1), 171–185.

Pepper, S. C. (1942). *World hypotheses*. Berkeley: University of California Press.

Piaget, J. (1952). *The origins of intelligence in children*. New York: Norton.

Piaget, J. (1967). *Six psychological studies*. New York: Random House.

Piaget, J. (1970). Piaget's theory. In P. H. Mussen (Ed.), *Carmichael's manual of child psychology* (3rd ed., Vol. 1, pp. 703–732). New York: Wiley.

Piaget, J. (1985). *The equilibration of cognitive structures*. Chicago: University of Chicago Press.

Pidwirny, M. (2006). The hypothetico-deductive method: Fundamentals of physical geography. Retrieved from *www.physicalgeography.net/fundamentals/3b.html*.

Pierce, C. S. (1992). *Reasoning and the logic of things: The Cambridge conference lectures of 1898*. Cambridge, MA: Harvard University Press.

Pigliucci, M., & Mueller, G. B. (2010a). Elements of an extended evolutionary synthesis. In M. Pigliucci & G. B. Mueller (Eds.), *Evolution: The extended synthesis* (pp. 1–17). Cambridge, MA: MIT Press.

Pigliucci, M., & Mueller, G. B. (Eds.). (2010b). *Evolution: The extended synthesis*. Cambridge, MA: MIT Press.

Popper, K. (1959). *The logic of scientific discovery*. London: Hutchinson.

Popper, K. (1963). *Conjectures and refutations*. London: Routledge & Kegan Paul.

Popper, K. (1970). Normal science and its dangers. In I. Lakatos & A. Musgrave (Eds.), *Criticism and the growth of knowledge* (pp. 51–58). New York: Cambridge University Press.

Prosch, H. (1964). *The genesis of twentieth century philosophy*. New York: Doubleday.

Putnam, H. (1983). *Realism and reason: Philosophical papers* (Vol. 3). New York: Cambridge University Press.

Putnam, H. (1987). *The many faces of realism*. Cambridge, UK: Cambridge University Press.

Putnam, H. (1994). *Words and life*. Cambridge, MA: Harvard University Press.

Quine, W. V. (1953). *From a logical point of view*. Cambridge, MA: Harvard University Press.

Randall, J. H. (1960). *Aristotle*. New York: Columbia University Press.

Redhead, M. (1990). Explanation. In D. Knowles (Ed.), *Explanation and its limits* (pp. 135–152). Cambridge, UK: Cambridge University Press.

Ricco, R., & Overton W. F. (2011). Dual systems competence ↔ procedural processing: A relational-developmental-systems approach to reasoning. *Developmental Review, 31*, 119–150.

Ricoeur, P. (1984). *Time and narrative* (Vol. 1; K. McLaughlin & D. Pellauer, Trans.). Chicago: University of Chicago Press.

Ricoeur, P. (1991). *From text to action: Essays in hermeneutics* (Vol. II, K. Blamey & J. B. Thompson, Trans.). Evanston, IL: Northwestern University Press.

Romesburg, H. C. (2009). *Best research practices*. Morrisville, NC: Lulu Enterprises.

Ross, W. D. (1959). *Aristotle*. Cleveland, OH: World Publishing.

Rowlands, M. (2010). *The new science of the mind: From extended mind to embodied phenomenology*. Cambridge, MA: MIT Press.

Rozeboom, W. W. (1997). Good science is abductive, not hypothetico-deductive. In L. L. Harlow, S. A. Mulaik, & J. H. Steiger (Eds.), *What if there were no significance tests?* (pp. 335–391). Mahwah, NJ: Erlbaum.

Russell, J. (1996). *Agency: Its role in mental development*. New York: Taylor & Francis.

Santostefano, S. (2010). Developmental psychopathology—self, embodiment, meaning: A holistic-systems perspective. In R. M. Lerner (Ed.-in-Chief) & W. F. Overton (Vol. Ed.), *Handbook of life-span development: Vol. 1. Cognition, biology, and methods* (pp. 792–836). Hoboken, NJ: Wiley.

Searle, J. (1992). *The rediscovery of the mind*. Cambridge, MA: MIT Press.

Serlin, R., & Lapsley, D. K. (1985). Rationality in psychological research: The good enough principle. *American Psychologist, 40*, 73–83.

Sheets-Johnstone, M. (1990). *The roots of thinking*. Philadelphia: Temple University Press.

Siegler, R. S. (1996). *Emerging minds: The process of change in children's thinking*. New York: Oxford University Press.

Siegler, R. S., & Crowley, K. (1991). The microgenetic method: A direct means for studying cognitive development. *American Psychologist, 46*, 606–620.

Siegler, R. S., & Crowley, K. (1992). Microgenetic methods revisited. *American Psychologist, 47*, 1241–1243.

Siegler, R. S., & Munakata, Y. (1993, Winter). Beyond the immaculate transition: Advances in the understanding of change. *SRCD Newsletter*.

Smith, L. B. (2005). Cognition as a dynamic system: Principles from embodiment. *Developmental Review, 25*, 278–298.

Sterba, S. K., & Bauer, D. J. (2010). Matching method with theory in person-oriented developmental psychopathology research. *Development and Psychopathology, 22*, 239–254.

Stewart, J., Gapenne, O., & Di Paolo, E. A. (2010). *Enaction: Toward a new paradigm for cognitive science*. Cambridge, MA: MIT Press.

Taylor, C. (1964). *The explanation of behavior*. New York: Humanities Press.

Taylor, C. (1995). *Philosophical arguments*. Cambridge, MA: Harvard University Press.

Thagard, P. (1988). *Computational philosophy of science*. Cambridge, MA: MIT Press.

Toulmin, S. (1953). *The philosophy of science*. New York: Harper & Row.

Toulmin, S. (1961). *Foresight and understanding*. New York: Harper & Row.

Turiel, E. (2010). Domain specificity in social interactions, social thought, and social development. *Child Development, 81*, 720–726.

Turkheimer, E. (2011). Still missing. *Research in Human Development, 8*(3–4), 227–241.
Urban, J. B., Osgood, N. D., & Mabry, P. L. (2011). Developmental systems science: Exploring the application of systems science methods to developmental science questions. *Research in Human Development, 8*(1), 1–25.
van der Maas, H. L. J., & Molenaar, P. C. M. (1992). Stagewise cognitive development: An application of catastrophe theory. *Psychological Review, 99*, 395–417.
van Geert, P. (1991). A dynamic systems model of cognitive and language growth. *Psychological Review, 98*, 3–53.
van Geert, P. (2003). Dynamic systems approaches and modeling of developmental processes. In J. Valsiner & K. J. Connolly (Eds.), *Handbook of developmental psychology* (pp. 640–672). London: Sage.
von Eye, A., Bergman, L. R., & Hsieh, C. (in press). Person-oriented approaches. In R. M. Lerner (Editor-in-Chief) & W. F. Overton & P. C. M. Molenaar (Eds.), *The handbook of child psychology and developmental science: Vol. 1. Theory and methods* (7th ed.). Hoboken, NJ: Wiley.
von Wright, G. H. (1971). *Explanation and understanding.* Ithaca, NY: Cornell University Press.
Wartofsky, M. (1968). *Conceptual foundations of scientific thought.* Toronto: Macmillan.
Werner, H. (1948). *Comparative psychology of mental development.* New York: International Universities Press.
Werner, H. (1957). The concept of development from a comparative and organismic point of view. In D. B. Harris (Ed.), *The concept of development: An issue in the study of human behavior* (pp. 125–148). Minneapolis: University of Minnesota Press.
West-Eberhard, M. J. (2003). *Developmental plasticity and evolution.* New York: Oxford University Press.
Winch, P. (1958). *The idea of a social science and its relation to philosophy.* London: Routledge & Kegan Paul.
Witherington, D. C. (2007). The dynamic systems approach as metatheory for developmental psychology. *Human Development, 50*, 127–153.
Witherington, D. C. (2011). Taking emergence seriously: The centrality of circular causality for dynamic systems approaches to development. *Human Development, 54*, 66–92.
Witherington, D. C., & Heying, S. (2013). Embodiment and agency: Toward a holistic synthesis for developmental science. *Advances in Child Development and Behavior, 44*, 161–192.
Wittgenstein, L. (1958). *Philosophical investigations* (G. E. M. Anscombe, Trans.) (3rd ed.). Englewood Cliffs, NJ: Prentice-Hall.
Zednik, C. (2011). The nature of dynamical explanation. *Philosophy of Science, 78*(2), 238–263.

CHAPTER 3

Relational Developmental Systems Theories of Positive Youth Development

Methodological Issues and Implications

G. John Geldhof, Edmond P. Bowers, Sara K. Johnson, Rachel Hershberg, Lacey Hilliard, Jacqueline V. Lerner, and Richard M. Lerner

Interest in the strengths of youth, the plasticity of human development, and the concept of resilience coalesced in the 1990s to foster the evolution of the positive youth development concept (PYD; Lerner, Lerner, & Benson, 2011; Lerner, Phelps, Forman, & Bowers, 2009). As discussed by Hamilton (1999), the concept of PYD can be understood in three interrelated but distinct ways: as a developmental process, as a philosophy or approach to youth programming, and as instances of youth programs and organizations focused on fostering healthy or positive development among youth. In this chapter, we focus on the idea of PYD as a developmental process, as this framework has inspired considerable research across the adolescent period (Lerner et al., in press).

In the decade following Hamilton's (1999) discussion of PYD, several models of the developmental process believed to be involved in PYD were used to frame descriptive or explanatory research across the adolescent period (Lerner et al., 2012). Although these models differ in important ways (Lerner et al., 2012), they all reflect ideas associated with relational developmental systems theory (RDST; e.g., Overton, 2010, 2011, 2012; Overton & Müller, 2012; see also Overton, Chapter 2, this volume). As such, these models carry with them methodological problematics—that is, theoretical and procedural issues that researchers must keep in mind when conducting RDST-derived research about PYD. The purpose of this chapter is to articulate these methodological problematics and their RDST roots, illustrate how existing PYD research has addressed them, and provide practical implications for developmental scientists committed to conducting RDST-driven research on human development in general and on PYD more specifically (Lerner, 2006).

RDST and the Roots of the PYD Perspective

The study of PYD as a developmental process is rooted in an approach to developmental science that seeks to describe, explain, and optimize intraindividual change, and interindividual differences in intraindividual change, across the life span (Baltes, Reese, & Nesselroade, 1977). The contemporary theoretical frame for such scholarship involves RDST models (Overton, 2010, and Chapter 2, this volume). Models reflecting RDST include Bronfenbrenner's bioecological theory (e.g., Bronfenbrenner & Morris, 2006), action theory models of intentional goal-directed behaviors (e.g., Baltes, 1997; Brandtstädter, 2006; Heckhausen, 1999), Elder's (1998) life course theory, Magnusson's (1999; Magnusson & Stattin, 2006) holistic person–context interaction theory (Sterba & Bauer, 2010), and the developmental systems formulations articulated by Ford and Lerner (1992) and Gottlieb (1998).

These models emphasize that the basic process of human development involves mutually influential relations between developing individuals and the multiple levels of their complex and changing contexts. These bidirectional individual ↔ context relations constitute the fundamental unit of analysis in the study of human development from a RDST perspective. Such relations regulate the pace, direction, and outcomes of development. When these "developmental regulations" involve individual ↔ context relations that benefit both the person and his or her ecology, they are considered "adaptive developmental regulations" (Brandtstädter, 2006).

One key assumption of RDSTs—and the use of these theories to understand both adolescent development in general and to frame the PYD concept of developmental processes more specifically—is that there is always change and at least some potential for systematic change (i.e., plasticity) across the life span (Baltes, Lindenberger, & Staudinger, 2006; Lerner, 1984). This potential for change represents a fundamental strength of human development. Of course, plasticity means that change for the better or worse can characterize any individual's developmental trajectory. Both RDST and the PYD perspective linked to it emphasize, however, that the developmental system is sufficiently diverse and complex such that researchers and practitioners may find some means to connect individual and context in ways that enhance the probability of change for the better and of promoting more positive features of human development (Baltes et al., 2006; Lerner, 2002, 2004; Lerner et al., 2009).

The adolescent period—the primary focus of much PYD research—is characterized by an enormous number of individual and contextual changes. For example, changes in the prefrontal cortex, increases in the interconnectivity among brain regions, and increases in dopamine levels provide both vulnerabilities of risk and opportunities for growth in cognitive control and self-regulation (Steinberg, 2010). At the same time, most youth in Western societies experience contextual changes, including changing schools (e.g., Eccles, 2004) and an increased relevance of peer pressure for risk taking (e.g., Gardner & Steinberg, 2005). For youth in what has been termed *third world* nations but, today, are more frequently termed *majority world* nations, the pace of social change may be historically unique, relatively (or even markedly) abrupt, and may be accompanied by political and socioeconomic stressors that, at some times and in some places, may be life-threatening (e.g., in the case of revolutions or wars; Larson, Wilson, & Rickman, 2009; Silbereisen, 2005; Silbereisen & Chen, 2010). Moreover, adolescents have the cognitive, behavioral, and social relational skills to contribute actively and often effectively to their own development (Lerner, 1982; Lerner & Busch-Rossnagel, 1981; Lerner & Walls, 1999).

Due to the plasticity of the adolescent period, a strength-based view of adolescents, such as the PYD perspective, has been increasingly used as the lens of choice for viewing development both within the United States (e.g., Lerner et al., 2009) and internationally (e.g., Gestsdóttir & Lerner, 2007; Silbereisen & Lerner, 2007). *Plasticity* means that young people are capable of systematic change in structure and function, and the core idea of the PYD perspective is that this potential constitutes a strength that, if aligned with resources that might promote adaptive, healthy functioning (i.e., contextual variables that have been termed *developmental ecological assets*; Benson, Scales, & Syvertsen, 2011; Theokas & Lerner, 2006), then youth may be put on a more positive trajectory across adolescence (Lerner, Almerigi, Theokas, & Lerner, 2005).

Using these ideas, researchers have developed various models of how PYD may occur through aligning the strengths of young people and the resources that exist in their ecologies (e.g., Benson et al., 2011; Damon, 2004; Eccles, 2004; Hamilton & Hamilton, 2009; Larson, 2000; Lerner, Phelps, et al., 2009; Masten, 2001; Spencer, 2006). The model of the PYD process used by Lerner and colleagues (e.g., Lerner, Lerner, et al., 2005; Lerner, Lerner, von Eye, et al., 2011) explicitly draws on the individual ↔ context relational conception within RDST, and it has received the most empirical support (Heck & Subramaniam, 2009). Accordingly, we focus on this model as we discuss the methodological problematics of RDST-informed PYD research and draw on research derived from other PYD models as appropriate.

The Five Cs Model of PYD and the 4-H Study of PYD

The Five Cs Model of PYD has been elaborated in the context of the 4-H Study of PYD (e.g., Bowers et al., 2010; Lerner, 2011; Lerner, Lerner, et al., 2005; Lerner, von Eye, et al., 2009; Lerner, von Eye, Lerner, Lewin-Bizan, & Bowers, 2010), a longitudinal study conducted by Lerner, Lerner, and colleagues and supported by a grant from the National 4-H Council. The purpose of the 4-H Study was to identify relationships between individuals and their ecologies that promote thriving, as well as those that may have a preventive effect in regard to risk/problem behaviors. Within the 4-H Study, *thriving* is conceptualized as the growth of attributes that mark a flourishing, healthy young person; these characteristics are termed the *Five Cs* of PYD: competence, confidence, character, connection, and caring (Eccles & Gootman, 2002;

Lerner, Lerner, et al., 2005; Roth & Brooks-Gunn, 2003a, 2003b). Table 3.1 lists definitions of the Five Cs that have been previously used.

The 4-H Study was designed to test the Five Cs Model of PYD by focusing on youth in their actual environments, rather than conducting randomized controlled trials. In these environments, youth and their parents—rather than research investigators—make decisions about how they spend their time. Accordingly, the 4-H Study represents one of the few empirical studies that examine the Five Cs and their relations with positive and problematic developmental outcomes as they occur in the actual lives of youth.

As noted earlier, a key hypothesis tested in this study was that if the strengths of youth can be aligned with the resources for positive growth, then young people's healthy development may be optimized (Lerner, 2004). In addition, given that positively developing youth should be involved in adaptive developmental regulations, a thriving young person should act to contribute to the context that is benefiting him or her; there should be contributions to self, family, community, and civil society (Jelicic, Bobek, Phelps, Lerner, & Lerner, 2007; Lerner, Lerner, et al., 2005; Zaff, Kawashima-Ginsberg, & Lin, 2011). In other words, if positive development rests on mutually beneficial relations between the adolescent and his or her ecology, then thriving youth should be positively engaged with, and act to enhance, their world and should be less prone to engage in problem behaviors.

The developmental process envisioned by Lerner and Lerner (e.g., Lerner, Lerner, et al., 2005) involves adaptive developmental regulations between the strengths of youth and the developmental assets present in their ecologies. These mutually beneficial individual ↔ context relations are depicted as being associated with PYD (and,

TABLE 3.1. Definitions of the Five Cs of Positive Youth Development

C	Definition
Competence	Positive view of one's actions in domain-specific areas, including social, academic, cognitive, and vocational. *Social competence* pertains to interpersonal skills (e.g., conflict resolution). *Cognitive competence* pertains to cognitive abilities (e.g., decision making). School grades, attendance, and test scores are part of *academic competence. Vocational competence* involves work habits and career choice explorations, including entrepreneurship.
Confidence	An internal sense of overall positive self-worth and self-efficacy; one's global self-regard, as opposed to domain-specific beliefs.
Connection	Positive bonds with people and institutions that are reflected in bidirectional exchanges between the individual and peers, family, school, and community in which both parties contribute to the relationship.
Character	Respect for societal and cultural rules, possession of standards for correct behaviors, a sense of right and wrong (morality), and integrity.
Caring	A sense of sympathy and empathy for others.

Note. Based on Lerner et al. (2005) and Roth and Brooks-Gunn (2003).

by extension, with the Five Cs) and the enhanced probability of youth contributions to their ecology and the reduced probability of problem behaviors. The outcomes of these adaptive developmental regulations feed back to the individual and his or her context, and in doing so, create a qualitatively new individual ↔ context relation, in the sense of their unique locations in time and place (Elder, 1998; Elder, Modell, & Parke, 1993); as such, these relations are substantively nonrecursive. These relations provide as well a basis for further adaptive developmental regulations. These adaptive developmental regulations and their positive and problematic sequelae exist within the broader ecology of human development. This ecology includes both cultural and historical (temporal) variation, and thus introduces change at all levels of organization within the developmental system (Bronfenbrenner & Morris, 2006; Elder, 1998). Such changes are manifested by intraindividual change, by interindividual differences in intraindividual change, and by normative and non-normative contextual variation (Baltes, Reese, & Nesselroade, 1977).

At present, the 4-H Study of Positive Youth Development represents the most comprehensive "test" of the Five Cs Model of PYD. The study began in the fall of 2002, and data were collected annually from participants during their 5th- through 12th-grade years and included information about strengths of youth as well as their contexts. Using these data, researchers have assessed how the individual strengths of adolescents interact with aspects of their families, schools, and communities to promote the development of the Five Cs and other positive developmental outcomes (e.g., active and engaged citizenship; Zaff et al., 2011), as well as diminish the likelihood of risks and problems. Given the comprehensiveness of the 4-H Study and its close ties to the Five Cs Model, we use examples from it throughout our discussion of the methodological problematics involved in conducting RDST-derived PYD research.

Problematics for RDST Research

We have thus far discussed PYD as a theoretical approach framed by RDST. We have not, however, discussed how PYD researchers actually apply the tenets of RDST when generating and testing research hypotheses. To preview a key point about the fit between theory and research within this area of scholarship, it remains the case that many developmental scientists argue for relational theories while simultaneously using data collection and analysis methods that are not consistent with the complexity and nuance that RDST implies. Therefore, in the remainder of this chapter we describe conceptual and methodological problematics that currently pertain to PYD research derived from RDST. In particular, we discuss implications related to the assumptions that (1) development involves interrelated changes in a complex, multilevel system; (2) developmental trajectories of all individuals remain relatively plastic across the life span; and (3) examining relations in the presence of such complexity requires multimethod integration. Faithfully conducting research that follows from RDST requires that theoretical ideas be actualized through methodological choices related to research design, data collection, and analysis methods. This obligation is characteristic, as Lerner and Overton (2008) note, of "good science—selecting all features of one's methodology based on the nature of the (theoretically predicated) questions asked" (p. 250).

Development Occurs in a Complex Person–Context System

The RDST approach to developmental science emphasizes holism as a fundamental guiding principle. In opposition to a fixed, atomistic reality composed of elements that preserve their identity regardless of context (Overton, 2010, 2012, and Chapter 2, this volume), holism views objects and events as necessarily related to the context in which they are embedded. The whole exists as an organized and self-organizing system of parts, each defined by its relations to other parts and to the whole itself (Overton, 2012, and Chapter 2, this volume). The key empirical question for developmental scientists interested in describing, explaining, and promoting positive human development in the context of a complex holistic system is therefore composed of five interrelated "whats": *What* attributes of *what* individuals, in relation to *what* contextual conditions, and at *what* points in ontogenetic, family or generational, and cohort or historical time can be integrated to promote *what* instances of positive human development?

Armed with appropriate RDST-informed research questions, researchers must make methodological decisions that acknowledge (1) that developmental systems are embedded (Overton, 2010, 2012), that is, they are characterized by holism; (2) that individuals in part produce (i.e., direct) their own ontogenetic development; and (3) that systematic plasticity is present across the life span. We provide more details concerning each of these ideas.

Focus on the System

Once researchers embrace the concepts underlying RDST, the types of questions that they ask must necessarily shift, as exemplified in the multicomponent "what" question. A research framework informed by RDST must include multiple elements that together account for the plasticity and dynamism that constitute ontogeny. For such research to match the complicated theory from which it derives, researchers must consider the complex and dynamic developmental system of which the developing person is a part.

Several studies derived from the 4-H Study illustrate how to frame and test research questions that recognize the bidirectional relationships between developing persons and their changing contexts. For example, Urban, Lewin-Bizan, and Lerner (2010) showed that both the strengths of youth and the resources of their contexts are involved in youth thriving. Urban and colleagues explored whether youth intentional self-regulatory skills moderated the effect of participation in out-of-school-time (OST) activities in predicting PYD and risk outcomes among adolescents in low-resource neighborhoods. These neighborhoods were classified as low-resource because, using census data, the authors found that the opportunities available to youth in their ecological contexts along dimensions of human resources, physical or institutional resources, collective activity, and accessibility were limited compared to other neighborhoods (see Theokas & Lerner, 2006, for more information on how this measure was derived). Urban and colleagues found that youth who reported the highest self-regulatory capacity benefited the most from involvement in OST activities. The strength of these relations was most evident in girls.

Future research should consider additional person- and context-level variables that might explain specific trajectories of youth development. For example, the findings of Urban and colleagues (2010) indicate sex differences, which might be explained more fully by accounting for additional person-level variables (e.g., age, race, socioeconomic status, religion, and household structure) in future studies. Although Urban and colleagues considered ontogenetic time, the relative influence of generational, cohort, or historical time might also explain why this evidence in support of the RDST-based PYD model seems to hold more for adolescent girls than for adolescent boys (Way, 2011). In addition, future research could account for more context-level variables, such as indicators of social capital and social networks. Such complexity implies that understanding how characteristics of individuals interact with aspects of their contexts requires a research focus on person ↔ context relations rather than simple aggregations of person- and context-level attributes.

Person ↔ Context Relations as Units of Analysis

Development is impacted by aspects of the individual and context, but the impact of any given personal or contextual characteristic can be interpreted only as part of a larger person ↔ context system. A specific level of competence may represent positive person ↔ context relations for some individuals in some contexts, but the same level of competence may represent neutral or even negative person ↔ context relations for other individuals or in other contexts (Masten & Coatsworth, 1998). In other words, person ↔ context relations will vary between individuals (Molenaar, 2007; Tobach & Greenberg, 1984). As such, the goal of RDST research is to capture and understand relations among the meaningful, adaptive person ↔ context relations (i.e., adaptive developmental regulations) that characterize development across diverse populations.

Accurately capturing the often idiographic nature of developmental regulations requires that researchers consider the contexts in which their participants are embedded, as well as which interactions with those contexts are adaptive. This problem can be tackled through idiographic research designs and analyses (see the section "Integrating Idiographic and Nomothetic Processes" for a more detailed discussion of this issue; see also Nesselroade & Molenaar, 2010, 2012), or they may alternatively be represented as a statistical interaction between self-reported measures of internal assets and objectively measured indices of contextual resources (e.g., Theokas et al., 2005).

Perhaps more appropriately, researchers can obtain and analyze nomothetic information through surveys that require participants to interpret survey questions idiographically (e.g., Nesselroade, Gerstorf, Hardy, & Ram, 2007). For instance, researchers interested in assessing participants' connection to their communities could obtain more useful information from a Likert-scaled item such as "How involved are you in your community?" than by averaging several Likert-scaled items about community service, connection to a religious group, or participation in school-related organizations. The first question allows flexibility in how participants interpret community involvement, whereas the later set of questions constrains the possible domains in which connection can occur. Asking both types of questions, however, would enable researchers to ascertain empirically the links between the idiographically phrased item and the domain-specific items.

The Five Cs Model of PYD presents one example of how person ↔ context relations can be the target of both theoretical and empirical consideration when incorporated in a nomothetic research design. The Five Cs Model emphasizes the importance of adaptive developmental regulations and discusses each C as a strength arising from person ↔ context relations. High levels of each C require not only the presence of a personal strength (e.g., ability) but also the successful application of strengths within each individual's unique context. These strengths represent broad, multidimensional domains that allow youth to display qualitatively distinct yet equifinal pathways to thriving. For example, high scores on the C of Competence require the successful application of personal strengths in one or more contexts. Although additive (e.g., high levels of competence in multiple domains leads to a higher Competence score than high levels in fewer domains), the 4-H Study's measure of Competence allows youth to display competent behavior in any combination of academic, social, and physical domains. Similarly, the C of Connection implies relationships between an individual and his or her context but allows individuals to display connection to different components and their context, such as their schools, communities, families, and peers. Each of the Five Cs, then, represents person ↔ context relations as the unit of analysis.

Individuals as Active Producers of Their Own Development

Developmental scientists aim to optimize contexts in ways that promote positive development, but recognize that individuals must also regulate their behavior in ways that take advantage of available resources. Developmental regulations represent the bidirectional ways individuals influence and are influenced by their contexts (e.g., Lerner, 2002). Individuals' influences on their developmental regulations (i.e., self-regulation) enable them to intentionally influence their own developmental outcomes (Brandtstädter & Lerner, 1999; Lerner, 2002). When individuals intentionally contribute to their development in such a way that successfully aligns their interests, desires, and needs with available contextual resources, their intentional self-regulation is viewed as adaptive (e.g., Baltes et al., 2006; Brandtstädter, 1998, 2006; Gestsdóttir & Lerner, 2008).

The 4-H Study has systematically investigated the processes through which adolescents adaptively and intentionally self-regulate, and thus actively contribute to and produce their developmental outcomes (e.g., Gestsdóttir & Lerner, 2007, 2008). Researchers involved in the 4-H Study have argued that a process of selecting goals, optimizing resources to achieve these goals, and adjusting expectations and strategies when they encounter obstacles toward achieving goals may explain how youth draw resources from their contexts in ways that positively influence their development (Gestsdóttir & Lerner, 2008). This process of selecting, optimizing, and compensating (SOC; e.g., Freund & Baltes, 2002) has been found to be a key asset for individuals in achieving future positive developmental outcomes (Baltes et al., 2006).

For this reason, the 4-H Study has used measures of SOC to assess self-regulation in youth. For example, Schmid, Phelps, and Lerner (2011) assessed whether measures of SOC and hopeful future expectations (another self-regulatory strength) predicted positive and negative trajectories of youth development across grades 7–9. They found that both SOC and hopeful future expectations were associated with positive developmental trajectories. That is, youth with higher self-regulatory strengths, such as hopeful future expectations and SOC, were more likely to have the most favorable trajectories of PYD, community contribution, and in turn, trajectories involving fewer depressive symptoms. Bowers and colleagues (2012) similarly found that hopeful future expectations mediated the relationship between the quantity and quality of positive relationships youth have with adults and the development of aspects of PYD. Together, these studies illustrate that intentional self-regulation and hope can make important contributions to positive development across adolescence, and thus potentially enable adolescents to direct their lives in meaningful ways.

As has been noted throughout the 4-H Study, the relationship between intentional self-regulation and PYD appears to apply to adolescents regardless of demographic differences. However, examining the ways diverse youth contribute to their own development and utilize the resources to which they have access will require that researchers gain more information from youth about the contextual resources to which they have access and how much they value these resources (Leventhal & Brooks-Gunn, 2000; Leventhal, Dupere, & Brooks-Gunn, 2009).

In interpreting these results, researchers must view the above findings as occurring during a specific developmental period, for a particular sample of adolescents living in a specific historical epoch. Person ↔ context relations intertwine individuals and contexts across time in highly complex ways. Acknowledging and accounting for temporal complexity both facilitate and complicate the tasks of developmental scientists, however, because temporal complexity presents its own set of methodological problematics (Lerner, Schwartz, & Phelps, 2009; Wohlwill, 1973).

The Temporality of Complex Developmental Processes

Complexity in development implies that the developmental trajectories of all individuals remain relatively plastic across the life span (Lerner, 1984). Research derived from RDST accordingly acknowledges the existence of inter- and intraindividual variability in development. For instance, the stated goal of the PYD perspective is to optimize the trajectories of all youth—an objective that depends on the presence of plasticity in intraindividual change. The acknowledgment of such plasticity impacts research derived from RDST in many ways. We highlight key ideas related to developmental plasticity that are especially important for PYD researchers to consider.

Predicting Developmental Phenomena Requires Change-Sensitive Measurement Tools

Development and its plasticity can only be examined using tools that are themselves sensitive to change. Although not surprising at first glance, this statement suggests that the tools used to study development must be able to detect changes over a period of interest, which potentially excludes many scales designed to display high test–retest stability. For instance, research derived from the 4-H Study often treats the Five Cs of PYD as indicators of a single higher-order latent construct (e.g., Bowers et al., 2010). This higher-order construct tends to show generally flat developmental trajectories (e.g., Schmid et al., 2011), which suggests that the higher-order PYD construct is stable. The indicators of PYD encompass a very wide array of constructs, and it is difficult to discern whether item- or subscale-level changes correspond to meaningful changes at the higher-order construct (i.e., PYD) level. More research is needed to examine these alternatives, but a higher-order PYD construct that is not sensitive to item-level changes may explain why research predicting change in global PYD is markedly absent from the literature (or produces only weak relations; e.g., Lewin-Bizan, Bowers, & Lerner, 2010).

Researchers must also be cognizant that the very structure, or qualitative meaning, of a scale or construct may vary across time and place (Elder, 1998). This possibility underscores the importance of quantitative invariance testing and qualitative exploration of ecological validity across time and place. Bowers and colleagues (2010) have found support for the invariance of the Five Cs of PYD across grades 8–10, for instance; yet, they also suggest that what constitutes PYD differs between early and middle adolescence. Although prior work suggests a relation between athletic competence and the C of Competence among early adolescents (Phelps et al., 2009), the results of Bowers and colleagues' (2010) study indicates that athletic competence does not indicate the C of Competence in middle adolescence.

Factorial invariance of a scale, however, does not necessarily mean that the construct of interest is itself invariant. In other words, the items in a particular scale may be invariant because of the strategy used to build the scale itself, or they may be invariant because the observed invariance truly reflects a property of the underlying construct. Many questionnaires are specifically designed to measure stable attributes, and so change-sensitive items are omitted during scale creation due to a lack of longitudinal reliability. Invariance in a scale that was specifically designed to be invariant over time says more about the scale's construction than about the target construct's actual meaning across the life span.

Qualitative research, which often includes interview or narrative data from participants (Denzin & Lincoln, 2005), can inform the development of a construct and/or the development of a quantitative measure for examining a particular construct across large populations of people. Qualitative research can also explore the presence of a construct at different points of development. For example, qualitative interviews often require participants to reflect on their current and past life experiences. The retrospective data garnered in this context provide another means through which time effects of particular phenomena can be approximated. For example, when participants are asked to think about how their behavior in high school differed from their behavior in college, information about developmental changes that occurred as a function of person ↔ context relations (e.g., participants' time-varying relationships with their teachers and peers) is being reported by the people who experienced these changes directly. Asking participants about changes they experienced and why the changes occurred often elicits a close examination of person ↔ context relations that may be hard to measure with quantitative scales, especially when the phenomenon of interest develops in a complex, nonlinear way.

Developmental Trajectories May Be Nonlinear

The trajectory of a plastic developmental system necessarily entails interactions within and among all levels of the system's integrated structure. Development can involve nonlinear interactions (e.g., quadratic relations, multiplicative interactions) or may even follow nonlinear functional forms. Figure 3.1 illustrates linear (A and B) and nonlinear (C and D) functional forms. From a quantitative perspective, linear models may be helpful for roughly approximating such complex development; in truth, however, development likely extends beyond additively concatenated relationships among variables (Little, 2013).

Researchers may therefore benefit from applying statistical techniques that explicitly assume nonlinearity, including many of the techniques presented in this handbook. Grimm and Ram (2009) similarly discuss the application of structured latent curve models (e.g., Blozis, 2004; Browne, 1993), the parameter estimates of which do not necessarily correspond to additive relationships. For instance, they provide an example that specifies Gompertz-shaped growth. The Gompertz function represents S-shaped growth (see also Figure 3.1D), which allows researchers to more accurately model developmental trajectories with lower and upper local asymptotes. Despite the many potential uses for such models in the PYD literature (e.g., modeling S-shaped

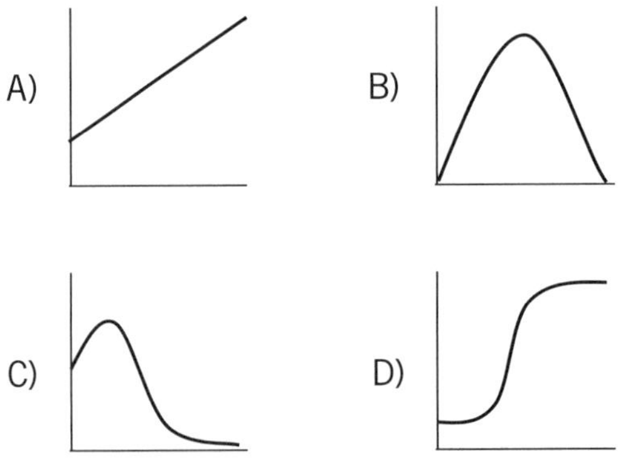

FIGURE 3.1. Examples of possible developmental trajectories: (A) linear, (B) inverse U-shaped, (C) nonlinear with a lower asymptote, and (D) S-shaped.

development of community contribution), such models have been markedly absent from the literature. Extending the application of nonlinear models such as these, and those discussed elsewhere in this volume, is therefore an important direction for future RDST-derived research.

Time Is a Proxy for Development

Adding another layer of complexity to RDST-derived approaches to development is that development is simultaneously affected by processes that occur on multiple, loosely coupled timescales. Modeling complex developmental trajectories requires that researchers explicitly account for the many ways that time can be manifested in a developmental system. As noted by several authors (e.g., Elder, 1998; Lerner, Schwartz, et al., 2009; Little, Card, Preacher, & McConnell, 2009; Wohlwill, 1973), the concept of time can mean many different things in relation to an individual's development. Time might represent chronological factors (e.g., age in years), generational changes (e.g., people changing from an F2 to an F1 generation), historical factors (e.g., the Great Depression), idiographic experiential factors (e.g., years in school), nomothetic episodic factors (e.g., months since September 11, 2001), or idiographic episodic factors (e.g., years since the onset of puberty), for instance. Because these instantiations of time are, of course, all involved in life course changes, developmental researchers must pay close attention to how they conceptualize, measure, and analyze development as a function of the multiple meanings of time (Wohlwill, 1973).

Disentangling the effects of various instantiations of developmental time requires careful methodological forethought, both in terms of study design and data analysis. For instance, Schaie (e.g., 1965) and Baltes (e.g., 1968) discuss multiple study designs that allow researchers to disentangle the integrated effects of chronological time, age, and birth cohort. Among these, the cohort sequential design longitudinally follows participants from multiple birth cohorts and is often heralded as a key method that not only helps researchers make inferences about age-related changes across and among cohorts (e.g., Baltes et al., 1977) but also allows them to study developmental change in an accelerated fashion (Collins, 2006). The emphasis of RDST on complexity and integration highlights the importance of implementing such sophisticated research methods.

The 4-H Study represents a form of cohort sequential design that replaces the traditional concept of a birth cohort with the concept of a test–retest control cohort. In other words, the 4-H Study followed individuals from a single birth cohort but added previously unmeasured participants in each wave to allow for the examination of possible retest effects. This design, however, confounds factors that impact human development with factors that specifically impacted development within this birth cohort. Generalizing findings from the 4-H Study beyond the single birth cohort examined thus requires additional research that examines alternative birth cohorts of various types (e.g., multiple birth cohorts across multiple cultural settings).

Acknowledging that different instantiations of time can impact development has implications for how data are analyzed and how hypotheses are tested. Researchers must ensure that they measure and analyze instantiations of time in a metric that is meaningful to the phenomenon of interest and at a rate that allows for the accurate representation of that phenomenon's development (Lerner, Schwartz, et al., 2009; Wohlwill, 1973). Although studies of PYD have explicitly considered multiple metrics for measuring the progression of development, PYD research has been less effective at specifying and measuring phenomena at intervals consistent with their anticipated rate of development. For instance, Lerner, Schwartz, and colleagues (2009) drew on data gathered as part of the 4-H Study of PYD to show the implications of treating development as a function of either age or pubertal status.

Like most large-scale longitudinal studies, the 4-H Study only assessed participants annually (e.g., Lerner, Lerner, et al., 2005). Annual assessments may be appropriate for examining some developmental phenomena, but the choice of annual measurement in the 4-H Study was apparently made for reasons of practicality and funding rather than a theory of the *x*-axis. That is, annual assessments were not made because of a theoretical specification of the rate or form of change. Data from the 4-H Study might accordingly be appropriate for examining the development of some constructs but may only offer an initial glimpse into phenomena that develop over intervals shorter than 1 year.

Understanding Complex Development Requires Multimethod Integration

The integrative and iterative nature of relational developmental science highlights the importance of triangulation, or the "attempt to map out, or explain more fully, the richness and complexity of human behavior by studying it from more than one standpoint" (Cohen, Manion, & Morrison, 2000, p. 254). Although several researchers have drawn attention to the benefits of rigorous mixed-methods research in developmental science (e.g., Yoshikawa, Weisner, Kalil, & Way, 2008), faithful adoption of these techniques has been slow. Moreover, although several studies of PYD include the collection of both quantitative and qualitative forms of data, the majority of these studies are dominated by one form of data collection and analysis over another. Most commonly, qualitative interview data are collected to "supplement" or illustrate substantial quantitative data collection and analyses (i.e., mixed methods "lite"; Greene, 2012).

Although this design can be informative, developmental scientists have yet to take full advantage of the array of mixed-methods designs available to them, several of which call for substantial and rigorous qualitative and quantitative data collection and analyses (Creswell & Plano Clark, 2011). We believe that approaching PYD research from an RDST framework requires developmental scientists to embrace a more pragmatic approach by collecting multiple forms of data while considering the ways data are integrated in the discussion of findings. Of the multiple mixed-methods designs identified in social science research, a convergent parallel mixed-methods design may hold significant promise for future research in developmental science. Informed by the paradigm of pragmatism (Creswell & Plano Clark, 2011), this design prevents researchers from becoming "the prisoner of a particular [research] method or technique" and from simply presenting findings derived through different methods alongside each other but discussing them separately (Robson, 1993, p. 291; see also Felizer, 2010).

Truly mixed-method designs require much more than simple triangulation of quantitative and qualitative methods. Next we discuss additional methodological dichotomizations that RDST researchers must fuse if they wish to obtain a broad understanding of any phenomena of interest.

Integrating Idiographic and Nomothetic Perspectives

Since Allport (e.g., 1942) introduced Windelbrandt's terms "idiographic" and "nomothetic" to psychology's vocabulary (see Marciel, 1977, and Holt, 1962, for reviews), researchers and theorists have debated whether the province of psychology is to study common (i.e., nomothetic) characteristics shared by all people or the idiosyncratic (i.e., idiographic) characteristics that make each person unique. Emmerich (1968) added a group differential focus to this discussion. In addition to arguing that each person is like all other people (the nomothetic approach in the Kluckholn & Murray, 1948, formulation) or that each person is like no other person (the idiographic approach in the Kluckholn & Murray, 1948, formulation), Emmerich added that each person is like only some other people (in the Kluckholn & Murray, 1948, formulation). As Holt (1962) commented over 50 years ago, the idiographic versus nomothetic debate is "one of the hardiest perennial weeds in psychology's conceptual garden" (p. 376) and indeed it remains a source of considerable discussion to date (e.g., Lamiell, 2009).

Marciel (1977) notes that the implications of this debate extend beyond psychology's conceptual purpose to include the specific methods researchers use to investigate research questions and test hypotheses. From a methodological perspective, RDST-based positions such as the PYD perspective take a middle road (see also Kluckholn & Murray, 1948) by acknowledging that developmental science requires a synthesis of idiographic, group differential, and nomothetic methods. In practice, unfortunately, such synthesis is rare. Nomothetic analyses have remained the primary tool in the methodological arsenals of developmental researchers interested in PYD (as well as across psychological research in general; Molenaar, 2004, 2007; Molenaar & Nesselroade, 2012; Nesselroade & Molenaar, 2010) from a quantitative perspective.

Most large-scale longitudinal studies of youth development are designed to address nomothetic issues. The 4-H Study is typical of these studies in that it relied on measuring a large sample of youth annually. This data collection schedule optimally facilitates nomothetic quantitative analyses such as growth curve analyses, and the data also may be used for the comparison of analytically derived, differential groups through techniques such as cluster analysis (e.g., Zarrett et al., 2009) and mixture modeling (e.g., Bowers, Gestsdóttir, Geldhof, Nikitin, & von Eye, 2011). These types of analyses provide important information about development; when applied to many participants across relatively few time points, however, they cannot provide truly idiographic information.

Data collection methods that support idiographic analyses involve collecting data from fewer individuals but across more occasions (e.g., Nesselroade & Molenaar, 2010). These types of data are commonly collected in several areas of psychological research, although they often are analyzed using nomothetic methods (e.g., growth curve analyses). For example, diary and experience sampling methods often are used in health and mental health research (e.g., Myin-Germeys et al., 2009), personality research (Conner, Tennen, Fleeson, & Barrett, 2009), and the study of relationships (Laurenceau & Bolger, 2005). Similar techniques have been used with adolescent samples (e.g., time use studies such as Larson & Verma, 1999) but are rarely used to study developmental phenomena from an idiographic perspective. These types of data, however, could provide valuable information about development from an RDST perspective. The use of truly idiographic quantitative analysis methods, such as dynamic factor analysis (Molenaar & Lo, 2012), the idiographic filter (Nesselroade et al., 2007), and recent integrations of these techniques (Molenaar & Nessleroade, 2012), could enable researchers to more fully understand the nature of developmental phenomena such as PYD and how these constructs change across adolescence.

Interindividual qualitative analyses can also provide valuable and rich information about groups of youth in a particular program, but they may fall short of promoting a truly idiographic and nuanced understanding of an individual's experiences of a phenomenon or developmental context. Within the broad range of qualitative data collection and analysis methods, however, are several person-centered techniques that could be used within RDST-informed PYD research. Narrative inquiry, for example, provides one way to examine the experiences of, and stories told by, "particular actors, in particular social places, at particular times" (Abott, 1992, p. 428). In addition to focusing on the particular, this idiographic approach allows researchers to highlight the diverse ways in which individuals produce their own development. Moreover, such an approach can lead to the possible discovery of developmental phenomena unique to a person or differential group. If, for instance, PYD research focused more on the experiences of minority youth in the United States, who often feel alienated from civic institutions and less politically efficacious than youth who are part of the majority and middle-class culture (Kirshner, 2009), we could possibly discern facets of PYD, such as critical consciousness, that may be an essential part of optimizing the positive development of all youth, and especially of marginalized youth in the United States and internationally (Hershberg & Lykes, 2012). This construct might be an important "C" of PYD that could encourage youth participation and contribution while simultaneously enabling researchers to identify forms of youth participation and contribution that are often overlooked in research—that is, organizing and activism—that have been identified as arenas wherein marginalized young people contribute to their communities, civic processes, and their own development by attacking social problems head on (Ginwright, Noguera, & Cammarota, 2006). In addition, inclusion of this potential "C" in the PYD model may elicit research regarding the already established Cs in the model and explain *why*, rather than simply conclude that marginalized youth reportedly experience low levels of them (e.g., Hart & Atkins, 2002). The issue of idiographic versus nomothetic measurement also has implications for researchers' choice of correlational versus experimental research designs, to which we turn next. Although both approaches offer some degree of flexibility between idiographic versus more nomothetic measurement, experimental designs explicitly control environmental conditions in order to uncover nomothetic laws of human development. As such, supplementing nomothetic findings with the findings from idiographic studies may require additional integration of experimental and correlational research designs.

Experimental versus Correlational Research Designs

Relational developmental science may represent a paradigm shift from traditional psychogenic ideas in psychology and the traditional focus in this field on experimental methods designed to disentangle cause from effect (Overton & Lerner, 2012). As Cattell (1966) notes, a great deal of psychological research has followed in the experimental tradition and has accordingly focused on data derived from rigidly controlled experiments.

Although not denying the importance of experimental methods and analyses, individuals represent more than simple summative aggregations of base-level components (e.g., neurons, personality traits). Stressing embodiment (Overton, 2010), relational developmental systems science points to the deep and complex relations that connect all possible units of analysis into a synthetic and developing whole. Relational developmental systems scientists therefore tend to favor integrative multivariate analyses and the simultaneous use of contextualized inductive and deductive analytical techniques over experimental methods that attempt to "wash out" individual differences and the influence of the ecologically valid environment. As such, relational developmental researchers often implement correlational methods derived from the traditions of Pearson and Galton (see Cattell, 1966). Common methods may emphasize partial and semipartial relations (e.g., multiple regression models) or treat multiple related items as imperfect indicators of underlying latent constructs (e.g., factor analysis, structural equation modeling).

The complexity of the developmental system, however, suggests that some relationships will be difficult to capture through such ecologically valid observational work. Research that draws information from the natural ecology may make it especially difficult to focus on subtle, nuanced, infrequently occurring, or internalized or "private" facets of the developmental system. Developmental scientists may therefore draw on the strengths of experimental methods to help ascertain such attributes involved in, as well as the outcomes of, PYD-focused programs and activities. For example, Tierney, Grossman, and Resch (1995) used an experimental design to evaluate the Big Brothers/Big Sisters program, a community-based mentoring intervention. The study addressed nine PYD constructs (including social, emotional, cognitive, and behavioral competencies, positive identity, and prosocial norms). Researchers randomly assigned participants to the intervention condition or a wait-list control group. The evaluation demonstrated positive results on behavioral and attitude measures. Although such uses of experimental design are insufficient to describe, explain, or optimize change in the relational developmental system, careful coordination of experimental and correlational designs might provide a more nuanced picture of the developmental system as a whole.

In short, RDST research has been generally approached from a correlational framework, with occasional experimental studies used to support correlational findings. Correlational analyses better capture development as a complex phenomenon; such studies also allow for more flexibility in hypothesis generation. Thus, rather than approaching data from a framework that specifies experimental deduction, correlational studies more easily allow for the dynamic interplay between induction and deduction, which we discuss below.

The Inductive–Hypothetico-Deductive Spiral

Just as prerelational theories often led scientists to artificially split parts from the theoretical whole (i.e., genes vs. environment; see Overton, 2010), pre–relational developmental systems researchers commonly emphasized the independence of inductive versus deductive logic when deriving and testing hypotheses. Taken to extremes, strict application of only inductive or deductive logic can respectively lead to verificationism or an unrealistically strict form of empiricism (Little, 2013). Neither extreme aligns well with the tenets of relational developmental science. Relational-developmental theories specify development as occurring through nonrecursive relations between separate components of a larger system, and from these theories it follows that the practice of relational developmental science progresses through iterative processes of induction and deduction (see Overton, 2012).

As Cattell (1966, p. 14) noted, most scientific advances begin "with dim, fleeting, and far-flung hypotheses, gleaned from the faint movement of straws in the wind." These inductively derived hypotheses become the subject of later deductive testing. Deductive testing may contradict such theories in their infancy and lead the scientist to dismiss them as arising from the observation of an interesting chance event. Other times, however, data from a first deductive test supports the initial hypothesis and provides fodder for additional inductive consideration and hypothesis refinement. These refinements lead to additional deductive testing, and so on, with each turn of this *inductive–hypothetico-deductive spiral* (Cattell, 1966) providing stronger and stronger evidence for the initial theory or for some alternative derived from it.

The development of the PYD perspective itself exemplifies this process. As mentioned earlier in this chapter, a new vision and vocabulary for discussing young people began to emerge in the early 1990s (Lerner, Almerigi, et al., 2005). These discussions were led not only by academic researchers from a variety of disciplines such as developmental science (e.g., Lerner, Brentano, Dowling, & Anderson, 2002) and community psychology (e.g., Trickett, Barone, & Buchanan, 1996), but also by practitioners in the field of youth development (e.g., Floyd & McKenna, 2003; Little, 1993; Pittman, Irby, & Ferber, 2001; Wheeler, 2003) and policymakers concerned with improving the life chances of diverse youth and their families (e.g., Cummings, 2003; Gore, 2003). These interests converged in the idea that youth are resources to be developed, rather than problems to be managed (Roth & Brooks-Gunn, 2003a, 2003b).

From these ideas, researchers began developing conceptual frameworks of PYD, such as the Five Cs Model (Lerner, 2004; Lerner, Almerigi, et al., 2005). Researchers then began efforts to "establish that the concept of PYD, as it had been discussed in the literature, had empirical reality, both in its purposed structure and covariation with other key individual and ecological variables" (Lerner, Lerner, et al., 2005, pp. 18–19). With data from the first wave of the 4-H Study of PYD, Lerner and colleagues found empirical evidence for five first-order latent factors, each corresponding to one of the Cs. Once the structure of PYD from this framework was empirically verified, researchers began to use the concept in models linking characteristics of PYD with other individual and ecological aspects of youth's lives.

This research has, in turn, sparked additional theorizing about the model. For example, Hershberg and Lykes (2012) have suggested that qualitative studies with minority youth may help discern processes involved in youth development that could be particularly salient to optimizing the development of youth in specific cultural settings, in the majority world, and/or in underresourced communities wherein they may be systematically marginalized from participating in activities typically recognized as indicating PYD (e.g., raising the concept of critical consciousness discussed earlier). This example also illustrates how qualitative and quantitative research can complement each other as theorizing and research in a particular area of human development moves up and down the inductive–hypothetico-deductive spiral. Historically, quantitative research has been associated with deductive approaches and qualitative research with inductive, but a thorough survey of methods within each of these traditions shows that there is tremendous diversity. Our conception of an RDST framework supports varying levels of induction and deduction within both qualitative and quantitative approaches, and indeed we argue that the full variety is necessary to move PYD research forward in a way that is consistent with RDST.

Conclusions and Future Directions

RDST models of human development such as the PYD perspective emphasize that changes in the mutually influential relationships between individuals and the multiple levels of their contexts constitute the basic process of human development (Overton, 2010, 2011, 2012, and Chapter 2, this volume; Overton & Müller, 2011). As the present chapter has detailed, the RDST framework entails several conceptual and methodology considerations. Over the past decade, the PYD perspective has been adopted by researchers studying adolescence, by practitioners in youth development, and by policymakers concerned with improving the lives of youth and their families (Lerner et al., 2012). Therefore, the tenets and assumptions of RDST models, and thus the PYD perspective, have implications for the work of researchers and practitioners interested in promoting the positive development of all young people, and for contributing to the promotion of social justice (Lerner & Overton, 2008). In particular, we have discussed three assumptions that are most relevant for this work: (1) development involves interrelated changes in a complex, multilevel system; (2) developmental trajectories of all individuals remain relatively plastic across the life span; and (3) examining relations in the presence of such complexity requires multimethod integration.

Taking these assumptions into account, to frame and test a set of questions within RDST models is a formidable challenge for researchers. Such framing requires multilevel, multivariate, and multimethod longitudinal research that must attend to both intraindividual change and interindividual differences in intraindividual change. Therefore, research derived from RDST entails considerable commitment, both in time and resources, of an interdisciplinary team of researchers. In addition, well-designed RDST research entails gaining the commitment and engagement of research participants in order to obtain the data necessary to test RDST-derived questions and hypotheses.

For example, in RDST research, bidirectional individual ↔ context relations constitute the fundamental unit of analysis, and the individual is an active producer of his or her own development. Therefore, rather than focus on the socially defined contexts (e.g., family, school, neighborhood) that mark much of the research on development, researchers should also collect information about the active individual's attributes in relation to the subjective social context that the individual deems important.

A concurrent challenge when deciding how to design and implement a study that appropriately addresses all of these issues is that researchers must also consider the political, financial, and academic climate within which a study is conducted. That is, research should not be conducted only *with* an RDST framework, but it should also be recognized that research is conducted *within* an RDST framework. What would be the most feasible, practical, and fundable research agenda, given the current policy and funding priorities? As systems science methods are meant to complement traditional research methods (Urban, Osgood, & Mabry, 2011; Urban et al., Chapter 4, this volume), would it make more sense to conduct a series of small studies that focus on particular variables within a broader framework, or would a more comprehensive study that accounts for multiple variables at multiple levels over multiple time points be the appropriate agenda to follow?

The goal of applied developmental science is to describe, explain, and optimize human development. Often, optimizing development involves designing and implementing research-based interventions; however, conducting intervention studies framed within a relational developmental systems approach to applied developmental science raises questions about the use of randomized control trials (RCTs) in tests of causality. There are several financial, practical, and ethical concerns inherent in conducting an RCT in an applied setting. In addition, there are also methodological and conceptual shortcomings in undertaking an RCT, such as limited external validity, contamination, endogeneity, and the infrequency of use of the Solomon four-group design, which involves two control groups other than the typical one involving pretest and posttest but no manipulation (i.e., there is a control group that involves no pretest but includes the manipulation and the posttest—as a control for the reactive effects of pretesting; and there is a control group that includes only the posttest—as a control for maturation; Solomon & Lessac, 1968). For example, an RCT is not appropriate when researchers seek to determine whether an intervention prevents rare events or when the intervention requires active participation, both of which would be the case for many PYD-derived interventions. An overarching challenge to conducting research in the current era is that many funding mechanisms require the use of RCTs unless there is a strong justification for a quasi-experimental design. The prevailing commitment to RCTs by some funding organizations adversely affects researchers who work in applied settings and who recognize that an RCT design may be inappropriate, impossible, or inadequate for the situation in which they conduct research. A PYD researcher must work to provide justification to overcome a mindset in both the research and practice communities that claims that RCTs are the "only way to be sure about cause and effect."

Fiduciary issues also arise in considering the work of practitioners as RDST-derived models become more popular with individuals working to enhance the positive growth of young people (Beets et al., 2009; Duerden, Witt, Fernandez, Jolliff, & Theriault, 2012; Kurtines et al., 2008). It is not always clear what particular developmental processes are explicitly used in the "philosophical" approaches to youth programming pertinent to PYD or in the particular instances of youth programs designed to foster PYD (Lerner, Bowers, et al., 2012). This uncertainty most likely occurs because most youth programs in the United States are not evaluated (cf. Catalano, Berglund, Ryan, Lonczak, & Hawkins, 1999) and, most critically, a theory of change and an evaluation design logic model (Weiss, 1972) are absent from most programs (e.g., see Roth, Brooks-Gunn, Murray, & Foster, 1998). These omissions may be a result of a lack of understanding about the essential need for such frames for programs, but they may also be a casualty of the limited budgets of youth-serving programs. Practitioners often choose to spend funds on programming elements in which youth participate rather than on scientifically rigorous evaluation. Even if funds were available to practitioners, programs framed by the PYD perspective would entail a level of complexity and oversight that can be identified in, let alone attained, by only a select few programs (e.g., Catalano, Haggerty, Oesterle, Fleming, & Hawkins, 2004; Flay & Allred, 2003; Kurtines et al., 2008). A review of these programs illustrates that university–community collaborations are critical for RDST-derived scholarship and practice success (Kurtines et al., 2008; Lerner & Simon, 1998).

Therefore, academic institutions as well as policymakers and funding institutions should work to establish and develop real collaborations with partners in the community. These partners, in turn, must recognize the importance of research-based designs and scientifically rigorous evaluation and also work to recruit and accept the resources available at academic institutions. Both sides must work to identify the mutually beneficial relations that can produce programs and research agendas that are practical, efficient, fundable—and thus sustainable (Lerner & Overton, 2008). With such researcher–community collaborations, RDST-based models such as the PYD perspective (Lerner et al., 2012) can enhance the quality of the information obtained about youth, can provide a more likely to-be-deployed basis for evidence-based practice, and may afford more, and more ecologically valid, contributions to promoting social justice for the diverse young people of the world.

ACKNOWLEDGMENTS

All authors other than G. John Goldhof and Jacqueline V. Lerner are members of the Institute for Applied Research in Youth Development, Tufts University. The preparation of this chapter was supported in part by grants from the John Templeton Foundation.

REFERENCES

Abott, A. (1992). "What do cases do?": Some notes on activity in sociological analysis. In C. C. Ragin & H. S. Backer (Eds.), *What is a case?: Exploring the foundation of social inquiry* (pp. 53–82). Cambridge, UK: Cambridge University Press.

Allport, G. W. (1942). The use of personal documents in psychological science [Monograph]. *Social Science Research Council Bulletin, 49.*

Baltes, P. B. (1968). Longitudinal and cross-sectional sequences in the study of age and generation effects. *Human Development, 11*, 145–171.

Baltes, P. B. (1997). On the incomplete architecture of human ontology: Selection, optimization, and compensation as foundation of developmental theory. *American Psychologist, 23*, 366–380.

Baltes, P. B., Lindenberger, U., & Staudinger, U. M. (2006). Lifespan theory in developmental psychology. In W. Damon & R. M. Lerner (Series Eds.) & R. M. Lerner (Vol. Ed.), *Handbook of child psychology: Vol. 1. Theoretical models of human development* (6th ed., pp. 569–664). Hoboken, NJ: Wiley.

Baltes, P. B., Reese, H. W., & Nesselroade, J. R. (1977). *Life-span developmental psychology: Introduction to research methods.* Monterey, CA: Brooks/Cole.

Beets, M. W., Flay, B. R., Vuchinich, S., Snyder, F. J., Acock, A., Li, K. K., et al. (2009). Use of a social and character development program to prevent substance use, violent behaviors, and sexual activity among elementary-school students in Hawaii. *American Journal of Public Health, 99*(8), 1438–1445.

Benson, P. L., Scales, P. C., & Syvertsen, A. K. (2011). The contribution of the developmental assets framework to positive youth development theory and practice. *Advances in Child Development and Behavior, 41*, 195–228.

Blozis, S. A. (2004). Structured latent curve models for the study of change in multivariate repeated measures. *Psychological Methods, 9*, 344–353.

Bowers, E. P., Geldhof, G. J., Schmid, K. L., Napolitano, C. M., Minor, K., & Lerner, J. V. (2012). Relationships with important nonparental adults and positive youth development: An examination of youth self-regulatory strengths as mediators. *Research in Human Development, 9*(4), 298–316.

Bowers, E. P., Gestsdóttir, S., Geldhof, G. J., Nikitin, J., & von Eye, A. (2011). Developmental trajectories of intentional self regulation in adolescence: Implications for positive and problematic development among diverse youth. *Journal of Adolescence, 34*, 1193–1206.

Bowers, E. P., Li, Y., Kiely, M. K., Brittian, A., Lerner, J. V., & Lerner, R. M. (2010). The five C's model of positive youth development: A longitudinal analysis of confirmatory factor structure and measurement invariance. *Journal of Youth and Adolescence, 39*(7), 720–735.

Brandtstädter, J. (1998). Action perspectives on human development. In W. Damon & R. M. Lerner (Series Eds.) & R. M. Lerner (Vol. Ed.), *Handbook of child psychology: Vol. 1. Theoretical models of human development* (5th ed., pp. 807–863). New York: Wiley.

Brandtstädter, J. (2006). Action perspectives on human development. In W. Damon & R. M. Lerner (Series Eds.) & R. M. Lerner (Vol. Ed.), *Handbook of child psychology: Vol. 1. Theoretical models of human development* (6th ed., pp. 516–568). Hoboken, NJ: Wiley.

Brandtstädter, J., & Lerner, R. M. (Eds.). (1999). *Action and self-development: Theory and research through the life-span*. Thousand Oaks, CA: Sage.

Bronfenbrenner, U., & Morris, P. A. (2006). The bioecological model of human development. In W. Damon & R. M. Lerner (Series Eds.) & R. M. Lerner (Vol. Ed.), *Handbook of child psychology: Vol. 1. Theoretical models of human development* (6th ed., pp. 793–828). Hoboken, NJ: Wiley.

Browne, M. W. (1993). Structured latent curve models. In C. M. Cuadras & C. R. Rao (Eds.), *Multivariate analysis: Future directions* (Vol. 2, pp. 171–198). Amsterdam: North-Holland.

Catalano, R. F., Berglund, M. L., Ryan, J. A. M., Lonczak, H. S., & Hawkins, J. D. (1999). *Positive youth development in the United States: Research findings on evaluations of youth development programs*. Washington, DC: U.S. Department of Health and Human Services.

Catalano, R. F., Haggerty, K. P., Oesterle, S., Fleming, C. B., & Hawkins, J. D. (2004). The importance of bonding to school for healthy development: Findings from the Social Development Research Group. *Journal of School Health, 74*, 252–261.

Cattell, R. B. (1966). Psychological theory and scientific method. In R. B. Cattell (Ed.), *Handbook of multivariate experimental psychology* (pp. 1–18). Chicago: Rand McNally.

Cohen, L., Manion, L., & Morrison, K. R. B. (2000). *Research methods in education* (5th ed.). London: Routledge.

Collins, L. M. (2006). Analysis of longitudinal data: The integration of theoretical model, temporal design, and statistical model. *Annual Review of Psychology, 57*, 505–528.

Conner, T. S., Tennen, H., Fleeson, W., & Barrett, L. F. (2009). Experience sampling methods: A modern idiographic approach to personality. *Social and Personality Psychology Compass, 3*, 1–22.

Creswell, J. W., & Plano Clark, V. (2011). *Designing and conducting mixed methods research* (2nd ed.). Thousand Oaks, CA: Sage.

Cummings, E. (2003). Foreword. In D. Wertlieb, F. Jacobs, & R. M. Lerner (Eds.), *Handbook of applied developmental science: Promoting positive child, adolescent, and family development through research, policies, and programs: Vol. 3. Promoting positive youth and family development: Community systems, citizenship, and civil society* (pp. ix–xi). Thousand Oaks, CA: Sage.

Damon, W. (2004). What is positive youth development? *Annals of the American Academy of Political and Social Science, 591*, 13–24.

Denzin, N. K., & Lincoln, Y. S. (2005). *Sage handbook of qualitative research* (3rd ed.). Thousand Oaks, CA: Sage.

Duerden, M. D., Witt, P. A., Fernandez, M., Jolliff, M., & Theriault, D. (2012). Measuring life skills: Standardizing the assessment of youth development indicators. *Journal of Youth Development, 7*, 99–117.

Eccles, J. S. (2004). Schools, academic motivation, and stage–environment fit. In. R. M. Lerner & L. Steinberg (Eds.), *Handbook of adolescent psychology* (Vol. 2, pp. 125–153). Hoboken, NJ: Wiley.

Eccles, J. S., & Gootman, J. (Eds.). (2002). *Community programs to promote youth development*. Washington, DC: National Academy Press.

Elder, G. H., Jr. (1998). The life course and human development. In W. Damon & R. M. Lerner

(Series Eds.) & R. M. Lerner (Vol. Ed.), *Handbook of child psychology: Vol. 1. Theoretical models of human development* (5th ed., pp. 939–991). New York: Wiley.

Elder, G. H., Jr., Modell, J., & Parke, R. D. (Eds.). (1993). *Children in time and place: Developmental and historical insights.* New York: Cambridge University Press.

Emmerich, W. (1968). Personality development and concepts of structure. *Child Development, 39*, 671–690.

Felizer, M. Y. (2010). Doing mixed methods research pragmatically: Implications for the rediscovery of pragmatism as a research paradigm. *Journal of Mixed Methods Research, 4*(1), 6–16.

Flay, B. R., & Allred, C. G. (2003). Long-term effects of the Positive Action program. *American Journal of Health Behavior, 27*(1), S6–S21.

Floyd, D. T., & McKenna, L. (2003). National youth serving organizations in the United States: Contributions to civil society. In R. M. Lerner, F. Jacobs, & D. Wertlieb (Eds.), *Handbook of applied developmental science: Promoting positive child, adolescent, and family development through research, policies, and programs: Vol. 3. Promoting positive youth and family development: Community systems, citizenship, and civil society* (pp. 11–26). Thousand Oaks, CA: Sage.

Ford, D. H., & Lerner, R. M. (1992). *Developmental systems theory: An integrative approach.* Newbury Park, CA: Sage.

Freund, A. M., & Baltes, P. B. (2002). Life-management strategies of selection, optimization, and compensation: Measurement by self-report and construct validity. *Journal of Personality and Social Psychology, 82*, 642–662.

Gardner, M., & Steinberg, L. (2005). Peer influence on risk-taking, risk preference, and risky decision-making in adolescence and adulthood: An experimental study. *Developmental Psychology, 41*, 625–635.

Gestsdóttir, S., & Lerner, R. M. (2007). Intentional self-regulation and positive youth development in early adolescence: Findings from the 4-H Study of Positive Youth Development. *Developmental Psychology, 43*(2), 508–521.

Gestsdóttir, S., & Lerner, R. M. (2008). Positive development in adolescence: The development and role of intentional self regulation. *Human Development, 51*, 202–224.

Ginwright, S., Noguera, P., & Cammarota, J. (2006). Youth activism in the urban community: Learning critical civic praxis with community organizations. *International Journal of Qualitative Studies in Education, 20*(6), 693–710.

Gore, A. (2003). Foreword. In R. M. Lerner & P. L. Benson (Eds.), *Developmental assets and asset-building communities: Implications for research, policy, and practice* (pp. xi–xii). Norwell, MA: Kluwer.

Gottlieb, G. (1998). Normally occurring environmental and behavioral influences on gene activity: From central dogma to probabilistic epigenesis. *Psychological Review, 105*, 792–802.

Greene, J. C. (2012). Engaging critical issues in social inquiry by mixing methods. *American Behavioral Scientist, 56*, 755–773.

Grimm, K. J., & Ram, N. (2009). Nonlinear growth models in Mplus and SAS. *Structural Equation Modeling, 16*, 676–701.

Hamilton, S. F. (1999). *A three-part definition of youth development.* Unpublished manuscript, Cornell University College of Human Ecology, Ithaca, NY.

Hamilton, S. F., & Hamilton, M. A. (2009). The transition to adulthood: Challenges of poverty and structural lag. In R. M. Lerner & L. Steinberg (Eds.), *Handbook of adolescent psychology: Vol. 2. Contextual influences on adolescent development* (3rd ed., pp. 492–526). Hoboken, NJ: Wiley.

Hart, D., & Atkins, R. (2002). Civic competence in urban youth. *Applied Developmental Science, 6*(4), 227–236.

Heck, K. E., & Subramaniam, A. (2009). *Youth development frameworks* [Monograph]. Davis, CA: 4-H Center for Youth Development, University of California.

Heckhausen, J. (1999). *Developmental regulation in adulthood: Age-normative and sociocultural constraints as adaptive challenges.* New York: Cambridge University Press.

Hershberg, R. M., & Lykes, M. B. (2012). Redefining family: Transnational girls narrate experiences of parental migration, detention, and deportation. *Forum Qualitative Sozialforschung/Forum: Qualitative Social Research, 14*(1), Art. 5.

Holt, R. R. (1962). Individuality and generalization in the psychology of personality. *Journal of Personality, 30*, 377–404.

Jelicic, H., Bobek, D., Phelps, E. D., Lerner, J. V., & Lerner, R. M. (2007). Using positive youth development to predict contribution and risk behaviors in early adolescence: Findings from the first two waves of the 4-H Study of Positive Youth Development. *International Journal of Behavioral Development, 31*(3), 263–273.

Kirshner, B. (2009). "Power in numbers": Youth organizing as a context for exploring civic identity. *Journal of Research on Adolescence, 19*(3), 414–440.

Kluckhohn, C., & Murray, H. A. (1948). *Personality in nature, society, and culture.* New York: Knopf.

Kurtines, W. M., Ferrer-Wreder, L., Berman, S. L., Cass Lorente, C., Silverman, W. K., & Montgomery, M. J. (2008). Introduction to special issue on promoting positive youth development: A developmental intervention approach—new developments in developmental theory, methods, and research. *Journal of Adolescent Research, 23*, 233–244.

Lamiell, J. T. (2009). Reviving person-centered inquiry in psychology: Why its erstwhile dormancy? In J. Valsiner, P. Molenaar, M. Lyra, & N. Chaudhary (Eds.), *Dynamic process methodology in the social and developmental sciences* (pp. 31–43). New York: Springer.

Larson, R. W. (2000). Toward a psychology of positive youth development. *American Psychologist, 55*(1), 170–183.

Larson, R. W., & Verma, S. (1999). How children and adolescents spend time across cultural settings of the world: Work, play, and developmental opportunities. *Psychological Bulletin, 125*, 701–736.

Larson, R. W., Wilson, S., & Rickman, A. (2009). Globalization, societal change, and adolescence across the world. In R. M. Lerner & L. Steinberg (Eds.), *Handbook of adolescent psychology: Vol. 2. Contextual influences on adolescent development* (3rd ed., pp. 590–622). New York: Wiley.

Laurenceau, J., & Bolger, N. (2005). Using diary methods to study marital and family processes. *Journal of Family Psychology, 19*(1), 86–97.

Lerner, J. V., Bowers, E. P., Minor, K., Lewin-Bizan, S., Boyd, M. J., Mueller, M. K., et al. (2012). Positive youth development: Processes, philosophies, and programs. In I. B. Weiner (Series Ed.) & R. M. Lerner, M. A. Easterbrooks, & J. Mistry (Vol. Eds.), *Handbook of psychology: Vol. 6. Developmental psychology* (2nd ed., pp. 365–392). Hoboken, NJ: Wiley.

Lerner, J. V., Phelps, E., Forman, Y., & Bowers, E. P. (2009). Positive youth development. In R. M. Lerner & L. Steinberg (Eds.), *Handbook of adolescent psychology: Vol. 1. Individual bases of adolescent development* (3rd ed., pp. 524–558). Hoboken, NJ: Wiley.

Lerner, R. M. (1982). Children and adolescents as producers of their own development. *Developmental Review, 2*, 342–370.

Lerner, R. M. (1984). *On the nature of human plasticity*. New York: Cambridge University Press.

Lerner, R. M. (2002). *Concepts and theories of human development* (3rd ed.). Mahwah, NJ: Erlbaum.

Lerner, R. M. (2004). *Liberty: Thriving and civic engagement among America's youth*. Thousand Oaks, CA: Sage.

Lerner, R. M. (2006). Developmental science, developmental systems, and contemporary theories of human development. In W. Damon & R. M. Lerner (Series Eds.) & R. M. Lerner (Vol. Ed.), *Handbook of child psychology: Vol. 1. Theoretical models of human development* (6th ed., pp. 1–17). Hoboken, NJ: Wiley.

Lerner, R. M., Almerigi, J., Theokas, C., & Lerner, J. V. (2005). Positive youth development: A view of the issues. *Journal of Early Adolescence, 25*(1), 10–16.

Lerner, R. M., Brentano, C., Dowling, E. M., & Anderson, P. M. (2002). Positive youth development: Thriving as the basis of personhood and civil society. In G. Noam (Series Ed.) & R. M. Lerner, C. S. Taylor, & A. von Eye (Vol. Eds.), *New directions for youth development: Theory, practice, research: Vol. 95. Pathways to positive development among diverse youth* (pp. 11–33). San Francisco: Jossey-Bass.

Lerner, R. M., & Busch-Rossnagel, N. A. (Eds.). (1981). *Individuals as producers of their development: A life-span perspective*. New York: Academic Press.

Lerner, R. M., Lerner, J. V., Almerigi, J., Theokas, C., Phelps, E., Gestsdóttir, S., et al. (2005). Positive youth development, participation in community youth development programs, and community contributions of fifth-grade adolescents: Findings from the first wave of the 4-H Study of Positive Youth Development. *Journal of Early Adolescence, 25*(1), 17–71.

Lerner, R. M., Lerner, J. V., & Benson, J. B. (2011). Research and applications for promoting thriving in adolescence: A view of the issues. In R. M. Lerner, J. V. Lerner, & J. B. Benson (Eds.), *Advances in child development and behavior: Vol. 42. Positive youth development: Research and applications for promoting thriving in adolescence* (pp. 1–16). Amsterdam: Elsevier.

Lerner, R. M., Lerner, J. V., von Eye, A., Bowers, E. P., & Lewin-Bizan, S. (2011). Individual and contextual bases of thriving in adolescence: A view of the issues. *Journal of Adolescence, 34*(6), 1107–1114.

Lerner, R. M., & Overton, W. F. (2008). Exemplifying the integrations of the relational developmental system: Synthesizing theory, research, and application to promote positive development and social justice. *Journal of Adolescent Research, 23*(3), 245–255.

Lerner, R. M., Schwartz, S. J., & Phelps, E. (2009). Problematics of time and timing in the longitudinal study of human development: Theoretical and methodological issues. *Human Development, 52*, 44–68.

Lerner, R. M., & Simon, L. A. K. (1998). The new American outreach university: Challenges and options. In R. M. Lerner & L. A. K. Simon (Eds.), *University–community collaborations for the twenty-first century: Outreach scholarship for youth and families* (pp. 3–23). New York: Garland.

Lerner, R. M., von Eye, A., Lerner, J. V., & Lewin-Bizan, S. (2009). Exploring the foundations and functions of adolescent thriving within the 4-H Study of Positive Youth Development: A view of the issues. *Journal of Applied Developmental Psychology, 30*(5), 567–570.

Lerner, R. M., von Eye, A., Lerner, J. V., Lewin-Bizan, S., & Bowers, E. P. (2010). Special issue introduction: The meaning and measurement of thriving—a view of the issues. *Journal of Youth and Adolescence, 39*(7), 707–719.

Lerner, R. M., & Walls, T. (1999). Revisiting individuals as producers of their development: From dynamic interactionism to developmental systems. In J. Brandtstädter & R. M. Lerner (Eds.), *Action and self-development: Theory and research through the life-span* (pp. 3–36). Thousand Oaks, CA: Sage.

Leventhal, T., & Brooks-Gunn, J. (2000). The neighborhoods they live in: The effects of neighborhood residence on child and adolescent outcomes. *Psychological Bulletin, 126*(2), 309–337.

Leventhal, T., Dupere, V., & Brooks-Gunn, J. (2009). Neighborhood influences on adolescent development. In R. M. Lerner & L. Steinberg (Eds.), *Handbook of adolescent psychology: Vol. 2. Contextual influences on adolescent development* (3rd ed., pp. 411–443). Hoboken, NJ: Wiley.

Lewin-Bizan, S., Bowers, E.P., & Lerner, R. M. (2010). One good thing leads to another: Cascades of positive youth development among American adolescents. *Development and Psychopathology, 22*, 759–770.

Little, R. R. (1993, March). *What's working for today's youth: The issues, the programs, and the learnings.* Paper presented at the Institute for Children, Youth, and Families Fellows' Colloquium, East Lansing, MI.

Little, T. D. (2013). *Longitudinal structural equation modeling.* New York: Guilford Press.

Little, T. D., Card, N. A., Preacher, K. J., & McConnell, E. (2009). Modeling longitudinal data from research on adolescence. In R. M. Lerner & L. Steinberg (Eds.), *Handbook of adolescent psychology: Vol. 1. Individual bases of adolescent development* (3rd ed., pp. 15–54). Hoboken, NJ: Wiley.

Magnusson, D. (1999). Individual development: Toward a developmental science. *Proceedings of the American Philosophical Society, 143*, 86–96.

Magnusson, D., & Stattin, H. (2006). The person in the environment: Towards a general model for scientific inquiry. In W. Damon & R. M. Lerner (Series Eds.) & R. M. Lerner (Vol. Ed.), *Handbook of child psychology: Vol. 1. Theoretical models of human development* (6th ed., pp. 400–464). Hoboken, NJ: Wiley.

Marciel, J. C. (1977). Implicit dimensions of idiography and nomothesis: A reformulation. *American Psychologist, 32*, 1046–1055.

Masten, A. S. (2001). Ordinary magic: Resilience processes in development. *American Psychologist, 56*(3), 227–238.

Masten, A. S., & Coatsworth, J. D. (1998). The development of competence in favorable and unfavorable environments: Lessons from research on successful children. *American Psychologist, 53*, 205–220.

Molenaar, P. C. M. (2004). A manifesto on psychology as a idiographic science: Bringing the person back into scientific psychology, this time forever. *Measurement, 2*, 201–218.

Molenaar, P. C. M. (2007). On the implications of the classical ergodic theorems: Analysis of developmental processes has to focus on intra-individual variation. *Developmental Psychobiology, 50*, 60–69.

Molenaar, P. C. M., & Lo, L. L. (2012). Dynamic factor analysis and control of developmental processes. In B. Laursen, T. D. Little, & N. A. Card (Eds.), *Handbook of developmental research methods* (pp. 333–349). New York: Guilford Press.

Molenaar, P. C. M., & Nesselroade, J. R. (2012). Merging the idiographic filter with dynamic factor analysis to model process. *Applied Developmental Science, 16*, 210–219.

Myin-Germeys, I., Oorschot, M., Collip, D., Lataster, J., Delespaul, P., & van Os, J. (2009). Experience sampling research in psychopathology: Opening the black box of daily life. *Psychological Medicine, 39*(9), 1533–1547.

Nesselroade, J. R., Gerstorf, D., Hardy, S. A., & Ram, N. (2007). Idiographic filters for psychological constructs. *Measurement, 5*, 217–235.

Nesselroade, J. R., & Molenaar, P. C. M. (2010). Emphasizing intraindividual variability in the study of development over the lifespan. In R. M. Lerner (Ed.-in-Chief) & W. F. Overton (Vol. Ed.), *Handbook of life-span development: Vol. 1. Cognition, biology, and methods* (pp. 30–54). New York: Wiley.

Overton, W. F. (2010). Life-span development: Concepts and issues. In R. M. Lerner (Ed.-in-Chief) & W. F. Overton (Vol. Ed.), *Handbook of life-span development: Vol. 1. Cognition, biology, and methods* (pp. 1–29). Hoboken, NJ: Wiley.

Overton, W. F. (2011). Relational developmental systems and quantitative behavior genetics: Alternative of parallel methodologies. *Research in Human Development, 8*(3–4), 258–263.

Overton, W. F. (2012). Evolving scientific paradigms: Retrospective and prospective. In L. L'Abate (Ed.), *The role of paradigms in theory construction* (pp. 31–65). New York: Springer.

Overton, W. F., & Lerner, R. M. (2012). Relational developmental systems: Paradigm for developmental science in the post-genomic era. *Behavioral and Brain Sciences, 35*(5), 375–376.

Overton, W. F., & Müller, U. (2012). Meta-theories, theories, and concepts in the study of development. In I. B. Weiner (Series Ed.) & R. M. Lerner, M. A. Easterbrooks, & J. Mistry (Vol. Eds.), *Handbook of psychology: Vol. 6. Developmental psychology* (pp. 19–58). New York: Wiley.

Phelps, E., Zimmerman, S., Warren, A. A., Jelicic, H., von Eye, A., & Lerner, R. M. (2009). The structure and developmental course of positive youth development (PYD) in early adolescence: Implications for theory and practice. *Journal of Applied Developmental Psychology, 30*(5), 571–584.

Pittman, K., Irby, M., & Ferber, T. (2001). Unfinished business: Further reflections on a decade of promoting youth development. In P. L. Benson & K. J. Pittman (Eds.), *Trends in youth development: Visions, realities and challenges* (pp. 4–50). Norwell, MA: Kluwer.

Robson, C. (1993). *Real world research: A resource for social scientists and practitioner-researchers.* Oxford, UK: Blackwell.

Roth, J. L., & Brooks-Gunn, J. (2003a). What exactly is a youth development program?: Answers from research and practice. *Applied Developmental Science, 7*, 94–111.

Roth, J. L., & Brooks-Gunn, J. (2003b). What is a youth development program?: Identification and defining principles. In F. Jacobs, D. Wertlieb, & R. M. Lerner (Eds.), *Handbook of applied developmental science: Promoting positive child, adolescent, and family development through research, policies, and programs: Vol. 2. Enhancing the life chances of youth and families: Public service systems and public policy perspectives* (pp. 197–223). Thousand Oaks, CA: Sage.

Roth, J. L., Brooks-Gunn, J., Murray, L., & Foster, W. (1998). Promoting healthy adolescents: Synthesis of youth development program evaluations. *Journal of Research on Adolescence, 8*, 423–459.

Schaie, K. W. (1965). A general model for the study of developmental problems. *Psychological Bulletin, 64*, 92–107.

Schmid, K., Phelps, E., & Lerner, R. M. (2011). Constructing positive futures: Modeling the relationship between adolescents' hopeful future expectations and intentional self regulation in predicting positive youth development. *Journal of Adolescence, 34*(6), 1127–1136.

Silbereisen, R. K. (2005). Social change and human development: Experiences from German unification. *International Journal of Behavioral Development, 29*, 2–13.

Silbereisen, R. K., & Chen, X. (Eds.). (2010). *Social change and human development: Concepts and results*. Thousand Oaks, CA: Sage.

Silbereisen, R. K., & Lerner, R. M. (Eds.). (2007). *Approaches to positive youth development*. London: Sage.

Solomon, R. L., & Lessac, M. S. (1968). A control group design for experimental studies of developmental processes. *Psychological Bulletin, 70*, 145–150.

Spencer, M. B. (2006). Phenomenological variant of ecological systems theory (PVEST): A human development synthesis applicable to diverse individuals and groups. In W. Damon & R. M. Lerner (Series Eds.) & R. M. Lerner (Vol. Ed.), *Handbook of child psychology: Vol. 6. Theoretical models of human development* (6th ed., pp. 829–894). Hoboken, NJ: Wiley.

Steinberg, L. (2010). A behavioral scientist looks at the science of adolescent brain development. *Brain and Cognition, 72*, 160–164.

Sterba, S. K., & Bauer, D. J. (2010). Matching method with theory in person-oriented developmental psychopathology research. *Development and Psychopathology, 22*, 239–254.

Theokas, C., Almerigi, J. B., Lerner, R. M., Dowling, E. M., Benson, P. L., Scales, P. C., et al. (2005). Conceptualizing and modeling individual and ecological asset components of thriving in early adolescence. *Journal of Early Adolescence, 25*(1), 113–143.

Theokas, C., & Lerner, R. M. (2006). Observed ecological assets in families, schools, and neighborhoods: Conceptualization, measurement, and relations with positive and negative developmental outcomes. *Applied Developmental Science, 10*(2), 61–74.

Tierney, J. P., Grossman, J. B., & Resch, N. L. (1995). *Making a difference: An impact study of Big Brothers/Big Sisters*. Philadelphia: Public/Private Ventures.

Tobach, E., & Greenberg, G. (1984). The significance of T. C. Schneirla's contribution to the concept of levels of integration. In G. Greenberg & E. Tobach (Eds.), *Behavioral evolution and integrative levels* (pp. 1–7). Hillsdale, NJ: Erlbaum.

Trickett, E. J., Barone, C., & Buchanan, R. M. (1996). Elaborating developmental contextualism in adolescent research and intervention: Paradigm contributions from community psychology. *Journal of Research on Adolescence, 6*, 245–269.

Urban, J. B., Lewin-Bizan, S., & Lerner, R. M. (2010). The role of intentional self regulation, lower neighborhood ecological assets, and activity involvement in youth developmental outcomes. *Journal of Youth and Adolescence, 39*(7), 783–800.

Urban, J. B., Osgood, N., & Mabry, P. (2011). Developmental systems science: Exploring the application of non-linear methods to developmental science questions. *Research in Human Development, 8*, 1–25.

Way, N. (2011). *Deep secrets: Boys' friendships and the crisis of connection*. Cambridge, MA: Harvard University Press.

Weiss, C. H. (1972). *Evaluation research: Methods for assessing program effectiveness*. Englewood Cliffs, NJ: Prentice-Hall.

Wheeler, W. (2003). Youth leadership for development: Civic activism as a component of youth development programming and a strategy for strengthening civil society. In R. M. Lerner, F. Jacobs, & D. Wertlieb (Eds.), *Handbook of applied developmental science: Promoting positive child, adolescent, and family development through research, policies, and programs: Vol. 2. Enhancing the life chances of youth and families: Public service systems and public policy perspectives* (pp. 491–505). Thousand Oaks, CA: Sage.

Wohlwill, J. F. (1973). *The study of behavioral development.* New York: Academic Press.

Yoshikawa, H., Weisner, T., Kalil, A., & Way, N. (2008). Mixing qualitative and quantitative research in developmental science: Uses and methodological choices. *Developmental Psychology, 44*(2), 344–354.

Zaff, J. F., Kawashima-Ginsberg, K., & Lin, E. S. (2011). Advances in civic engagement research: Issues of civic measures and civic context. *Advances in Child Development and Behavior, 41*, 271–306.

Zarrett, N., Fay, K., Carrano, J., Li, Y., Phelps, E., & Lerner, R. M. (2009). More than child's play: Variable- and pattern-centered approaches for examining effects of sports participation on youth development. *Developmental Psychology, 45*(2), 368–382.

CHAPTER 4

Developmental Systems Science

Extending Developmental Science with Systems Science Methodologies

Jennifer Brown Urban, Nathaniel Osgood, Janet Okamoto, Patricia Mabry, and Kristen Hassmiller Lich

Developmental systems science refers to the application of systems science methodologies (e.g., social network analysis, system dynamics, and agent-based modeling) to developmental science questions, particularly those derived from a developmental systems theoretical perspective. The term *developmental systems science* deliberately combines *developmental science* with *systems science* and is meant to reflect the joining together of these two fields. *Developmental science* is an approach to the study of human development that emphasizes multidisciplinary and systemic thinking and includes the spectrum from basic to applied forms of inquiry (Lerner, 2006). *Systems science* refers to a family of methodologies that enables the study of complex problems and typically involves modeling and simulation. This chapter aims to present a compelling rationale for the application of systems science methods in developmental science as well as an introduction to three specific systems science methodologies. This chapter builds on our previous work in this area (Urban, Osgood, & Mabry, 2011), by giving a much richer and more detailed description of the selected systems science methods and by including a brief review of systems science applications in developmental science.

Developmental science is at the cutting edge of thinking and research in the study of human development (Lerner, 2006), though developmental science, as a field, is still relatively young. Historically, those interested in the study of human development operated through the lens of developmental psychology (Lerner, 2006). In recent years, a multidisciplinary (and occasionally interdisciplinary) approach to the study of human development has gained prominence and along with it, a more holistic and systemic conceptualization of human development has emerged. Thus, *developmental science* is now the more appropriate term for this area of study.

Along with a move toward expanding the disciplines involved in the study of development, there have been fundamental shifts in our thinking and theorizing about human development. In particular, the emphasis on Cartesian splits (e.g., nature vs. nurture, continuity vs. discontinuity) has been replaced with a relational metanarrative that supports an inclusive conceptualization of development that privileges neither biological nor sociocultural influences on development (Overton, 1998, 2006, and Chapter 2, this volume). As the field of developmental science has moved toward embracing relational metatheories of development, developmental systems theories, in particular, have gained prominence.

For several decades developmental science has embraced a theoretical orientation that reflects a "systems thinking" orientation. Developmental systems theories include bioecological systems theory (Bronfenbrenner & Morris, 2006), developmental systems theory (Ford & Lerner, 1992), probabilistic epigenesis (Gottlieb, 1992, 1998; Gottlieb, Wahlsten, & Lickliter, 2006), dynamic systems theory (Thelen & Smith, 2006), and holistic person–context interaction theory (Magnusson & Stattin, 2006). These approaches share several defining features (Lerner, 2006), the first of which is a fundamental rejection of a dichotomous and reductionist view of development. Therefore, it is not surprising that they also share a multidisciplinary (and occasionally interdisciplinary) perspective on development. Additionally, they all focus on the integration of multiple levels of organization from the microlevel (e.g., cells) to the macrolevel (e.g., society). These multiple levels do not operate in isolation; rather, development is viewed as emerging from bidirectional individual–context *interactions* (or *coactions*; Gottlieb et al., 2006). These theoretical approaches also share a common emphasis on temporality and the plasticity (i.e., capacity for change) of the developmental system. This potential for plasticity, however, is constrained and regulated by individual–context interactions and may vary over developmental time. Despite the relative nature of this developmental plasticity, the potential for change is considered a fundamental *strength* of human development. As such, all of these theoretical approaches reject a deficit conceptualization of development and instead emphasize the *promotion* of positive development. For example, this emphasis could include a focus on developing and testing interventions in the form of policies or programs that align the strengths of an individual with the resources available in their context (Lerner, 2006).

As a class, developmental systems theories hold tremendous promise for the study of human development. However, until now, the methods used have been insufficient to support the testing of these theories. Existing statistical techniques do not readily allow for the testing of many of the defining theoretical concepts presented in developmental systems theories. For example, reciprocal relationships, or feedback mechanisms, are difficult to model using standard statistical methods. Similarly, although hierarchical linear modeling can account for the nonindependence of multilevel systems, these statistical approaches are not able to adequately address the relational structure of hierarchically embedded systems. The fundamental challenge, then, is to develop or discover analytic methods capable of handling developmental complexity.

Although there has been a handful of efforts aimed at addressing this problem (e.g., Fogel, 2011; Hollenstein, 2007; Lewis, 2005; Molenaar & Newell, 2010; Molenaar, Sinclair, Rovine, Ram, & Corneal, 2009; Newell & Molenaar, 1998; Spencer & Perone, 2008; van der Maas et al., 2006; van Geert, 2011), it seems that much could be gained by adopting systems science methodologies, which are expressly designed to handle complexity and include the involvement of stakeholders.

Interestingly, developmental science has a rich theoretical history that draws from the systems science literature base (e.g., Bertalanffy, 1969; Capra, 1977, 1982). Although developmental systems theories share a common theoretical background with systems science, developmental scientists have not fully capitalized on this shared etiology. However, the field is now well positioned to capitalize on recent technological advances, including access to software, as well as a general shift within the scientific community toward a focus on interdisciplinary teams that provide developmental scientists with access to researchers with systems science modeling expertise (e.g., computer scientists, engineers, mathematicians). The application of systems science methodologies holds tremendous promise for the study of human development, particularly for developmental scientists who operate from a developmental systems theoretical perspective.

The next section addresses the specific methodological challenges developmental scientists face and discusses how systems science methodologies can address those issues. This is followed by an overview of, and introduction to, three of the preeminent systems science methods: social network analysis, system dynamics, and agent-based modeling. Specifically, we discuss how each of the methods relates to developmental science, when it is most appropriate to use each method, and how each method relates to other analysis tools. Lastly, we briefly discuss current uses of systems science methods in developmental science and related fields as well as potential future applications.

Systems Science Methodologies Can Address Methodological Challenges in Developmental Science

Systems science is a family of methodologies that is generally characterized by modeling and simulation and includes social network analysis, system dynamics, and agent-based modeling. Systems science methods grew out of a variety of disciplines, including engineering, mathematics, physics, computer science, sociology, and operations research. Within the behavioral and social sciences, these methodologies have primarily been employed in the fields of public health, sociology, business, and economics. However, frequently research in these areas is conducted by interdisciplinary teams comprised of both content experts and systems science modelers. Systems science methods are not intended to replace more traditional linear statistical approaches to research. Rather, they provide an innovative complement to the methodological arsenal, particularly when the focal research question involves the study of complexity.

Systems science methods are particularly well suited for analyzing a variety of complex phenomena commonly encountered in developmental science. These include *threshold behavior*, which refers to the often dramatic relationship changes that occur when some threshold is reached. For example, the development of secondary sex characteristics is triggered when hormonal levels reach a certain level. The *worse-before-better phenomenon*, whereby a problem does not get better or it gets worse before an intervention has a chance to work its way through the system, is another example of a nonlinear relationship that systems science methods can capture.

Systems science methods also enable the study of complex problems while simultaneously taking into account the big picture that addresses developmental scientists' interest in studying the integration of multiple levels of organization. Specifically, developmental scientists are interested in studying and understanding the bidirectional individual ↔ context relations. Systems science methods allow researchers to examine the dynamic interrelationships of variables that occur at multiple levels of analysis (e.g., from cells to society) simultaneously (often through causal feedback processes), while also studying the impact on the behavior of the system as a whole over time (Midgley, 2003).

By definition, developmental scientists are interested in change over time and specifically the concepts of temporality and plasticity. Systems science methods address these concepts by (1) capturing the dynamic behavior of the system over time and potentially highlighting opportunities for systematic change, (2) illuminating time-delayed effects that can aid in foreseeing unintended consequences, and (3) allowing for the observation of emergent properties (e.g., when individual behavior leads to an aggregate outcome). These are fundamental characteristics of complex systems and are at the heart of the study of human development. Systems science tools provide us with a means of understanding such emergent phenomena at a particular level and with straightforward techniques for building models of these phenomena that directly incorporate cross-level interactions. For example, an adolescent's capacity for self-regulation may be an individual-level phenomenon, but the expression of that capacity is affected not only by the behavior of other individuals (e.g., peers), but also by factors at the family, school, and neighborhood levels (e.g., the availability of resources in the community for actualizing that potential).

Developmental scientists are also interested in the promotion of positive development through a search for effective interventions (e.g., policies and programs). Systems science methods are very well suited for addressing these more applied research questions that seek to understand and improve the human condition. Reductionist methods tend to fall short in their ability to address these more practical research questions, since they often require the use of controlled conditions that may be either impossible or unethical to implement. Systems science methods are particularly valuable in simulating these situations as they can be used to generate possible alternative futures in order to simulate the impact of various policy decisions before they are ever put into practice (Sterman, 2006).

The complexity of problems in which developmental scientists are interested requires an interdisciplinary approach, and systems science methods facilitate the integration of data and knowledge from multiple disciplines. For example, consider the multiple disciplines involved in studying the long-term impact of an early intervention such as Head Start (e.g., psychology, economics, sociology, urban planning, policy analysis, education, health/nutrition). Methods that enable integration across this wide array of fields, such as systems science, are needed.

In general, systems science methodologies are particularly useful when the research questions relate to theory generation/exploration (which includes evidence synthesis, the proposal of theoretical constructs, interactions among those constructs to be tested, and hypothesis generation) and policy analysis/decision making (Urban et al., 2011). They are also occasionally used for theory testing, which includes empirically testing competing hypotheses. The various systems science methodologies highlight different concepts and handle them differently. The choice of method will depend primarily on the questions being asked. Sometimes a combination of multiple systems science methods is desirable and/or a combination of systems science method(s) with more traditional statistical approaches. The next sections outline three of the leading systems science methods: social network analysis, system dynamics, and agent-based modeling.

Overview of Social Network Analysis

Network science primarily focuses on the patterns of the relationships between actors and tends to be grounded in systematic empirical data. *Social network analysis* also relies heavily on graphic imagery and on the use of mathematical and/or computational models. Wasserman and Faust (2008) outlined basic principles of network analysis, stating that network analysis (1) refers to a set of actors and the ties among them; (2) examines characteristics of actors that arise from structural or relational processes or focuses on the relational systems themselves; (3) includes examination of the relationships among units in a study; (4) is primarily tasked with understanding the properties of the social structural environment and how these structural properties influence observed characteristics; and (5) focuses primarily on relational ties among actors, whereas attributes of actors are secondary.

Use of network analysis in the social sciences has increased exponentially in the past two decades. The dramatic growth in the field of network science has encompassed study and application across a wide array of disciplines and phenomena, including public health, computer and information sciences, operations research, biology, genetics, physics, engineering, and mathematics. The life sciences, physical sciences, and hybrid disciplines such as neuroscience have also prominently applied network methods, for example, in mapping structures in the brain (e.g., Sporns, 2011). A large part of the appeal of social network analysis is the focus on examining and further understanding the interplay between actors (e.g., people, genes) and the context in which actors operate.

Social network analysis provides a nice entry and introduction to systems science methodologies, as there is a grounding in empirical data that is familiar to those trained in traditional statistical methodologies, while also taking into account the "big picture" view of the system (or network, in this case) and how the individual components of the system (i.e., the network) interact and are interdependent with each other. The foundation of social network analysis is multidisciplinary and traces its roots back to graph theory in mathematics (e.g., Scott, 2000; Wasserman & Faust, 1994). It was Moreno's (1934) study of teenage runaways using "sociograms" (i.e., network maps) in the early part of the 20th century that is often considered one of the early examples of the application of modern social network analysis and a key step forward in the development of a network science. Many of the techniques and descriptive measures of networks were developed in the fields of sociology and anthropology. More recently—and paired with recent advances in computing power—statistics, computational mathematics, and the computer sciences have made significant contributions in the form of new models and methods along with faster and easier-to-use analytic programs.

Some of the earliest antecedents of social network analysis were studies conducted by researchers interested in developmental issues (see Bott, 1928; Hagman, 1933; Wellman, 1926), so a rich history links these two fields. The intersection between studies of human development and network science, however, has remained a small research niche to date. Some of the earliest work examining what would later be called *social networks* examined social development in early childhood and served as key historical points of reference in network science.

Jacob Moreno's work (*Who Shall Survive?* [1934] in particular) is often cited as some of the originating groundwork of modern network science. He studied children from kindergarten to high school and examined various childhood phenomena, including the runaways mentioned above. He is also credited with popularizing use of the sociogram as a visual means of conveying a network of relationships. Moreno's work in linking social developmental questions with social network research inspired Bronfenbrenner, the father of ecological systems theory (Bronfenbrenner, 1979, 1999), to examine human development from a network perspective (for further detail, see Hanish & Rodkin, 2007).

Despite the strong early ties, the advancement of developmental network science stalled after the initial excitement produced by the work of Moreno, Bronfenbrenner, and their collaborators—though to be completely accurate, so did network science as a whole, until around the 1970s (see Freeman, 2004, for review). This gap could be attributed to many factors, but one strong possibility is the relative lack of computational power.

Since the dawn of the personal computer and especially as computational power has grown, network science has advanced at an astonishing pace and is more accessible to the developmental scientist than ever before. Social network analysis in particular enables developmental researchers who focus on interpersonal relationships to move beyond the scope of discrete dyadic relations and examine them in the larger social context in which they operate.

Newman (2003) has identified four somewhat broad categories of network analysis: social networks, information networks, technological networks, and biological networks. Among these four categories, *social* network analysis has been historically the most relevant network analysis domain to the study of human development. The central tenet of *social* network analysis is that the structure of the social relations in which an individual is embedded affects his or her behavior.

Core Concepts in Social Network Analysis

In the language of networks, members of a network are called *actors* and are also referred to as *nodes* and sometimes *vertices* (the singular form is *vertex*). Relationships between two actors are called a *tie* or *edge*. A representation of all such connections or ties is called a *network* and is referred to by mathematicians as a *graph*. Network graphs can be considered *directed* or *undirected*. Directed graphs contain ties that imply direction to the relationship; that is, the relationship between nodes is not necessarily mutual. For example, in a friendship network, node *A* may nominate *B* as being a friend, without *B* nominating *A* as a friend. If all ties can be considered reciprocal—that is, if *A* is connected to *B*, then *B* is by definition connected to *A*—then the network is undirected. For example, a network of cousins would imply that if *A* is a cousin of *B*, then by definition B must be a cousin of A. Directed networks are more commonly encountered than undirected networks; in directed networks, the node generating a tie is called the *ego*, and the receiver of the tie is an *alter*.

Levels of Social Network Analysis

There are different levels of analysis in network science that correspond with different metrics. At the *individual* level, the focus is on properties (both absolute and relative) of single actors, links, or incidences. Analysis often examines relationships between attributes (intrapersonal characteristics) of *ego* and individual measures of network status. That is, at the individual level, properties associated with individual nodes—such as how many ties each one has or how many connections make up the shortest route from one node to another—are the focus of the analysis.

The second level of analysis is the *group*. For this overview, we include dyads (ties and relationships between two actors) and triads (ties and relationships between three actors) in this level, though argument could be made that these should be separate analytic levels due to unique metrics that can be calculated for each. This level of analysis also begins to consider the characteristics of the tie itself. At this level, one is interested in classifying the elements of a network and properties of subnetworks (e.g., actor equivalence classes and cluster identification). Identification of subgroups, cliques, and components of a network is a main focus of this level of analysis. Identification can be straightforward in small networks, but as the size of the network increases, this becomes more difficult (for a discussion of the approaches to identifying subgroups within a large-scale network, see Newman, 2012).

At the *network* level, one is interested in properties of the overall network, such as connectivity or balance. For many situations the attributes of the individual nodes are insufficient, and what is of interest is how the group is structured and the attributes of the group as a whole.

Egocentric and Sociometric Levels of Analysis

Another way to distinguish between levels of analysis is to classify them as egocentric or sociometric. An *egocentric* network approach considers only the direct linkages of a given node. In practical terms, this relies on self-reports from ego (the individual of interest) about his or her relationship network, whether it be a friendship, sexual relationship, a kinship, or a coauthoring relationship. A *sociometric* approach considers a full set of actors with relationships to each other and the linkages among them; this is called a *complete network*. It is difficult, if not impossible, to capture a truly complete network (think of all your friends, all of their friends, all of the friends' friends, and so on) unless the network is *bounded*. A bounded network is one that has limits to it and usually follows a natural boundary. For example, a friendship network *within a school* is an example of a bounded network because the network is limited by the boundary of the school. Friends that do not attend the identified school are not considered part of the network.

Individual and Relational Models

There are also two common *types* of network models: individual and relational models. O'Malley and Marsden (2008) have identified *individual*-level models as focusing on an individual-level outcome wherein network data are used to define explanatory variables and *relational*-level models as focusing on the relationships between individuals in a network. Therefore, individual-level models focus on the attributes of an individual, which include network position and composition, and how those attributes are related to some individual outcome. Relational models make inferences, not about an individual outcome, but about the ties linking individuals and the structure of those ties (or the network). Individual-level models resemble standard regression models that examine an outcome from an individual using network metrics as independent variables. In relational-level models, the outcomes are a measure of the relationships (or ties) between individuals in the network and are used to account for the structure of the network (e.g., exponential random graph models; see Goodreau, Kitts, & Morris, 2009).

Individual-level models are helpful because they allow for the inclusion of measures of interpersonal influence, such as an individual's network position or the attributes of those to whom an individual is connected. Relational models are important because they allow for the modeling of the relationships in a network simultaneously, taking into account the interdependencies among them. This type of modeling is important for identifying structural properties of networks.

Basic Network Measures and Metrics

Since the primary purpose of this overview is to provide a general background and introduction to social network analysis for developmental science, only the most basic descriptions of a selection of the various network measures are given. (For more detailed descriptions of these metrics from a developmental research methods perspective, see Gest & Kinderman, 2012, or one of the many introductory textbooks on network analysis, e.g., Scott, 2000; Wasserman & Faust, 1994, 2008). Table 4.1 provides brief definitions of the most common network measures at the different levels of analysis.

When Is It Appropriate to Use Social Network Analysis?

Social network analysis is particularly well suited for exploratory work. It allows for the examination of the "big picture" view and is useful in illuminating patterns across individuals. It is also helpful in wading through large datasets or when there are multiple levels of influence proposed by the theoretical model. Social network analysis is

TABLE 4.1. Description of Common Network Measures by Level of Analysis

Level of analysis	Network measure	Definition
Individual	Degree Centrality	The simplest centrality measure; a simple count of the actor's total number of social ties. *Indegree* refers to incoming ties, and *outdegree* refers to outgoing ties.
	Betweenness Centrality	The extent to which a node lies between or connects two nodes that would not otherwise be connected
	Closeness Centrality	How close a node is to all other nodes in the network
	Structural Equivalence	The extent to which nodes plays a similar role within the network, described as having the same pattern of connections to other nodes
Group or subgroup	Modularity	A measure of the strength of division of a network into modules. Networks with high modularity have dense connections between the nodes within modules but sparse connections between nodes in different modules.
	Clique	Groups are identified as *cliques* if every individual is directly tied to every other individual in a subgroup
	Transitivity	The extent to which two nodes that are connected to the same third node are also connected to each other, or the probability that the "friends of my friends are also *my* friends" (can also be computed for the complete network)
Network	Density	The proportion of pairs of nodes that have ties of a given type (existing ties in a network/all possible ties in the network)
	Reciprocity	The proportion of all directional ties that are reciprocated
	Centralization	The extent to which the network revolves around one or a few nodes
	Diameter	The longest of all geodesics (the shortest path between two nodes)
	Clustering	The extent to which a network has clumps (smaller subgroups within the network with close ties to each other)

also suited to studying complex relational questions or processes that go beyond what can be gathered from self-report alone.

As noted earlier, network data can also be treated as covariates or explanatory variables in more traditional statistical techniques. In addition, these data can provide detailed visualizations that can be used to complement traditional tables and graphs and are also useful in identifying patterns in data.

When Is It Not Appropriate to Use Social Network Analysis?

First, it may be best to start discussing the challenges and limitations of social network analysis by noting when it is appropriate, or not appropriate, to use the *different* network analysis models and types. The model, type, level of analysis, and so on, should be chosen based on the best "fit" for the research question. Not all social network analysis is the same, and using an egocentric measure when the question requires examination at the sociometric level, and vice versa, will likely result in frustration at best. As mentioned above, there are some fundamental differences in the ways that social network analysis can be used, and finding which best fits the research question is critical.

From a methodological viewpoint, collecting network data can be challenging. There are sampling issues to consider. For example, are there resources available to collect data on a complete network? Is the complete network even accessible? These issues are usually highly related to issues of resources, but even with sufficient resources, the quality of the data collected can have a large impact. For example, nonresponse and missing data can have different meanings with network data, and handling or adjusting for missing data can require different procedures than for traditional methods. The basic assumption of independence also does not usually hold with network data.

Just finding someone with the training and expertise to conduct analyses is often a challenge. This may be the most common limiting factor for social network analysis to date. However, with greater accessibility to textbooks, courses, trainings, workshops, and so on, than ever before, this factor is slowly being remedied. Software is also easier to use and, for basic analyses, no longer requires a lot of advanced knowledge of programming languages (see Batagelj & Mrvar, 1998; Borgatti, Everett, & Freeman, 2002).

How Does Social Network Analysis Relate to Other Tools?

Network data are also well suited to serve as a complement to other methodologies and techniques. Networks are everywhere, so social network analysis is a useful way to consider context. As noted earlier, network data can be treated as another variable in more traditional statistical techniques. These data are also often utilized in other systems science techniques (e.g., agent-based modeling; see Marouslis et al., 2010; Skvoritz, 2003). Network data fit well with multilevel models of human behavior and provide detailed visualizations of data, which can be incredibly useful in exploratory work or with large and/or complex datasets.

In the *Handbook of Developmental Research Methods*, Gest and Kinderman (2012) provide a detailed overview of social network methods and applications to developmental science. They suggest several avenues of future exploration, particularly ways to "improve designs and methods to support causal inferences" in network analysis. These avenues encompass the need for longitudinal designs in order to more carefully study the issues around *influence* and *selection* in peer networks; the examination of contextual confounds that affect members of a network in the same way; a better understanding of how transmission occurs in peer networks and the mechanisms that facilitate this transmission; and the need for matched control groups.

The evolution of social network analysis again provides developmental studies with a useful methodology that allows for a more detailed study of social processes that goes beyond the "self" perspective. Recent computational advances, easier-to-use statistical and visualization software, and greater accessibility to information, texts, and tutorials will hopefully serve to widen the scope and breadth of this intersection further. The social environment remains important throughout the life course yet differs and changes across developmental periods; network analysis provides a practical and informative methodology that is uniquely equipped to examine the impact of this environment on human development.

Overview of System Dynamics Modeling

System dynamics modeling is a compartmental or aggregate modeling method that can accommodate both qualitative and quantitative data to help conceptualize, describe, analyze, and manage complex systems (Meadows, 2008). System dynamics modeling delivers insight and value from the very earliest stages (typically using qualitative tools) to the final stages (using rigorous quantitative methods). By focusing heavily on the modeling process itself, system dynamics approaches have created a powerful set of techniques for working with diverse interdisciplinary stakeholders. This includes tools that allow teams to easily articulate and sketch qualitative and semiquantitative knowledge to advance discussion, compare "mental models" of a situation, enhance collaboration, and critique and refine knowledge. In a seminal book, Vennix (1996) articulated and formalized participatory model building and group model elicitation approaches (Randers, 1977; Roberts, 1977; Rohrbaugh, 1992; Senge, 1990; Stenberg, 1980; Vennix, Scheper, & Willems, 1993; Weil, 1980; Winch, 1993).

System dynamics is frequently confused with the cognate field of dynamical systems, which has also been increasingly applied in developmental science (Thelen & Smith, 2006; van Geert, 2011). In addition to the confusing resemblance in name, the approaches do share many similarities. Both system dynamics and dynamical systems approaches are concerned with characterizing and studying the temporal behavior of systems. Another similarity concerns the formalisms used for characterizing system behavior. Whereas modern dynamical systems approaches make use of a wide range of approaches for specifying model behavior (including stochastic and delay differential equations), ordinary differential equations (ODEs) and their discrete time analogue (difference equations) are the most classic mechanisms and remain very widely used as dynamical systems formalisms. ODEs and difference equations also serve as the mathematical formalisms underlying the overwhelming majority of quantitative system dynamics models.

Despite these similarities in the analytic tools employed, the two domains are fundamentally distinguished by the questions that they pose and their scope. Reflecting its focus on improving the *management* of systems of which humans are an integral part, system dynamics seeks to realize sustainable change via understanding and changing people's mental models of a system within which they are embedded, so that they are empowered to behave more judiciously and coordinate more effectively, at less cross-purposes to each other and to the system's natural behaviors—and even realize a new system design that is less vulnerable and more effective. Reflecting this objective—and rather than simply focusing on the model as a mathematical artifact that describes "how things work out there"—system dynamics traditionally focuses on stakeholders as embedded in the system, and places conscious emphasis on facilitating all stages of the modeling process, so as to enhance the opportunity for insight by stakeholders, to secure buy-in and sense of ownership of the model and its outputs. To support this ownership goal, system dynamics is distinguished among dynamic modeling approaches by offering a broad, stakeholder-centric methodology that seeks to address the human theater in which modeling takes place, including—but not limited to—stakeholder-facing "learning laboratories" or "action labs" using simulation models, techniques and detailed scripts for participatory group model building, and qualitative and widely accessible mapping tools and graphical representations that aid in the elicitation, sharing, and analysis of mental models.

By contrast, the primary focus of most dynamical systems research lies squarely in the analytic domain, and places comparatively more emphasis on mathematical characterization of the behavior of complex systems—for example, identifying tipping points and phase transitions in complex systems and their governing criteria (e.g., conditions related to the basic reproductive constant [Anderson & May, 1991] in infectious disease modeling that determines whether an introduced entity will lead to an outbreak or die out), the location and stability of equilibria, proving general invariants of system behavior under wide ranges of parameter values, and so on. Although many system dynamics practitioners do use such methods to secure insight into model behavior, most tend to rely on simulation, rather than mathematical analysis, for insight.

Core Concepts in System Dynamics

In system dynamics, model formulation and analysis are based on a set of simple, yet versatile and universal, building blocks. The focus is on elements related to system state: stocks and flows. A *stock* represents an accumulation of something—for example, a capacity to self-regulate, a degree of risk perception related to some activity for a single person or for a segment of the population, people with a particular health risk, and so on. The value of a stock specifies the size of that accumulation (e.g., the number of people at the specified time point who have the identified health risk). Stocks collectively express the "memory" or "state" of a system, and are the factors underlying system inertia and delays. In an aggregate system dynamics model depicting a population, the members of the population are divided according to the combination of attributes (static or dynamic) with which they are associated. For each combination of attributes, we record the count of individuals who are characterized by that specific combination. For example, we might have stocks counting the number of individuals who are "never," "current," and "former" smokers, or the numbers of males and females (respectively) at each of several different levels of education. Alternatively, a model seeking to capture population dynamics might have stocks associated with infants, children, youth, and adults. Although less common in health research historically, system dynamics models can also be articulated at an individual level, with stocks representing factors such as immune memory (Vickers & Osgood, 2007), individual weight (Hamid, 2009; Rahmandad & Sabounchi, 2011), degree of belief (Karanfil, Moore, Finley, Brown, & Glass, 2011), or level of perceived risk (Gonzalez & Sawicka, 2003). Within a *stock-and-flow* diagram, a stock is depicted using a rectangle, suggesting the accumulation that it represents.

Stocks change due to *flows* with which they are associated. A flow typically connects two stocks, flowing out of one and into the other. In contrast to a stock (whose value specifies the quantity accumulated), a flow is associated with a value measured in terms of a *rate*—a certain quantity per unit time. Stocks and flows can be connected in a variety of ways. Figure 4.1 shows a simple "aging chain" formulation, in which people enter the world as "infants." This is the first stock, with new infants accumulating in the stock as they are born. Infants then pass into childhood (leaving the infant stock and entering the childhood stock), and eventually move into adulthood (the third stock). Death is an additional stock (not depicted) that people from any other stock can move to at any time, but they cannot move out of this stock. Note that although each of the stocks represents individuals, the model does not model each individual separately. What is tracked is the size of the group (or accumulation) within each stock at any given time. An individual will typically remain in each stock for a period of time, and then transition (treated as an instantaneous event) through a *flow* to a new stock. Whereas the stocks here would be counted in terms of people, the flows might be quantified as people per unit time. The arrangements of stocks and flows are typically not so uniform.

FIGURE 4.1. An "aging chain" of stocks and flows.

Figure 4.2 shows a slightly more involved complex of stocks and flows, one exhibiting cycles (involving pregnancy), and multiple paths between stocks (e.g., from the normoglycemic state to the diabetic state). In addition to stocks and flows, system dynamics models typically also include supporting variables in the form of model parameters (values that are static or whose dynamics are predetermined; e.g., constants or specified time series) and "auxiliary values" (convenient names for the result of applying a formula involving model variables, with the naming allowing for reuse elsewhere in the model).

The elements within system dynamics models are connected by hypothesized causal relationships and are themselves included in feedbacks that are frequently the foremost drivers for high-level system behavior. Such elements serve as the foundation for model formulation and analysis from early qualitative stages of modeling to the point where a team is working with and reasoning about the behavior of a fully quantitative model. Dependency of one variable on another is shown in a stock-and-flow diagram using directional arrows. If variable *B* depends on variable *A*, the arrow is shown as traveling from the representation of variable *A* to that of variable *B*.

The most fundamental relationships are those linking stocks and flows. The special nature of the relationship between a flow and a stock and that relationship's central importance in the system's dynamics has led to special diagrammatic conventions. In comparison with a stock (shown in a rectangle and analogized with a reservoir or bathtub), flows are shown as "pipes" into or out of a stock. The rate of change in a given stock depends on the net value of the flows to which that stock is connected. When the sum of the rate of the outflows of a stock exceeds the sum of the rate of

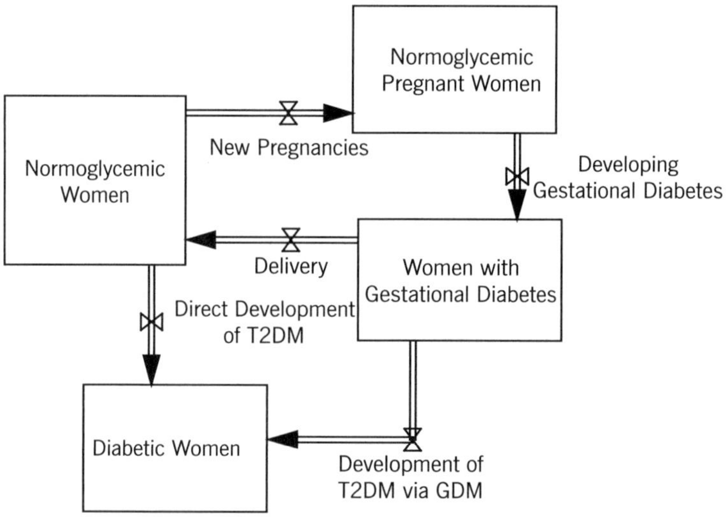

FIGURE 4.2. A more involved set of pathways of stocks and flows. GDM, gestational diabetes mellitus; T2DM, type 2 diabetes mellitus.

the inflows (e.g., water is draining out of a bathtub faster than it is arriving via the faucet), the stock (level of water in the bathtub) will decline over time. In the reverse case (e.g., water is flowing into a bathtub via the faucet faster than it is draining), the stock will rise. When the sum of the rate of the inflows exactly equals the sum of the rate of the outflows (e.g., water is flowing into a bathtub at the same rate that it is draining), the stock's value will neither rise nor fall but remain in stasis. The special diagrammatic representation used to distinguish stocks and flows helps remind us of the special semantics of this relationship. For example, although a stock's evolution will depend on an incoming or outgoing flow, even a sudden and pronounced change in that flow will not lead to an instantaneous, pronounced change in the stock. Just as doubling the speed with which water flows into a bathtub will not instantaneously fill it (but will instead just affect the rate with which it is filling), a change in the rate of a flow will affect only the *rate of change* of the connected stocks, rather than their immediate level. Recognizing the nature of stocks and flows in a system can help caution us against raising unrealistic expectations regarding how soon the impact of an intervention will be felt.

Although a stock's rate of change is driven by its flows, the value of those flows at any given moment in time ultimately depends on the values of the stocks at that point in time, plus any other model parameters, such as constant values estimated from datasets or studies. The exact dependence of a flow upon the stock(s) on which it depends will vary from context to context, and is typically based on hypothesized causal connections between factors. This dependence of a flow on other variables in the model is schematically depicted by placing a symbol for a "valve" on the "pipe" associated with a flow, with incoming arrows from the other variables on which it depends.

Feedback

Stocks and flows provide an extremely general and powerful way of describing and analyzing the behavior of complex systems. However, to understand the primary drivers of system behavior, it is important to appeal to the interaction of stocks and flows in the context of *feedbacks*. A feedback is a structural feature of causal connections in which a change to a stock will lead to a cascading series of changes in other factors that ripple around to affect the flow into that stock, either reinforcing (and further amplifying) or "pushing back against" (moving in the opposite direction of and thus dampening) the original change. Feedbacks come in two broad forms: Reinforcing (or "positive") feedbacks and balancing (or "negative") feedbacks. Reinforcing feedbacks are frequently the driver of observed rapid changes in system behavior and include those underlying "vicious cycles" such as those involved in the development of addiction or substance abuse and violence. For example, an individual who experiments with addictive substances may experience mood disorders that lead to difficulty in maintaining employment and to deteriorating relationships. The resulting adverse circumstances may raise stress levels in that individual, thereby elevating his or her substance use and further worsening the cycle. Although reinforcing feedbacks are often associated with "vicious cycles," the same structure frequently permits them to operate as virtuous cycles. For example, positive changes in self-regulation may reinforce the strength of relationships, allow for enhanced social capital, and increase trust, all of which permit greater access to learning resources and a feeling of security that reduces stress levels and builds an individual's sense of what he or she can accomplish—thereby reinforcing the original increase in self-regulation.

By contrast, balancing feedbacks are associated with stasis and resistance to change. Sometimes (e.g., for homeostasis and self-regulation) this is desirable; in other contexts (e.g., policy resistance or "lock-in" effects associated with poverty or addiction), such resistance to change is deleterious. A fundamental type of balancing feedback is that associated with self-regulation. Individuals may observe the consequences of their behavior, judge the effectiveness of those consequences in light of their goals, and adjust their behavior accordingly. In different individuals and circumstances, this balancing feedback will operate with different levels of strength and different delays (differences readily captured by variations in model parameters), but the structure is relatively consistent. Similar feedbacks operate at the interpersonal level (e.g., with a tutor attempting to adapt to a child's learning style) as well as at the policy level (albeit with far longer time delays and varying effectiveness).

System Structure and Causal Loop Diagrams

From the earliest phases of model conceptualization, it is helpful to sketch a group's understanding of the basic causal pathways operating within a system—briefly putting aside the quantitative nature of those dependencies—in a manner that is transparent to all stakeholders. Engaging stakeholders in the characterization of high-level hypotheses provides opportunities for group brainstorming about the scope of the model and the nature of key interactions in the system. However, it also offers value after a model has been developed as a way of keeping a high-level understanding of relationships in mind. Whereas later stages of model development focus on quantitative data and relationships (with formulae describing the rate of flows, initial values, and parameters of a model), stock-and-flow graphical representations can be used independent of such quantitative information. Stock-and-flow diagrams also form the conceptual basis for two additional types of qualitative diagrams: *system structure* diagrams and *causal loop* diagrams.

A system structure diagram further augments a stock-and-flow diagram with a depiction of the *polarities* associated with relationships. Specifically, consider a visual link from variable *A* to variable *B*, indicating a dependency of variable *B* on variable *A*. A system structure diagram labels that link with a "+" sign if an increase in variable *A* would lead (all other things remaining equal) to an increase in variable *B* (compared to the value it would have held had *A* retained its original value). By contrast, this link would be labeled with a "–" sign if that increase in *A* would lead (all other things remaining equal) to a decrease in *B*, compared to the value that it would have held had *A* not changed its value. With links labeled in this fashion, we can, by extension, derive the polarities associated with a specific pathway between two variables that are otherwise not directly connected. A bit of reflection will demonstrate that such a pathway offers a net positive polarity if and only if there is an *even* number (potentially 0) of "–" signs along that pathway. Conversely, the pathway will be associated with a net negative polarity if and only if there is an *odd* number of "–" signs along the pathway. The most important application of classifying path polarities is for feedback loops. When a given loop is associated with a positive polarity, it is classified as a reinforcing (positive) feedback. When a loop is associated with a negative polarity, it is classified as a balancing (negative) feedback. Reflecting the critical impact that feedbacks exert on system behavior, the convention in system structure diagrams is to label the polarities and directions of important feedback loops and to give them succinct names.

Particularly early in the modeling process, modelers secure many of the benefits of the system structure diagram with a simpler causal loop diagram. A causal loop diagram (see Figure 4.3) is similar to a system structure diagram, except it does not visually distinguish stocks and flows. Within causal loop diagrams, identical conventions are used for labeling links and loops, but variables are simply labeled with a brief name. No attempt is made in such diagrams to distinguish flows and stocks from other sorts of variables. While extensive experience with the system dynamics modeling process has demonstrated the effectiveness of system structure diagrams in working with diverse stakeholders, causal loop diagrams are even simpler for nontechnical stakeholders to understand.

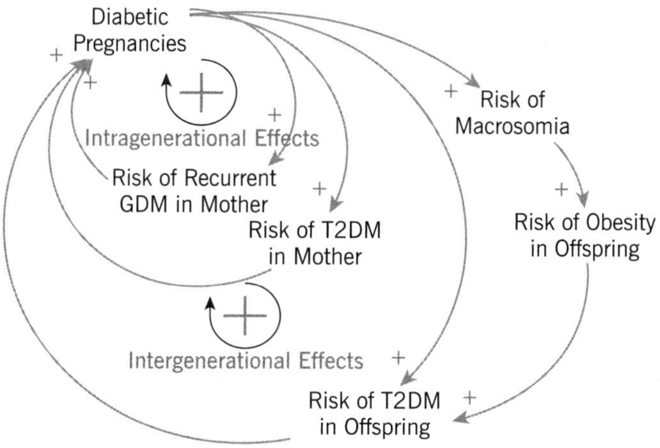

FIGURE 4.3. An example of a causal loop diagram that illustrates two reinforcing feedbacks that are labeled with words as well as their polarity. The figure depicts several vicious cycles hypothesized to be operating in the link between gestational diabetes mellitus (GDM) and Type 2 diabetes mellitus (T2DM). *Intra*generational effects play out within a single generation and specifically involve elevated risk of T2DM and future GDM on the part of a mother, following her development of GDM during a pregnancy. By contrast, *inter*generational effects reflect the heightened risks that offspring exposed to a diabetic intrauterine milieu will develop T2DM during their lifetime, or (for female offspring) GDM during pregnancies—thus perpetuating the "vicious cycle" (Osgood et al., 2011).

Mathematical Foundations

In contrast to many other dynamic modeling methods (e.g., agent-based modeling and discrete event modeling), the modeling tools used in system dynamics have transparent mathematical underpinnings that support straightforward mathematical analysis. Specifically, system dynamics models can be directly transliterated into ordinary differential equations—a framework widely used to describe dynamic systems in diverse branches of science and engineering. In this framework, stocks correspond directly to state variables, and the sum of the flows for a stock represent the derivative (rate of change) of that state variable with respect to time. For modeling teams containing the necessary expertise, this approach permits powerful analyses that can identify (1) general patterns of modeling behavior over a wide variety of contexts (Anderson & May, 1991), (2) properties of long-term system evolution, (3) the resiliency of a system in the face of internal or external disturbance (Vickers, Zhang, & Osgood, 2009), and (4) degree of leverage exerted by different parameters (Goncalves, 2006; Guneralp, 2005; Kampmann & Oliva, 2006; Saleh, Davidsen, & Bayoumi, 2005; Zhang & Osgood, 2010). It should be cautioned that application of certain analyses is limited to smaller models, but others readily accommodate models of any size. Developmental scientists seeking to learn more about system dynamics are advised to consult the foundational textbook on the subject (Sterman, 2000).

When Is It Appropriate to Use System Dynamics Modeling?

System dynamics modeling is particularly useful when feedback effects may be playing an important role, such as via "vicious cycles," "virtuous cycles," system inertia or resistance to change, and in cases where a system is characterized by one or more series of stages of progression (e.g., stages of cognitive development). As perhaps the only extant simulation technique that provides a full set of qualitative and quantitative tools and articulated participatory model building techniques that actively support the delivery of value and insight for stakeholders and the modeling team from the earliest stages of the modeling process, system dynamics offers a particular attraction for those seeking a continuous process of engagement with diverse stakeholders, including those lacking technical background. System dynamics is appropriate when knowledge concerning a system is largely qualitative, admits to divergent hypotheses regarding causal mechanisms, or where quantitative knowledge or data are so fragmentary as to be incapable of supporting a full quantitative model in the short run. In such cases, qualitative methods (optionally deployed in participatory sessions) can offer great value as a means of succinctly capturing and expressing different hypotheses, and working toward a common ground, including an articulation of empirical study or data collection priorities.

System dynamics models can be formulated at either an aggregate or individual level. When only aggregate data or aggregate knowledge of dynamics is available, system dynamics modeling confers particular advantage. By virtue of their clear mathematical structure (which affords use of refined numeric methods) and deterministic character, system dynamics models can typically yield results very quickly—frequently rapidly enough to be used interactively. Such rapid execution permits powerful use of models in compelling "learning laboratories" and "management flight simulators" (Diehl, 1994; Sterman, 1992), which aid stakeholders in securing an intuition for model behavior and allows for rich exploration of policy options.

The formal mathematical basis for system dynamics models also makes deep mathematical analysis possible. Such tools can be used to analyze whether the model behavior is likely to resist change in response to disturbance from inside or outside the model (e.g., if the external economy or neighborhood dynamics were to change), or whether such outside disturbances could precipitate pronounced changes in model behavior. The transparent basis of system dynamics models on deeply studied and widely used mathematics allows them to be used with a variety of control theoretical techniques to identify high-leverage parameters and highly effective policies that are adaptive to external conditions.

When Is It Not Appropriate to Use System Dynamics Modeling?

System dynamics modeling is not appropriate unless there is a dynamic problem being investigated. Aggregate system dynamics models are not appropriate when there is a desire to flexibly experiment with changing which heterogeneity is considered for a given actor in the model, where it is a recognized priority to include many (e.g., more than five) dimensions of heterogeneity (Osgood, 2004), or where one needs to maintain continuous variables or a large number of discrete categories with respect to some of those dimensions of heterogeneity. A particularly notable type of heterogeneity that is difficult to capture in aggregate system dynamics models relates to longitudinal information on a particular actor (such as might be required for simulating history-contingent interventions or to match model output to longitudinal data). Other types of heterogeneity of particular note are spatial and network contexts. Although aggregate models can readily capture very simple aspects of discrete spatial context and less structured ("random") mixing within and between subcategories of individuals, such models offer poor support for investigations that require representing continuous or irregular spatial environments or depiction of network structure. Moreover, although it is possible to readily incorporate some stochastic factors into system dynamics models, if understanding statistical variability of model results is a high priority, and if one is seeking to capture stochastics with regard to a large amount of behavior, other naturally stochastic forms of dynamic modeling (e.g., agent-based modeling or discrete event simulation) may offer a better fit than system dynamics modeling.

How Does System Dynamics Modeling Relate to Other Tools?

Parameter estimation for both system dynamics and agent-based models employs statistical estimation techniques such as multivariate regression (to help estimate the influence of one variable on another), survival analysis, and competing risks methods to estimate hazard rates and their dependence on risk factors. Statistical tools assess goodness of fit of system dynamics model results against empirical data during calibration and model validation. Calibration itself—a process undertaken with both system dynamics and agent-based models in which less well-known model parameters are adjusted so as to offer the best fit to historic data—bears a certain similarity to nonlinear regression, with the proviso that there is no explicit "functional form" given for the behavior of the model. Instead, this overall model behavior is merely implicit in the set of rules given for updating the state of the model (including states of agents and the state of the environment) over time. Often modelers make use of aggregate (and often fragmented) cross-sectional data, though longitudinal and individual-level data can also be used for these purposes. Statistical tools are quite useful in improving analysis of system dynamics simulation output in support of improving understanding of system behavior, and in improving model calibration and validation efforts. Although population-level aggregate system dynamics models do not distinguish individuals as such (and thus provide more limited integration with multilevel modeling), such models can provide powerful support for informing certain aspects of study design (e.g., study duration required to see an effect, or training resources that must be mobilized for effective intervention) and a framework for evaluating certain types of statistical tools (particularly those using repeated aggregated cross-sectional measures).

Overview of Agent-Based Modeling

The purpose of all of the systems science modeling methodologies is to create simplified replicas of the real world ("models") that are easier to understand and manipulate than the complex world on which they are modeled. Agent-based modeling (ABM) is a dynamic modeling technique that leverages the computational power of computers to provide fine-grained insights into real-world systems behaviors by creating models that attempt to capture the behaviors of the individuals/entities that comprise the systems. It is the emphasis on capturing and modeling the behavior of the individual agents in the system that sets ABM apart from most other systems science methodologies. Microsimulation techniques also model the behavior of each individual; what distinguishes them is that ABM features agents that interact with one another (and often alter their behavior based on those interactions), whereas in a microsimulation the individual agents do not have the ability to interact with one another. Agent-based modeling is useful when dealing with systems in which there are clear discrete components, referred to as *agents*. For example, an agent-based model seeking to understand and help prevent or control substance abuse in a high school population might include agents to represent students, parents, and teachers. Agents are typically assigned attributes (e.g., age, income, number of friends, hair color, genotype) and rules governing their behavior. Attributes typically include both *parameters*—factors that are invariant for a given agent, or change only infrequently (e.g., ethnicity, gender, the network of relations)—and *state*—properties that change over time (e.g., age, attitudes, norms, knowledge, location, health conditions, skills, knowledge, capacity to self-regulate, history of substance abuse). A modeler describes agent behavior via *rules*—typically probabilistic—that govern how the various components of an agent's state evolve over time. For example, such rules might specify how agents' attitudes are shaped by those around them, or how agents' states of health evolve according to current and past behavioral risk factors and characteristics of their surroundings. The attributes, and even the rules governing behavior, can differ across individuals. It is important to note that in cases where the group of interest is fairly homogeneous, ABM is often not needed.

Core Concepts in ABM

Aspects of an agent's surroundings are abstracted in what is known as the *environment.* An environment might, for example, provide a stylized description of an agent's location in a physical two- or three-dimensional environment (e.g., classrooms within a school, a school or home in a neighborhood) or in a social network similar to those discussed in the network analysis section of this chapter (e.g., networks connecting peers or family members). The environment can further be associated with rules that govern its behavior or evolution; for example, the school could be closed over certain time windows, the economy could be associated with business cycles, seasonality effects could govern the accessibility of the outdoor environment for recreation, or illicit substances could be more or less readily available outside of the school.

Stochasticity

ABMs are typically stochastic. That is, even given identical assumptions (captured in the form of the rules, parameter values, and initial state), two particular simulations (*realizations*) of a model will typically yield varying results. Such variation reflects the stochastic nature of the rules governing agent and environmental dynamics. By virtue of their flexibility in accumulating information regarding an agent's history that is arbitrarily detailed and structured (e.g., as a record of past duration or even all instances of substance abuse, or past contacts with detoxification programs or peer counselors), ABMs offer very powerful capacity to support history-dependent interventions (e.g., detoxifications that offer special handling of individuals with a history of inhalant abuse, or programs matching peer counselors and at-risk individuals who have similar backgrounds).

Micro-Level Rules Determine Macro-Level Behavior

ABMs often exhibit behavior referred to as *emergence*—that is, the interactions between various components at the microlevel (characteristics of the agents, rules and probabilities for how agents change or interact) lead to behavior very different from the behavior of any one component in isolation or from the simple sum of the behavior of those components. Although such emergent patterns of stocks and flows are seen with behavior over time in system dynamics models, within ABM such patterns are additionally seen across space, across networks, and—perhaps most notably—across multiple scales. Specifically, to the degree that an ABM captures different levels of analysis in the environment—such as schools and their neighborhoods within which agents circulate, or one or more networks (e.g., of peers and family)—quite distinct dynamic patterns may emerge at each of these levels, capturing the evolution of phenomena that might traditionally be examined using multilevel statistical models, such as hierarchical linear models. For a more detailed introduction to ABM, interested readers are referred to a recent textbook on the subject (Railsback & Grimm, 2011).

When Is It Appropriate to Use ABM?

ABMs merit particular consideration when (1) a dynamic problem exhibits pronounced heterogeneity (or need to flexibly explore the impact of different aspects of heterogeneity), (2) agent behavior depends strongly on the particulars of the surrounding local context for that individual, or (3) when the research question focuses on complex components of individual-level history or of learning effects that also depend on that individual environment or history, or phenomena that differ over individual context. A related important motivation for use of ABM applies when potential interventions differentially affect factors that are more easily captured within an ABM, such as characteristics of individual context (e.g., geographic or network position), diverse heterogeneity, or individual history. For example, if we wished to examine the impact of interventions targeting individuals who have a history of abusing more than three distinct substances, or focus peer counseling on individuals with distinct network positions, we could conduct such studies more easily with ABM than with classical aggregate-level modeling (as would be commonly used in system dynamics approaches). Another important motivation for ABM use is the plethora of developmental data that are available at an individual level. When we seek to match a model's outcome to such data, the individual-level perspective of ABM can offer distinct advantages. For example, ABMs that accumulate history information can be readily matched during model calibration against longitudinal data.

An additional important attraction for ABM is the ability to capture the different emergent behaviors seen at different scales of analysis—for example, at the individual, family, school, and neighborhood levels. Like most modeling projects, ABM initiatives take place in a "human theater" that involves diverse stakeholders. When stakeholder understanding of the dynamic phenomenon to be understood or changed can be articulated and formulated in a model most comfortably or readily at an individual level, or when stakeholders find it preferable to understand or validate against model results at an individual level, ABMs warrant particular attention.

When Is It Not Appropriate to Use ABM?

ABM tools are versatile and powerful, but they are not well suited to application in many contexts. Perhaps the most fundamental factor is whether there is a genuinely complex dynamic problem at hand—a problem in which a complex series of interactions among a set of factors gives rise to behavior over time that a researcher is seeking to either understand or modify. A second broad criterion for applicability is whether the system is composed of recognizable, discrete units.

ABM projects typically look to the model itself as the defining source of project value. This focus places a premium on the conceptualization, formulation, implementation, debugging, and validation of such a model. Of these steps, the formulation, implementation, and debugging steps for ABMs currently demand a level of technical familiarity and comfort with some basics of object-oriented programming in languages such as Java or Logo. In some cases, the computational burden of the model may be sufficiently large that the routine operation of it will also require some technical support (e.g., to run it on a computer cluster).

When knowledge and data concentrate on aggregate-level phenomena, and where there is a deep lack of knowledge of individual-level phenomena that give rise to such high-level behavior, building an individual-level model will often not be a fruitful path to quantitative insight. In such cases, creation of an individual-based model may be such an exercise in speculation that it is not yet worth pursuing for quantitative findings. However, if what is being sought is an exploratory tool for hypothesizing about factors occurring at the individual level, highly stylized ABMs may offer value.

Sometimes a model that affords analysis on a scenario-by-scenario basis is viewed as being too specific. Instead, what is being sought is something much more general—for example, a broad understanding of how the system will behave under a wide variety of circumstances, the long-term behavior of the system and points of stasis associated with a system, and the robustness of that behavior in response to small changes caused by factors not directly incorporated in the model (e.g., assumptions regarding the surrounding economic context, or the level of drug dealing seen in the surrounding society). In such contexts, modelers traditionally turn to mathematical analysis (Anderson & May, 1991). Unfortunately, ABMs are currently relatively poor in supporting such analysis, due to the typical absence of mathematically precise specifications for such models, and the challenges in conducting formal analysis of general-purpose stochastic systems. Other dynamic modeling techniques, such as system dynamics, offer much greater support for such analyses by providing more mathematically tractable formulations.

How Does ABM Relate to Other Tools?

ABMs can be used synergistically both with traditional tools and with other methods from the system sciences. As with system dynamics, statistical analyses are commonly used for estimating model parameters and for assessing correspondence of model results with observed data (e.g., during calibration), but there are textured differences and unique opportunities associated with ABMs. Given the stochastic character of ABMs, statistical tools—and particularly hypothesis-testing tools—are used widely with respect to model outcomes. The capacity of ABMs to accumulate longitudinal information on individual agents opens up opportunities for evaluation of model dynamics against longitudinal empirical data using approaches such as survival analysis. ABM is also distinguished by virtue of its ability to support "simulation studies" to evaluate statistical methods applied to individual-level data, which can be transparently applied to the individual agents. Much more so than their aggregate cousins, ABMs are also highly suited to evaluating elements of study design by simulating study operation over time in a synthetic population, data collection processes (and attended measurement error), loss to follow-up, and so on.

An important distinguishing strength of ABMs is their capacity to straightforwardly support analysis at multiple scales. For example, a given model can represent not only individuals, but also groupings of those individuals such as in families, schools, neighborhoods, and cities. Multilevel modeling techniques (e.g., hierarchical linear models, multilevel logistic regression, multilevel longitudinal approaches) can provide valuable estimates for parameter values at these different levels. Using such tools in conjunction with ABMs can help improve model calibration and validation by increasing understanding of the degree to which observed multilevel statistical patterns are consistent with the synthetic multilevel patterns observed within the ABM.

Systems Science in Developmental Science: Current and Future Directions

The following sections of this chapter focus on how systems science methodologies are currently being employed in developmental science and related fields. We also explore potential future applications of systems science methods in developmental science. Publication of systems science papers in the developmental science literature has been uneven across methods: a substantial number of papers featuring social network analysis has been published, whereas many fewer papers have featured ABMs and system dynamics. Moreover, much of the system dynamics and ABM work is very recent.

Systems Science Applications in Developmental Science

Most of the social network analysis work published in the developmental science literature does not attempt to explore development per se (i.e., change over time), but rather is aimed at understanding the social networks within a specific age group thought to be important in development (e.g., adolescents, children). This may be, at least in part, a reflection of the current state of the field, as most of the published social network analysis work features analysis of networks at a single point in time, rather than longitudinally across time.

Although we cannot cite all of the developmental science studies that have utilized social network analysis we highlight some particularly noteworthy contributions. Valente and colleagues have explored the friendship networks of adolescents with regard to a variety of health phenomena, including obesity (Valente, Fujimoto, Chou, & Spruijt-Metz, 2009), smoking (Hall & Valente, 2007; Valente, Unger, Ritt-Olson, Cen, & Anderson Johnson, 2006), and drinking (Fujimoto & Valente, 2012). Many groups have explored the social networks of those who use or abuse substances or alcohol (recent contributions include Lakon, Hipp, & Timberlake, 2010; Mason, Mennis, & Schmidt, 2011; for review of social network analysis use in studying adolescent smoking, see Seo & Huang, 2012). Recent work by Dishion and Tipsord (2011) reviews the literature on how peer networks influence child and adolescent social-emotional development and deviant behavior, and some of the work reviewed features social network analysis. Fujisawa, Kutsukake, and Hsegawa (2009) have used social network analysis to explore social relationships among preschoolers. Schaefer, Simpkins, Vest, and Price (2011) examined the importance of extracurricular activities to adolescent friendships using social network analysis. Mouttapa and colleagues (Mouttapa, Valente, Gallaher, Rohrbach, & Unger, 2004) used social network analysis to examine bullying (bullies, victims, and aggressive victims—children who are both bullies and victims).

There are very few publications of system dynamics models in the developmental science literature. We did find an exploratory investigation by Hovmand and colleagues (Hovmand & Ford, 2009; Hovmand, Ford, Flom, & Kyriakakis, 2009) in which the authors applied both qualitative system dynamics diagramming and quantitative simulation modeling to understand victim arrests associated with domestic violence, and to understand the life course of abusive relationships. In another ongoing project relevant to developmental science, Hovmand (personal communication, April 5, 2012) led a team organizing and facilitating a set of four workshops in which a group of locally diverse elementary, middle, and high school students and young adults (independently) used innovative group model-building methods to diagram the key causal determinants of youth weight based on the "Connection Circles Exercise" (Quaden, Ticotsky, & Lyneis, 2006).

In another project, Osgood, Dyck, and Grassmann (2011) used a system dynamics model to investigate the link between gestational diabetes mellitus (GDM) and type II diabetes mellitus (T2DM), which are thought to have distinct etiologies, as a possible driver for the increase in T2DM prevalence observed in Canadian Aboriginal communities. The basic dynamic hypotheses investigated are depicted in Figure 4.3.

Given the high degree of heterogeneity observed in developmental science, the nested and interwoven contexts in which an individual is embedded, and the multifaceted character of the individual state, the complexity of behavior, and the importance of learning and individual history in shaping behavior change, individual-level system dynamics models offer much to recommend them. While not yet applied to developmental science topics, a growing class of system dynamics models focus on individual-level dynamics such as for obesity (Hamid, 2009; Karanfil et al., 2011; Rahmandad & Sabounchi, 2011) and multiscale models of communicable disease dynamics (Vickers & Osgood, 2007), with characteristics of different individuals maintained in different stocks. Although aggregate models can yield much insight, such individual-level system dynamics models afford a depth of description of individual dynamics and a flexibility that make them particularly well suited to modeling in developmental science.

With regard to ABM, very little work has been published in the developmental science literature. Of note is Orr and Evans's (2011) agent-based model that was developed to understand diffusion of sexual initiation among adolescents. Thomas (2012) used ABM to evaluate three intervention strategies aimed at reducing unintended pregnancy: improved access to family planning via Medicaid, mass-media campaigns to promote condom use, and teen prevention programs informed by evidence-based practices. While all three interventions showed potential for cost savings, the Medicaid expansion scenarios imposed the greatest cost in the model, but also exhibited the highest benefit–cost ratios—largely reflecting their close targeting of those in greatest need of assistance in supporting contraception costs. Other ABM work in developmental science appears to be limited to a call for the use of these methods (Schlesingera & Parisi, 2001).

Future Directions for Developmental Systems Science

One area of methodological development is in hybridizing the different modeling methods. The goal of such work is often to overcome limitations of any single method and to get "the best of all worlds." Partly reflecting their joint focus at the individual level and capacity for representing complex network structures, ABM and social network analysis offer many opportunities for cross-leveraging. Social network analysis data can be a rich source of contextual information about relationships—for example, places of contact, type or closeness of relationship, and so on—than are likely to be initially captured within an ABM. By identifying important patterns in the empirical data, social network analysis can help identify priority areas for investigation using the ABM, which can allow for exploring possible dynamic hypotheses to explain patterns observed in social network analysis. ABM can also be used to develop hypotheses regarding network structure that were not empirically observed, thereby helping to prioritize future empirical social network analysis studies.

The use of system dynamics and ABMs offers complementary insights and rich sets of synergies that incorporate elements drawn from both methods. One way in which the two techniques can be combined in the same modeling process is by drawing on system dynamics methods for early stages of the process and ABM later in the process. For example, a modeling project might draw on system dynamics methods for group model building and qualitative mapping and yet build an ABM when it comes to quantitative model construction.

Agent-based methods can also complement system dynamics models through the specification of partial or complete agent behavior using the language of stocks and flows. Stocks provide a powerful way of describing continuous states associated with each agent, and flows support the characterization of the ways in which each state evolves over time. Agent populations can also be embedded within system dynamics models. Changes in those agent populations—such as shifts in agent behavior, in agent mortality, or in agent natality—could then drive flows in those models between stocks that summarize population-wide state or in computing other quantities of interest.

There is also incredible potential for the use of dynamic and longitudinal network analysis. Indeed, further development and refinement of statistical and computational methods to better understand how networks change and evolve over time may be the key issue for developmental network analysis moving into the future. Much of the recent research in network science has focused on the development and refinement of ways to study *dynamic* networks (see Veenstra & Steglich, 2012, for a detailed overview of dynamic networks from a developmental perspective). Development of these models and techniques has been a central advancement in network science in recent years and remains a burgeoning area of growth in the field. Being able to statistically model information *temporally* and *dynamically* is critical for developmental science.

The actor-oriented model of network evolution proposed by Snijders (1996, 2001, 2005) is one such method that continues to rapidly evolve and provide additional solutions to past limitations of the more static network analysis models. Recent developments of the actor-oriented model allow for partial imputation of missing ties over time (Snijders, Steglich, Schweinberger, & Huisman, 2007), and they also allow actors to join or leave the network over time (Huisman & Snijders, 2003; Snijders, 2005). Development of the SIENA software (Snijders et al., 2007) to analyze these longitudinal, actor-oriented models has also been a key factor in the rapid growth and dissemination of these methods. An area related to longitudinally evolving networks is network co-evolution; that is, when actors in one network are also actors in another network (e.g., children are members of a peer network and each child also has a family network), and each network is evolving simultaneously, with the possibility for each network to influence the other (e.g., Chen, Marathe, & Marathe, 2010; Steglich et al., 2010).

Although there are relatively few large datasets that have longitudinal network data, it is important to point out accessible sources of this type of data. One is the National Longitudinal Study of Adolescent Health (Add Health) dataset (see *www.cpc.unc.edu/projects/lifecourse/research_projects/add_health*). Also, the popularity of social media such as Facebook presents an enormous opportunity for longitudinal social network data. However, it comes with significant challenges, including ethical and privacy concerns (for a good discussion of these issues, see Liu, Krishnamurthy, Gummadi, & Mislove, 2011).

Conclusions

This chapter has promoted developmental systems science and, more specifically, it introduced developmental scientists to systems science methodologies, including an introduction to three specific methodologies: social network analysis, system dynamics modeling, and agent-based modeling. We expect that some readers will be inspired by this chapter to learn more and to incorporate systems science into their own work. There are at least two ways to do this. One is to become a full-fledged expert by pursuing an academic degree, postdoctoral study, or otherwise receive in-depth training. Another option is to collaborate with someone who has systems science expertise. Either way, we expect many readers will want to know what they can do to learn more. Various resources are available through the Office of Behavioral and Social Sciences Research at the National Institutes of Health. Resources include a series of introductory videocasts on systems science, in-person training courses and online training materials, and a listserv for obtaining current information on articles of interest, funding opportunities, conferences, training, and the like. To obtain information, write to chapter author Patricia Mabry at mabryp@od.nih.gov.

Systems science methodologies hold tremendous promise for developmental science scholars. Developmental systems science is a newly emerging field that should be embraced particularly by those interested in testing theoretical constructs posited by developmental systems theories. Applying systems science tools is the next logical step in advancing the field of developmental science.

AUTHOR NOTE

The content of this chapter is solely the responsibility of the authors and does not necessarily represent the official views of the Office of Behavioral and Social Sciences Research or the National Institutes of Health.

REFERENCES

Anderson, R. M., & May, R. M. (1991). *Infectious diseases of humans: Dynamics and control.* New York: Oxford University Press.

Batagelj, V., & Mrvar, A. (1998). Pajek: Program for large network analysis. *Connections, 21*(2), 47–57.

Bertalanffy, L. (1969). *General systems theory.* New York: George Braziller.

Borgatti, S. P., Everett, M. G., & Freeman, L. C. (2002). *Ucinet for Windows: Software for social network analysis.* Harvard, MA: Analytic Technologies.

Bott, H. (1928). Observation of play activities in a nursery school. *Genetic Psychology Monographs, 4,* 44–88.

Bronfenbrenner, U. (1979). *The ecology of human development.* Cambridge, MA: Harvard University Press.

Bronfenbrenner, U. (1999). Environment in developmental perspective: Theoretical and operational models. In S. L. Friedman & T. D. Wachs (Eds.), *Measuring environment across the life span: Emerging methods and concepts* (pp. 3–29). Washington, DC: American Psychological Association.

Bronfenbrenner, U., & Morris, P. A. (2006). The bioecological model of human development. In W. Damon & R. M. Lerner (Eds.), *Handbook of child psychology* (6th ed., Vol. 1, pp. 793–828). Hoboken, NJ: Wiley.

Capra, F. (1977). *The tao of physics.* New York: Bantam.

Capra, F. (1982). *The turning point: Science, society, and the rising culture.* New York: Bantam.

Chen, J., Marathe, A., & Marathe, M. (2010, March). *Coevolution of epidemics, social networks and individual behaviors: A case study.* Proceedings of the third International Conference on Social Computing, Behavioral Modeling, and Prediction (SBP10), Bethesda, MD.

Diehl, E. (1994). Managerial microworlds as learning support tools. In J. D. W. Morecroft & J. D. Sterman (Eds.), *Modeling for learning organizations* (pp. 327–338). Portland, OR: Productivity Press.

Dishion, T. J., & Tipsord, J. M. (2011). Peer contagion in child and adolescent social and emotional development. *Annual Reviews in Psychology, 62*, 189–214.

Fogel, A. (2011). Theoretical and applied dynamic systems research in developmental science. *Child Development Perspectives, 5*(4), 267–272.

Ford, D. H., & Lerner, R. M. (1992). *Developmental systems theory: An integrative approach.* Newbury Park, CA: Sage.

Freeman, L. C. (2004). *The development of social network analysis: A study in the sociology of science.* Vancouver, BC, Canada: Booksurge.

Fujimoto, K., & Valente, T. W. (2012). Decomposing the components of friendship and friends' influence on adolescent drinking and smoking. *Journal of Adolescent Health, 51*, 136–143.

Fujisawa, K. K., Kutsukake, N., & Hasegawa, T. (2009). Social network analyses of positive and negative relationships among Japanese preschool classmates. *International Journal of Behavioral Development, 33*, 193–201.

Gest, S. D., & Kindermann, T. A. (2012). Analysis of static social networks and their developmental effects. In B. Laursen, T. D. Little, & N. A. Card (Eds.), *Handbook of developmental research methods* (pp. 598–618). New York: Guilford Press.

Goncalves, P. (2006). *Eigenvalue and eigenvector analysis of dynamic systems.* Proceedings of the 24th International Conference of the System Dynamics Society, Nijmegen, Netherlands.

Gonzalez, J. J., & Sawicka, A. (2003, July). *The role of learning and risk perception in compliance.* Paper presented at the 21st International Conference of the System Dynamics Society, New York.

Goodreau, S. M., Kitts, J. A., & Morris, M. (2009). Birds of a feather, or friend of a friend?: Using exponential random graph models to investigate adolescent social networks. *Demography, 46*(1), 103–125.

Gottlieb, G. (1992). *Individual development and evolution: The genesis of novel behavior.* New York: Oxford University Press.

Gottlieb, G. (1998). Normally occurring environmental and behavioral influences on gene activity: From central dogma to probabilistic epigenesis. *Psychological Review, 105*(4), 792–802.

Gottlieb, G., Wahlsten, D., & Lickliter, R. (2006). The significance of biology for human development: A developmental psychobiological systems view. In W. Damon & R. M. Lerner (Eds.), *Handbook of child psychology* (Vol. 1, pp. 210–257). Hoboken, NJ: Wiley.

Guneralp, B. (2005, July). *Progress in eigenvalue elasticity analysis as a coherent loop dominance analysis tool.* Proceedings of the 23rd International Conference of the System Dynamics Society, Boston.

Hagman, E. P. (1933). The companionships of preschool children. *University of Iowa Studies in Child Welfare, 7*, 10–69.

Hall, J., & Valente, T. W. (2007). Adolescent smoking networks: The effects of influence and selection on future smoking. *Addictive Behaviors, 32*(12), 3054–3059.

Hamid, T. K. A. (2009). *Thinking in circles about obesity: Applying systems thinking to weight management.* New York: Springer.

Hanish, L. D., & Rodkin, P. C. (2007). Bridging children's social development and social network analysis. *New Directions for Child and Adolescent Development, 118*, 1–8.

Hollenstein, T. (2007). State space grids: Analyzing dynamics across development. *International Journal of Behavioral Development, 31*(4), 384–396.

Hovmand, P. S., & Ford, D. N. (2009). Sequence and timing of three community interventions to domestic violence. *American Journal of Community Psychology, 44*(3–4), 261–272.

Hovmand, P. S., Ford, D. N., Flom, I., & Kyriakakis, S. (2009). Victims arrested for domestic violence: Unintended consequences of arrest policies. *System Dynamics Review, 25*(3), 161–181.

Huisman, M., & Snijders, T. (2003). Statistical analysis of longitudinal network data with changing composition. *Sociological Methods and Research, 32*, 253–287.

Kampmann, C. E., & Oliva, R. (2006). Loop eigenvalue elasticity analysis: Three case studies. *System Dynamics Review, 22*(2), 141–162.

Karanfil, Ö., Moore, T. M., Finley, P., Brown, T. Z. A., & Glass, R. (2011, July). *A multi-scale paradigm to design obesity policy: Exploring the integration of individual-based modeling and system dynamics*. Paper presented at the 29th International Conference of the System Dynamics Society, Washington, DC.

Lakon, C. M., Hipp, J. R., & Timberlake, D. S. (2010). The social context of adolescent smoking: A systems perspective. *American Journal of Public Health, 100*(7), 1218–1228.

Lerner, R. M. (2006). Developmental science, developmental systems, and contemporary theories of human development. In R. M. Lerner (Ed.), *Theoretical models of human development* (6th ed., Vol. 1, pp. 1–17). Hoboken, NJ: Wiley.

Lewis, M. D. (2005). Bridging emotion theory and neurobiology through dynamic system modeling. *Behavioral and Brain Science, 28*, 169–245.

Liu, Y., Krishnamurthy, B., Gummadi, K. P., & Mislove, A. (2011, July). *Analyzing Facebook privacy settings: User expectations vs. reality*. Paper presented at the Internet Measurement Conference, Berlin.

Magnusson, D., & Stattin, H. (2006). The person in context: A holistic-interactionistic approach. In W. Damon & R. M. Lerner (Eds.), *Handbook of child psychology* (6th ed., Vol. 1, pp. 400–464). Hoboken, NJ: Wiley.

Marouslis, S., Guimera, R., Petry, H., Stringer, M. J., Gomez, L. M., Amaral, L. A. N., et al. (2010). Complex systems view of educational policy research. *Science, 330*, 38–39.

Mason, M. J., Mennis, J., & Schmidt, C. D. (2011). A social operational model of urban adolescents' tobacco and substance use: A mediational analysis. *Journal of Adolescence, 34*(5), 1055–1063.

Meadows, D. (2008). *Thinking in systems: A primer.* White River Junction, VT: Chelsea Green.

Midgley, G. (Ed.). (2003). *Systems thinking.* Thousand Oaks, CA: Sage.

Molenaar, P. C. M., & Newell, K. M. (Eds.). (2010). *Individual pathways of change: Statistical models for analyzing learning and development.* Washington, DC: American Psychological Association.

Molenaar, P. C. M., Sinclair, K. O., Rovine, M. J., Ram, N., & Corneal, S. E. (2009). Analyzing developmental processes on an individual level using nonstationary time series modeling. *Developmental Psychology, 45*(1), 260–271.

Moreno, J. L. (1934). *Who shall survive?: A new approach to the problem of human interrelations.* Washington, DC: Nervous and Mental Disease.

Mouttapa, M., Valente, T., Gallaher, P., Rohrbach, L. A., & Unger, J. B. (2004). Social network predictors of bullying and victimization. *Adolescence, 315*(21), 315–335.

Newell, K. M., & Molenaar, P. C. M. (Eds.). (1998). *Applications of nonlinear dynamics to developmental process modeling.* Mahwah, NJ: Erlbaum.

Newman, M. E. J. (2003). The structure and function of complex networks. *SIAM Review, 45*(2), 167–256.

Newman, M. E. J. (2012). Communities, modules and large-scale structure in networks. *Nature Physics, 8*, 25–31.

O'Malley, A. J., & Marsden, P. V. (2008). The analysis of social networks. *Health Services and Outcomes Research Methodology, 8*(4), 222–269.

Orr, M. G., & Evans, C. R. (2011). Understanding long-term diffusion dynamics in the prevalence of adolescent sexual initiation: A first investigation using agent-based modeling. *Research in Human Development, 8*(1), 48–66.

Osgood, N. (2004, July). *Representing heterogeneity in complex feedback system modeling: Computational resource and error scaling.* Paper presented at the 22nd International Conference of the System Dynamics Society, Oxford, UK.

Osgood, N., Dyck, R., & Grassmann, W. (2011). Inter- and intra-generational impact of gestational diabetes on the epidemic of type 2 diabetes. *American Journal of Public Health, 101*(1), 173–179.

Overton, W. F. (1998). Developmental psychology: Philosophy, concepts, and methodology. In R. M. Lerner (Ed.), *Theoretical models of human development* (5th ed., Vol. 1, pp. 107–188). Hoboken, NJ: Wiley.

Overton, W. F. (2006). Developmental psychology: Philosophy, concepts, methodology. In R. M. Lerner (Ed.), *Theoretical models of human development* (6th ed., Vol. 1, pp. 18–88). Hoboken, NJ: Wiley.

Pettitt, D. J., & Ovanovic, L. (2007). The vicious cycle of diabetes and pregnancy. *Current Diabetes Reports, 7*(4), 295–297.

Quaden, R., Ticotsky, A., & Lyneis, D. (2006). Do you want fries with that?: Learning about connection circles. *The Shape of Change.* Retrieved from *www.clexchange.org/ftp/documents/x-curricular/CC2010–11Shape10FriesConnexnSF.pdf.*

Rahmandad, H., & Sabounchi, N. (2011, January). *Building and estimating a dynamic model of weight gain and loss covering both childhood and adulthood.* Paper presented at the System Dynamics Winter Conference, Austin, TX.

Railsback, S. F., & Grimm, V. (2011). *Agent-based and individual-based modeling: A practical introduction.* Princeton, NJ: Princeton University Press.

Randers, J. (1977). *The potential in simulation of macro social processes, or how to be a useful builder of simulation models.* Oslo, Norway: Gruppen for Ressurstudier.

Roberts, E. (1977). Strategies for effective implementation of complex corporate models. *Interfaces, 8*(1), Part 1.

Rohrbaugh, J. W. (1992). Collective challenges and collective accomplishments. In R. P. Bostron, R. T. Watson, & S. T. Kinney (Eds.), *Computer augmented team work: A guided tour* (pp. 299–324). New York: Van Nostrand Reinhold.

Saleh, M., Davidsen, P., & Bayoumi, K. (2005, July). *A comprehensive eigenvalue analysis of system dynamics models.* Paper presented at the 23rd International Conference of the System Dynamics Society, Boston.

Schaefer, D. R., Simpkins, S. D., Vest, A. E., & Price, C. D. (2011). The contribution of extracurricular activities to adolescent friendships: New insights through social network analysis. *Developmental Psychology, 47*(4), 1141–1152.

Schlesingera, M., & Parisi, D. (2001). The agent-based approach: A new direction for computational models of development. *Developmental Review, 21*(1), 121–146.

Scott, J. (2000). *Social network analysis: A handbook.* Thousand Oaks, CA: Sage.

Senge, P. M. (1990). *The fifth discipline: The art and practice of the learning organization.* New York: Doubleday.

Seo, D. L., & Huang, Y. (2012). Systematic review of social network analysis in adolescent cigarette smoking behavior. *Journal of School Health, 82*(1), 21–27.

Skvoritz, J. (2003). Complexity theory and models for social networks. *Complexity, 8*(1), 47–55.

Snijders, T. (1996). Stochastic actor-oriented models for network change. *Journal of Mathematical Sociology, 21*, 149–172.

Snijders, T. (2001). The statistical evaluation of social network dynamics. In M. E. Sobel & M. P. Becker (Eds.), *Sociological methodology* (pp. 361–395). Malden, MA: Basil Blackwell.

Snijders, T. (2005). Models for longitudinal social network data. In P. Carrington, J. Scott, & S. Wasserman (Eds.), *Models and methods in social network analysis* (pp. 215–247). Cambridge, UK: Cambridge University Press.

Snijders, T. A. B., Steglich, C. E. G., Schweinberger, M., & Huisman, M. (2007). *Manual for SIENA version 3*. Groningen, The Netherlands: University of Groningen.

Spencer, J. P., & Perone, S. (2008). Defending qualitative change: The view from dynamical systems theory. *Child Development, 79*(6), 1639–1647.

Sporns, O. (2011). *Networks of the brain*. Cambridge, MA: MIT Press.

Steglich, C., Snijders, T. A. B., & Pearson, M. (2010). Dynamic networks and behavior: Separating selection from influence. *Sociological Methodology, 40*(1), 329–393.

Stenberg, L. (1980). A modeling procedure for public policy. In J. Randers (Ed.), *Elements of the system dynamics method*. Cambridge, MA: MIT Press.

Sterman, J. (1992). Flight simulators for management education. *OR/MS Today, 19*(5), 40.

Sterman, J. (2000). *Business dynamics: Systems thinking and modeling for a complex world*. Boston: Irwin/McGraw-Hill.

Sterman, J. (2006). Learning from evidence in a complex world. *American Journal of Public Health, 96*(3), 505–514.

Thelen, E., & Smith, L. B. (2006). Dynamic systems theories. In W. Damon & R. M. Lerner (Eds.), *Handbook of child psychology* (6th ed., Vol. 1, pp. 258–312). Hoboken, NJ: Wiley.

Thomas, A. (2012). Three strategies to prevent unintended pregnancies. *Journal of Policy Analysis and Management, 31*(2), 280–311.

Urban, J. B., Osgood, N., & Mabry, P. L. (2011). Developmental systems science: Exploring the application of systems science methods to developmental science questions. *Research in Human Development, 8*(1), 1–25.

Valente, T. W., Fujimoto, K., Chou, C. P., & Spruijt-Metz, D. (2009). Adolescent affiliations and adiposity: A social network analysis of friendships and obesity. *Journal of Adolescent Health, 45*(2), 202–204.

Valente, T. W., Unger, J. B., Ritt-Olson, A., Cen, S. Y., & Anderson Johnson, C. (2006). The interaction of curriculum type and implementation method on 1-year smoking outcomes in a school-based prevention program. *Health Education Research, 21*(3), 315–324.

van der Maas, H. L. J., Dolan, C. V., Grasman, R. P. P. P., Wicherts, J. M., Huizenga, H. M., & Raijmakers, M. E. J. (2006). A dynamical model of general intelligence: The positive manifold of intelligence by mutualism. *Psychological Review, 113*(4), 842–861.

van Geert, P. (2011). The contribution of complex dynamic systems to development. *Child Development Perspectives, 5*(4), 273–278.

Veenstra, R., & Steglich, C. (2012). Actor-based model for network and behavior dynamics. In B. Laursen, T. D. Little, & N. A. Card (Eds.), *Handbook of developmental research methods* (pp. 598–618). New York: Guilford Press.

Vennix, J. A. M. (1996). *Group model-building: Facilitating team learning using system dynamics*. Chichester, UK: Wiley.

Vennix, J. A. M., Scheper, W., & Willems, R. (1993, July). *Group model-building: What does the client think of it?* Paper presented at the International System Dynamics Conference, Cancun, Mexico.

Vickers, D. M., & Osgood, N. (2007). A unified framework of immunological and epidemiological dynamics for the spread of viral infections in a simple network-based population. *Theoretical Biology and Medical Modelling, 4*, 49.

Vickers, D. M., Zhang, Q., & Osgood, N. D. (2009). Immunobiological outcomes of repeated chlamydial infection from two models of within-host population dynamics. *PLoS One, 4*(9), e6886.

Wasserman, S., & Faust, K. (1994). *Social network analysis.* Cambridge, UK: Cambridge University Press.

Wasserman, S., & Faust, K. (2008). *Social network analysis: Methods and applications.* Cambridge, UK: Cambridge University Press.

Weil, H. B. (1980). The evolution of an approach for achieving implemented results from system dynamics projects. In J. Randers (Ed.), *Elements of the system dynamics method* (pp. 271–291). Cambridge, MA: MIT Press.

Wellman, B. (1926). The school child's choice of companions. *Journal of Educational Research, 14*(715), 126–132.

Winch, G. W. (1993). Consensus building in the planning process: Benefits from a "hard" modeling approach. *System Dynamics Review, 9*(3), 287–300.

Zhang, Q., & Osgood, N. (2010, December). *Summary function elasticity analysis for an individual-based system dynamics model.* Paper presented at the 2010 Winter Simulation Conference, Baltimore.

PART III

EPIGENETIC DEVELOPMENT AND EVOLUTION

CHAPTER 5

How Development Directs Evolution

Mae-Wan Ho

Almost 35 years ago, Ho and Saunders (1979) proposed the then outrageous idea that the intrinsic dynamics of developmental processes are the source of nonrandom variations that *directs* evolutionary change in the face of new environmental challenges; and the resulting evolutionary novelties are reinforced in successive generations through epigenetic mechanisms, *independently of natural selection*. Our proposal has held up well against subsequent research findings and is all the more relevant in view of the numerous molecular mechanisms discovered in epigenetic inheritance (Ho, 2009c, 2009d) that could transmit developmental novelties to subsequent generations.

We have demonstrated how the nonlinear dynamics of living processes predict the major features of macroevolution such as "punctuated equilibria" (long period of stasis interrupted by abrupt changes); large changes from small critical disturbances, and discontinuous changes from continuously varying parameters; and why the macroevolution of form and function is decoupled from the microevolution of gene sequences. We showed that the same (nonrandom) developmental changes are repeatedly produced by specific environmental stimuli. Furthermore, we demonstrated how general mathematical models can account for all the developmental transformations experimentally produced, which can make strong evolutionary predictions and offer a natural taxonomy based on the predicted transformations. However, neither the epigenetic mechanisms nor the dynamics of developmental processes are taken into account in the recent studies on evolution and development.

The totality of research findings gives no support to the neo-Darwinian theory of evolution by the natural selection of random genetic mutations, nor to any theory ascribing putative differences in human attributes predominantly to genes. The overwhelming determinants of health and behavior are social and environmental. Heredity is distributed over the seamless web of nested organism–environment interrelationships extending from the social and ecological to the genetic and epigenetic. Consequently, there is no separation between development and evolution, and the organism actively participates in shaping its own development as well as the evolutionary future of the entire ecological community of which it is part.

"Epigenetic" Then and Now

The term *epigenetic*, as used today in *epigenetic inheritance*, refers to effects that do not involve DNA base sequence changes, but only the chemical modifications of DNA or histone proteins in chromatin (complex of DNA and protein that make up chromosomes in the nucleus of cells) that alter gene expression states. Epigenetic inheritance has been defined (Bird, 2007, p. 398) as "the structural adaptation of chromosomal regions so as to register signal or perpetuate altered activity states." But such a definition is rapidly becoming obsolete (Ho, 2009a, 2009b, 2009c, 2009d, 2009e, 2009f, 2009g). In reality, epigenetic modifications encompass a great variety of mechanisms. They act during and after transcription, and at translation of genetic messages; they can even rewrite genomic DNA (see Ho, 2009c). Hence the distinction between genetic and epigenetic is increasingly blurred.

Epigenetic, as originally used, was derived from *epigenesis*, the theory that organisms are not *preformed* in the germ cells, but come into being through a process of development in which the environment plays a formative role. Most evolutionists have used *epigenetic* to mean hereditary influences arising from environmental effects in the course of development.

Evolution: Lamarck versus Darwin

Evolution refers to the natural (as opposed to supernatural) origin and transformation of organisms on earth throughout geological history to the present day. The first comprehensive *general* theory of evolution—that evolution has occurred—was proposed by Jean Baptiste de Lamarck in his book published more than 200 years ago (1809; see Barthelemy-Madaule, 1982; Burkhardt, 1977; Ho, 1983). It was a *uniformitarian* theory in that causes proposed to be operating in the past are seen as the same as those that can be observed at present. The theory postulated the spontaneous generation of the living from the nonliving and unlimited transformation over time, which gave rise to whole kingdoms of organisms beginning from a single origin of life. In addition, Lamarck proposed special mechanisms whereby new species could evolve through changes in how the organism relates to its environment in *pursuing its basic needs*—changes that produce new characteristics that become inherited after many generations. These special mechanisms are use and disuse—*use* enhances and reinforces the development of the organs or tissues, whereas *disuse* results in atrophy—and the "inheritance of acquired characters"—the transmission to subsequent generations of the tendency to develop certain new characteristics that the organism has acquired in its own development.

Thus, Lamarck was also responsible for the first epigenetic theory of evolution, in which development is seen as playing a key role in initiating the evolutionary change, and specific epigenetic mechanisms transmit the change and reinforce it in subsequent generations (see Ho, 1983, 1984a, 1984b).

Darwin's (1859) special theory of evolution by natural selection states that, given the organisms' capability to reproduce more of their numbers than the environment can support (and there are heritable variations), then, within a population, individuals with the more favorable variations would survive to reproduce their kind at the expense of those with less favorable variations. The ensuing competition and "struggle for life" results in the "survival of the fittest," so the species will become better adapted to its environment, and if the environment changes in time, there will be a gradual but definite "transmutation" of the species. Thus, nature effectively "selects" the fittest in the same way that artificial selection by plant and animal breeders ensures that the best or the most desirable characters are bred and preserved. In both cases, new varieties are created after some generations.

In addition to natural selection, Darwin invoked the effects of use and disuse and the inheritance of acquired characters in the transmutation of species. However, those Lamarckian ideas do not fit into the theory of natural selection, and Darwin's followers all regard the lack of a theory of heredity and variation as the weakest link in the argument for natural selection (Ho, 1986).

In my book *Genetic Engineering Dream or Nightmare?: The Brave New World of Bad Science and Big Business* (Ho, 1998), I have described in detail how Darwin's followers created the "neo-Darwinian synthesis" by expurgating Darwin's Lamarckian tendencies, including his theory of pangenesis. Darwin's theory of pangenesis had actually received a great deal of support (see Ho, 2009b). The rediscovery of Mendel at the turn of the last century provided evidence that particulate genes controlling the characteristics of organisms are passed on unchanged, except for rare random mutations. This process fits in perfectly with August Weismann's (1834–1914) discovery of the material basis of heredity as the "germplasm" in germ cells that becomes separate from the rest of the animal's body early in development to ensure its protection from environmental influences. Development is therefore irrelevant to evolution. We now know that Weissman's theory is wrong and that there are numerous exceptions to Mendelian inheritance. Nevertheless, Darwinism was promptly reinterpreted according to the gene theory in the neo-Darwinian synthesis from the 1930s up to the 1950s and 1960s.

As the result of the neo-Darwinian synthesis, evolution is seen as occurring strictly by the natural selection of *random* gene mutations, or changes in base sequences of DNA; those that happen to increase reproductive fitness are selected at the expense of the others that do not.

Evolution, Development, and Heredity

The theories of evolution, development, and heredity are closely intertwined. Just as evolutionists needed a theory of heredity, so plant breeders in the 18th century who inspired Mendel's discovery of genetics were motivated by the question as to whether new species could evolve from existing ones. In accounting for change or transformation, it is also necessary to locate where the constancy or stability that constitutes heredity resides. In order to explain the evolution of form and function, development (epigenesis) is central, as Lamarck clearly grasped. In contrast, Darwin and neo-Darwinists see new variations arising at *random* in the sense that they bear no direct relationship to the environment; those that happen to be adaptive are selected, whereas the rest are eliminated. The theory of natural selection is essentially preformist, and development is seen as playing little or no role in determining evolutionary change (Ho, 1984a, 1984b, 1987).

A number of different epigenetic theories of evolution has been proposed since Lamarck, some predating the neo-Darwinian synthesis. A common starting point for all epigenetic theories is the developmental flexibility of all organisms. In particular, it has been observed that artificially induced developmental modifications often resemble (*phenocopy*) those existing naturally in related geographical races or species and that appear to be genetically determined. Thus, it seems reasonable to assume that evolutionary novelties first arose as developmental modifications, which somehow became stably inherited (or not, as the case may be) in subsequent generations.

Epigenetic Reorganization Initiates Evolutionary Change

Early proponents of epigenetic theory included James Mark Baldwin (1896), who suggested that modifications arising in organisms developing in a new environment produce "organic selection" forces internal to the organisms, which stabilize the modification in subsequent generations. Another notable figure, Richard Goldschmidt (1940), proposed that evolutionary novelties arise through *macromutations* that produce "hopeful monsters" that can initiate new species. In his defense, he pointed out that monsters could be hopeful because of the inherent *organization* of the biological system that tends to "make sense" of the mutation. Following Goldschmidt, Søren Løvtrup (1974) advocated a similar theory of macromutations for the origin of major taxonomic groups of organisms, such as phyla.

But random mutations—changes in the DNA—that generate hopeful monsters must be hopelessly rare, and to make things worse, major taxonomic groups tend to appear suddenly in clusters, as "adaptive radiations" (Gould & Eldredge, 1972), rather than isolated at different geological times.

The extraordinarily rich fossil finds of the Cambrian "explosion" responsible for most of the major animal phyla are a prime example of evolution occurring in bursts of "adaptive radiation" followed by relatively long periods of stasis, or "punctuated equilibria" (Gould & Eldredge, 1972). Furthermore, evolution does seem to proceed top-down, from phyla to subphyla, classes, orders, and so on (Valentine, 2004), rather than the converse, as predicted by Darwinian and neo-Darwinian natural selection of small random mutations, and crucially, all the evidence indicates that macroevolution is decoupled from molecular or microevolution (more on this later).

These considerations suggest that adaptive radiations involve major novelties arising from *epigenetic reorganization* provoked by large environmental changes or changes in the organisms' way of life, which also seem to coincide with adaptive radiations. For example, oxygen is very important for the evolution of complex organisms, and the Cambrian explosion between 1 billion and 0.5 billion years ago is believed to have been triggered by the rapid increase in atmospheric oxygen levels from a low of about 15% to the current level of about 20% (see Ho, 2009i).

"Evo-Devo" Still Blinded by a "Genetic Program" of Development

In a sense, there is nothing new about the current revival of "evo-devo" (Blumberg, 2009; Brakefield, 2006; Carroll, 2005; Coyne, 2009; Gilbert, 2003). It is still dominated by the idea, going back at least 20 years, that genes control development in a "genetic program" of gene regulation and interaction (Coyne, 2009), and that large evolutionary changes in body pattern are the result of changes in gene regulation due to natural selection. There is still no recognition that the *patterns* themselves and the biological *forms* need to be explained in their own right, independently of whether natural selection operates or not, and independently of the action of specific genes (Ho, 1986; Ho & Saunders, 1979; Saunders, 1984; Webster & Goodwin, 1982). Not surprisingly, there is still little or no recognition that epigenetic and nongenetic environmental influences can give rise to large alterations in form and function.

In a brilliant critique of the genetic determinist approach to behavior, Gottlieb (1998) deconstructed the idea that genes determine body pattern by pointing to the very different expression patterns of the same *Hox* genes in the fruit fly, the centipede, and the Onychophora. *Hox* (homeotic) genes are supposed to control segmental patterning during development; instead, the same genes appear to be simply responding to different patterning processes in the different animals. There is decidedly no homology of genes that corresponds to the homology of biological structures. This same theme emerged in a comprehensive review of segmentation in arthropods by Peel, Chipman, and Akam (2005), which showed that different groups have distinct modes of segmentation and divergent genetic mechanisms.

In a study of left–right asymmetry in development, Palmer (2004) analyzed the genes determining vertebrate heart asymmetry, known as *the nodal signaling cascade*. Like the molecular mechanisms that define the anteroposterior and dorsoventral axes, the nodal signaling cascade includes a "curious mix" of conserved and divergent elements, with nonconserved elements greatly outnumbering the conserved elements. He stated: "Above all, one conclusion is inescapable: The molecular pathway underlying the binary asymmetry switch that yields a left-sided heart, an unambiguously homologous state in all vertebrates, includes more divergent than conserved elements. Developmental system drift may therefore be the rule rather than the exception as development evolves" (p. 828). The idea that natural selection of developmental genes accounts for the evolution of form is obsolete; it is time to move on.

Some researchers now despair of trying to explain pattern formation with complicated computational networks of genes that pass for "systems biology." Kondo and Miura (2010) stated that "the behaviour of such systems often defies immediate or intuitive understanding, [and] it becomes almost impossible to make a meaningful prediction" (p. 1616). For that reason, they are turning back to mathematical reaction–diffusion models (more on this later).

Waddington's Theory of Canalization and Genetic Assimilation

The most influential figure among the "epigenetic evolutionists" was Conrad H. Waddington (1905–1975), who attempted to accommodate "pseudo-Lamarckian" phenomena within neo-Darwinism in his theory of genetic assimilation. Like all Darwinian and neo-Darwinian evolutionists, he wanted to explain the origin of *adaptive* characters, that is, characters that seem to fit the functions they serve.

Waddington (1957) conceptualized the flexibility and plasticity of development, as well as the capacity of development to regulate against disturbances, using his famous term *epigenetic landscape*, a general metaphor for the nonlinear dynamics of the developmental process (Saunders, 1990). The developmental paths of tissues and cells are constrained or *canalized* to "flow" along certain valleys due to the "pull" or force exerted on the landscape by the various gene products that define the fluid topography of the landscape (see Figure 6.1, p. 156). Thus, certain paths along valley floors will branch off from one another to be separated by hills (thresholds) so that different developmental results (alternative attractors) can be reached from the same starting point. However, some branches may rejoin further on, so that different paths will nevertheless lead to the same developmental result. Genetic or environmental disturbances tend to "push" development from its normal pathway across the threshold to another pathway. Alternatively, other valleys (developmental pathways) or hills (thresholds) may be formed due to changes in the topography of the epigenetic landscape itself.

The significance of the conceptual epigenetic landscape is that its topography is determined by *all* of the genes whose actions are inextricably interlinked, and is not immediately dependent on specific alleles of particular genes (Ho & Saunders, 1979). This model is in accord with what we know about metabolism and the epigenetic system, particularly as revealed by the new genetics (more on this later). It also effectively decouples the evolution of the organism, of form and function, from alleles of specific genes and explains the notable lack of correlation between morphological and genetic differences between species (Lowenstein, 1986).

Waddington proposed that a new phenotype arises when the environment changes so that development proceeds to a new pathway in the epigenetic landscape, or else a remodeling of the epigenetic landscape itself takes place (both of which are possible from what we now know about epigenetic processes at the molecular level). Thereafter, the new phenotype becomes reinforced or "canalized" through natural selection for modifier genes so that a more or less uniform phenotype results from a range of environmental stimuli, and later, the phenotype is "genetically assimilated," such that it occurs in the absence of the original environmental stimulus. Waddington (1953) and others carried out experiments showing that artificial selection for the bithorax phenocopy in *Drosophila*, induced by ether exposure during early embryogenesis, resulted in canalization and genetic assimilation.

Ho and Saunders's Epigenetic Theory of Evolution

The first distinctive feature of our epigenetic theory of evolution (Ho & Saunders, 1979, 1982, 1984) is that neo-Darwinian natural selection plays little or no role, based on evidence suggesting, on the one hand, that most genetic changes are irrelevant to the evolution of organisms, and on the other, that a relative *lack* of natural selection may be the prerequisite for major evolutionary change.

The second feature is that the intrinsic dynamics of the epigenetic system—developmental dynamics—is determined not so much by gene interactions as by *physical and chemical forces* of nonlinear complex systems in general, which are amenable to mathematical description (Saunders, 1984, 1992a), especially in terms of catastrophe theory (Saunders, 1980, and Chapter 6, this volume). (This hypothesis has recently been generalized to stochastic systems by Wagenmakers, Molenaar, Grasman, Hartelman, and van der Maas, 2005.) That is why, contrary to the neo-Darwinian view, variations in the phenotype that arise during development in response to new environments are *nonrandom* and *repeatable*.

We proposed, therefore, that the intrinsic dynamical structure of the epigenetic system is the source of nonrandom variations that *direct* evolutionary change in the face of new environmental challenges. These evolutionary novelties are reinforced (canalized) in subsequent generations through cytoplasmic/epigenetic mechanisms, *independently of natural selection*.

When a population of organisms experiences a new environment *or adopts a new behavior*, the following sequence of events is envisaged:

1. A novel response arises during development in *a large proportion, if not all*, of the organisms in a population experiencing a new environment, due to the intrinsic dynamics of the epigenetic system. In the case of a new behavior initiated by a single individual in a social group, the behavior can also spread quite rapidly. For example, Kawai (1962) found that the new habit of washing sweet potatoes in the sea initiated by a young female had spread to the entire troop of wild monkeys on Koshima Island in Japan within 9 years. Doubtlessly, this behavior may also have triggered developmental changes in the monkey's brain.
2. This response, which has been demonstrated in experiments in our laboratory (more on this below), is canalized in successive generations through epigenetic mechanisms *independent of natural selection*.
3. After some generations, the response *may* become genetically assimilated, in

that it arises even in the absence of the stimulus. As in Waddington's epigenetic landscape, this assimilation could entail a change in the topography to bias the original branch point in favor of the new pathway, so that the new phenotype will persist in the absence of the environmental stimulus. Random genetic mutations could also be involved.

Corroborations of Ho and Saunders's Epigenetic Theory

Since we proposed our theory, we have obtained important empirical and theoretical corroboration. We questioned Waddington's assumption that selection of (modifier) genes is necessary for canalization and genetic assimilation, and in a series of experiments, Ho, Tucker, Keeley, and Saunders (1983) demonstrated that canalization occurred in the *absence* of selection *for* the new character. We showed that successive generations of ether treatment during early embryonic development in *Drosophila* increased the frequency of the bithorax phenocopy in the adults, without selecting *for* the phenocopy. If anything, the phenocopy was almost certainly selected *against*, as it obviously interfered with flight and other normal functions. We had identified a case of "epigenetic inheritance" of a maladaptive character, consistent with recent findings in "epigenetic toxicology" in which the toxic effects of exposure to environmental pollutants were transmitted to grandchildren (Ho, 2009f).

At least one study of the fossil record (Palmer, 2004) provided evidence that left–right asymmetry in animals and plants may have originated as phenotypic novelties that became genetically assimilated subsequently.

We stipulated that genetic assimilation is not a necessary part of the response to change (Ho & Saunders, 1979), as it would preserve the important property of developmental flexibility or "adaptability." In retrospect, this reasoning has proved correct. We now know that maternal behavior, long regarded as genetically inherited and instinctive, is actually associated with epigenetic gene markings made during early infancy that are erased at every generation, yet perpetuated indefinitely from mother to daughter, parent to offspring (Ho, 2009a, 2011b). More than that, it is an epigenetic system that enables foster parents to literally influence their adopted child biologically.

An important motivation for focusing on development for evolutionary change is that developmental changes are far from random or arbitrary (Ho & Saunders, 1979, 1982, 1984; Saunders, 1984; Webster & Goodwin, 1982). Developmental changes are determined by dynamic processes, independently of the action of specific genes.

Physicochemical Forces and Flows in Growth and Form

Patterns are generated everywhere in the physical world where no genes are involved, and many of the patterns closely resemble those found in the living world. It is the dynamics of physical and chemical forces and flows that generate patterns and forms, as Scottish biologist and mathematician D'Arcy Thompson (1860–1948) so beautifully argued in his classic book, *On Growth and Form* (Thompson, 1917/1961). Closer to our time, Alan Turing (1912–1954), English mathematician, logician, code breaker, and computer pioneer, is also well known for his work on morphogenesis (Turing, 1952). Turing's reaction-diffusion model shows, for the first time, how patterns can arise *spontaneously* in an initially homogeneous domain—precisely the problem of how patterns can form in a featureless egg in development (Turing, 1952; Saunders, 1992a, 1993). The Turing model inspired much work on pattern formation in biological systems before it got lost in the proliferating thicket of genes that "control pattern formation."

The complex nonlinear dynamics of the developmental process have been explored mathematically in greater detail (Saunders, 1984, 1989, 1990, 1992a), and their evolutionary consequences made explicit. For example, the nonlinear dynamics of development account for "punctuated equilibria" (Eldredge & Gould, 1972). They also show how large organized changes can occur in response to a relatively small disturbance, and how continuously varying environmental parameters can nevertheless precipitate discontinuous phenotypic change (see Saunders, 1990, and Chapter 6, this volume).

The physical and chemical forces that organize living systems were the subject of my book *The Rainbow and the Worm: The Physics of Organisms* (Ho, 1993, 1998, 2008a, 2008c), now in its third enlarged edition. The book presents evidence that cells and organisms are liquid crystalline, with water the most important constituent of the liquid crystalline matrix. I pointed out that electrical polarities determine the alignment of the liquid crystals and hence the major body axes. Furthermore, electrodynamical forces acting on liquid crystal mesophases may play a key role in pattern formation and morphogenesis. As consistent with this hypothesis, we demonstrated dramatic effects with brief exposures of early *Drosophila* embryos to very weak static magnetic fields; the segmental body pattern of the larva that emerged 24 hours later were transformed into helices (Ho et al., 1992).

Recently, there has been a revival of interest in electrodynamical processes in development, as changes in membrane potential and the establishment of ionic currents and endogenous electric fields appear to determine polarities long before the relevant genes are expressed (Ho, 2011a). This and other evidence suggest that electrodynamical processes are involved in pattern formation via the liquid crystalline cortex of the egg and epithelial cells in regeneration. The connection to genes is hence indirect and nonlinear.

In contrast, developmental geneticists generally assume that diffusion gradients of special "morphogens" determine body pattern by providing "positional information" for particular genes to "interpret." For example, in *Drosophila*, on which the most complete genetic analysis of development has been carried out (Nusslein-Volhard, 2006), the maternal gene product bicoid is identified as the morphogen; its anteroposterior gradient serves to initiate the cascade of "combinatorial regulation" of genes that eventually gives rise to the complete body pattern. The difficulty is that very few molecules, if any, diffuse freely in the liquid crystalline matrix, and the bicoid protein is no exception. If anything, it appears that a gradient of transcription/translation and degradation is actively maintained in the embryo during several cycles of synchronous nuclear divisions (Gibson, 2007; Gregor et al., 2007), by an as-yet unknown patterning process, which may well be electrodynamical.

Status of Ho and Saunders's Epigenetic Theory

To summarize the present status of Ho–Saunders's epigenetic theory, I want to stress again that natural selection plays little or no role in evolution, especially in the evolution of major novelties or macroevolution (except in the negative sense of eliminating deleterious mutations with large effects) for three reasons. First, the epigenetic (developmental and nongenetic) novelties produced in response to new environments are common to most, if not all, individuals in a population, and would swamp out residual effects due to genetic variation. Second, the fluidity of the genome—the constant interaction between genome and environment, the epigenetic markings of genes, and the blurring between genetic and epigenetic—makes clear that organism and environment are inseparable; hence there can be no selection of any static, preformed variant that is independent, or random, with respect to the selective environment. Third, the physical and chemical forces and flows that *generate* biological patterns and forms are independent of natural selection and require their own explanations (see also Ho, 2011a).

Our theory has been corroborated since it was first proposed, as follows:

- Canalization of bithorax phenocopy independent of natural selection (Ho et al., 1983) experimentally demonstrated.
- Epigenetic toxicology shows that harmful effects are transmitted across generations (reviewed in Ho, 2009f).
- There is evidence for genetic assimilation in the evolution of left–right asymmetry and other morphologies (Palmer, 2004).
- Genetic assimilation need not occur; for example, stable epigenetic inheritance of maternal behavior is nevertheless reversed in cross-adoption studies in rats (reviewed in Ho, 2009a, 2011b).
- Complex nonlinear dynamics do exhibit properties of the epigenetic landscape and account for major features of macroevolution (Saunders, 1990, Chapter 6, this volume).
- Electrodynamical forces are involved in pattern formation, with only nonlinear connections to genes (Ho, 1993, 1998, 2008b, 2011a; Ho et al., 1992).

Next I show why the dynamics that generate patterns and forms are much more than weak "developmental constraints" to natural selection, and then I address the "neutral mutation hypothesis," the proposal that most, if not all, DNA base changes during evolution are due to random genetic drift decoupled from the evolution of organisms.

Rational Taxonomy Based on the Generative Dynamics of Biological Form

The dynamics of developmental (epigenetic) processes, being amenable to mathematical description, provide a powerful perspective for understanding the development and the evolution of form. That perspective is the basis of "structuralism in biology" (Goodwin, Webster, & Sibatani, 1989; Webster & Goodwin, 1982) or, more accurately in our view, "process structuralism" (Ho, 1984b, 1988b; Ho & Saunders, 1984; Saunders, 1984, 1989, 1992b, 1993).

The developmental dynamics define a set of possible transformations that is highly constrained, so that particular transformations may be *predictably* linked to specific environmental stimuli. The fundamental importance of development for evolution is that evolutionary transformations can ultimately be understood in terms of developmental transformations that can be empirically investigated, and which in turn provide us with the criteria for a rational taxonomy, a natural system of classification based on the generative dynamics of form. I describe two examples: the segmentation defects in *Drosophila* larva produced by exposing early embryos to ether vapor, and phyllotaxis, the arrangement of leaves around the stem.

The segmentation pattern of the first instar *Drosophila* larva is determined during early embryogenesis. In the course of our studies on the bithorax phenocopy, we discovered that brief exposures to ether vapor also produced characteristic defects in the segmental pattern that reflected a dynamic process arrested at different stages (Ho, Matheson, Saunders, Goodwin, & Smallcombe, 1987). These defects phenocopy *all* the major genetic mutants identified, and the most general model of successive bifurcation could produce all the observed defects, giving a rational taxonomy of both the observed and yet to be observed forms (Ho, 1990; Ho & Saunders, 1993). This rational taxonomy, based on generative dynamics, differs from one based on genealogy or similarity of DNA, and interestingly, also differs significantly from one based on cladistic analysis (Ho, 1990). Saunders and Ho (1995) subsequently produced a mathematical model of reliable segmentation based on successive bifurcation.

Figure 5.1 (Ho, 1990; Ho & Saunders, 1993) is a transformational "tree" of the range of segmental patterns obtained *during development*. The main sequence, going up the trunk of the tree, is the normal transformational pathway, which progressively divides up the body into domains, ending up with 16 body segments of the normal larva. All the rest (with solid outlines) are transformations in which the process of dividing up the body has been arrested at different positions in the body. The patterns with dashed outlines are hypothetical forms, not yet observed, connecting actual transformations.

This transformational tree reveals how different forms are related to one another; how superficially similar forms are far apart on the tree, whereas forms that look most different are neighbors. It is the most parsimonious tree relating all the forms.

More importantly, the ontogenetic transformation tree predicts the possible forms that can be obtained in evolution (phylogeny), most likely by going up the sequence

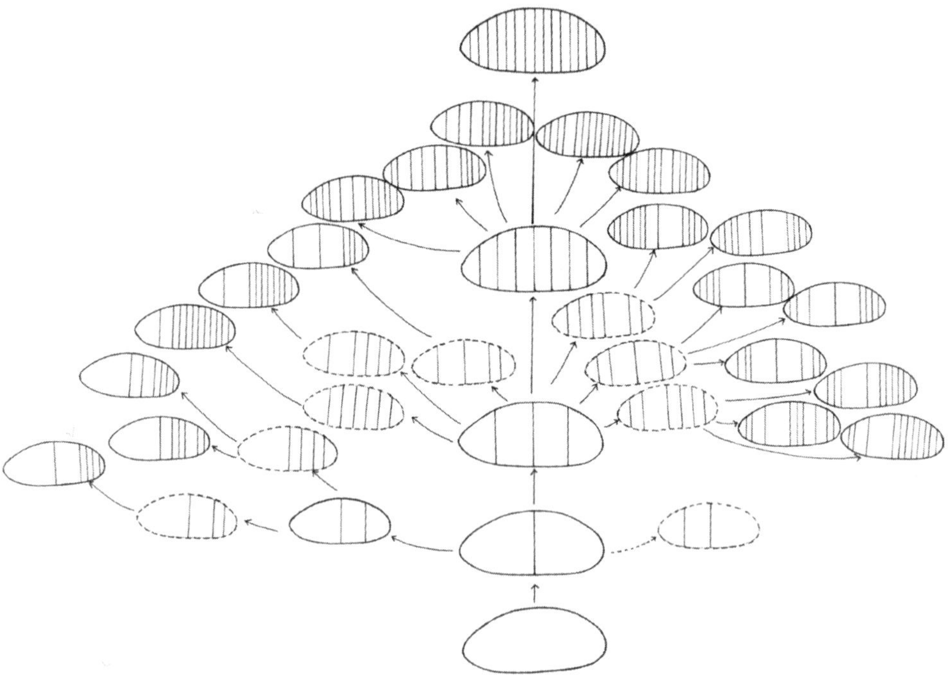

FIGURE 5.1. Transformation tree of body patterns in fruit fly larvae based on a model of successive bifurcation.

of successive bifurcations, but occasional reversals to simpler forms could also take place. This is why phylogeny appears to recapitulate ontogeny (Gould, 1977), though actually *it does not*; ontogeny and phylogeny are simply related through the dynamics of the generic processes generating form.

A *natural* system of classification results from the tree. The 24 actual forms or species are classified hierarchically into one family with two orders, the first order containing three genera, and the second order, eight genera. The forms not yet found (depicted in dashed lines in Figure 5.1) would also fit neatly in the natural system of classification should they be discovered in future. There are 676 possible forms according to the dynamic model of successive bifurcation. If all the body segments were free to vary independently, the number of possible forms would have been 2^{16}, or more than 60,000. This demonstrates how highly the generative dynamics can constrain the possible forms, and why incidental parallelisms are rife in evolution (Ho, 1984a; Ho & Saunders, 1982).

Obviously, the scheme proposed is an oversimplification, which is why eight hypothetical intermediates (represented in dashed outlines) were not actually observed. The actual process itself may well predict many less possible forms.

In the second example, we produced a transformation tree for all possible ways leaves are arranged around the stem in plants (Figure 5.2; Ho & Saunders, 1994), based on the generic and robust dynamics that generate the patterns, discovered by French mathematical physicists Douady and Couder (1992). The discovery caused quite a stir in France, as leaf arrangement, or *phyllotaxis*, has been a long-standing problem in biology, ever since Alan Turing drew attention to how the spiral patterns of leaves around the stem conform to the Fibonacci sequence (Saunders, 1984, 1989, 1992a, 1993). Many neo-Darwinian "just-so stories" have been invented over the years to account for different leaf arrangements in terms of "selective advantage," all of which have been proven irrelevant in one stroke. The power of dynamics—the syntax of form—is that it predicts the set of possible transformations, *excluding all others*. This syntax also tells us how the possible forms are related by transformation (Ho, 2008c).

It is not known if all the possible forms in Figure 5.2 actually exist in nature. The main Fibonacci sequence with a divergence angle of 137.5° is in the middle row (marked with *), where the successive number increasingly approaches the golden ratio. At the bottom is an alternative Fibonacci sequence with a divergence angle of 99.5° (marked with **). Like the transformation tree in Figure 5.1, it makes very definite predictions concerning neighboring transformations. Thus, parastichies 8,11 and 9,11 (secondary spirals; the numbers indicate spirals to the right and left, respectively, starting from the center of the flower or top of the cone), despite their apparent similarity, are quite far apart on the tree, whereas the neighboring parastichies 8,13 and 13,21 appear superficially very different. As the tree is also an ontogenetic tree, it predicts that plants such as the Canadian pine (*Pinus resinosia*), with parastichies 8,13 in the cone, goes through all of the main sequence in development. We do not know if that is true, but we did find that the leaf shoot bearing the cone has 3,5 parastichies.

For the same reasons, we would predict that the decussate arrangement 2,2 is the earliest divergence from the main Fibonacci sequence, followed by the alternative Fibonacci sequence beginning with 1,3. Phylogenetic transformations are strictly

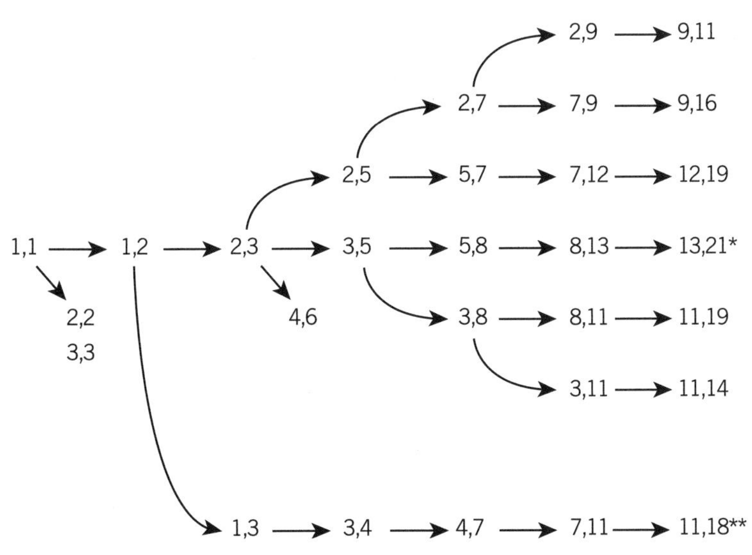

FIGURE 5.2. The transformation tree of possible phyllotaxis patterns.

predicted. For example, one would not expect an ancestor of a plant with parastichies 8,13 to have had parastichies 7,12, or even 2,5, but most likely, 5,8.

The mathematics of spiral phyllotaxis turns up everywhere in nature. The 2011 Nobel prize for chemistry went to Dan Schechtman for the discovery of crystal structures with the golden proportion (Fernholm, 2011).

The dynamics of the processes are subject to contingent "complexification" (or simplification) in the course of evolution, by virtue of the lived experience of the organisms themselves. Nevertheless, these dynamics are highly constrained when it comes to pattern formation. It has become clear that directed genetic changes in given environments are just as nonrandom as morphological changes, and hence, possibly subject to comparable systemic constraints (Ho, 1987) (more on this later).

Natural Selection and Molecular Evolution

Molecular evolution, the study of how proteins and nucleic acid sequences evolve in different species, has been dominated by the neutralist–selectionist controversy that continues to the present day.

Motoo Kimura (1924–1994) was best known for his neutral theory of molecular evolution (Kimura, 1968), which proposed that most of the amino acid and base changes in evolution resulted from random genetic drift of neutral mutations—that is, mutations that did not influence the "fitness" of the organisms. In fact, he did not deny that natural selection could be operating, only that it was not reflected in the evolution of molecules. In effect, molecular evolution appears to be decoupled from the evolution of organisms—which, at least, is consistent with all other observations indicating the lack of simple translations between genes and phenotype, and is an independent corroboration of Waddington's (1957) concept of the epigenetic landscape.

The neutral mutation theory was inspired by earlier discovery that when the amino acid or DNA base sequences of genes in different organisms were compared, they diverged in an apparently linear fashion, according to the time at which the organisms shared a common ancestor. This discovery gave rise to the "molecular clock" hypothesis (Margoliash, 1963; Zuckerkandl & Pauling, 1962), which stated that the rate of amino acid or nucleotide substitution is approximately constant per year over evolutionary time and among different species (Lowenstein, 1986). As more data became available, the molecular clock hypothesis ran into trouble. Although there is a correlation between genetic distance and time of divergence, such correlation is not universal and indeed is often violated.

Numerous studies on extant organisms show that mutation rates are far from constant (e.g., Huang, 2009). For example, genetic differences between two subpopulations of medaka fish that had diverged for about 4 million years are three times that as between two primate species, humans and chimpanzees, which are thought to have split 5–7 million years ago. Genetic distances measured on genealogical timescales of less than 1 million years are often an order of magnitude *larger* than those on geological timescales of more than a million years.

TABLE 5.1. Genetic Distance and Estimated Divergence Time

	Percent identity				Divergence time (MYBP)
	PRDM2	BTK	CytC	GCA1A	
H. sapiens versus *D. rerio*	39	61	80	66	450
H. sapiens versus *X. laevis*		55	85	75	360
H. sapiens versus *G. gallus*	71	85	87	81	310
H. sapiens versus *M. musculus*	91	98	91	91	91
F. rubripes versus *D. rerio*	45				420
		71			400
			89		200
				91	91

Note. Adapted from Huang (2009). Reprinted with permission from the author.

To illustrate the paradox, four randomly selected genes in different species were compared for their similarity (i.e., percent identity). All four genes behave as good clocks in macroevolution from fish *(D. rerio*, zebrafish), to frog (*X. laevis*, African clawed toad), to bird (*G. gallus*, red jungle fowl), to mouse (*M. musculus*), and human (*H. sapiens*). However, they gave wildly contradictory timing at lower levels (see Table 5.1). When different species of fish were compared with each other, *F. rubripes* (puffer fish) versus *D. rerio*, divergence time ranged from 91 to 420 MYBP (million years before present).

Epigenetic Complexity versus Genetic Diversity, Macroevolution versus Microevolution

Huang (2009) proposed that an inverse relationship exists between genetic diversity and epigenetic complexity: Multicellular organisms differentiated into tissues and cells are epigenetically complex and can tolerate less genetic variation (germline DNA mutation), whereas single-celled organisms, being epigenetically simple, can tolerate more. Consequently, each level of epigenetic complexity will reach its maximum level of variations. This simple theory explains the major features of evolution, including the paradox of an overestimate of divergence times when some gene sequences in lower taxonomic levels are compared (see Table 5.1).

Humans are undoubtedly the most epigenetically complex species, but in terms of the number of genes, humans have only roughly 1.6 times the genes of a fruit fly and about the same as the mouse or fish. However, the number of certain enzymes responsible for epigenetic gene organization, such as the PRDM subfamily of histone methyltransferases, increases during metazoan evolution from 0 in bacteria, yeasts, and plants, to 2 in worms, 3 in insects, 7 in sea urchins, 15 in fishes, 16 in rodents, and 17 in primates. The core histone genes H2A, H2B, H3, and H4 have been duplicated in humans but not chimpanzees, and the number of genes for microRNA (which play key regulatory functions) correlates well with organism complexity. Complex organisms also show complex gene expression patterns: 94% of human genes have alternative products or alternative splicing compared to only 10% in the nematode *C. elegans*.

According to Huang (2009), organisms undergo epigenetic changes or genetic mutations in a certain range in accordance with their epigenetic complexity. More significantly, epigenetic complexity change is almost, by definition, macroevolution, whereas genetic changes due to mutations causing minor variations in phenotypes that do not affect the epigenetic programs are microevolution. Microevolution is a continuous process of accumulating mutations (see Table 5.2).

Macroevolution from simple to complex organisms is associated with a punctuational increase in epigenetic complexity and in turn a punctuational loss in genetic diversity. From a common ancestor, the genetic distance between two splitting descendants may gradually increase with time until a maximum is reached, and then it remains constant thereafter.

The maximum genetic diversity hypothesis predicts that if time is long enough for genetic distance to reach the maximum, then the genetic distance between two genera of the same family should be similar to that between two families or orders or phyla. That was demonstrated to be the case for a very old group such as fungi; in contrast, the molecular clock hypothesis predicts that the genetic distance between two fungi genera of the same family should be *smaller* than that between families, and still smaller than that between orders, and so on. His hypothesis, Huang (2009) claims, explains top-down evolution, which is also consistent with the epigenetic origin of evolutionary novelties (see earlier) and the decoupling of macroevolution from the microevolution of genetic distance.

Continuity between Epigenetic and Genetic Changes

Huang's (2009) theory does explain a lot and could, in principle, resolve nearly all the major paradoxes in molecular evolution, except perhaps the widely different rates of divergence between different genes within the same organism. More importantly, I

TABLE 5.2. Shi Huang's (2009) Theory of Molecular Evolution

- There is an inverse relationship between genetic diversity and epigenetic complexity.
- Multicellular organisms are epigenetically complex and tolerate less genetic diversity than simple single-celled organisms.
- Each level of epigenetic complexity has its maximum level of genetic diversity.
- From a common ancestor, the genetic distance between two splitting descendants gradually increases with time until a maximum is reached and remains constant thereafter; if time is long enough for genetic distance to reach the maximum, then genetic distance between two genera of the same family would be similar to that between two families or orders or phyla.
- Macroevolution involves epigenetic changes and is decoupled from microevolution of genes.

believe that Huang's hypothesis that epigenetically complex organisms are less tolerant of genetic or germline diversity is incomplete because the level of germline diversity is *actively maintained*. A key feature of epigenetics in complex organisms is that they have become more efficient at generating the sequence diversity required at the precise local somatic level (Ho, 2009i) and, incidentally, also more efficient at reducing it at the germline level through mechanisms such as gene conversion and concerted evolution (Ho, 2004a, 2004b, 2004c, 2004d), all part of the death of the central dogma of molecular biology that has been happening since the 1980s.

Epigenetic processes such as RNA editing, alternative and trans-splicing, exonization, and somatic hypermutation can generate huge sequence diversity wherever and whenever required (Ho, 2009i). Some of those processes, coupled with reverse translation, are powerful mechanisms for generating sequence diversity that can be tested by function within the individual organism and then used to overwrite the germline sequence(s). I have reviewed these mechanisms in some detail elsewhere (Ho, 2009c), including a range of evidence indicating that mutations are far from random, with the organism choosing when and how to mutate or not to mutate at all (Ho, 2004a).

DNA recoding—rewriting genome DNA—appears to be a central feature of both the immune and nervous systems. DNA recoding is involved at the level of establishing neuronal identity and neuronal connectivity during development, learning, and brain regeneration. And it appears that the brain, like the immune system, also changes according to experience. Mattick and Mehler (2008) suggest that the potential recoding of DNA in nerve cells (and similarly in immune cells) might be primarily a mechanism whereby productive or learned changes induced by RNA editing are *rewritten* back to DNA via RNA-directed DNA repair. This effectively fixes the altered genetic message once a particular neural circuitry and epigenetic state have been established (see Ho, 2009i). Steele (2008) has proposed a similar RNA-directed recoding of DNA for the immune system. Unlike Steele, Mattick and Mehler (2008) fall short of proposing that the RNA-templated recoding of the genome and the associated structural and functional adaptations could be transmitted to the next generation. This transmission could be crucial for brain evolution in primates leading up to humans, so that the gains made by successive generations could be accumulated (Ho, 2009i).

If the analogy with the immune system holds, then as suggested by Steele and others, edited RNA messages or their reverse transcribed DNA counterparts could become inherited via sperm (Ho, 2009e; Steele, 1981). "Sperm-mediated gene transfer" is well documented as a process whereby new genetic traits are transmitted to the next generation by the uptake of DNA or RNA by spermatozoa and delivery to the oocytes at fertilization.

Macroevolution therefore involves epigenetic and epigenetically directed genetic changes, and is decoupled from the random microevolutionary accumulation of base sequence changes. These processes (reviewed in greater detail in Ho, 2009c) are part and parcel of the fluid genome (Ho, 2003), a molecular "dance of life" that is necessary for survival (e.g., see Ho, 2008a). In summary, there is continuity between macroevolutionary epigenetic and genetic changes due to fluid genome processes (see Table 5.3).

TABLE 5.3. Continuity of Macroevolutionary Epigenetic and Genetic Changes

- Epigenetically complex organisms are more efficient at generating genetic diversity locally and as required, by directed mutation, exonization, somatic hypermutations, RNA editing, alternative splicing, trans-splicing, etc.
- Local diversity-generating mechanisms feed back to rewrite genomic and germline DNA with functionally tested gene sequences, especially in the brain and the immune system.
- Epigenetic complex organisms actively decrease genomic and germline diversity by gene conversion, concerted evolution, and sperm-mediated gene transfer.
- Macroevolution involves epigenetic and epigenetically directed genetic changes and is decoupled from random microevolutionary accumulation of base sequence changes.

Note. Based on Ho (2003, 2004a, 2004b, 2004c, 2004d, 2009c).

Heredity and Evolution in Light of the New Genetics and Epigenetics

How should we view heredity in light of the new genetics and epigenetics? Where does heredity reside if the genome itself is dynamic and fluid? Clearly, heredity does not reside solely in the DNA of the genome.

Ten years since the announcement of the human genome sequence has brought little progress in understanding life, health, or disease. Herculean efforts to locate the genes responsible for common diseases have yielded next to nothing (Ho, 2010b), not surprisingly at all, given the fluidity of the genome and the associated complexity of epigenetic mechanisms.

It has been clear to some of us since before the Human Genome Project was conceived, and copiously corroborated by the findings since, that *heredity resides in an epigenetic state, a dynamic equilibrium between genetic–epigenetic and other cellular processes*. But heredity does not end at the boundary of cells or organisms. As organisms engage their environments in a web of mutual feedback interrelationships, they transform or maintain their environments, which are also passed on to subsequent generations as home ranges and other cultural artifacts (Ho, 1984a, 1986; Ho & Saunders, 1982). Embedded between organisms and their environment are social habits and traditions, an inseparable part of the entire dynamical complex that gives rise to the stability of the developmental process, and which we recognize as heredity (Ho, 1984b, 1986, 1988a). Heredity is thus distributed over the whole web of organism–environment interrelationships, wherein changes and adjustments are constantly taking place, propagating through all space–time in the maintenance of the whole, and some of these changes may involve genomic DNA. Thus, the fluidity of the genome is a *necessary* part of the dynamic stability, for genes must also be able to change as appropriate to the system *as a whole* (see Figure 5.3).

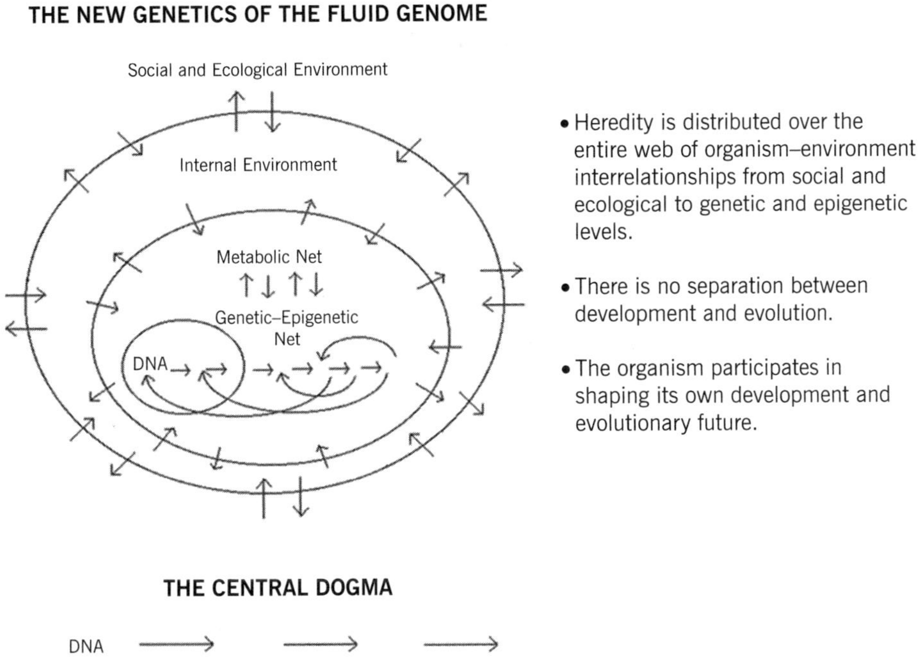

FIGURE 5.3. The new genetics of the fluid genome versus the central dogma.

Conclusions

The epigenetic approach fully reaffirms the fundamentally holistic nature of life and discredits any theory ascribing putative group differences in human attributes to genes (Ho, 2010a). It also gives no justification to *simplistic, mechanistic* ideas about arbitrary effects arising from use and disuse or the inheritance of acquired characteristics. It does not lead to any kind of determinism, environmental or genetic. Organisms are, above all, complex, nonlinear, dynamical systems (Saunders, 1992b) and, as such, they have regions of stability and instability that enable them to maintain homeostasis or to adapt to change, or not, as the case may be. The appearance of novelties and of mass extinctions alike in evolutionary history are but two sides of the same coin; we cannot be complacent about the capacity of organisms to adapt to any and all environmental insults that are perpetrated, the most pressing of which is anthropogenic global warming. The dynamics of the developmental process ultimately hold the key to heredity and evolution, by determining the sorts of changes that can occur and in the resilience to certain perturbations and susceptibility to others. Our knowledge in this crucial area is urgently required.

What are the implications for evolution? Just as interaction and selection cannot be separated, nor can variation (or mutation) and selection, for the "selective" regimen may itself cause specific epigenetic variations or "directed" mutations. The organism experiences its environment in one continuous nested *process*, adjusting and changing, leaving imprints in its epigenetic system, its genome, as well as on the environment, all of which are passed on to subsequent generations. Thus, *there is no separation between development and evolution*. In that way, the organism actively participates in shaping its own development as well as the evolution of its ecological and social community. We do hold the future in our hands; it is precious, be careful.

REFERENCES

Baldwin, J. M. (1896). A new factor in evolution. *American Naturalist, 30*, 441–451, 536–553.

Barthelemy-Madaule, M. (1982). *Lamarck the mythical precursor* (M. H. Shank, Trans.). Cambridge, MA: MIT Press.

Bird, A. (2007). Perceptions of epigenetics. *Nature, 447*, 396–398.

Blumberg, M. S. (2009). *Freaks of nature: What anomalies tell us about development and evolution*. Oxford, UK: Oxford University Press.

Brakefield, P. M. (2005). Bringing evo-devo to life. *PLoS Biology, 3*, 1693–1695.

Brakefield, P. M. (2006). Evo-devo and constraints on selection. *Trends in Ecology and Evolution, 21*, 362–367.

Burkhardt, R. (1977). *The spirit of systems*. Cambridge, MA: Harvard University Press.

Carroll, S. B. (2005). *Endless forms most beautiful: The new science of evo devo*. New York: Norton.

Coyne, J. A. (2009). Evolution's challenge to genetics. *Nature, 457*, 382–383.

Darwin, C. (1859). *On the origin of species by means of natural selection or the preservation of favoured races in the struggle for life*. London: Murray.

de Lamarck, J. B. (1809). *Philosophie zoologique*. Paris.

Douady, S., & Couder, Y. (1992). Phyllotaxis as a physical self-organized growth process. *Physical Review Letters, 68*, 2098–2101.

Eldredge, N., & Gould, S. J. (1972). Punctuated equilibria: An alternative to phyletic gradualism. In T. J. M. Schopf (Ed.), *Models in paleobiology* (pp. 82–115). New York: Freeman.

Fernholm, A. (2011). *The Nobel Prize in Chemistry 2011. Crystals of golden proportions*. Stockholm: Royal Swedish Academy of Sciences. Retrieved from *www.kva.se/Documents/Priser/Nobel/2011/kemi/pop_ke_en_11.pdf*.

Gibson, M. C. (2007). Bicoid by the numbers: Quantifying a morphogen gradient. *Cell, 130*, 14–16.

Gilbert, S. F. (2003). The morphogenesis of evolutionary developmental biology. *International Journal of Developmental Biology, 47*, 467–477.

Goldschmidt, R. B. (1940). *The material basis of heredity*. New Haven, CT: Yale University Press.

Goodwin, B. C., Webster, G., & Sibatani, A. (Eds.). (1989). *Dynamic structures in biology*. Edinburgh, UK: Edinburgh University Press.

Gottlieb, G. (1998). Normally occurring environmental and behavioral influences on gene activity: From central dogma to probabilistic epigenetics. *Psychological Review, 105*, 792–802.

Gould, S. J. (1977). *Ontogeny and phylogeny*. Cambridge, MA: Harvard University Press.

Gould, S. J., & Eldredge, N. (1972). Punctuated equilibria: An alternative to phyletic gradualism.

In T. J. M. Schopf (Ed.), *Models in paleobiology* (pp. 82–115). San Francisco: Freeman, Cooper.
Gregor, T., Wieschaus, E. F., McGregor, A. P., Bialek, W., & Tank, D. W. (2007). Stability and nuclear dynamics of the bicoid morphogen gradient. *Cell, 130*, 141–152.
Ho, M. W. (1983). Lamarck the mythical precursor: A book review. *Paleontological Association Circular, 14*, 10–11.
Ho, M. W. (1984a). Environment and heredity in development and evolution. In M. W. Ho & P. T. Saunders (Eds.), *Beyond neo-Darwinism. An introduction to the new evolutionary paradigm* (pp. 267–290). London: Academic Press.
Ho, M. W. (1984b). Where does biological form come from? *Rivista di Biologia, 77*, 147–179.
Ho, M. W. (1986). Heredity as process. *Rivista di Biologia, 79*, 407–447.
Ho, M. W. (1987). Evolution by process, not by consequence: Implications of the new molelcular genetics on development and evolution. *International Journal of Comparative Psychology, 1*, 3–27.
Ho, M. W. (1988a). Genetic fitness and natural selection: Myth or metaphor. In *Proceedings of the 3rd Schneirla Conference* (pp. 85–111). Mahwah, NJ: Erlbaum.
Ho, M. W. (1988b). How rational can rational morphology be?: A post-Darwinian rational taxonomy based on the structuralism of process. *Rivista di Biologia, 81*, 11–55.
Ho, M. W. (1990). An exercise in rational taxonomy. *Journal of Theoretical Biology, 147*, 43–57.
Ho, M. W. (1998). *Genetic engineering dream or nightmare: The brave new world of bad science and big business.* New York: Macmillan.
Ho, M. W. (2003). *Living with the fluid genome.* London: ISIS/TWN.
Ho, M. W. (2004a). Are ultra-conserved elements indispensable? *Science in Society, 24*, 5.
Ho, M. W. (2004b). How to keep in concert. *Science in Society, 24*, 12–13.
Ho, M. W. (2004c). Subverting the genetic text. *Science in Society, 24*, 6–9.
Ho, M. W. (2004d). To mutate or not to mutate. *Science in Society, 24*, 9–10.
Ho, M. W. (2008a). GM is dangerous and futile. *Science in Society, 40*, 4–8.
Ho, M. W. (2008b). *The rainbow and the worm: The physics of organisms* (2nd ed.). Singapore: World Scientific.
Ho, M. W. (2008c). In search of the sublime, significant form in science and art. *Science in Society, 39*, 4–11.
Ho, M. W. (2009a). Caring mothers strike fatal blow against genetic determinism. *Science in Society, 41*, 6–9.
Ho, M. W. (2009b). Darwin's pangenesis: The hidden history of genetics and the dangers of GMOs. *Science in Society, 42*, 42–45.
Ho, M. W. (2009c). Development and evolution revisited. In K. E. Hood, C. T. Halpern, G. Greenberg, & R. M. Lerner (Eds.), *Handbook of developmental science, behavior, and genetics* (pp. 61–109). New York: Blackwell.
Ho, M. W. (2009d). Epigenetic inheritance: "What genes remember." *Science in Society, 41*, 4–5.
Ho, M. W. (2009e). Epigenetic inheritance through sperm cells: The Lamarckian dimension in evolution. *Science in Society, 42*, 40–42.
Ho, M. W. (2009f). Epigenetic toxicology. *Science in Society, 41*, 13–15.
Ho, M. W. (2009g). From genomics to epigenomics. *Science in Society, 41*, 10–12.
Ho, M. W. (2009h). Living with oxygen. *Science in Society, 43*, 9–12.
Ho, M. W. (2009i). Rewriting the genetic text in human brain development and evolution. *Science in Society, 41*, 16–19.

Ho, M. W. (2010a). Celebrating the uses of human genome diversity and dissecting the controversies. *Science in Society, 48*, 26–29.

Ho, M. W. (2010b). Ten years of the human genome: Reams of data and no progress in sight. *Science in Society, 48*, 22–25.

Ho, M. W. (2011a). Genes don't generate body patterns. *Science in Society, 52*, 16–19.

Ho, M. W. (2011b). Nurturing nature. In S. Krimsky & J. Gruber (Eds.), *Genetic explanations: Sense and nonsense* (pp. 256–269). Cambridge, MA: Harvard University Press.

Ho, M. W., Matheson, A., Saunders, P. T., Goodwin, B. C., & Smallcombe, A. (1987). Ether-induced segmentation disturbances in *Drosophila melanogaster. Roux's Archives of Developmental Biology, 196*, 511–524.

Ho, M. W., & Saunders, P. T. (1979). Beyond neo-Darwinism: An epigenetic approach to evolution. *Journal of Theoretical Biology, 78*, 673–691.

Ho, M. W., & Saunders, P. T. (1982). Epigenetic approach to the evolution of organisms—with notes on its relevance to social and cultural evolution. In H. C. Plotkin (Ed.), *Learning, development, and culture: Essays in evolutionary epistemology* (pp. 343–361). Chichester, UK: Wiley.

Ho, M. W., & Saunders, P. T. (Eds.). (1984). *Beyond neo-Darwinism: Introduction to the new evolutionary paradigm*. London: Academic Press.

Ho, M. W., & Saunders, P. T. (1993). Rational taxonomy and the natural system with particular reference to segmentation. *Acta Biotheoretica, 41*, 289–304.

Ho, M. W., & Saunders, P. T. (1994). Rational taxonomy and the natural system: Segmentation and phyllotaxis. In R. W. Scotland, D. J. Siebert, & D. M. Williams (Eds.), *Models in phylogeny reconstruction* (pp. 113–124). Oxford, UK: Systematics Association.

Ho, M. W., Stone, T. A., Jerman, I., Bolton, J., Bolton, H., Goodwin, B. C., et al. (1992). Brief exposure to weak static magnetic fields during early embryogenesis cause cuticular pattern abnormalities in *Drosophila* larvae. *Physics in Medicine and Biology, 37*, 1171–1179.

Ho, M. W., Tucker, C., Keeley, D., & Saunders, P. T. (1983). Effects of successive generations of ether treatment on penetrance and expression of the bithorax phenocopy. *Journal of Experimental Zoology, 225*, 1–12.

Huang, S. (2009). Inverse relationship between genetic diversity and epigenetic complexity. *Nature Precedings*. Retrieved from *http://precedings.nature.com/documents/1751/version/2/files/npre20091751-2.pdf.*

Kawai, M. (1962, November). *On the newly-acquired behaviours of the natural troop of Japanese monkeys on Koshima Island.* Abstracts of the papers read in the seventh annual meeting of the Society for Primate Researches, Japan Monkey Centre, Inuyama, Japan. Available at *www.springerlink.com/content/w85j34568088u82q*.

Kimura, M. (1968) Evolutionary rate at the molecular level. *Nature, 217*, 624–626.

Kondo, S., & Miura, T. (2010). Reaction-diffusion model as a framework for understanding biological pattern formation. *Science, 329*, 1616–1620.

Løvtrup, S. (1974). *Epigenetics*. London: Wiley.

Lowenstein, J. M. (1986). Molecular phylogenetics. *Annual Review of Earth Planet Sciences, 14*, 71–83.

Margoliash, E. (1963). Primary structure and evolution of cytochrome c. *Proceedings of the National Academy of Sciences, 50*, 672–679.

Mattick, J. S., & Mehler, M. F. (2008). RNA editing, DNA recoding and the evolution of human cognition. *Trends in Neuroscience, 31*, 227–233.

Nusslein-Volhard, C. (2006, August 28). *Gradients that organize embryo development: A few crucial molecular signals give rise to chemical gradients that organize the developing embryo*. Paper presented at the Max Planck Institute for Developmental Biology. Available at *www.eb.tuebingen.mpg.de/departments/3-genetics/christiane-nusslein-volhard/gradients-that-organize-embryo-development.*

Palmer, A. R. (2004). Symmetry breaking and the evolution of development. *Science, 306*, 828–833.

Peel, A. D., Chipman, A. D., & Akam, M. (2005). Arthropod segmentation: Beyond the *Drosophila* paradigm. *Nature Reviews Genetics, 6*, 905–916.

Saunders, P. T. (1980). *An Introduction to Catastrophe Theory*. Cambridge, UK: Cambridge University Press.

Saunders, P. T. (1984). Development and evolution. In M. W. Ho & P. T. Saunders (Eds.), *Beyond neo-Darwinism: An introdution to the new evolutionary paradigm* (pp. 243–263). London: Academic Press.

Saunders, P. T. (1989). Mathematics, structuralism and the formal cause in biology. In B. C. Goodwin, G. C. Webster, & A. Sibatani (Eds.), *Dynamic structures in biology* (pp. 107–120). Edinburgh, UK: Edinburgh University Press.

Saunders, P. T. (1990). The epigenetic landscape and evolution. *Biological Journal of the Linnean Society, 39*, 125–134.

Saunders, P. T. (Ed.). (1992a). *Alan Turing's collected works: Morphogenesis*. Amsterdam: Elsevier.

Saunders, P. T. (1992b). The organism as a dynamical system. In W. Stein & F. J. Varela (Eds.), *Thinking about biology* (pp. 41–63). Reading, MA: Addison-Wesley.

Saunders, P. T. (1993). Alan Turing and biology. *IEEE Annals of the History of Computing, 15*, 33–36.

Saunders, P. T., & Ho, M. W. (1995). Reliable segmentation by successive bifurcation. *Bulletin of Mathematical Biology, 57*, 539–556.

Steele, E. J. (1981). *Somatic selection and adaptive evolution: On the inheritance of acquired characters*. Chicago: University of Chicago Press.

Steele, E. J. (2008). Reflections on the state of play in somatic hypermutation. *Molecular Immunology, 45*, 2723–2726.

Thompson, D. W. (1961). *On growth and form* (abridged ed.; J. T. Bonner, Ed.). Cambridge, UK: Cambridge University Press. (Original work published 1917)

Turing, A. (1952). The chemical basis of morphogenesis. *Philosophical Transactions, B, 237*, 37–72.

Valentine, J. W. (2004). *On the origin of phyla*. Chicago: University of Chicago Press.

Waddington, C. H. (1953). Genetic assimilation of an acquired character. *Evolution, 7*, 118–126.

Waddington, C. H. (1957). *The strategy of the genes*. London: Allen & Unwin.

Wagenmakers, E. J., Molenaar, P. C. M., Grasman, R. P. P. P., Hartelman, P. A. I., & van der Maas, H. L. J. (2005). Transformation invariant stochastic catastrophe theory. *Physica D, 211*, 263–276.

Webster, G., & Goodwin, B. C. (1982). The origin of species: A structuralist approach. *Journal of Social and Biological Structures, 5*, 15–47.

Zuckerkandl, E., & Pauling, L. B. (1962). Molecular disease, evolution, and genetic heterogeneity. In M. Kasha & B. Pullman (Eds.), *Horizons in biochemistry* (pp. 189–225). New York: Academic Press.

CHAPTER 6

Dynamical Systems, the Epigenetic Landscape, and Punctuated Equilibria

Peter T. Saunders

The most commonly used methods of applied mathematics, which involve writing down and solving differential equations, are far less effective in the biological and social sciences. There are seldom adequate quantitative data available, and even if there were, the number of equations required to describe a large complex system such as an organism or an economy would clearly be unmanageable.

One way forward is to draw on the comparatively new branch of mathematics known as dynamical systems, which deals chiefly with the qualitative behavior of classes of equations and systems of equations rather than seeking the solutions of particular ones. Not only can this approach yield results where the traditional method cannot, qualitative results about classes can be more useful than numerical results about one particular exemplar that may or may not be typical of the class as a whole.

Waddington's (1940) *epigenetic landscape* can be regarded as an early example of this approach. He intended it as a metaphor, but in the light of contemporary mathematics, we can see it is more than that. It is a visualization of some important typical properties of nonlinear dynamical systems. This means that inferences we can draw from it have a solid basis. They are essentially statements that, if we were able to write down and solve the differential equations that describe the system we are studying, these are properties we would expect to find.

An example described in this chapter is the prediction that in a complex system, large changes are likely to occur relatively abruptly rather than by the gradual accumulation of small ones. In biology, this implies that microevolution and macroevolution are essentially different processes. We therefore expect to observe the mode of evolution described by Eldredge and Gould (1972) as *punctuated equilibria*.

The Stability of Development

It is now about 70 years since Waddington (1940) first drew attention to some important general features of the developmental process in organisms. An embryo does not need an absolutely perfect environment to develop properly, and it can survive many small disturbances and even some large ones. Two embryos do not have to be clones to turn into very similar adults. The organisms that are the result of this process can be divided into distinct species rather than forming a continuous spectrum of forms.

The process is stable, though not in the same way as a ball at the bottom of a cup. An embryo that is perturbed will not return to the state that it was in before. If it can recover at all, it will continue to develop, in most cases eventually reaching more or less the state it would have attained had it been left alone. What is stable is not the state of the embryo at any one time, but its pathway of development.

Waddington introduced the words *homeorhesis* (similar flow) to describe the property of returning to the developmental trajectory when perturbed and *chreod* (necessary path) for the trajectory itself. He also used the term *canalization* to describe the property that development typically can proceed to one or more of a restricted number of alternative end states rather than to a broad spectrum.

These properties are clearly advantageous, for without them few if any organisms would develop properly. The reason we observe them, however, is simply that they are characteristic of complex nonlinear dynamical systems. Such a system, if it has any stable states at all, generally has a much richer sort of stability than a linear one. It will typically have more than one equilibrium, just as nonlinear algebraic equations often have many solutions (a nondegenerate linear system has, at most, one). Instead of just equilibrium points, it can have stable trajectories; the *limit cycle* is the most familiar of these, but the trajectories do not have to be cycles (Saunders, 1993).

Almost all the complex systems we observe are significantly nonlinear. They are also bound to be in some sense stable, for if they were not, we would not consider them as systems. It follows that we may expect to observe these properties in almost any highly complex system we are studying.

Homeorhesis, chreods, and canalization are nothing but a biologist's terms to describe the stable trajectories we expect to find in any such system, providing it is sufficiently complex—which organisms certainly are. It may be surprising that there are organisms at all, but given that there are, it is not at all surprising that they have these properties.

The Epigenetic Landscape

To help explain his ideas, Waddington portrayed the developmental system as a mountainous terrain with the valleys representing possible developmental pathways (Figure 6.1). The precise shape of the landscape depends on a complex network of guy ropes beneath it. The ropes stand for the effects of the genes, and the complexity of the

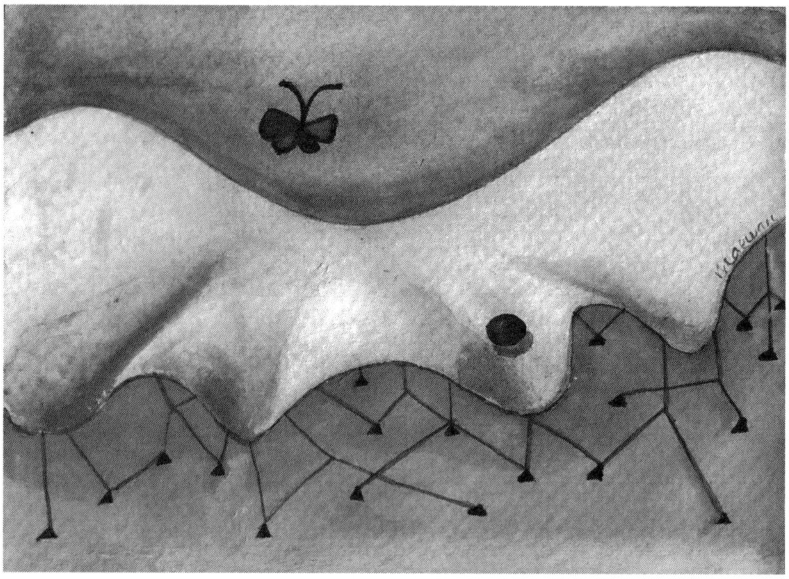

FIGURE 6.1. The epigenetic landscape. The butterfly symbolizes the role of catastrophe theory, among other things.

network is to remind us that although genes influence development, they do so in very complicated ways.

A ball rolls down the landscape, and the path it follows stands for the actual development of the organism. An environmental perturbation is represented either by the ball being pushed up the side of the valley or by a small, temporary change in the shape of the valley itself, as if the landscape were a tent and someone had kicked it. In either case, if the disturbance is not too great, then the ball returns to the valley bottom—not at the point where it was disturbed but somewhere further down. The system thus exhibits *homeorhesis.*

If the ball is disturbed so much that it is forced out of the valley, it will probably reach a dead end, but it is also possible that it will pass over a watershed and then continue down a different valley just as stable as the original one. This point too illustrates a property that is actually observed: Sometimes when an embryo is perturbed, it neither dies nor returns to normal development but switches to an alternative pathway that leads to the production of a viable, though significantly different, organism.

A mutation is represented by a change in the tension or position of a guy rope, which may (though on account of the complexity need not) bring about a permanent change in the shape of a valley. Most of these will be relatively minor, and the ball will keep to its original course. Mutations are most likely to have significant effects if they disturb the landscape near where one valley divides into two. At such a point even a small alteration in the topography can be enough to send the ball down a different path. Thus a small genetic change can bring about a large change in the phenotype. Equally, an environmental perturbation, represented by something disturbing the ball or temporarily distorting the landscape, is also likely to have a significant effect only if it occurs near a natural fork.

The model also suggests that environmental perturbations and mutations can have similar effects: They are simply different ways of diverting development into the same alternative pathway. This point is observed in the common phenomenon of *phenocopying*. If, for example, genetically normal *Drosophila* embryos are immersed for a short time in ether vapour, some of them will develop into adult flies that closely resemble the *bithorax* mutant (e.g., Ho, Tucker, Keeley, & Saunders, 1983; Waddington, 1956). Another example, by the way, is the well-known deformities caused by thalidomide. There is a rare mutation that brings about the same effect.

Phenocopying is largely ignored by biologists, but it actually has a lot to tell us. Above all, it is convincing evidence for the existence of alternative pathways. If such pathways did not exist, it would be very difficult to explain how it is that so often a mutation and an environmental perturbation can have the same effect.

The complexity of the network of guy ropes means that changing the position or tension of any one rope is unlikely to alter the landscape very much. Hence most such changes will have little or no influence on the end result, unless they affect the landscape near a junction between two valleys. This is of course what is observed: There is typically a great deal of genetic variation within populations where most individuals have nearly identical phenotypes.

The exceptions will mostly be near the end of the process, where the sides of valleys are flattening out and where the ball may not have enough time to return to its original trajectory. These changes will, however, be relatively minor because all the major features of the organism have already been determined.

It is, of course, on these minor changes that evolutionists concentrate their attention: the beaks of finches, the color of the peppered moth, things like that. This focus does not, however, address the real question, which is how the major changes came about. It is one thing to understand how the color of a moth changed over time; even some creationists are willing to accept this account. But how did there come to be moths in the first place?

In principle, large changes could be just long sequences of small ones, each giving a small, competitive advantage. This occurrence would require a succession of small deformations of the landscape, each moving the endpoint a little bit until it had reached a significantly different position. Although this scenario is possible in principle, it seems unlikely. If pulling a little on a rope shifts the landscape in the appropriate direction at first, after a while the tensions from the other ropes to which it is linked will stop it from continuing the same effect. And at the same time, the complexity of the network means that other parts of the landscape may also be affected, and in ways that were not intended.

Again, what the model illustrates are real properties. Even artificial selection seems inevitably to run up against limits. Luther Burbank (1931), the great plant breeder, famously remarked that he could get plums of many different sizes but not as small as a pea or as big as a grapefruit. Both *pleiotropy* (one gene affecting more than one character) and *genetic linkage* (the fact that genes that are physically close together on a chromosome tend to be passed on together) will oppose strong directional selection.

But a small change in the landscape can have a large effect on the organism if it occurs just where it can divert the ball down a different path. No other part of the landscape would have to be affected significantly. What is more, since it is only a matter of a diversion into an existing path, not the creation of a new one, we would expect that any one of a number of different small alterations in the network of ropes would do.

The model thus suggests not only that a single mutation can bring about a large phenotypic change but also that different single mutations can have the same effect. Thus, evolutionary change can occur without the same mutation occurring more or less simultaneously in a large number of organisms. (As Haldane, 1927, pointed out many years ago, a mutation that occurs in only one individual is unlikely to survive, even if it confers a significant selective advantage. It is more likely to be eliminated by "genetic drift," that is, by chance. This point is customarily ignored in arguments that begin, "Suppose there were a mutation that caused some individual to. . . . ")

Punctuated Equilibria

The model predicts that large changes will happen rapidly rather than by successions of small ones. Macroevolution is *not* just a lot of microevolution added up. The mode of evolution suggested by the epigenetic landscape is thus precisely that of *punctuated equilibria*, the idea that was put forward by Eldredge and Gould (1972) and has caused controversy ever since.

The problem is that if the organisms we see today have come into being through a long process of descent with gradual modification, we would expect this process to be reflected in the fossil record. It should look something like Figure 6.2a, with some measure of complexity plotted against time. In fact, it does not. The record is embarrassingly full of gaps, which opponents of the theory of evolution, from Cuvier to the present day, have always cited as evidence against the theory.

This situation would not be such a problem if what we saw were steady, gradual progress but with some bits missing (Figure 6.2b). Unfortunately for the theory, what we actually see is a succession of long periods within which very little seems to have happened, and with no intermediate forms between them (Figure 6.2c).

Despite this evidence, Darwinists have always insisted that evolution is gradual and that the missing intermediates *must* have existed. In the *Origin of Species*, we read "natura non facit saltum" (nature does not make leaps) and over a century later Dawkins (1985) writes, "Gradualism is of the essence. In the context of the fight against creationism, gradualism is more or less synonymous with evolution itself. If you throw out gradualism you throw out the very thing that makes evolution more plausible than creation" (p. 683).

Why are evolutionists so committed to gradualism when that is not what the evidence shows? Why offer the creationists such an easy target? Would it not be better to

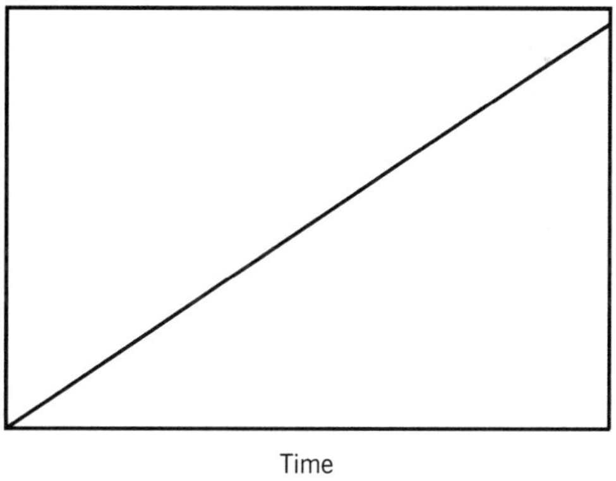

FIGURE 6.2a. What the fossil record should look like if evolution were gradual and most of the fossils have been found.

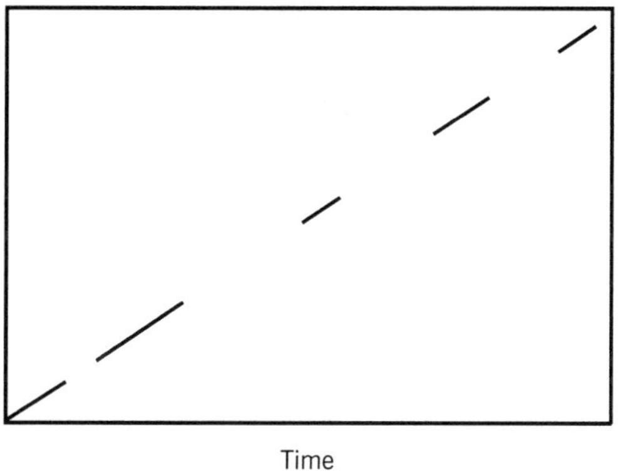

FIGURE 6.2b. What the fossil record should look like if evolution were gradual, but many of the fossils have not been found.

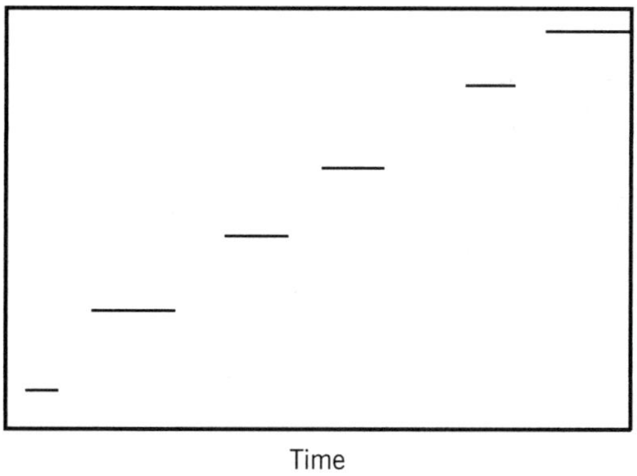

FIGURE 6.2c. What the fossil record actually looks like.

make the case for evolution on other, stronger grounds, such as carbon dating or the observation that the deeper we dig, the simpler the organisms we find, and leave the question of whether evolution has been gradual to be decided later?

Not for neo-Darwinists, because for them it is an article of faith that evolution occurs by the natural selection of random variations. What is more, the variations are random not in the weak sense that they are not preferentially in the direction in which they are needed, but in the strong sense that practically nothing can be said about them. They have to be because otherwise natural selection is not "the only direction giving force in evolution" (Mayr, 1980, p. 42) and that idea would be heresy.

If the variations are random in the strong sense, they have to be small because it must be next to impossible that a large strongly random variation could be anything but disastrous for the organism in which it occurs. And that is why evolution must be gradual, whatever the evidence may say. If, however, we concentrate less on the genes and think instead about how organisms develop and how the process can be affected by mutations or other perturbations, then we realize that *phenotypic* variations are not strongly random and the problem disappears.

In fact, neo-Darwinists have found a way of explaining away the gaps in the fossil record, and especially why all the interesting evolution seems to have taken place just when no one was looking. The major changes, they tell us, must have occurred by small groups becoming isolated, evolving rapidly, and then returning to out-compete the main population. There were indeed intermediates, but we do not find their fossils partly because the group that was doing all the evolving was very small and partly because they were not in the same place as the population whose history we are trying to trace.

But the gaps in the fossil record and punctuated equilibria are not awkward facts that require special explanation. On the contrary, they are precisely what we should expect.

Large Changes

Waddington (1940) intended the epigenetic landscape as a metaphor, and he even suggested that it would be useful to write down a large set of differential equations and work with those. In fact, he was closer to the modern approach than he could have known because mathematicians now think of a dynamical system as a flow on a manifold, which is really what the landscape is, albeit in far too few dimensions. We can, however, see how abrupt changes could occur by considering the following simple equation, originally put forward by Lewis, Slack, and Wolpert (1977):

$$\frac{dg}{dt} = S + \frac{g^2}{1+g^2} - 0.4g$$

Here, g is a substance that will determine some feature that we are concerned with and which, following Turing (1952), we refer to as a "morphogen," and S is a "signal substance" whose concentration might be determined by some gene. To find the equilibrium states, we have only to solve the equation $dg/dt = 0$, that is:

$$2g^3 - 5(S+1)g^2 + 2g - 5S = 0$$

This is a cubic in *g*, so it has three roots. For some values of *S*, all three are real; for others, only one is real and the other two are a complex conjugate pair. Figure 6.3 shows the equilibrium values of *g* as a function of *S*. Where there is only one real value of *g* for a given *S*, the equilibrium is stable; where there are three, the two outer ones are stable and the middle one is unstable and consequently never observed.

We can base a number of different scenarios on this equation, but the simplest is to suppose that *S* has a fixed value in any individual organism and *g* is initially zero. Then providing that *S* is less than the critical value $S_c = 0.0418$, *g* will reach an equilibrium value between 0 and about 0.25. If *S* increases over time—for example, due to a number of mutations—so does the equilibrium value of *g*, and as long as *S* remains below the critical value, the change in *g* will be small.

If, however, $S > S_c$, then *g* will reach equilibrium at about 2. So any evolutionary change in *S* that takes it from below the critical value to above it will bring about a large change in *g* and then in anything that *g* determines. Note that the size of the change in *g* depends hardly at all on the size of the change in *S*. All that matters is

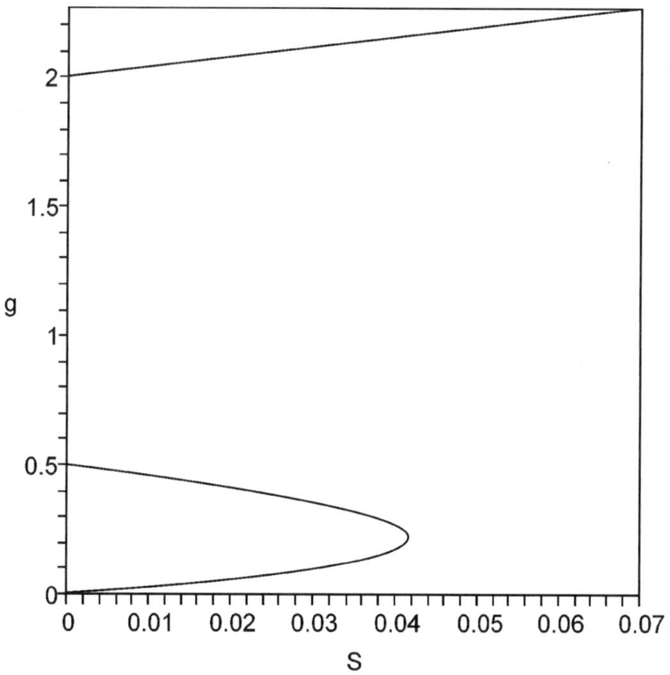

FIGURE 6.3. Equilibrium concentration of the morphogen *g* as a function of the concentration of the signal substance *S*.

that it takes S above S_c. It is just like the epigenetic landscape, where what matters is whether or not a perturbation takes the ball over a watershed. And as in the epigenetic landscape, the large change in S is to another predictable stable equilibrium state, not to some random value or to a collapse.

This model illustrates how a plausible chemical reaction scheme can produce a sudden jump, but we have no reason to suppose that this particular reaction is involved in any discontinuity in development or evolution. It certainly is not involved in all of them. What happens if some different mechanism is involved?

To deal with the general case, we turn to catastrophe theory (Thom, 1972). This theory tells us that if a variable, call it X, undergoes a sudden change from one equilibrium state to another, then however complex the system is, there will typically be two "control variables," usually denoted by U and V, involved in the change, not just one (i.e., S) as in the model of Lewis, Slack, and Wolpert (1977), and not more, as you might have expected. Instead of a curve of equilibria, there will therefore be a two-dimensional surface of equilibria (Figure 6.4). And from that surface we can infer the sorts of behaviour we expect from the system, without knowing the underlying mechanism (Saunders, 1980).

It may seem surprising that there are only two control variables, but in general these variables are not physical variables but combinations of physical variables. They are like principal components, or *eigenvariables*. Think how we can speed up or slow down a reaction by varying the amount of one or more reagents, adding a catalyst, changing the temperature or the pressure, and so on. All these different factors thus combine to form a single control variable. This scenario means that although there may be a considerable number of physical variables that can affect the process, the number of possible patterns of discontinuities (or, as Thom [1972] called them, *elementary*

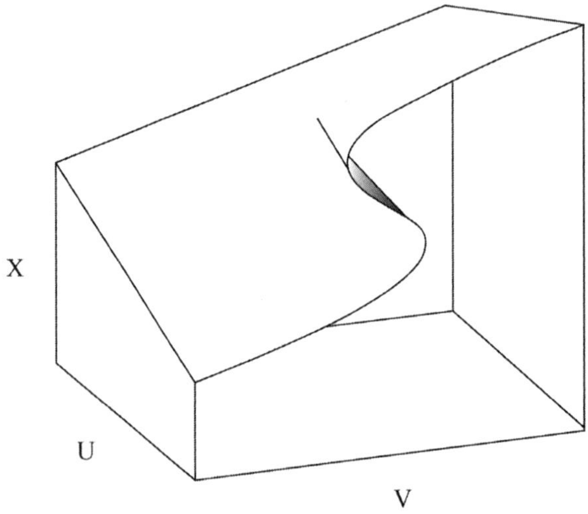

FIGURE 6.4. Equilibrium surface of the canonical cusp catastrophe.

catastrophes) is small. And in this sort of discussion, it is the qualitative patterns, not the details, that concern us. If we do want to know the details, then we have to work with the physical variables and equations. But having used catastrophe theory to establish the qualitative picture, we will know the sort of thing we are looking for, and this can be a vital clue. Bazin and Saunders (1978) and Saunders (1980) present an example.

You can see that if you hold one of the control variables, *U*, constant, you get a curve (Figure 6.5) that is very much like Figure 6.3, though the latter does not have the full *S*-shape because the figure is cut off at $S = 0$. The second variable increases the range of behaviors of the system (Saunders & Ho, 1985), but even if we hold it constant for the sake of simplicity, the model still allows us to understand some otherwise mysterious phenomena. For example, we can see how a number of different mutations, influencing the same reaction in different ways or by different amounts, or influencing different reactions, could each take V across V_c and bring about *precisely the same phenotypic change*. This example provides a real basis for the epigenetic landscape. It also helps solve the Haldane problem that an advantageous mutation that occurs in only one individual is very unlikely to survive.

We can now see why phenocopies should exist; it is simply because the set of physical variables that contribute to the control variable V can include both genetic and environmental components. To mention just one example, many butterfly collectors try to find a specimen of every genetic variety of a species. Every collector knows

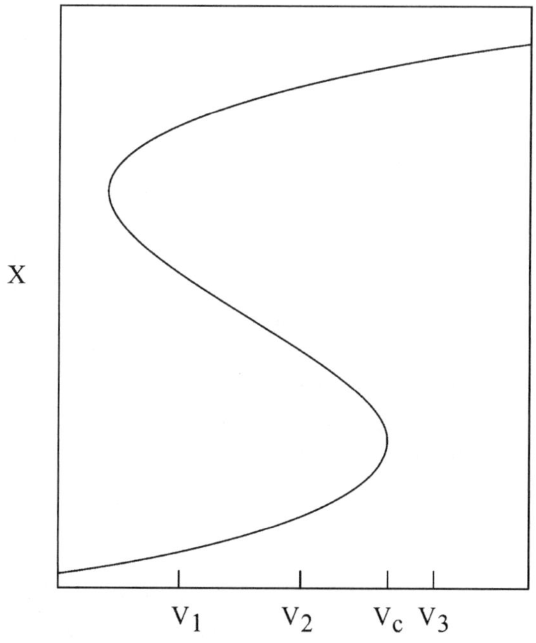

FIGURE 6.5. A model for genetic assimilation.

that you can produce many of these varieties by chilling the pupae for a while, but it is considered cheating.

Genetic Assimilation

Flat fish like the flounder and the sole have both eyes on the same side of the head. That's obviously an advantage for a fish that swims on its side along the bottom of the sea, but how did the other eye get to the upper surface? Neo-Darwinist theory tells us that evolution occurs by long sequences of small changes, each giving the individual in which it occurs an advantage over its competitors. But what conceivable selective advantage could there have been in the bottom eye moving just a tiny bit forward as the first step toward being on the top?

Darwin was challenged on this point by Mivart (1871), and it was one of the objections to his theory that he discussed in the sixth edition of the *Origin of Species* (Darwin, 1872, p. 173). His suggested explanation was that the eye moved around gradually through generations of fish straining to move their eyes forward so that they could see something with them. This Lamarckian account has been an embarrassment to Darwin's successors, but they have not found a better one to replace it.

The model we are discussing in this section makes it easy to see how such a change could come about. For the sake of argument, let us imagine that some trait is determined by the concentration of a single morphogen, X. The low value corresponds to the current trait and the higher value to the more advantageous one. (See Figure 6.5.)

Suppose that the original level of the signal substance is V_1 and that the most any single mutation can do is to raise it to V_2, which is still less than the critical value V_c. If one does occur, it will seem to confer no selective advantage, and the chance that another will occur in the same individual or even in the same line will be very small indeed. So the desired trait is very unlikely to evolve, even though it might be very useful to the organism if it did.

On the other hand, suppose that V can also be increased by some environmental perturbation. Then any mutation that increased V would be advantageous because it would make it more likely that such a perturbation would raise V to some $V_3 > V_c$ and the trait would appear. There would therefore be selection for any mutation that led to an increase in V. Eventually a large proportion of individuals would have such a mutation, and so it would not be at all unlikely that a second one would occur in some of them. The trait would then appear without the stimulus from the environment. This scenario is how a change can come about even if it requires a number of mutations, none of which, by itself, produces a phenotype that is any better adapted to the environment than the rest of the population.

The process of genetic assimilation was first proposed by Waddington (1942) but has never been fully accepted by evolutionists. Their problem is that even though it involves only processes that are known to happen, it amounts to the inheritance of acquired characters, and this type of inheritance is anathema to neo-Darwinists. Darwin himself took a different view.

A Quite Different Example

Over 55 years ago, economist William Phillips (1958) observed that if he plotted the inflation rate against the rate of unemployment, the points lay approximately on a curve (now called the *Phillips curve*). When inflation was high, unemployment was low, and vice versa. This observation, and theoretical work that was later based on it, implied that there is a tradeoff between inflation and unemployment. Governments could affect both, but not independently, because the economy can only move in one direction or the other along the curve.

The Phillips model fit the observations quite well for some time, but in the early 1970s it appeared to break down. A scatter plot of inflation against unemployment (more precisely, unemployment in the previous year) for the period 1953–1992 seems to show no correlation of any kind (Figure 6.6). If, however, the data are grouped into three separate time periods—1955–1971, 1974–1985, and 1985–1992—then the points in each of these groups lie approximately on a line with negative slope. There appear to be three separate Phillips curves with relatively sudden transitions from one to another; a formal analysis indicates there may be five (see, e.g., Abel & Bernanke, 1995).

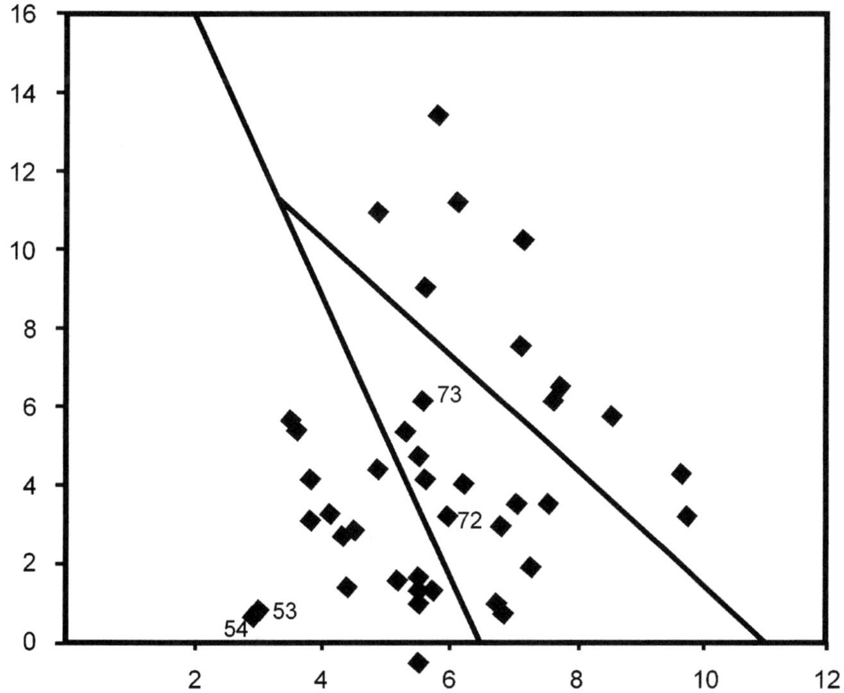

FIGURE 6.6. The U.S. inflation rate plotted against the unemployment rate in the previous year for the period 1953–1992. The points in the lower left group are pre-1972, the points in the upper right group are between 1974 and 1984, and those in the middle group are post-1984. The exceptions—the points for 1953, 1954, 1972, and 1973—are labeled. Data from the U.S. Bureau of Labor Statistics.

The system thus appears to have several basins of attraction. What can happen within each of these is constrained by the dynamic in the way described by a Phillips curve, but the basin itself—that is, which curve the system is on—is determined on other grounds: for example, by history, by government measures, or by events outside the system or at least outside the part of the system included in the model. That large nonlinear systems can have multiple equilibria warns us that the claim that there is no alternative may be overly pessimistic—or overly optimistic if we are comfortable where we are.

Conclusions

The key property of organisms that we have used here is that they are complex nonlinear systems. There are, of course, many other examples of such systems, and we should be able to make inferences about these too. For example, Saunders and Skar (2001) have argued that the archetype, an important but ill-defined term in Jungian thought, can be explained in terms of self-organization. (Bénard convection and hurricanes are examples of self-organization in physical systems.)

It may seem surprising that we can infer so much from the mere observation that the system we are interested in is complex and nonlinear, but our intuition has been developed over many years of studying linear systems. We all know that real systems are seldom strictly linear, but the usual strategy has been to treat them as though they were and then add on the nonlinear effects as (hopefully) small corrections. We assume, in effect, that the systems we are analyzing (and, by an obvious if unjustified extension, systems in general) are what we might call "essentially linear"—that is, their most important properties are what they would be if the systems were actually linear.

So, for example, we expect effects to be more or less proportional to causes, that the whole is approximately the sum of the parts, and that structures do not form without a template or some force or forces directly related to the shape that appears. Exceptions to all of these have been known for years: for example, Bénard convection was first studied in 1900 and the abrupt collapse of an overloaded structure was analyzed by Euler in 1744 (and the proverb about the straw that broke the camel's back is much older than that), but they are how we expect systems to behave.

Recent developments in mathematics and in computing have made the study of nonlinear systems a major research area in mathematics. We are learning more and more about phenomena such as symmetry breaking, catastrophes, chaos, and self-organization. These phenomena do not occur in all nonlinear systems, but when they do, we do not have to seek special explanations for them. Up to now, these new ideas have appeared mostly in mathematical modeling, as we might expect. The example of punctuated equilibria demonstrates how they can be used more widely, as they become part of an expanded intuition about systems.

REFERENCES

Abel, A. B., & Bernanke, B. S. (1995). *Macroeconomics* (2nd ed.). Reading, MA: Addison-Wesley.

Bazin, M. J., & Saunders, P. T. (1978). Determination of critical variables in a microbial predator prey system by catastrophe theory. *Nature, 275,* 52–54.

Burbank, L. (1931). *The harvest of the years.* New York: Houghton Mifflin.

Darwin, C. R. (1872). *The origin of species by means of natural selection, or the preservation of favoured races in the struggle for life* (6th ed.). London: John Murray.

Dawkins, R. (1985). What was all the fuss about? (Review of *Time Frames* by N. Eldredge). *Nature, 316,* 683–684.

Eldredge, N., & Gould, S. J. (1972). Punctuated equilibria: An alternative to phyletic gradualism. In T. J. M. Schopf (Ed.), *Models in paleobiology* (pp. 82–115). San Francisco: Freeman Cooper.

Haldane, J. B. S. (1927). The mathematical theory of natural and artificial selection. *Proceedings of the Cambridge Philosophical Society, 23,* 838–844.

Ho, M. W., Tucker, C., Keeley, D., & Saunders, P. T. (1983). Effects of successive generations of ether treatment on penetrance and expression of the bithorax phenocopy in *Drosophila melanogaster. Journal of Experimental Zoology, 225,* 357–368.

Lewis, J., Slack, J. M., & Wolpert, L. (1977). Thresholds in development. *Journal of Theoretical Biology, 65,* 579–590.

Mayr, E. (1980). Some thoughts on the evolution of the evolutionary synthesis. In E. Mayr & W. B. Provine (Eds.), *The evolutionary synthesis: Perspectives on the unification of biology* (pp. 1–48). Cambridge, MA: Harvard University Press.

Mivart, St. G. J. (1871). *On the genesis of species.* London: Macmillan.

Phillips, W. (1958). The relationship between unemployment and the rate of change of money wages in the United Kingdom 1861–1957. *Economica, 25,* 283–299.

Saunders, P. T. (1980). *An introduction to catastrophe theory.* Cambridge, UK: Cambridge University Press.

Saunders, P. T. (1993). The organism as a dynamical system. In W. Stein & F. J. Varela (Eds.), *Thinking about biology: Vol. III. SFI studies in the sciences of complexity* (pp. 41–63). Reading, MA: Addison-Wesley.

Saunders, P. T., & Ho, M. W. (1985). Primary and secondary waves in prepattern formation. *Journal of Theoretical Biology, 114,* 491–504.

Saunders, P. T., & Skar, P. (2001). Archetypes, complexes and self-organisation. *Journal of Analytical Psychology, 46,* 305–323.

Thom, R. (1972). *Stabilité structurelle et morphogénèse.* Reading, MA: Benjamin. English translation by D. H. Fowler (1975). *Structural stability and morphogenesis.* Reading, MA: Benjamin. (For a more accessible approach, see Saunders, 1980.)

Turing, A. M. (1952). The chemical basis of morphogenesis. *Philosophical Transactions of the Royal Society of London, B237,* 37–72.

Waddington, C. H. (1940). *Organisers and genes.* Cambridge, UK: Cambridge University Press.

Waddington, C. H. (1942). The epigenotype. *Endeavour, 1,* 18–20.

Waddington, C. H. (1956). Genetic assimilation of the *bithorax* phenotype. *Evolution, 10,* 1–13.

PART IV

NEURAL NETWORKS AND DEVELOPMENT

CHAPTER 7

Nonlinear Epigenetic Variance in Developmental Processes

Maartje E. J. Raijmakers, Kees Jan Kan, Annemie Ploeger, and Han L. J. van der Maas

In behavior genetic studies phenotypic variance is linearly decomposed into components attributable to two general latent sources, the genetic and the environmental. Genetic variance can be decomposed into additive, dominance, and epistatic genetic variance, whereas environmental variance can be decomposed into shared environmental variance and nonshared environmental variance. Shared influences are attributable to the environment that cohabiting individuals share. In humans, this decomposition is carried out using twins (e.g., Plomin, DeFries, McClearn, & McGuffin, 2008). In these twin studies it is assumed that genotypic and environmental factors reflect the underlying mechanisms causing phenotypic individual differences (e.g., Plomin et al., 2008). This assumption is not without problems.

First, a limiting feature of the behavior genetic methodology is that the causal interpretation pertains to phenotypic individual differences, and not to phenotypes themselves (Dolan & Molenaar, 1995; Lewontin, 1974; Oyama, 1985). Second, the validity of the methodology may be undermined by the limitations of the statistical model employed to carry out the decomposition of phenotypic variance. For instance, genotype–environment interaction, genotype–environment covariance, and assortative mating are not taken into account in the standard linear model (but see Plomin et al., 2008). Third, as we argue, the search for specific genetic and environmental variables may be limited in that an important part of phenotypic variance may be due to nonlinear (epigenetic) processes (Molenaar, Boomsma, & Dolan, 1993; Zuk, Hechter, Sunyaev, & Lander, 2012). These effects will appear unsystematic and are therefore difficult to distinguish from specific environmental effects and measurement error.

The aim of this chapter is to study the third limitation: the role of nonlinear epigenetic processes as a source of phenotypic variance. We review the relevant literature and present ample evidence in support of these processes. In addition, we present the results of computer simulations. The implications and consequences of the presence of nonlinear epigenetic variance are particularly interesting in the light of the largely unsuccessful attempts to identify specific nonshared environmental influences (e.g., see Turkheimer & Waldron, 2000).

Nonlinear Epigenetic Processes during Development

Developmental processes can be conceived of as the outcome of some nonlinear dynamical system (Guckenheimer & Holmes, 2002; van der Maas & Molenaar, 1992). In contrast to linear systems, nonlinear systems may show a disproportional relationship between cause and effect (Arnold, Afrajmovich, Il'yashenko, & Shil'nikov, 1994). This disproportion implies that large influences sometimes have small or limited effects, whereas small causes (e.g., initial changes or differences) may have large effects. Hence, almost similar initial conditions in the same system may produce different outcomes.

Sensitivity to initial conditions can result into two forms of unpredictability (Arnold et al., 1994). In some nonlinear systems the result is *seemingly random behavior*. Such behavior is referred to as *chaotic*. In other nonlinear systems the result is a *sudden qualitative change in behavior* when a small, smooth change is made to a parameter. In mathematics such a change is termed a *bifurcation*. Bifurcations are a general characteristic of nonlinear systems in physics (Prigogine, 1980), biology (Meinhardt, 1982), and psychology (e.g., Kelso, 1995; Ploeger, van der Maas, & Raijmakers, 2008; van der Maas & Molenaar, 1992). In physics they are called *phase transitions*; in biology and psychology they are often called *stage transitions*. We use the term *bifurcation*.

Nonlinear systems can attain order, structure, and stability by a process known as *self-organization* (e.g., Camazine, 2001). Numerous instances of self-organization have been found in both nonliving and living systems. Examples are the formation of stripes in sand dunes and patterns on skins, coats, and shells (Camazine, 2001; Kan, Boomsma, Dolan, & van der Maas, 2012; Meinhardt, 1982). Self-organization also occurs in morphogenesis: Organs such as the lungs, the brain, the cardiac muscle network, and the blood circulatory system display a very deep and structured branching. The notion that the formation of these organs involves self-organization is supported by the fact that the total amount of information stored in the genome is far too small to prescribe these structures in any detail (Benno, 1990; Camazine, 2001; Molenaar et al., 1993; Stent, 1978). In other words, self-organizing processes may be required to explain the process of ontogenesis.

In this context, epigenetics and epigenesis are important concepts. There is a close correspondence between self-organization and epigenesis: "Epigenesis constitutes an instance of a self-organizing developmental process" (Molenaar & Raijmakers, 2000, p. 45). In other words, "Epigenesis may be regarded as the theory of self-organization as applied to ontogenetic phenomena" (Beloussov, 2006, p. 1165). Traditionally, *epigenetics* refers to the study of such processes; that is, "the way genes and their products bring the phenotype into being" (Jablonka & Lamb, 2002, p. 82; see also Waddington, 1957). Nowadays, with the greater understanding of the molecular mechanisms that control gene activity during embryonic development and cell differentiation, epigenetics is defined as "the study of mitotically and/or meiotically heritable changes in gene function that cannot be explained by changes in DNA sequence" (Jablonka & Lamb, 2002, p. 87).

Thus both traditional and molecular biological epigenetics focus on alternative developmental pathways. Nonlinear mechanisms initiate autonomous growth processes leading to structure and pattern (e.g., brain structure). However, the environment, the epigenetic process itself, and their interactions tend to perturb development, possibly resulting in different developmental pathways. That is, at critical points (i.e., bifurcation points) these perturbations may cause development to follow a different pathway, resulting in variations in structure or pattern (Waddington, 1957). Fascinating examples of divergent developmental pathways are given by Gärtner (1990). In the next section we discuss this work.

Empirical Evidence of Sources of Unsystematic Phenotypic Variance

Gärtner (1990) reviewed the decades-long efforts in his own laboratory to minimize the variance of biological traits in laboratory animals by standardization of environmental and genetic conditions. The variance in a number of traits, such as morphological, biochemical, and other quantitative traits, of highly inbred rats held under strict environmental control, was compared with the variance in the same traits of rats living in a natural wild setting. The variance in inbred rats was not substantially lower than the variance in wild rats. Apparently, neither the postnatal environment nor genes explained the phenotypic variance in this context.

Later research was performed to assess environmental variance directly. To this end, eight-cell stage mice embryos were divided, thereby creating monozygotic (MZ) twin pairs. Each twin pair was transplanted into the uterus of the same foster mother, which raised both twins. The coefficients of variation appeared to be similar in both groups. In another experiment with Friesian cattle, an additional comparison was made with a group of divided embryos transferred into, and raised by, different uterine foster mothers. A large amount (70–97%) of the phenotypic variance remained unexplained. From these studies, Gärtner (1990) concluded that the remaining variance had to be attributed to a component other than genetic and (external) environmental influences. Gärtner's work has been reprinted in a special issue of the *International Journal of Epidemiology* with commentaries and reply (see Ferrie, 2012).

Other interesting results concern studies with cloned organisms. Archer, Friend, Piedrahita, Nevill, and Walker (2003) compared cloned pigs with naturally bred controls on several phenotypic bodily traits as well as blood parameters. Analysis of the degree of phenotypic variance between clones and controls showed that cloned animals displayed equally high or even increased variance in several traits.

Yet another kind of variation reflects the poor ability of organisms to develop the same phenotype under the same environmental conditions. Bilateral organisms can show intraindividual variation known as *fluctuating asymmetry*. For example, Stige, Slagsvold, and Vøllestad (2005) repeatedly measured feather length and color patterns of wings and tail of the pied flycatcher. The degree of feather asymmetry persisted from nestling stage to adulthood, and even across moults. Genetic analyses revealed that the heritability of the asymmetry was almost nonexistent, and that shared environmental factors had little influence. The researchers concluded that asymmetry is possibly determined by (stochastic) events during early stages of development, permanently affecting the features of the feathers. In humans, similar variations emerge at a very young age and remain stable throughout the life span. Furthermore, the direction of these asymmetries is not under genetic control. Fink, Manning, Neave, and Grammer (2004) argued that facial asymmetries might be the results of perturbations in uterine conditions (e.g., hormone levels) in the first stages of embryonic development.

A related phenomenon, which counts as another indication of developmental instability, is the development of minor physical anomalies. Several studies have demonstrated the existence of atypical, nongenetically determined asymmetries in human brain anatomy (e.g., Eckert et al., 2002; Steinmetz, 1996; Thoma, Yeo, Gangestad, Lewine, & Davis, 2002; Wright, Sham, Murray, Weinberger, & Bullmore, 2002). For example, Eckert and colleagues (2002) examined the heritability of asymmetry in the planum temporale in 27 MZ and 13 dizygotic (DZ) male twins. Magnetic resonance imaging (MRI) measurements revealed significant dissimilarities in gyral and sulcal features between MZ twins. The authors hypothesized that intrauterine effects lead to divergent morphological development.

The unresolved variance originating in the process of development may not be limited to biological traits; it may well characterize psychological traits as well.

The Role of Nonshared Environment in Behavior Genetic Psychological Research

Behavior genetic research has revealed a number of consistent, common findings in cognitive abilities, personality, social attitudes, psychological interests, and psychopathology (e.g., Turkheimer, 2000). First, a substantial amount of phenotypic variance can be attributed to the genetic component. Second, shared environmental variance is substantially smaller than variance explained by genetic effects. Third, a large portion of the phenotypic variance is attributed to the component nonshared environment. Fourth, estimated heritabilities commonly increase with development, whereas environmentabilities decrease. However, the relative contributions of the shared component and nonshared component change over development: If present, the contribution of the shared component decreases significantly, often down to values close to zero. Hence, ultimately, all environmental variance is accounted for by the nonshared component. Finally, MZ twins are often more than twice as similar as DZ twins. However, recent papers by Charney (2012) and Lickliter (2012) indicate that across development MZ twins share increasingly less of their genotype variance. Usually this is interpreted as evidence for genetic dominance or epistasis (Bouchard & McGue, 2003; Turkheimer, 2004).

The nonshared environmental variance ranges from about 20% in cognitive abilities in adults to about 60% in personality, phobias, and some social attitudes (Bouchard, 2004; Bouchard & McGue, 2003; McGue & Bouchard, 1998; Plomin et al., 2008). A meta-analysis of specific effects of objectively defined nonshared environmental variables (e.g., differences in sibling, teacher, and peer relationships) indicated that in behavior genetic studies the largest part of nonshared environmental variance is left unexplained (Turkheimer & Waldron, 2000). The median percentage nonshared variance accounted for by these influences was no more than 2%.

In part, the ubiquitous nonshared environmental variance may in fact be due to nonlinear influences. Since the outcome of such processes appears unsystematic, these influences are hard to detect—which may explain why attempts to identify such influences have been largely unsuccessful (Turkheimer & Waldron, 2000). A part of the unsystematic phenotypic variance may actually be the result of nonlinear self-organizing epigenetic processes (Molenaar et al., 1993; Zuk et al., 2012). Simulation studies of nonlinear epigenetic processes have been used to investigate the role and effects of such processes on behavior genetic analyses (Eaves, Kirk, Martin, & Russell, 1999; Molenaar & Raijmakers, 1999). Their implications are discussed in the next section.

Computer Simulations of Nonlinear Epigenetic Variance

Both Eaves and colleagues (1999) and Molenaar and Raijmakers (1999) simulated nonlinear epigenetic variance in development with the discrete logistic equation as a simple model of development. This equation is known to display chaotic behavior, making it a good model to study the effects of sensitivity to initial conditions on twin correlations. Quantitative behavior genetic analyses revealed that, with time, the genetically induced correlations between the parameters had been concealed in the observed output: Observed heritabilities decreased to zero (Molenaar & Raijmakers, 1999).

From a theoretical and practical biological point of view, the model considered by Molenaar and Raijmakers (1999) and by Eaves and colleagues (1999) is quite limited mainly because, over time, chaotic processes will yield observed heritabilities of zero, which is not observed in behavior genetic research. Hence, the role of chaotic epigenetic processes in development may be quite limited, as Eaves and colleagues argue. The possibility remains, however, that other types of nonlinear processes (e.g., bifurcating systems) give rise to a different picture.

Turkheimer and Gottesman (1996) considered a nonlinear model of development that does display bifurcating behavior. Genotype, environment, and phenotype were modeled as locations (dots) in two-dimensional space. Whereas genotype was simulated as a fixed location, environment and phenotype changed dynamically according to a set of simple rules. Turkheimer (2000) conducted a twin simulation experiment based on this model. A significant part of phenotypic variance could be accounted for by genotypic variance, but none of it could be explained by the variance in environmentally determined starting locations. Furthermore, phenotypic DZ within-pair correlations were substantially smaller than MZ within-pair correlations. However, the DZ correlations were not close to zero, as in the studies of Eaves and colleagues (1999) and Molenaar and Raijmakers (1999). Turkheimer concluded: "Phenotype at any moment in development is the cumulative result of an organism's developmental history, encompassing genotype, environment, and all the complexities of their epigenetic interactions" (p. 184).

We studied the role of nonlinear epigenetic processes using a model of neurite outgrowth (Kan, Ploeger, Raijmakers, Dolan, & van der Maas, 2009). This model shows bifurcating, but not chaotic, behavior and is biologically realistic; time is modeled as continuous, the timescale is defined, and the system's output (the state of an organism) represents concrete characteristics (e.g., membrane potentials). Furthermore, previous computer simulations demonstrated results that are consistent with empirically observed phenomena, such as a temporary overproduction of synapses (see van Oss & van Ooyen, 1997). With this simulation study, we reconsidered the plausibility of nonlinear epigenetic processes as a source of unsystematic phenotypic variance. The consequences of the presence of such processes for behavior genetic analyses and the search for specific genes and environmental influences are discussed shortly.

Modeling Network Development

During development neurons are assembled into functional network structures (van Oss & van Ooyen, 1997). One important factor determining neuronal morphology and network formation is a mutual influence between network activity, on the one hand, and neuronal form, connectivity, and function, on the other. The network is able to generate patterns of activity, thereby changing the organization of the network, which in turn leads to an alteration of activity patterns. One example of a model of activity-dependent neural network development is the two-cell model of van Oss and van Ooyen.

Neurite outgrowth concerns the development of axons and dendrites; that is, the development of connections between neuronal cells. van Oss and van Ooyen (1997) used their two-cell model to investigate the effects of the combination between activity-dependent neurite outgrowth and inhibition. They showed that this combination may lead to multistability, which they associated with normal and pathological end states of network development. Because this model has the characteristics that are considered to underlie epigenetic variance, we considered it to be a good candidate to generate twin data. A mathematical description of the model and its behavior is given in the appendix of Kan and colleagues (2009). Here we give an informal description.

The two-cell model contains one excitatory and one inhibitory unit or cell (Figure 7.1). This model is based on the behavior of populations of neurons and was developed to replace the numerous differential equations (the network model) with a simpler model that still would display the characteristics and behavior of this network model. Comparable two-neuron models have been used to study, for example, working memory (Kirillov, Myre, & Woodward, 1991).

The two-cell model of van Oss and van Ooyen (1997) describes activity-dependent neurite outgrowth, one of the many dynamical processes involved in shaping neuronal morphology. Excitatory and inhibitory inputs take the membrane potential toward a finite maximum and minimum potential, respectively. The excitatory unit is connected to itself (with weight w_{xx}). For simplicity, reciprocal inhibition between inhibitory neurons is not modeled (weight $w_{yy} = 0$). The connection between the inhibitory unit and the excitatory unit is modeled as symmetric (weight $w_{xy} = w_{yx}$). We retained these assumptions to ensure that the bifurcation diagrams (explained below) could describe the model.

A unit's output is the mean firing rate, which is modeled as a sigmoidal function of the membrane potential. Parameters determine the steepness of the function as well as a low subthreshold firing rate. The latter can be interpreted as spontaneous activity as a result of synaptic noise or fluctuations in membrane potential. Furthermore, increasing neurite outgrowth is thought to imply increasing connection strength. In turn, as connection strength increases, the amount of input that a cell receives through this connection also increases. Hence, the system is characterized by feedback. Finally, the dynamics of the connection strength are considerably slower than those of neuronal activity. Therefore, connection strength can be considered as a slowly varying parameter.

Using their model, van Oss and van Ooyen (1997) investigated occurrences of bifurcations in neuronal development. Such occurrences would imply that individual networks are able to develop via different pathways, dependent on initial conditions. The researchers were interested in the behavior of the network as a function of the parameters that determine the membrane potential at which neurite outgrowth is 0 (ε), and the relative strength of the inhibitory connection (p). They established the presence of bifurcations, multistability, and rich dynamic behavior, including oscillations and transient jumps. Due to the relative simplicity of the model, van Oss and van

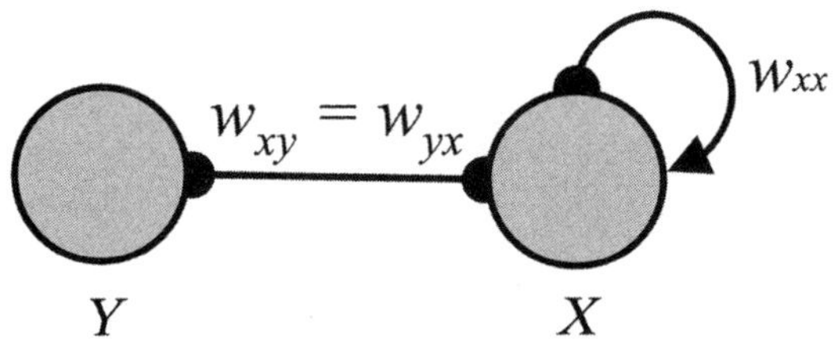

FIGURE 7.1. The two-cell model, where unit Y is interpreted as a population of inhibitory neurons and unit X as a population of excitatory neurons. X is connected to itself with weight w_{xx}. In contrast, Y is not connected to itself (mutual inhibition is not modeled). The connection between X and Y is modeled as symmetric ($w_{xy} = w_{yx}$). In the simulations, $w = w_{xx} = w_{xy} / p$.

Ooyen could construct a bifurcation diagram in which areas in the parameter plane (ε, p) are associated with qualitatively different dynamic behavior of the model. These areas are demarcated by bifurcations.

The model has several essential characteristics that illustrate the theoretical point that epigenetic processes constitute a source of variance. First, due to the reciprocal interactions and the sigmoid activation functions, the model is nonlinear. The specific properties of the model that may give rise to epigenetic variance are the occurrence of bifurcations (the possibility that small differences in parameter values result in qualitatively different phenotypes) and the presence of multistability (initial differences can result in different developmental pathways). The existence of qualitatively different possible phenotypes, given one genotype, implies that members of an MZ twin pair may follow different developmental pathways. Second, the model displays properties of self-organization. Third, the model describes development as a deterministic process. This characteristic is important theoretically because it means that the sensitivity to initial conditions does not arise from any randomness—that is, the epigenetic variance is not attributable to an external random process that introduces noise into the system.

Using the two-cell model of van Oss and van Ooyen (1997), we aimed to simulate twin data in the form of time series. These time series were generated under various assumptions concerning starting values, parameter values, and degree of genetic control in order to create nonlinear epigenetic variance.

Twin Simulations Applying the Two-Cell Model

Method

The two-cell model in the simulated twin study is interpreted as follows. The relevant phenotypic traits are x, the time-averaged membrane potential of the excitatory unit; y, the time-averaged membrane potential of the inhibitory cell; and w, the connection strength between excitatory neurons. We made the variances of the parameters ε and p attributable to genetic and environmental individual differences within the population. That is, individual differences in ε and p satisfy the standard behavior genetic model (see above). Thus, ε and p are considered as underlying variables causing individual differences in the development of phenotypic traits x, y, and w. Starting values of x, y, and w were set at 0. At the same time, ε and p are the control variables of the two-cell model for which bifurcations may occur in the model.

Phenotypic time series were generated for 200 MZ and 200 DZ twin pairs using a simple additive model:

$$\varepsilon_i = \varepsilon_{g,i} \sqrt{a} + \varepsilon_{c,i} \sqrt{b} + \varepsilon_{e,i} \sqrt{(1 - a - b)} + m_\varepsilon$$

and

$$p_i = p_{g,i} \sqrt{a} + p_{c,i} \sqrt{b} + p_{e,i} \sqrt{(1 - a - b)} + m_p$$

with $0 \leq a \leq 1$, $0 \leq b \leq 1$, and $a + b \leq 1$, where ε_i and p_i are the individual values of the parameters ε and p, respectively, and m_ε and m_p are the sample means of the parameters. The subscripts g, c, and e refer to the additive genetic, shared environmental, and nonshared environmental components of a parameter, with regression coefficients $\sqrt{a}$, $\sqrt{b}$, and $\sqrt{(1 - a - b)}$, respectively. Assuming that the additive genetic and environmental components of a parameter are uncorrelated and sampled from the standard normal distributions, heritability h^2 equals a, common environmentability c^2 equals b, and unique environmentability e^2 equals $(1 - a - b)$. MZ twin correlations of the genetic, shared, and nonshared environmental components equal 1 (100% additive genetic resemblance), 1, and 0, respectively. The same procedure was followed to generate 200 DZ twins, except the DZ twin correlations of the genetic components equal .5 in this case (on average 50% additive genetic resemblance).

In simulating the data, we chose a heritability of .5 for both parameters p and ε. Furthermore, referring to the epigenetic picture and to nonlinearity, initial conditions may be very similar. However, in real biological systems exactly equal initial conditions are impossible. Therefore, we set unique environmentability to a low, but nonzero, value (i.e., to .01). As a consequence, common environmentability was set to a value of .49. This procedure implies that in the absence of nonlinearity, standard behavior genetic analyses of the phenotypic measures should be expected to yield a heritability estimate of .5, a common environmentability of .49, and a unique environmentability of .01. However, given the occurence of bifurcations and sensitivity to initial conditions, additional (i.e., epigenetic) variance is expected to arise.

The differential equations were numerically solved using the variable time step integrator Lsoda (Petzoldt, 2003). Subsequently, phenotypic twin correlations of x, y, and w were obtained at time points $t = 0$ to 20,000 (by $t = 20{,}000$ all networks have reached their final stable states). We derived alternative phenotypic measures to cope with latency shifts due to oscillations. These are the mean of the signal over a certain period of time and the relative power. This mean signal is a smoothed signal, where high-frequency oscillations have been filtered out. The power of a signal is a time average of energy (energy per unit time), where energy is defined by the area under the signal. Relative power is the power of one frequency band relative to that of other frequencies. In this manner we could distinguish (almost) nonoscillating signals (frequencies < 0.1) from oscillating ones (frequencies ≥ 0.1). In order to calculate the two alternative phenotypic measures, time was divided into 40 equal intervals: [(0, 500), (500, 1,000), . . ., (19,500, 20,000)]. Twin correlations were calculated in each time window.

In simulation 1 we chose region 5b within the (ε, p) plane as the region in which variance was to be created in parameters ε and p (see Appendix in Kan et al., 2009). In this region one point attractor and two stable limit cycles exist. Such a limit cycle can be viewed as a switching between two states. With respect to epigenetics, we note that the system may follow multiple developmental pathways. However, which path is followed depends on the parameter values.

The means of parameters ε and p were set at .52, and .42, respectively. This parameter combination is located approximately in the center of region 5b. In this way, the variance of the parameters could be taken as large as possible (.0005 for both parameters), so we could minimize restriction of range. In the interest of biological plausibility, nonvarying parameter values were set identical to those in the study of van Oss and van Ooyen (1997); that is, $h = .1$, $\tau = .5$, $\alpha = .1$, and $b = .00005$. An exception was made for the value of q, which was set at .05. This parameter determines the outgrowth rate.

In simulation 2 we chose region 5a as parameter region. Region 5a differs only from region 5b in that the only attractor is the point attractor. Hence, multistability is not present in 5a. We set the parameter means of ε and p to values of .8, and .3, respectively. The variances of parameters ε and p as well as other parameter values were identical to the values used in simulation 1. We carried out simulation 2 to compare the differences between the amount of unexplained variance in case of multistability with the amount of unexplained variance in case of monostability. This difference can be regarded as the amount of nonlinear epigenetic variance.

Results

To determine discordance between twins, we used the relative power of x. A twin pair shows discordant behavior if the one member of a twin pair displays oscillating behavior, whereas the other member does not. Twin correlations were based on the filtered (smoothed) signals. The filtered signals of x and y still showed some oscillating behavior. As explained above, oscillations may lead to latency shifts, which yield lower observed twin correlations, which may affect estimates of heritability. In order to minimize such effects, we mainly concentrate on the analyses of the filtered signal of w.

As is common in behavior genetic research, we decomposed phenotypic variance into latent factors by fitting an ACE twin model, where A stands for additive genetic effects, C for common or shared environmental effects, and E for nonshared environmental effects. We did so over each of the 40 time intervals. In addition, in each simulation two stepwise regression analyses were carried out on the filtered signal of w over the last time interval ($19{,}500 \leq t < 20{,}000$). To describe the amount of variance explained by the model, we calculated the adjusted squared multiple correlation coefficient (R^2_{adjusted}) between all of the predictors and the dependent variable. We calculated the amount of unexplained variance as $(1 - R^2_{\text{adjusted}}) * 100\%$. In the regression model, the predictors were the variables p_g, p_c, p_e, ε_g, ε_c, and ε_e. As above, the dependent variable was the filtered signal of w. In the first analysis, the linear regression model (henceforth referred to as "G + E model") includes solely main effects for all independent variables. In addition to main effects, the second regression model (henceforth referred to as "G × E model") includes interaction terms for all possible combinations for all independent variables.

Simulation 1: Subregion 5b

In simulation 1, multiple stable attractors were present. Simulations using the two-cell model gave rise to bifurcations and multistability. The presence of bifurcations implied that small differences in parameter values could result in qualitatively distinct behavior: differences in developmental pathways and differences in final stable states (i.e., oscillating vs. nonoscillating behavior). Twin members could thus follow discordant developmental pathways. At the end state, 27.5% of the MZ twin pairs and 33% of the DZ twin pairs showed discordant behavior. In addition, even in the absence of noise, with time a substantial and increasing amount of unsystematic (nonlinear epigenetic) variance arose, which was attributed to the nonshared component in a linear behavior genetic analysis.

The effect of bifurcating processes was a lowering of twin correlations over time. Figure 7.2 illustrates the development of these correlations with respect to the filtered signal of *w*, together with the development of observed heritabilities and environmentabilities. As one can see, relatively high initial correlations (MZ correlations) are affected less than relatively low initial correlations (DZ correlations). In other words, compared to DZ twin similarity, MZ twin similarity was better preserved. As a consequence, observed common environmentabilities decreased to zero, whereas heritabilities and unique environmentabilities increased over time. Although not depicted, the development of twin correlations of the filtered signals of *x* and *y* gave qualitatively the same picture.

Fitting an ACE model to the twin covariance matrices of the filtered signal of *w* at the end state yielded good fits (χ^2 (3) = 0.495, p = .91). The variance attributed to the nonshared environmental component increased substantially (to 36.6%). The variance attributed to genetic effects also increased (to 63.4%), whereas estimated shared

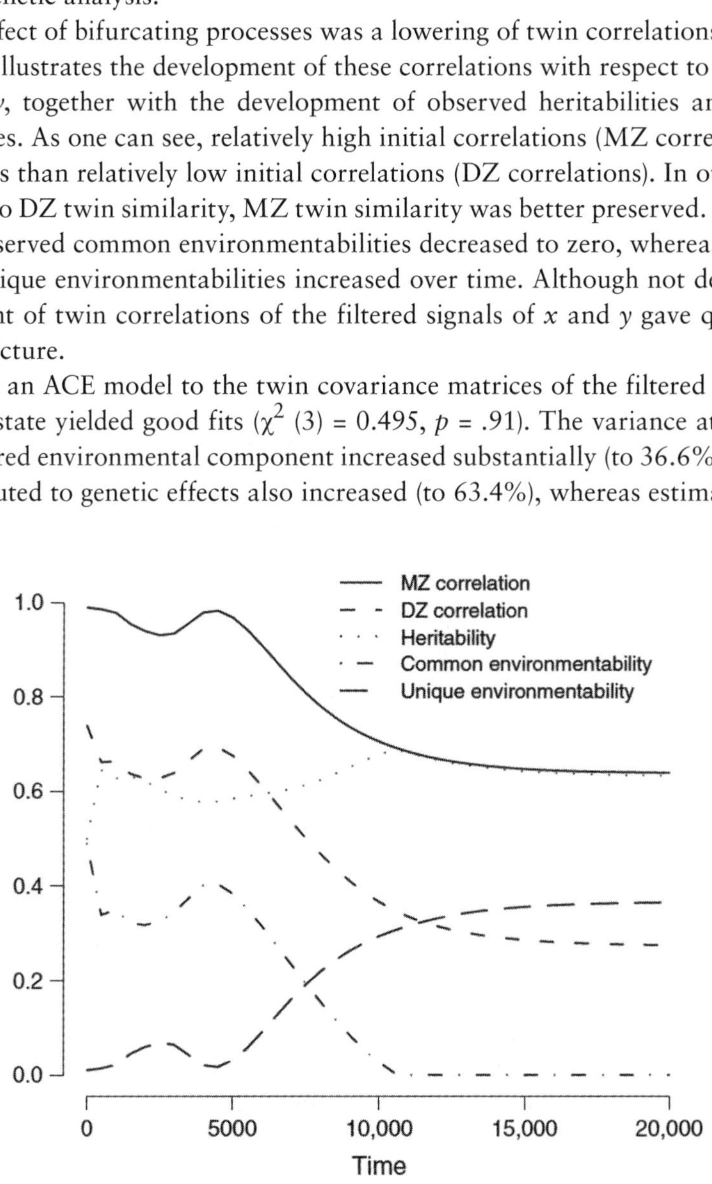

FIGURE 7.2. The development of MZ and DZ twin correlations, observed heritabilities, and observed environmentabilities of the filtered signal of *w* in case of multistability.

environmental variance disappeared (to 0.00%). As expected, the shared environment component (C) could be dropped without any reduction in fit relative to the ACE model (χ^2 (4) = 0.495, p = .974).

These results are in concordance with those of the stepwise regression analyses: In the G + E regression model on the filtered signal of w at the end state, a large part of the variance remained unexplained (58.1% in MZ twins and 63.3% in DZ twins). The model that included interaction terms (G × E model) did little to decrease the unexplained variance (53.1% in the MZ twins and 62.5% in the DZ twins). In other words, G × E interaction did little to account for the residual variance. Thus, nonlinear epigenetic variance was not subsumed under genotype–environment interaction.

As in latent variable modeling, in the regression analyses the genetic components (p_g and ε_g) explained a significant part of the variance in the dependent variable, the filtered signal of w. However, in contrast to latent variable modeling, in the regression analyses the shared environmental components could explain a significant part of the variance. Hence, we may conclude that the effects of shared environmental influences were not detectable in a standard twin design.

Simulation 2: Subregion 5a

In simulation 2, a single stable (point) attractor existed. Simulations using the two-cell model showed the implications of the absence of bifurcations and of multistability. Differences in parameter values did not result in qualitatively different phenotypes or different developmental pathways. As a consequence, discordance between twins was zero; nonlinear epigenetic variance did not arise. Behavior genetic models yielded good fits and (nearly) all variance was explained.

Figure 7.3 illustrates the development of phenotypic twin correlations of the filtered signal of w, together with the observed heritabilities and environmentabilities. Twin correlations remained stable over time. As a result, observed heritabilities and environmentabilities accurately reflected the initial genetic and environmental structure.

Fitting an ACE model on the filtered signal of w at the end state yielded good fits (χ^2 (3) = 0.00, p = .99). The percentage of variance attributed to genetic effects was 47.5%, whereas the percentages of variance attributed to the shared and nonshared environmental components were 51.7% and 0.90%, respectively. These results are in concordance with the results from the regression analyses: The G + E regression model on the filtered signal of w at the end state could explain nearly all of the variance (99.95% for the MZ twin data and 99.94% for the DZ twin data).

In the light of these results, we conclude that nearly all unexplained variance in simulation 1 could be attributed to nonlinear epigenetic variance. The unexplained variance was mainly the result of the disproportional relationships between causes and effects within the nonlinear process. This variance was subsumed under the nonshared environmental component in the linear behavior genetic model. Furthermore, nonlinear epigenetic variance is distinct from genotype–environment interaction. Last, the true genetic and environmental causes of phenotypic variance were not detected in the standard twin design.

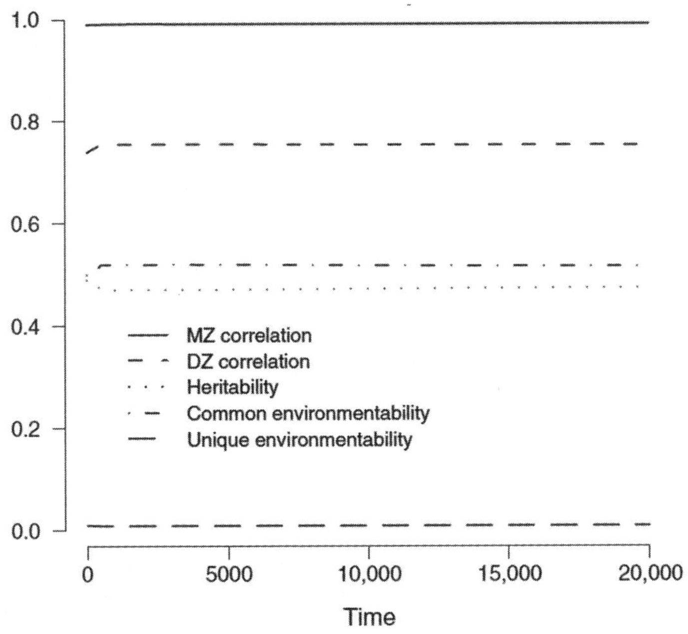

FIGURE 7.3. The development of MZ and DZ twin correlations, observed heritabilities, and observed environmentabilities of the filtered signal of *w* in case of monostability.

Conclusions

In the present chapter we first reviewed the literature on empirical evidence that suggests that a substantial portion of phenotypic variance is due to nonlinear (epigenetic) processes during ontogenesis. In part, such processes may be responsible for the ubiquitous nonshared environmental variance in human psychology. Second, we simulated twin data using a realistic biological model of neurite outgrowth that displayed the characteristics that were hypothesized to underlie nonlinear epigenetic variance. The results of our simulations demonstrate that the existence of multiple attractors due to bifurcations gives rise to phenotypic variance in simulated neuronal development. This variance cannot be attributed to genetic or environmental variance in the parameters, to measurement error, or to any other external random process that introduced noise into the system. Hence, the variance can be interpreted as nonlinear epigenetic variance. In a standard behavior genetic (twin) analysis, this type of variance is subsumed under the nonshared environmental component. The depressing effects of nonlinear epigenetic variance were greater on DZ than on the MZ twin correlation. As a consequence, heritabilities appeared to increase, and shared environmentabilities appeared to decrease to zero.

Our results are in concordance with the empirical evidence we reviewed, in which a substantial portion of phenotypic variance, provisionally attributed to unshared (nonshared) environmental influences, actually remains unexplained. The variance may be due to nonlinear sources, which probably include endogenous molecular or cellular processes (Figure 7.4). The hypothesis that nonlinear epigenetic processes may represent an appreciable source of variance may help to explain why nonshared effects are extremely hard to identify, and why systematic attempts thus far have had only limited success (Turkheimer & Waldron, 2000; Zuk et al., 2012).

The results of the simulations are also consistent with the common findings from behavior genetic studies in psychology. That is, over time the estimated relative contribution of nonshared variance to phenotypic variance increased, whereas shared variance decreased to zero. As a consequence, estimated heritabilities increased, which is often observed in longitudinal twin studies (e.g., Bouchard & McGue, 2003). Finally, monozygotic twins were more than twice as similar as dizygotic twins, which is also frequently observed (Turkheimer, 2004).

Our interpretation of nonshared environmental variance as consisting in part of nonlinear epigenetic variance is consistent with Molenaar and colleagues (1993), who

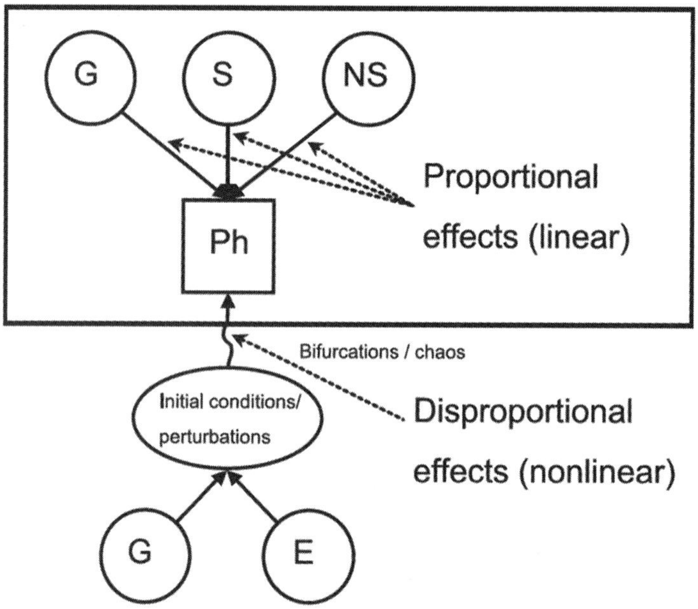

FIGURE 7.4. Phenotype (Ph) as a function of linear and nonlinear influences. In a linear behavior genetic model, the environmental component is decomposed into shared environment (S) and nonshared environment (NS). Error variance is subsumed under the nonshared component. In contrast to the modeled linear (proportional) effects, chaotic and bifurcating (epigenetic) processes will have unsystematic (disproportional) effects. These will be difficult to distinguish from error. The genetic (G) and environmental (E) influences on initial conditions or perturbations will be hard to detect. Although the ultimate influences on phenotype are only genetic and environmental, nonlinear (epigenetic) processes constitute a distinct and independent source of variance.

consider nonlinear epigenetic processes as a third source of variance, alongside genetic and environmental influences. According to the definition of nonshared variance in behavior genetics, nonlinear epigenetic variance is subsumed under the nonshared environmental component, even though initial conditions may be determined mainly by genetic effects and shared environmental influences.

ACKNOWLEDGMENTS

This chapter is a modified version of Kan, Ploeger, Raijmakers, Dolan, and van der Maas (2009). Adapted with permission from John Wiley & Sons, Inc. Figures 7.1–7.4 are reprinted with permission from John Wiley & Sons, Inc.

REFERENCES

Archer, G. S., Friend, T. H., Piedrahita, J., Nevill, C. H., & Walker, S. (2003). Behavioural variation among cloned pigs. *Applied Animal Behaviour Science, 82*, 151–161.

Arnold, V. I., Afrajmovich, V. S., Il'yashenko, Y. S., & Shil'nikov, L. P. (1994). *Bifurcation theory and catastrophe theory.* Berlin: Springer-Verlag.

Beloussov, L. V. (2006). A morphomechanical aspect of epigenesis. *Russian Journal of Genetics, 42*(9), 966–969.

Benno, R. H. (1990). Development of the nervous system: Genetics, epigenetics and phylogenetics. In M. E. Hahn, J. K. Hewitt, N. D. Henderson, & R. H. Benno (Eds.), *Developmental behaviour genetics: Neural, biometrical, and evolutionary approaches* (pp. 113–143). New York: Oxford University Press.

Bouchard, T. J. (2004). Genetic influence on human psychological traits. *Current Directions in Psychological Science, 13*(4), 148–151.

Bouchard, T. J., & McGue, M. (2003). Genetic and environmental influences on human psychological differences. *Journal of Neurobiology, 54*(1), 4–45.

Camazine, S. (2001). *Self-organization in biological systems.* Princeton, NJ: Princeton University Press.

Charney, E. (2012). Behavior genetics and post genomics. *Behavioral and Brain Sciences, 35*, 331–358.

Dolan, C. V., & Molenaar, P. C. M. (1995). A note on the scope of developmental behaviour genetics. *International Journal of Behavioral Development, 18*, 749–760.

Eaves, L. J., Kirk, K. M., Martin, N. G., & Russell, R. J. (1999). Some implications of chaos theory for the genetic analysis of human development and variation. *Twin Research, 2*(1), 43–48.

Eckert, M. A., Leonard, C. M., Molloy, E. A., Blumenthal, J., Zijdenbos, A., & Giedd, J. N. (2002). The epigenesis of planum temporale asymmetry in twins. *Cerebral Cortex, 12*, 749–755.

Ferrie, J. E. (2012). Incidents, incidence and golden eggs. *International Journal of Epidemiology, 41*, 329–332.

Fink, B., Manning, J. T., Neave, N., & Grammer, K. (2004). Second to fourth digit ratio and facial asymmetry. *Evolution and Human Behaviour, 25*(2), 125–132.

Gärtner, K. (1990). A third component causing random variability beside environment and genotype: A reason for the limited success of a 30 year long effort to standardize laboratory animals? *Laboratory Animals*, *24*(1), 71–77.
Guckenheimer, J., & Holmes, P. (2002). *Nonlinear oscillations, dynamical systems and bifurcations of vector fields.* New York: Springer.
Jablonka, E., & Lamb, M. J. (2002). The changing concept of epigenetics. *Annals of the New York Academy of Sciences*, *981*, 82–96.
Kan, J. K., Ploeger, A., Raijmakers, M. E. J., Dolan, C. V., & van der Maas, H. L. J. (2009). Nonlinear epigenetic variance: Review and simulations. *Developmental Science, 13*(1), 11–27.
Kan, K. J., Boomsma, D. I., Dolan, C. V., & van der Maas, H. L. (2012). Commentary: The presence of bifurcations as a "third component of individual differences": Implications for quantitative (behaviour) genetics. *International Journal of Epidemiology, 41*(2), 346–351.
Kelso, J. A. S. (1995). *Dynamic patterns: The self-organization of brain and behavior.* Cambridge, MA: MIT Press.
Kirillov, A. B., Myre, C. D., & Woodward, D. J. (1991). Bistable neurons and memory patterns in the inhibitory-feedback model inspired by neostriatum. *Society for Neuroscience Abstracts*, *17*, 51.
Lewontin, R. (1974). The analysis of variance and the analysis of causes. *American Journal of Human Genetics*, *26*, 400–411.
Lickliter, R. (2012). From gene activity to behavior (and back again): Commentary. *Behavioral and Brain Sciences, 35*, 369–370.
McGue, M., & Bouchard, T. J. (1998). Genetic and environmental influences on human behavioral differences. *Annual Review of Neuroscience*, *21*, 1–24.
Meinhardt, H. (1982). *Models of biological pattern formation.* London: Academic Press.
Molenaar, P. C. M., Boomsma, D. I., & Dolan, C. V. (1993). A third source of developmental differences. *Behaviour Genetics*, *23*, 519–524.
Molenaar, P. C. M., & Raijmakers, M. E. J. (1999). Additional aspects of third source variation for the genetic analysis of human development and behavior: A commentary on Eaves et al. *Twin Research*, *2*(1), 49–52.
Molenaar, P. C. M., & Raijmakers, M. E. J. (2000). A causal interpretation of Piaget's theory of cognitive development: Reflections on the relationship between epigenesis and nonlinear dynamics. *New Ideas in Psychology*, *18*(1), 41–55.
Oyama, S. (1985). *The ontogeny of information: Developmental systems and evolution.* New York: Cambridge University Press.
Petzoldt, T. (2003). R as a simulation platform in ecological modeling. *R News*, *3*, 8–16.
Ploeger, A., van der Maas, H. L. J., & Raijmakers, M. E. J. (2008). Is evolutionary psychology a metatheory for psychology?: A discussion of four major issues in psychology from an evolutionary developmental perspective. *Psychological Inquiry*, *19*, 1–18.
Plomin, R., DeFries, J. C., McClearn, G. E., & McGuffin, P. (2008). *Behavioral genetics* (4th ed.). New York: Worth.
Prigogine, I. (1980). *From being to becoming: Time and complexity in the physical sciences.* San Francisco: Freeman.
Steinmetz, H. (1996). Structure, function and cerebral asymmetry: In vivo morphometry of the planum temporale. *Neuroscience and Biobehavioural Reviews*, *20*, 587–591.
Stent, G. S. (1978). *Paradoxes of progress.* San Francisco: Freeman.
Stige, L. C., Slagsvold, T., & Vøllestad, L. A. (2005). Individual fluctuating asymmetry in pied

flycatchers (*Ficedula hypoleuca*) persists across moults, but is not heritable and not related to fitness. *Evolutionary Ecology Research, 7*, 381–406.

Thoma, R., Yeo, R. A., Gangestad, S. W., Lewine, J., & Davis, J. (2002). Fluctuating asymmetry and the human brain. *Laterality, 7*, 45–58.

Turkheimer, E. (2000). Three laws of behaviour genetics and what they mean. *Current Directions in Psychological Science, 9*(5), 160–164.

Turkheimer, E. (2004). Spinach and ice cream: Why environmental social science is so difficult. In L. F. DiLalla & I. I. Gottesman (Eds.), *Behavior genetics principles: Perspectives in development, personality and psychopathology* (pp. 161–189). Washington, DC: American Psychological Association.

Turkheimer, E., & Gottesman, I. I. (1996). Simulating the dynamics of genes and environment. *Development and Psychopathology, 8*(4), 667–677.

Turkheimer, E., & Waldron, M. (2000). Non-shared environment: A theoretical, methodological, and quantitative review. *Psychological Bulletin, 126*(1), 78–108.

van der Maas, H. L. J., & Molenaar, P. C. M. (1992). Stage-wise cognitive development: An application of catastrophe theory. *Psychological Review, 99*(3), 395–417.

van Oss, C., & van Ooyen, A. (1997). Effects of inhibition on neural network development through activity-dependent neurite outgrowth. *Journal of Theoretical Biology, 185*, 263–280.

Waddington, C. H. (1957). *The strategy of the genes*. London: Allen & Unwin.

Wright, I. C., Sham, P., Murray, R. M., Weinberger, D. R., & Bullmore, E. T. (2002). Genetic contributions to regional variability in human brain structure: Methods and preliminary results. *NeuroImage, 17*, 256–271.

Zuk, O., Hechter, E., Sunyaev, S. R., & Lander, E. S. (2012). The mystery of missing heritability: Genetic interactions create phantom heritability. *Proceedings of the National Academy of Science USA, 109*, 1193–1198.

CHAPTER 8

Dynamical Systems Thinking

From Metaphor to Neural Theory

Gregor Schöner

The Fundamental Tension in Development between Stability and Flexibility

As parents, we experience our children as unique individuals. When we have more than one child, we marvel at how each child is special from the very first day. On the other hand, our entire educational attitude is based on the assumption that children's experiences shape their development and that the environment to which they are exposed may make a difference in how they will develop.

As developmental scientists we have seen, time and again, how infants, toddlers, and children go through recognizable stages of development, formalized by developmental milestones. These stages invite accounts in which development resembles a program from which competences unfold in a fixed sequence. Growth processes in developmental biology provide the metaphor for how molecular processes could lead to maturation of the nervous system that may drive behavioral and cognitive development.

At the same time, there can be no doubt that the environment to which children are exposed matters and that their individual behavioral history influences the process of development. This observation invites accounts based on learning, in which the nervous system of the developing child is shaped by the activity induced by ongoing behavior and stimulation.

The Self-Organization Metaphor

One way developmental scientists have addressed this tension between the apparent stability of the developmental process and its flexibility in response to variable environments and experiential histories is by invoking the concept of self-organization. In a self-organizing system, the flexible and individual processes of development may emerge from the confluence of various forces, while the inherent organizational principles would hold the developmental process on track. Exactly how that may work requires closer examination.

This idea has been promoted as a metaphor for development that has inspired a research program into possible mechanisms of emergence, of how environmental factors contribute to development, how individual differences persist, and how the ensemble of these influences jointly control development ("soft causation") (Elman et al., 1997; Spencer et al., 2006; Thelen & Smith, 1994). This chapter reviews this metaphor, criticizes it, and proposes a specific direction in which this metaphor can be transformed into an operational theory of the development of behavior and cognition.

The perspective of self-organization was developed, in part, in opposition to an apparent alternative; that is, to an account of development that is based on maturation and innateness. In such an account the molecular machinery of growth guarantees the reproducible patterns of development and also explains how individual differences arise from different starting conditions, largely determined by genetic factors, and remain stable over the course of development (Fodor, 1981; for critiques, see Molenaar, 1986; Quartz, 1993).

Interestingly, the self-organization metaphor connects to older theoretical ideas about growth processes. In fact, within developmental biology an analogous tension exists between two metaphors, one based on information processing and programming, the other on self-organization and emergence. The fine-grained molecular machinery of growth engages gene expression, which may be thought of as the core driver of morphogenesis, which is the formation of macroscopic shape from microscopic processes. At this molecular level, information-processing metaphors such as *program, lock and key*, and *reading* or *writing* of genetic information, and so on, are commonly invoked. These metaphors have their own interesting conceptual history (Fox Keller, 2002). Molecular biologists resonated with these concepts as these seemed to fit to their experimental tools.

In an older view, morphogenesis had been thought of as a form of pattern formation. This older view dates back to Waddington's famous epigenetic landscape (Waddington, 1953; see also Ho, Chapter 5, and Newell & Liu, Chapter 12, this volume and Figure 8.1), often invoked but somewhat inconsequential to modern developmental biology. The metaphor suggests that "forces" shape the form of an organism, which then emerges from a process of equilibration of these forces. Over development, the landscape of forces and the associated equilibria become increasingly complex, leading to a more and more differentiated organism. This metaphor resonates with the famous mathematical model of pattern formation of Alan Turing (1952), in which the interaction of diffusion and chemical reaction kinetics generates concentration patterns. The structure of these patterns is encoded in the parameters of the chemical and diffusion dynamics that are hypothesized to drive growth. That and similar models remained essentially metaphorical too, as they did not make contact with experimentally accessible problems in development.

This changed somewhat in the 1980s, when experimental evidence was obtained that the substrate of such pattern formation, promoters of the growth of cell membranes, are at work in development. Mathematical models of pattern formation in morphogenesis (Gierer & Meinhardt, 1972; Murray, 2002) were based on the diffusion and chemical reaction kinetics of an activator and inhibitor of growth. The formation of chemical concentration patterns then creates a skeleton for growth processes. The theory was based on the mathematical framework of nonlinear dynamical systems, typically in the form of partial differential equations that undergo instabilities (Haken, 1983; Murray, 2002). The theory can account for features of morphogenesis, such as the qualitative form of misformation. For instance, an individual may have an extra finger, but the overall shape of the fingers and the hand is preserved. In the mathematical models, this is a property of the self-organization processes on which growth is based. The boundary conditions for a piece of tissue only allow for a discrete set of possible patterns. Among these, one typically emerges as the stable solution, but a perturbation or change of conditions may induce a neighboring pattern to "win" the competition (Murray, 2002). In fact, more generally, the study of malformations may provide provocative insight into development (Blumberg, 2010).

So, in a sense, in the domain of morphogenesis, the metaphor of Waddington has been shaped into mathematical models that have some traction empirically and are even, to some extent, reductionistic in nature. Is this the sort of transition from metaphor to theory at which we aim for an understanding of the development of behavior and cognition? Perhaps not. Even after the discovery of promoters, the mathematical theory of morphogenesis has had very little impact on the research program of developmental biologists, largely because the methods of molecular biology that have been most successful experimentally do not match the kinds of questions the theory promotes (Fox Keller, 2002). (This may begin to change as computational biology brings the mathematics of dynamical self-organization to the molecular level; see Harrison, 2011, for a survey).

Conversely, conceptually challenging and interesting questions that flow out of the metaphor of self-organization are not addressed by the mathematical theory. In particular, the metaphor of emergence suggests that growth processes are open to multiple forces that may literally help shape the organism. This idea has been at the core of the research program of developmental systems theory (Gottlieb, 2001; see also Griffiths & Gray, 2005, for a history of thought), which focuses on nonobvious interactions between the emerging function of a developing organism and the growth processes taking place through the environment, and modulating and mediating developmental processes.

Although the mathematics of self-organization may, in principle, be able to address such issues, models have not really played a critical role in this field. This is probably again due to a mismatch in levels of description, with the substrate of the mathematical theory being quite removed from the level of observation at which developmental systems theory seeks evidence for environmental influences.

From this history of ideas in developmental biology and morphogenesis, an important lesson can be derived for how the metaphor of self-organization can be transformed into theory. Clearly, that transformation requires more than mere mathematization. Theoretical concepts must relate to the level of description at which development is characterized experimentally and must be able to articulate the role of the various factors found to impact on developmental processes.

The Dynamical Systems Metaphor for the Development of Behavior and Cognition

The metaphor of pattern formation and the associated mathematical concepts of dynamical systems theory have been used to talk about how behavior and cognition may emerge from learning. Thelen and Smith (1994), for instance, interpreted Waddington's epigenetic landscape in terms of emergent behavioral and cognitive competences. At any particular stage of development, a landscape is a potential function, whose minima represent behaviors that can be realized at that particular stage of development (Figure 8.1). Because the minima are thought to arise from the different forces that shape the behavioral landscape, they are not fixed competences, but emerge as a function of environmental conditions and of an individual's inner state; they are "soft-assembled" in the moment.

Over development, the landscape is postulated to evolve, becoming increasingly differentiated and complex to reflect the increasing number of adaptive behaviors that

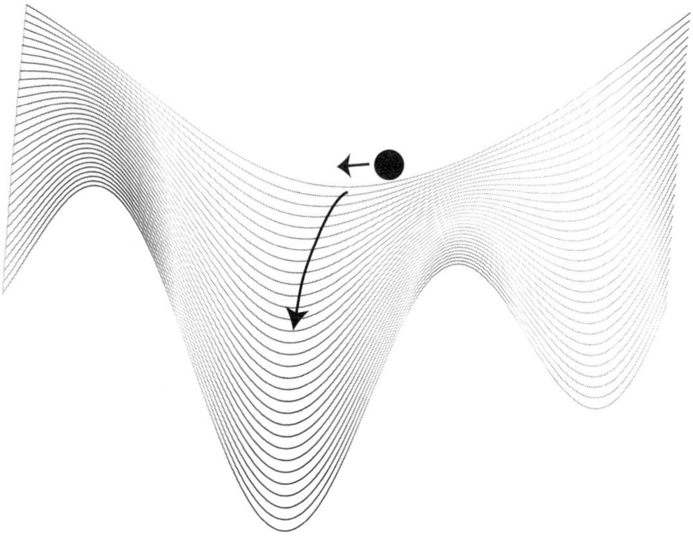

FIGURE 8.1. The epigenetic landscape visualizes the emergence of behaviors within a dimension (horizontal) as a potential landscape (vertical) evolves over time (back to front). The system (black ball) moves to minima of the potential on the fast timescale of behavior (short arrow). These minima may change and new minima may appear on the slow timescale of development (long arrow).

are the new competences acquired over development. In the metaphor, the behaviors expressed by the system contribute to the developmental processes. One way to visualize this in the metaphor is to think of the system as moving within the landscape and carving out new minima. In this picture, individually different routes in development may emerge from the way ongoing behavior drives the developmental process.

This metaphor has important strengths. First, it promotes a process account of development by directing attention to how behavior unfolds in time under the influence of various factors. This perspective stands in contrast to the emphasis on description and normalization that a maturational account suggests by decoupling development from the moment-to-moment experience of an individual. Process accounts are inherently stronger and more naturally open to intervention.

Along the same line, the metaphor emphasizes that environmental conditions must be taken into account to understand how competences emerge during development. When environmental conditions are favorable, a child may be able to perform at a more mature level than when conditions are more difficult. Thus, during development, the environmental constraints required for a child to display a particular competence may relax. This is an important insight.

The metaphor draws attention to learning as the core process of development. How experience matters for development is the central research question that the metaphor proposes for developmental science. The shaping of development by experience amplifies the role of environmental conditions: Because the environment may support, to varying degrees, the emergence of skilled behavior, it influences what experiences are accessible to the child and thus impact on the developmental process. Scaffolding, in which caregivers structure the environment to facilitate the emergence of new skills, makes immediate sense in this perspective (Lipscomb, Swanson, & West, 2004). Educators and parents are effective by enabling a child to gather his or her own experience.

Finally, the metaphor emphasizes individual differences and views them not as fixed traits, but as the result of the developmental process. Because experience shapes development, differences in experience have long-lasting impact and may lead to individual paths of development. This explains how individual differences may persist without attributing them necessarily to fixed structural differences. In fact, the metaphor suggests ways in which individual differences may emerge and then be amplified during development. Again, the implications for educators and parents are broad.

On a closer look, however, the dynamical systems metaphor is unclear and fuzzy in many ways. Consider the image of the epigenetic landscape again. Implicit in this image is a notion of a timescale: the slower timescale of development and the faster timescale of behavior. At a given moment during development, the different minima of the landscape represent the behavioral repertoire accessible at this time. How are these behaviors described? Are they embedded within a single space (illustrated along one dimension, but surely spanning multiple dimensions)? Are all behaviors equally accessible or does the distance between them matter? What are the processes of selecting a particular behavior? The metaphor is quite fuzzy about how behavior unfolds at any given moment during development.

The metaphor emphasizes that behavior is generated by an embodied and situated organism. However, how behavior is responsive to environmental conditions is obscured in the metaphor. Possible behaviors do not sit somewhere as minima, waiting for the state of the system to "fall" into the well. Instead, possible behaviors emerge "on the fly" in response to environmental conditions and the inner state of the system. For example, *looking* emerges in response to interesting things to look at and is strongly cued by salient visual transients. *Reaching* emerges when there are attractive objects within reach. Previous reaches affect new reaches.

The metaphor derives strengths from the idea that behaviors are shaped by the joint action of forces, the very notion of dynamics, from which stable states emerge. The landscape illustrates this notion through the minima to which the system is attracted. The time needed to move to the nearest minimum is really the timescale of behavior (Newell, Liu, & Meyer-Kress, 2009). But how can we conceive of these forces? How does the influence of the environment and inner states impact on the stability of a behavior? Behaviors may vary in stability. When environmental conditions change, behaviors may lose stability (Schöner & Kelso, 1988). So clearly, the shape of the minima vary with context. The dependence of the stability of behaviors on many factors, including the environment, intention, and attention, is an important dimension of development that needs to be made more explicit.

Relatedly, selecting a particular behavior involves active choice or decision processes. An infant may look, grasp, look away, or persist. This flexibility is not visualized in the metaphor, and the underlying process of decision making is another dimension of behavior that has its own dynamics and timescale (Newell et al., 2009). This point is critical because in many instances it is this very flexibility that develops. The *A*-not-*B* error is a signature of a lack of flexibility when the goal of a reaching act is changed. Developmentally, this flexibility is an achievement. Similarly, younger infants tend toward stereotypy—the inflexible persistence of behavioral patterns—and over development become increasingly flexible. Looking behavior is a well-studied example, in which very long fixations may occur early on, whereas gaze shifts become increasingly nimble over development (Colombo, 2001).

Finally, the metaphor is suggestive of overt motor behavior as the primary mark of competence, realized while the system is sitting in a minimum of the landscape. Cognition entails inner states that affect overt motor behavior. Through working memory, for instance, perceptual or motor processes at one point in time impact on decisions at a later time. This is one factor in explaining how older infants succeed in overcoming perseveration in the *A*-not-*B* paradigm. Older infants are capable of stabilizing a motor intention to reach for the *B* location during the delay. In the simple form suggested by the potential landscape, the dynamical systems metaphor does not provide a substrate for such inner states. Such inner states may have their own dynamics—for instance, in the process of memory formation—that live at an intermediate timescale, longer than behavior in the here and now, and shorter than the timescale over which development takes place.

So let us turn to the slower timescale of development on which the landscape itself evolves. The core hypothesis is that the developmental process is driven by experience, that is, by the behaviors (and inner states) that are realized on the faster timescale of behavior. How this happens remains vague in the metaphor. This is the major frontier, of course, for a process account of development. Here are a few questions that illustrate how much remains to be explored.

- What is the substrate for learning and development? Inner states are not obvious in the simple form of the dynamical systems metaphor. Connectionism proposes that change occurs in how the inner states are connected to the sensory and motor surfaces. But there may also be change in just the (recurrent) connectivity internal to the neural networks that drive behavior.
- What information is used to drive the learning process?
- Is learning largely driven by the statistics of afferent input, to which unsupervised learning rules are sensitive?
- Does it require a sensory feedback signal or reinforcement?
- Is the system open to learning in any of its subsystems, or is there an inherent organization of learning, in which one system is the focus of learning for some time and that focus is then shifted to the next system when a given level of performance has been reached?
- How does the system maintain a coherent learning process across the different episodes of experiencing a particular aspect of behavior or of stimulation?
- How does it autonomously control its learning processes?
- How do the learning processes actually lead to differentiation?
- Why does learning a new skill not unlearn a former skill? Is there a need for relearning old skills to stabilize them?

Clearly, at this point the self-organization metaphor has only scratched the surface of what a true process account of learning and development must account for. By way of summary, here are the three main challenges for a transition from the self-organization metaphor to dynamical systems theory of development.

- First, we need a much better, concrete account for what happens at the timescale of behavior. This account would include a systematic method to characterize behavior as the performance of a competence. This method must address how the environment—how stimuli and perception—shapes behavior and how ongoing behavior may be coupled with perception, but may also be conditioned by inner states, a particular mode of behavior that classically was called "task" and is all the more challenging to define for infants and children. And this method must characterize cognitive processes as a form of behavior. Not every behavioral state leads to immediate and overt motor action.

- Second, we need to account for flexibility and cognition that includes a systematic process understanding of how decisions are made in the here and now, and how those decisions are linked to sensory information and may generate motor consequences. This also requires a substrate for representations and an account of the processes through which sensory and motor processes interface with representations and create, update, or remove memories.

- Third, we need a much more profound approach toward the processes of learning that are hypothesized to take place during development. Connectionism and dynamical systems thinking have only given first hints at the autonomous learning processes in which infants, toddlers, children, and even adults are permanently engaged.

From the Dynamical Systems Metaphor to Dynamic Field Theory as a Theoretical Framework for Understanding Development

What does the transition from metaphor to theory entail? This section provides a rough sketch and introduces the main concepts of dynamic field theory (DFT; for reviews, see Schöner, 2008). In the next section, we consider the same ideas at a more pedestrian pace by using two worked examples to make things concrete and simple.

We begin with the issue of how to characterize behavior in a way that is open to sensory and motor processes, but may also reach into cognition. The answer, perhaps not surprisingly, comes from the neural level of description. We know that neuronal activity in the central nervous system is determined in part by the pattern of connectivity from the sensory surfaces to the neurons in question. Their tuning curves to parameters of sensory stimulation delineate the extent of this influence. In this sense, a neuron "stands for" the stimulus condition to which it is sensitive. Conversely, at the output level the forward connectivity from a neuron to the motor surface determines the motor "meaning" of that neuron's activity. Tuning to movement parameters estimates the extent of the motoric specificity of neural activation.

We also know that tuning curves throughout the higher nervous system, in the cortex, but also in such subcortical structures as the colliculus or the lateral geniculate nucleus, are relatively broad, so that when any single stimulus is presented or any individual motor act prepared, a considerable portion of the relevant neural populations is activated. Evidence for such population coding of perceptual and motor parameters comes from various sources that include the broad tuning curves of cortical neurons to behavioral parameters (Georgopoulos, Schwartz, & Kettner, 1986; Young & Yamane, 1992), the contribution that all activated neurons make to behavioral decisions (e.g., Groh, Bron, & Newsome, 1997; Lee, Rohrer, & Sparks, 1988), and the patterns of correlation between neurons and behavior (see contributions to Kriegeskorte & Kreima, 2012).

These facts support the proposal that DFT makes for the substrate of behavior and cognition: The activation of neuronal populations stands for particular perceptual states or motoric outcomes. These activation patterns need not be traced back to how they are distributed within the neural networks of the brain. Instead, we can think of these patterns as being defined directly in terms of the sensory and motor states to which the populations are sensitive. Thus, for instance, a distribution of activation may be defined over retinal visual space, over color space, or over the space of movement direction of visual objects. Similarly, a distribution of activation may be defined over the space of possible movement directions, movement extent, or amount of force generated. These distributions are the activation fields on which DFT is based.

Peaks of activation are then instances of a perceptual or motor state (see Figure 8.2). Generating such a peak amounts to falling into one of the minima of the potential landscape (see Figure 8.1). Note, however, that the peaks are not all standing there and waiting to be selected. In fact, typically a single peak or a small number of peaks may be activated. And peaks are tunable: They can be localized at different positions along the behavioral dimensions for which they stand, representing different perceptual or motor outcomes. Peaks can directly drive motor behaviors, or they can drive other representations to build other peaks. The processes of peak formation, peak updating, and peak deletion are the neural dynamics of neuronal populations. Input plays an important role in these processes. Environmental condition may impact on activation fields by generating patterns of input that come through the sensory surfaces.

How the forward connectivity determines input patterns is the central topic of connectionist modeling, and DFT shares this principle with connectionism. In connectionism, cascades of layered input stages may lead to increasing abstraction in sensory representations. The motor side is much less well understood, in large part because it is not well described by the feedforward conception. In DFT, even perceptual events are not purely input driven. Peaks of activation are postulated to be stable states, attractors of the neural dynamics, so that they can be continuously coupled to sensory information. Recurrent connectivity within the neural population stabilizes such peaks against decay and against competing inputs. This is based on the principle of local excitation, according to which similarly tuned neurons are exciting each other, and global inhibition, according to which differently tuned neurons are effectively inhibiting each other (through interneurons).

A consequence of this principle of the stabilization of activation peaks is that such peaks are separated from nonpeak patterns of activation by an instability, in which input-driven patterns of activation reach a critical level of activation, at which the detection decision is made. The instability comes from the nonlinear properties of neural

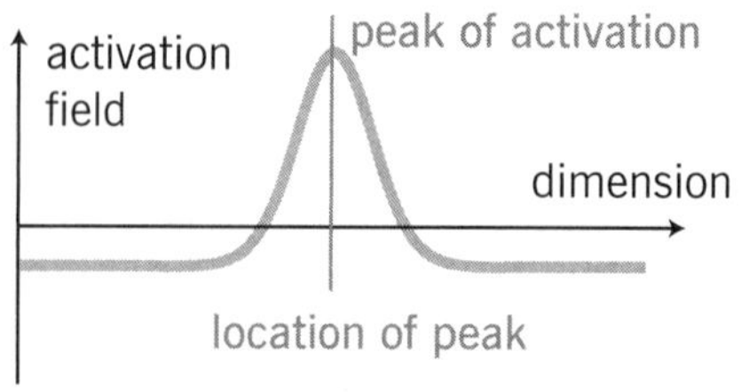

FIGURE 8.2. An activation field defined over a perceptual or motor dimension has a peak whose location specifies a particular value along that dimension.

dynamics, in which only sufficiently activated neurons in a population contribute to neural interaction, as described by a sigmoidal function. The concept of self-stabilized peaks of activation, sustained by neural interaction, is a central tenet of DFT, in which it departs from standard connectionist thinking (although there are connectionist models that entail the same kind of instabilities, e.g., Usher & McCelland, 2001). We consider the detection instability again in the next section (see Figure 8.5).

Activation peaks are units of representation in DFT. In whichever way a peak was induced, it resists change and impacts on whatever downstream neural networks or motor systems it projects. When a peak is induced by a single localized input, the decision to detect this input resists change: It persists even if the input is weakened or noisy. When multiple localized inputs are presented, an activation field may make a selection decision by generating a single peak located over the input that arose first or was strongest. That selection decision resists change (up to a point) when, for instance, distractor inputs impinge on the system. In fact, under appropriate dynamic conditions for the activation dynamics, a peak may persist after all localized input has been removed. That persistence provides an account for working memory of metric information, represented by the location of the peak in the activation field.

The stability of activation peaks has important consequences for learning processes. The detection instability may, essentially, amplify small inhomogeneities in a neural population into a macroscopic decision represented by a peak. It may be sufficient, for instance, to just push global activation into a field, without any specific information about the location to activate; that is, no specific information about the desired state or behavior. As soon as a field location creates enough activation so that the activation engages the sigmoidal function and interaction sets in, an entire peak is generated, essentially pulling itself up through its own excitatory interaction. The locations at which this is possible are dictated by inhomogeneities—that is, by small, competitive advantages for field locations through prior activation or strengthened synaptic connections to any input structures.

Learning Processes Are Also a Form of Neural Dynamics in DFT

The concept of self-stabilized peaks lowers demands on learning process. In conventional, largely feedforward neural networks, learning must shape synaptic connectivity sufficiently to make a difference at the output level, for instance, to generate a new "winner." In DFT, even small biases may be sufficient to lead to the formation of new peaks. As result, even very simple learning mechanisms, such as a memory trace of prior patterns of activation, may have a significant impact on a neural dynamics.

Such memory traces are, in a sense, a zeroth order learning mechanism because they are sensitive only to the history of activation, not to correlations between patterns of input and patterns of output. Hebbian learning, which is sensitive to such correlations, may, of course, also shape the field and determine which activation patterns are most easily induced.

Both the laying down of memory traces and the strengthening of connections according to a Hebbian rule can be understood as a form of dynamics at a slower time-scale (Erlhagen & Schöner, 2002; Schöner, 2008). These dynamics converge toward the learned state, which is a stable state of the learning rules. The learning dynamics are driven by the ongoing activation in the system. This is how experience matters and shapes the future network.

Understanding the dynamics of learning deeply is the theoretical frontier in our understanding of development. In spite of much work by connectionist modelers and the demonstration of the power of the memory trace within DFT, most of the fundamental questions about learning as a process have not been addressed. One big question is how learning is organized autonomously, so that a reproducible sequence of learning phases emerges from the experience of infants and children. How is consistency created across the varying experiences over the course of a day or of weeks, with continuing progress on a particular front at a particular stage of development? What role does reward play, and what are intrinsic rewards?

Worked Examples of DFT

Neural Grounding of Fields and Neural Dynamics

DFT is based on the hypothesis that neural processes control behavior through the activity of populations of neurons. To be specific, consider the preparation of a movement of the hand toward an object or location. Recording from the motor and premotor cortical areas of macaque monkeys, researchers have found populations of neurons whose firing is modulated in time, reflecting the movement task (Georgopoulos et al., 1986). These neurons begin to fire as the movement is prepared, and then stop firing at some point early in the actual hand movement. Moreover, the firing rate of the neurons depends on the spatial direction of the hand movement. That dependence is captured by the tuning curve, depicted schematically in Figure 8.3, which represents the neural firing rate as a function of movement direction. The tuning curves of most cortical neurons are single-humped and broad and can be approximated by a cosine function centered on a "preferred movement direction" of each neuron. The cosine function is the broadest smooth period function of direction, so tuning is broad in an objective sense. By implication, whenever a single movement direction is specified, a large ensemble of neurons fires—all those whose preferred direction lies within a range of about 180 degrees from the specified direction. Approximately half of the entire population of neurons tuned to movement direction is active when any specific movement is being prepared. Similarly broad patterns of activation have been observed in many other cortical and subcortical areas.

The activation of a population of neurons that is tuned to a task can be visualized as the formation of a peak in a field of neural activation. In some cortical areas such as the visual and other primary sensory cortices, neighboring neurons tend to have similar tuning curves. In such cases, neurons activated in the task will tend to be neighbors. The peak of activation can be thought of as sitting on the cortical surface, over which the neural population is distributed. In other areas, such as the motor and premotor cortex, neighboring neurons may have quite different tuning curves. The neurons activated in a task may be broadly distributed over the cortical area. We can still visualize the activation pattern as the formation of a peak, however, if we reorder the neurons in a virtual (or functional) space, in which neurons are put next to each other if they have similar tuning curves. The concept of a distribution of population activation (DPA) does this job. Each neuron contributes its entire tuning curve to a distribution of activation defined over the task space (Figure 8.3). In the example, the activation is thought of as distributed over the space of movement directions of the hand. When a specific movement is prepared, a peak in this activation field is positioned over that movement direction, which is currently specified (Figure 8.4).

As a movement is prepared, a peak arises in response to inputs that specify the upcoming motor act. In the laboratory, visual cues to the movement target may be provided at some point, but a participant may also anticipate and preactivate movements that have been frequently elicited in a given task setting (Erlhagen & Schöner, 2002). Such prior activation can be induced by the layout of the environment, in which a few graspable or reachable objects may be visible in the work space. It may also arise out of learned motor habits, in which previous reaching acts leave memory traces near the movement directions of those previous reaches. In each individual reaching incident, one particular movement direction must be selected from among the preactivated locations in the field.

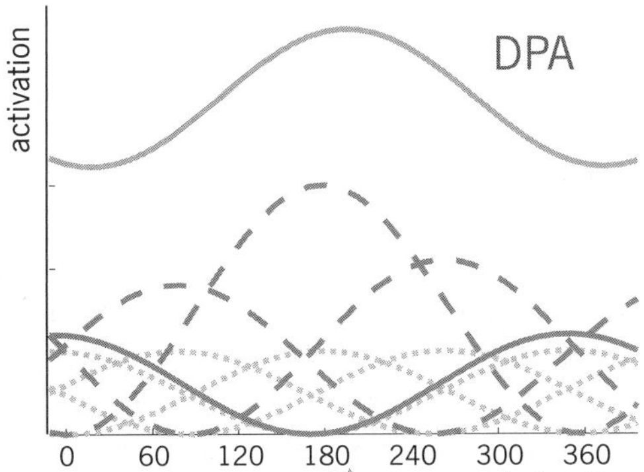

FIGURE 8.3. From tuning curves (schematically shown as dotted lines) of neurons in motor and premotor cortex to the direction of a hand movement, the distribution of population activation (DPA; schematically shown as a solid line) can be constructed by weighting the tuning curves with the firing rate currently observed (dashed line). In the illustration, the movement indicated by the arrow was prepared, leading to a current firing pattern in which neurons tuned to directions close to that value, firing more and thus weighing more than neurons tuned to other directions. As a result, the DPA develops a single peak centered on the currently prepared movement direction. Redrawn from Bastian, Schöner, and Riehle (2003).

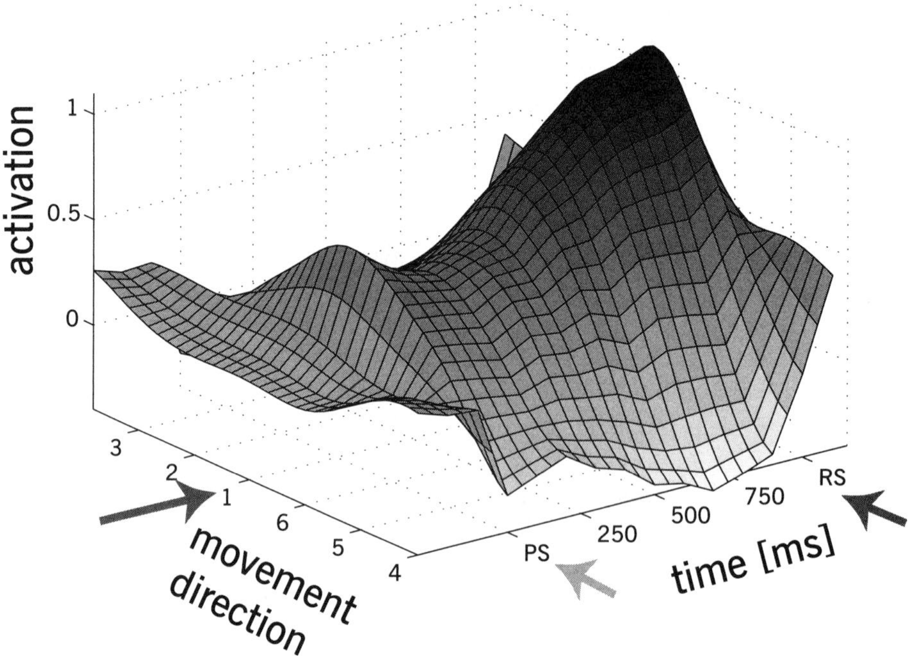

FIGURE 8.4. The DPA is obtained as a function of time by using the current firing rate from about 100 neurons in the motor cortex when a movement direction is first cued at "PS" and then the "go" signal is given at "RS." Data from Bastian, Schöner, and Riehle (2003). An optimal linear estimator was used to smooth the representation, as described in Erlhagen, Bastian, Jancke, Riehle, and Schöner (1999).

Selection decisions require competition within the neural population. All movement directions cannot become activated at the same time. Inhibition within the population is a neural mechanism that may bring about such competition. Inhibition is generic, of course, within cortical and subcortical neuronal networks. Excitatory coupling may also exist, and there is a typical gradient in which neurons with similar tuning curves tend to have net excitatory coupling, whereas neurons with dissimilar tuning curves tend to have net inhibitory coupling. In the network dynamics of neural populations this pattern of interaction stabilizes localized patterns of activation—peaks or blobs or bubbles depending on how many dimensions we envisage. The excitatory interaction among similarly tuned neurons stabilizes a localized pattern of activity against decay, as the neurons contributing to the localized patterns push each other up toward higher activation. The inhibitory interaction among more dissimilarly tuned neurons counterbalances this excitatory influence, which would tend to activate the neighborhood of localized patterns of activation, leading to activation spreading away from the localized pattern. Inhibition thus stabilizes peaks against diffusive decay. Localized peaks of activation are thus the stable states, or attractors, of the neural dynamics that may emerge under appropriate circumstances.

In DFT we abstract from the details of neural population activity and tuning curves. We think of distributions of neural activation defined directly over the functionally relevant spaces, here, the space that consists of all possible directions in space of hand movement. Patterns of localized activation are then peaks that are localized in these functional spaces, and that location specifies values along these dimensions. A movement that has been prepared is thus a peak of activation positioned at the direction in which the movement will be made. Connectivity downstream from the neural population captured by the dynamic field brings about that movement. That is how the tuning curves of the neurons that contribute to the field come about. For sensory representations, analogously, connectivity from the sensory surface to the neurons that contribute to the dynamic field brings about the sensory event encoded by a localized peak of activation, again, as captured by the tuning curves of the contributing neurons.

The A*-Not-*B *Paradigm*

We have seen how neural dynamics are grounded in the population activity of the brain. How do we use the concepts of DFT to understand how behavior and cognition unfold in the here and now? To answer this question in the concrete, let us have a look at what happens when an infant is induced to reach toward objects in Piaget's *A*-not-*B* paradigm (Figure 8.5). The infant would be seated in an infant chair and you, the experimenter, would interact with the baby.

At some point you retrieve an attractive toy from somewhere and show it to the infant by bringing it into his or her field of view. Even very young infants will direct their gaze to the toy. Infants around 5–6 months of age have learned to reach. Such infants will often initiate a reach for the toy when you move the toy close enough so that it is reachable. How would we think of such a reaching act within DFT (Thelen, Schöner, Scheier, & Smith, 2001)? A reach is initiated when a peak of activation arises in a dynamic field that represents prepared movements. Because the location of the peak determines the orientation of the reach, the peak should be positioned over the movement direction in which the object lies.

Sensory information about this object comes from the visual system. Extracting from the visual system information about the direction of a reach to a visible object actually requires a whole sequence of neural events, which are not modeled in any detail in this simplest description. For instance, neural mechanisms generate a shift of attention to the visual location of salient stimuli, usually accompanied by a shift of gaze to foveate the object. This entails a perceptual process of figure–ground segmentation in which the visual system focuses on the object. Extracting movement parameters from visual information about the object further requires transforming from visual, head-centered coordinates to body-centered coordinates and taking the initial position of the hand into account. All this is summarized in a simplest DFT description of a reach by assuming that a localized pattern of input is provided by the visual system to the dynamic field representing the movement direction of a prepared hand movement.

So before the infant's attention has been captured by the object, no localized input is provided to the movement direction field. When the infant attends to the object, such localized input is present and promotes the building of an activation peak localized over the movement direction in which the object lies. In DFT, the decision to reach amounts to the building of an activation peak in the relevant dynamic field. As the peak is built, could other visual objects, shifts of attention, or other movement intentions weaken the input and prevent the peak from remaining stable? Yes, in principle, this could happen. In DFT the peak is stabilized, however, to some extent. It is stabilized against building competing peaks at other locations, in which other distractor objects lie, by inhibitory interaction. It is stabilized against decay by local excitatory interaction, so a wavering of attention will not necessarily lead to a decay of the peak. A decision to reach is stabilized in the sense that the sensory input required to keep a peak of activation stable is less than the sensory information required to first bring about a peak.

In DFT models the induction of a peak by localized input involves the detection of instability, in which a peak-less pattern of activation becomes unstable, and the system switches to a peaked pattern of activation that is stable. This is illustrated in Figure 8.5. What happens is that below a threshold (conventionally defined as "zero" activation), neurons in the population do not interact. This reflects that only sufficiently activated (i.e., spiking) neurons impact on their postsynaptic targets. So excitatory and inhibitory interaction do not come into play until activation somewhere in the field exceeds that threshold. When that happens, local excitatory interaction drives up

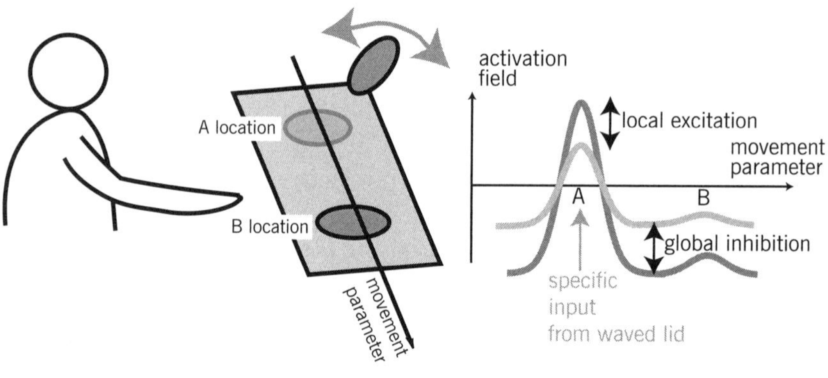

FIGURE 8.5. *Left*: An infant reaching to one of two cued locations, "A" or "B," at which lids are placed on a box. On an "A" trial, the experimenter attracts the attention of the infant to the "A" location, for instance, by waving a lid in front of the infant and then setting that lid down on the "A" location. *Right*: An activation field defined over the movement parameter "direction" accounts for the motor plans that the infant is generating. The waving action leads to localized ("specific") input to the field at the "A" location. The lid visible at the "B" location receives smaller input. Sufficiently strong, specific input at "A" induces a suprathreshold peak of activation. Local excitatory interaction leads to higher levels of activation within the peak than dictated by input. Global inhibitory interaction suppresses activation below resting level elsewhere in the field.

activation around the location where the threshold was first pierced. A suprathreshold peak grows. At the same time, inhibitory projections from that same region of the field begin to affect other regions, suppressing the field at other, alternative reaching directions. Once a peak has been generated, it is stabilized by the excitatory interaction within the peak region and inhibitory projection from the peak onto the rest of the field. Even when input is weakened, the peak may resist because it is, in a sense, self-stabilized by excitatory interaction.

Not every object in the visual array will automatically generate a reach. Infants are not little reaching machines that reach no matter what. For one thing, even as an infant attends to the object, he or she will often initiate a reach only when the object is pushed close enough to be in reaching space. Thus, other factors than the localized input that specifies the movement direction may influence whether or not a peak is generated. These factors can be captured as nonlocalized or as constant inputs to the field. For instance, seeing that the object is reachable, an encouragement by the parent to take the object or an endogenous intention to reach may provide a global boost to the dynamic field of movement direction. Such global input may push the activation field through the threshold somewhere, most likely where localized input is already present but was not sufficient by itself to induce a peak. In DFT we call this the *boost-driven detection instability*, and it is illustrated in Figure 8.6.

Another factor that may support initiating a reach is learning—that is, building the habit to reach. If we play a little game with the infant, presenting an object several times in a row, the infant often willingly reaches for the object. The infant may now need less of a cue to initiate a reach. This is exploited by experimenters, who often use an initial training regimen in which they push an object closer to the infant on the first few trials. In fact, in Piaget's *A*-not-*B* experiment, the lid on the "A" trough is often left a bit off kilter, closer to the infant on the first few "A" trials to encourage the infant to reach toward the "A" location. I explain this experiment a little better below.

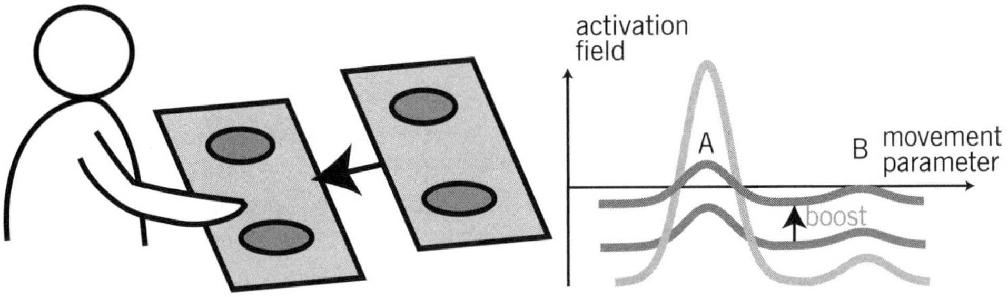

FIGURE 8.6. Pushing the box with the two lids into the reaching space of the infant is modeled by applying a homogeneous excitatory input, a *boost*, to the entire field. This boost may push activation at the "A" or the "B" location through the detection instability and induce a peak there. The field is preactivated at these locations due to residual activation from a peak that decayed during the delay, due to the memory trace of earlier reaches to a location, or due to the visual structure of the scene with lids visible at the two locations.

A habit to reach toward a specific location can be described in DFT as coming from an additional source of localized input that does not originate from the visual array, but from a memory trace of past reaches. Classically, habits have been viewed as powerful ways to reduce the amount of sensory specification required to bring about an action. William James (1890) famously speculated about the pervasiveness and power of habits, which he described as "ease of action" and considered the most common and basic form of learning. The neural mechanisms underlying habit formation are well studied in biological psychology (Yin & Knowlton, 2006), but little is known about how human infants acquire habits. In any case, in the *A*-not-*B* paradigm, a very small number of reaches is sufficient to induce a reaching habit, so this is a fast and robust form of learning. It is, perhaps, unclear at this time if the emergence of a memory trace of previous reaches over a few trials is truly akin to the formation of behavioral habits, as observed in paradigms of instrumental conditioning.

How about a hidden object? This is already quite a complex scenario. Presumably, the infant "knows" about the hidden object because earlier he or she saw the object, until it was hidden in a trough in a box, say, as in the typical *A*-not-*B* experiment. The trough was then covered with a lid, so the toy is hidden under the lid. When the object was shown, did the infant prepare a movement to reach for it? Perhaps not if the project was outside its reaching range. But there was some localized input that induced some localized activation, which is now decaying because the input is no longer available. How do we elicit the reach to the hidden toy? By moving the hiding location into the reach space of the infant. The lid is an object, so the reach may be induced just by that—by the visual stimulus of a reachable object. The hidden toy is discovered, becomes visible, and may elicit further object-oriented action. So it isn't necessarily the case that reaching for a hidden toy requires a true memory of the toy.

In the *A*-not-*B* experiment (Piaget, 1954; Wellman, Cross, & Bartsch, 1986) there are two locations, typically a box with two troughs, each covered with a lid. When the toy is first hidden under the "A" lid, the infant does select the "A" lid for his or her reach in most cases. This means that the previous action of attracting attention to the "A" location—of hiding an attractive toy there and putting down the lid over that location—left some activation trace that biases the competition among the two reaching locations, the two lids, toward the "A" location.

In many *A*-not-*B* experiments, the two lids are not equally spaced on the first few training trials; as noted previously, the "A" lid is left closer to the infant than the "B" lid, inviting a reach toward "A." This may bias the decision independently of any remaining activation from previous events.

Once the reach to "A" has been elicited several times over a few "A" trials, a memory trace solidly biases the motor decision toward the "A" location each time the box is pushed into the infant's reaching space. By the time the hiding location is switched to the "B" location, the bias toward "A" from the memory trace may be strong enough to overcome any remaining activation trace from the stimulation at "B" when the toy is hidden. The *A*-not-*B* perseverative error ensues.

The hallmark of the *A*-not-*B* error is the critical role of the delay. When the box is pushed toward the infant right after the toy is hidden at "B," the infant is likely to reach correctly toward "B." Young infants make the error even at short delays, whereas older infants tolerate increasingly longer delays. The delay directly controls the tradeoff between the decaying activation induced by the stimulus at "B" when the toy is hidden there and the bias to "A" that the motor habit, the memory trace at "B," imposes.

Presumably, the decay of activation at "B" is slower in older infants because they may engage in neural interaction more easily, better stabilizing the activation at "B" against decay. (The possibility that older infants may sustain a peak at "B" over long delays as a form of working memory for a planned action requires an account for why they still reach only when the box is pushed into their reaching space. It is likely that understanding the initiation of reaching requires a more refined account, in any case.)

Clearly, the *A*-not-*B* paradigm does not probe reaching to hidden objects per se. Rather, it probes how flexible infants are in switching to a new pattern of activation that must be stabilized against decay. The memory trace is a slow and inflexible form of stabilization of motor behavior, whereas the neural interaction within a neural field is a faster form of stabilization that may thus flexibly respond to changed demands.

Is the toy then actually needed? A toyless variant of the paradigm demonstrates that perseverative reaching emerges even in a purely sensory–motor version of the task (Smith, Thelen, Titzer, & McLin, 1999). So this is at heart a sensory–motor decision task in which different factors that bias the selection decision are varied. The developmental trajectory is one in which increasingly, the system is less dependent on current sensory input to bring about the selection decision.

Are the lids really needed? In a lidless version of the task (Spencer, Smith, & Thelen, 2001), the toys are hidden in sand, leaving no visible trace of the hiding location. Perseverative errors now show up as metric errors in which the searching location is biased metrically toward the "A" location. This can be understood in a different picture, in which a peak is sustained during the delay, but drifts under the influence of the memory trace. This metric bias occurs in much older children. Toddlers of 4 years, for instance, show a strong metric bias consistent with perseveration. The localized perceptual input at the locations marked by the lids in the conventional *A*-not-*B* paradigm locks peaks of activation in place. The absence of such visible perceptual markers makes the field more sensitive to other inputs or to internal inhomogeneities. This is how the memory trace of past reaches may have observable consequences in 4-year-olds in the sandbox, whereas infants older than about a year no longer show perseveration in the conventional *A*-not-*B* paradigm.

The sandbox reaches into spatial working memory (Schutte & Spencer, 2009). The hiding location must be remembered as just that, a location along a metric continuum. In DFT, working memory for metric dimensions arises from the capacity of neural interaction to sustain peaks of activation once the inducing localized input has been removed. This is the most common neural account for working memory and is consistent with the observation of sustained firing of neurons in many cortical structures (Fuster, 1995). Excitatory and inhibitory connectivity within a neural population is critical to enable self-sustaining neural activation patterns. Input from the sensory surface alone is not sufficient. The stronger such intrapopulation connectivity, the stabler the sustained activation patterns. Stable patterns of sustained activation may resist distractor inputs more effectively and are less easily biased by such inputs.

Across a broad set of experimental paradigms, the developmental trajectory has been found to shift from the dominant role of feedforward sensory inputs into neural representations toward a dominant role of intrapopulation recurrent connectivity (Schutte & Spencer, 2009; Schutte, Spencer, & Schöner, 2003; Simmering, Schutte, & Spencer, 2008). Activation patterns in the brains of younger infants seem to be more strongly dictated by inputs from the sensory surfaces and by fixed habits (modeled as inputs from memory traces). During development, activation patterns are increasingly controlled by input from within the neural population, whereas the role of external inputs is reduced. This developmental trend has been called the *spatial precision hypothesis* (Schutte, Spencer, & Schöner, 2003) because it manifests in increasing precision with which spatial (or other metric) information can be retained and processed. This same tradeoff between the influence of inputs, including a form of memory (latent memory implicit in connections), and the influence of recurrent connectivity (active memory) is at the core of a connectionist account of perseverative reaching (Munakata, McClelland, Johnson, & Siegler, 1997) that is largely equivalent to DFT.

The Habituation Paradigm

How about perception? The neural concepts at work are the same as for the motor domain: The space of possible percepts is spanned by feature values such as retinal location, visual orientation, movement direction, or color (see Figure 8.7 for an example). Cortical neurons are sensitive to such feature dimensions as indicated by their tuning curves. Any given stimulus evokes a distribution of population activation that may be construed as a peak localized within the feature spaces. Such a peak is then the neural representation of a percept, whose location represents an estimate of the feature value while the amplitude may reflect other stimulus properties such as contrast.

Perceptual peaks arise as a sensory stimulus impinges on the sensory surface. The forward connectivity from the sensory surface to the neural representation is largely responsible for the feature extraction, which determines where a peak is induced. The peak itself is a stable state of the perceptual field, stabilized by interaction. The detection instability, at which the peak arises, separates the preperceptual from the perceptual state. Typically, when the sensory array is sufficiently complex, a perceptual detection decision entails at the same time a selection decision in which one particular perceptual object or one particular perceptual dimension is brought into the foreground. Featural salience influences which object is thus selected. Infants are particularly sensitive to movement, for instance. In DFT, this selection decision amounts to generating a peak localized over one stimulated location, while inhibiting activation over other stimulated locations within a feature dimension.

This is a simple picture, in which percepts are attentional blobs in feature space, with a small number of estimated feature values. Even this simple-minded picture may account for much of what is known about infant perception and its development. In particular, the metrics along the feature dimensions matter for the perceptual

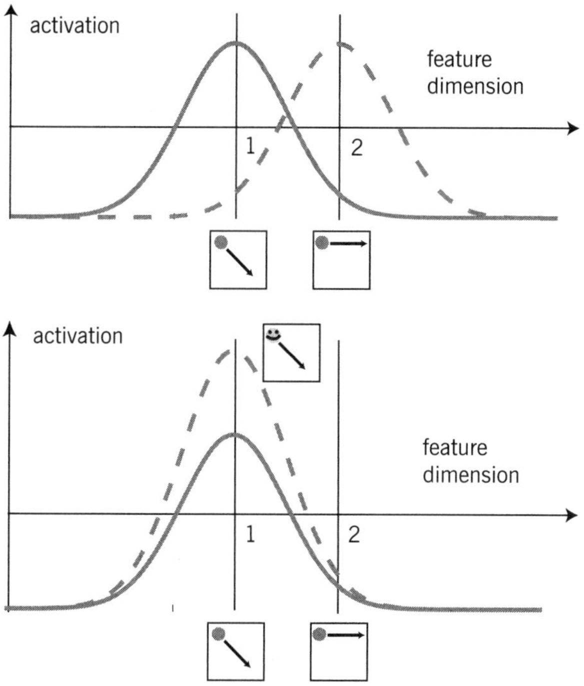

FIGURE 8.7. A perceptual activation field defined over a feature dimension, in this case, the direction of visual motion. An object moving in different directions is represented by a peak localized over the corresponding location in the field (*top*: solid line for diagonal motion, dashed line for horizontal motion). The luminance and complexity of the stimulus are reflected in the amplitude of the induced activation peak, higher for a bright stimulus with inner structure (*bottom*: dashed line) than for a plain stimulus (*bottom*: solid line).

experience of infants: Metric differences along feature dimensions separate familiar from novel stimuli, and development manifests in the metric differences required to shift from familiar to novel.

Perception is much harder to observe in infants than overt motor action, of course. The tools of modern psychophysics are not readily applied. One common way of indirectly assessing the state of visual perceptual representations is to register patterns of looking (Cohen, 1972). This method is based, of course, on the well-documented principles of perceptual habituation, in which the probability of continuing to fixate on an unchanged stimulus decreases as a function of the total looking time (Colombo & Mitchell, 2009). This more indirect assessment of the state of the perceptual system requires some extra machinery in DFT to link activation peaks in feature spaces to looking behavior.

Does decreased looking reflect a decrease or an increase in perceptual activation? Any dependence of "dwell time" on prior perceptual experience is used as an index of perceptual processing (Colombo, 2001). *Familiarity preference* is the tendency to look more at a similar or identical stimulus to the one previously looked at; *novelty preference* is the tendency to look more at a dissimilar stimulus. An account for how perceptual activation is linked to looking within DFT (Schöner & Thelen, 2006) is illustrated in Figure 8.8. A first assumption is that fixation is strengthened by perceptual activation. Thus, whenever a visual stimulus is perceived, the induced activation keeps the infant fixating on it. Perceptual activation thus promotes continued looking, accounting for the familiarity preference early during looking (Roder, Bushnell, & Sasseville, 2000). The excitatory link from visual activation to fixation is consistent with prolonged saccadic response times when a visual stimulus continues to be visible at fixation (overlap) than when it is removed (gap) in adults (Kopecz, 1995).

A second assumption is that perceptual activation drives the buildup of inhibition that is represented in a separate field in the model. Increasing inhibition, in turn, weakens perceptual activation and thus promotes looking away, accounting for novelty preference later during looking. Such inhibition is a well-known form of adult visual perception, wherein it is observed as an increase in perceptual threshold after prolonged experience of a particular percept, a phenomenon known as *selective adaptation* (see Hock, Schöner, & Hochstein, 1996, for the subtle issue of whether conscious perception is required to induce selective adaptation).

Together, the two assumptions lead to an account of the typical pattern of looking in infants exposed to a visual stimulus—an early increase followed by a decrease in looking time that is referred to as *visual habituation* (Colombo, Frick, & Gorman, 1997). The early increase is not always observed for reasons that can be understood. The DFT model also predicts how habituation is modulated by stimulus strength and prior levels of activation, which account for the difference between "fast" and "slow" habituators (Schöner & Thelen, 2006).

The DFT model also explains the pattern of dishabituation routinely used to assess infants' visual perception and cognition (Kaplan & Werner, 1986). Figure 8.8 illustrates a typical paradigm. In a first phase the infant is exposed repeatedly to a habituation stimulus, here a figure moving either horizontally or diagonally, until the infant has *habituated*, that is, until the time the infant looks at the stimulus has fallen from the initial level to a criterion level. In the following test phase, the infant is presented with new stimuli that differ in some way from the habituation stimulus. In the example the motion direction may change or the moving object may change color and have an additional smiley face added to it. If the change in stimulus induces renewed looking, called *dishabituation*, then this is evidence that the infant detected the change of stimulus and thus discriminated the test from the habituation stimulus.

The stimuli shown in Figure 8.9 were used in an unpublished experiment performed in Esther Thelen's lab by Connie Jing Feng, Melissa Clearfield, and myself. Overall, ninety-six 16-month-old infants participated, 24 in each of the four stimulus conditions shown. The typical stimulus-controlled procedure was used, in which an attention-grabbing stimulus (chime with a concentric moving pattern) attracted the infant's gaze to the display, which was then switched to the current stimulus until the child looked away for more than 2 seconds or the trial lasted more than 120 seconds. The habituation phase lasted at least 4 trials and a maximum of 20 trials. It was aborted when the mean looking time for three consecutive trials fell below 50% of the

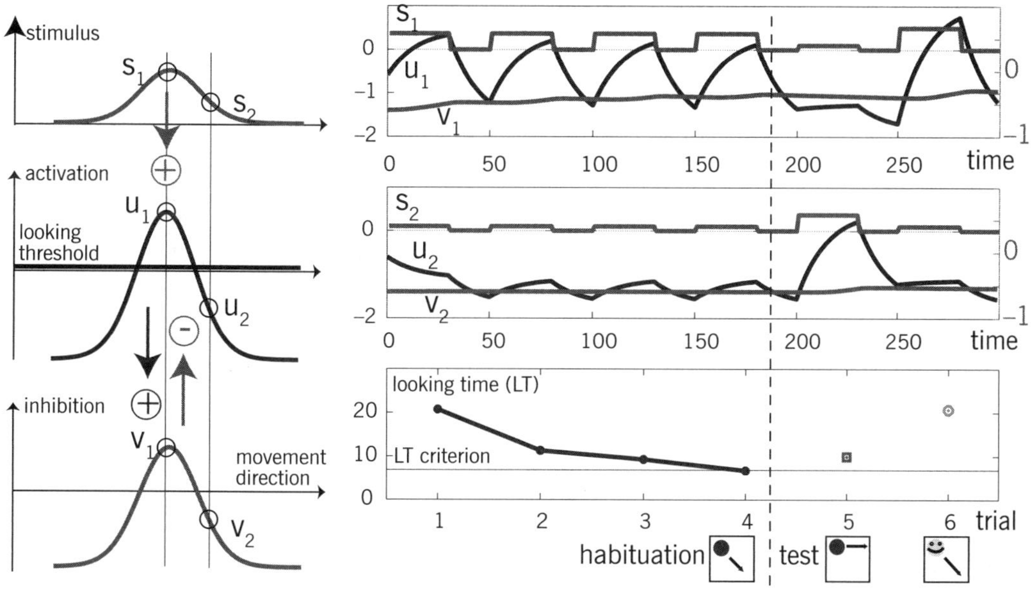

FIGURE 8.8. *Left*: In a model of visual habituation (Schöner & Thelen, 2006), an activation field (*middle*) defined over a relevant feature dimension (here, the direction of visual motion) receives input from the sensory surface (stimulus on top). The activation field drives an inhibitory field (*bottom*), which conversely inhibits the activation field. Levels of activation above a looking threshold promote fixation on the stimulus. *Right*: The neural dynamics of the activation and inhibition fields are simulated at two locations along the feature dimensions corresponding to diagonal (index 1) and horizontal motion (index 2). The moving objects are shown repeatedly (illustrated by the boxcar stimulus trace for S1 and S2). Whenever a matching stimulus is presented, activation at the corresponding location is driven up. For instance, while a diagonally moving object is presented (habituation phase), activation at location 1 (*top panel*) is driven up, leading to looking at the moving object (*bottom panel*). Activation decays when the stimulus is removed. Across stimulus presentations, inhibition increases whenever there is sufficient activation, leading to a reduction of activation and looking time during the habituation phase. In the test phase, dishabituation occurs when a new stimulus (horizontal motion, *second from right*) is shown or when the stimulus receives added strength (diagonal motion with brighter and more complex object, *far right*).

looking time during the first three trials. Figure 8.10 shows the mean looking times across conditions together with error bars and simulation results.

The pattern of looking in these experiments shows dishabituation when the movement direction changes from diagonal to horizontal (e.g., in condition 1A for the first test stimulus) or from horizontal to diagonal (e.g, in condition 2A for the first test stimulus). In the DFT model, such dishabituation arises when the test stimulus differs sufficiently along the perceptual dimension from the habituation stimulus such that it activates a different location in the perceptual field. That new location is not yet inhibited by previous perceptual experience and may therefore build up a strong activation peak that supports renewed looking. This account for dishabituation is symmetrical: Either direction of change along the perceptual dimension leads to dishabituation as long as the distance along that dimension is sufficiently large. This symmetrical form of dishabituation is thus truly a signature of discrimination.

The results also show dishabituation that is asymmetrical. A change from the simple to the bright and complex object moving in the same direction leads to renewed looking (e.g., in condition 1B, first test). The reverse change from a bright, complex to the simple object (e.g., in condition 4B, first test) does not lead to dishabituation. In the DFT model, such asymmetrical dishabituation reflects a change in input strength that leads to renewed looking through higher perceptual activation only when input strength is increased, not when it is decreased. In fact, the DFT model predicts shorter looking time than during habituation—the only mismatch between the model and this experiment (on condition 4). (In the experiment such a negative difference is not possible because the attention-grabbing stimulus is presented until the infant looks, so looking times cannot fall below a lower bound. The attention-grabbing stimulus is not modeled.)

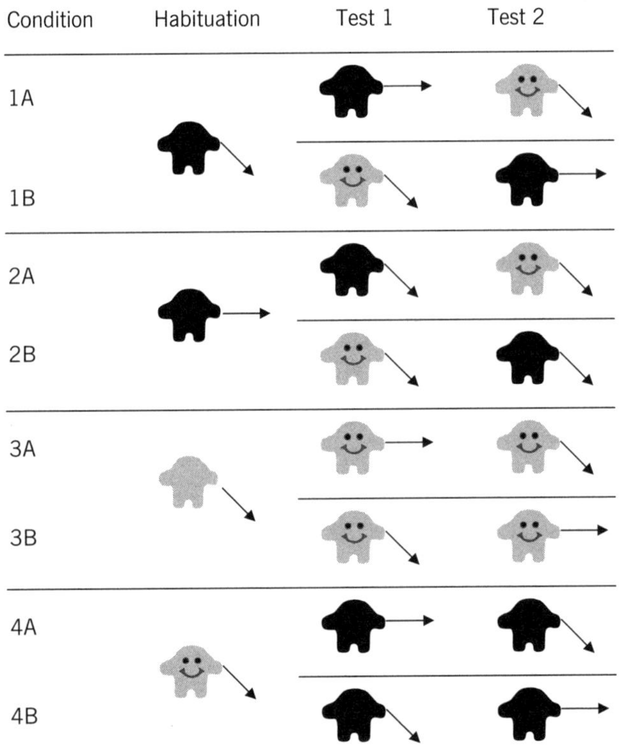

FIGURE 8.9. Stimuli used in a habituation experiment. The stimuli involved a moving schematic figure that differed between the habituation and test phases by movement direction (either horizontal or vertical) and complexity (either plain blue or bright yellow with a smiley face).

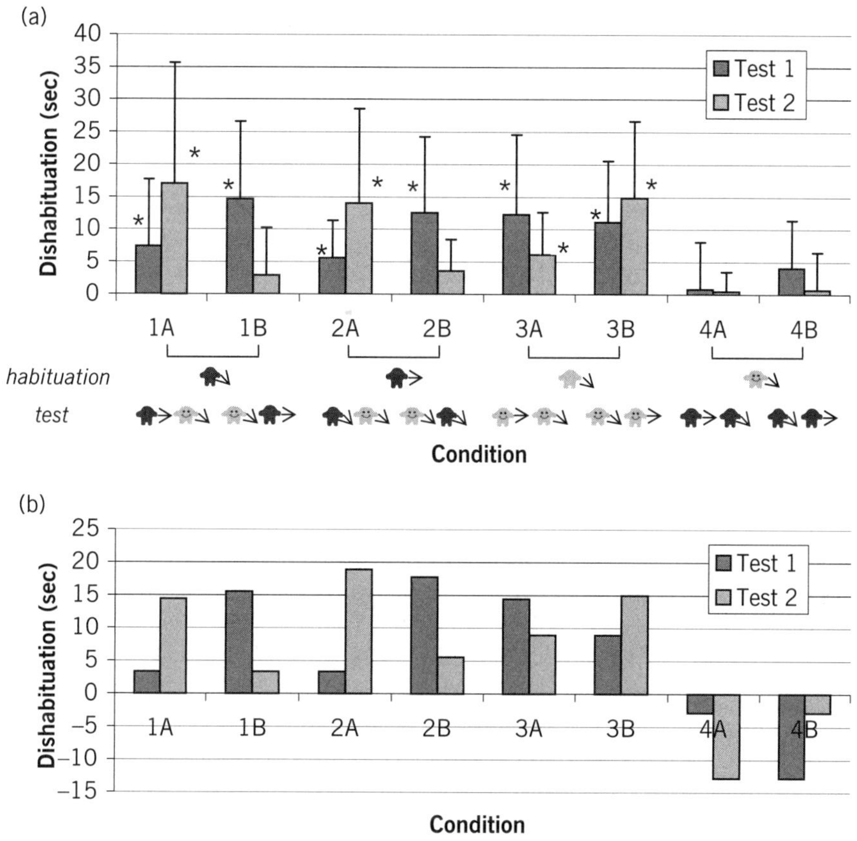

FIGURE 8.10. Mean looking time during the test phase for the four experimental conditions shown in Figure 8.9. Experimental results are shown on top (a). Significant dishabituation is marked by an asterisk (*). Results from model simulations are show on bottom (b).

Over development, perceptual discrimination generally improves (Kaplan & Werner, 1986). In DFT this improvement is consistent with a generalization of the spatial precision hypothesis, according to which activation peaks become sharper and stabler over development. As a result, they will overlap less. Shorter displacements of the test stimulus along the feature dimension are required to induce new activation peaks that do not share inhibition with the habituation stimulus.

A variant of the habituation paradigm has been extensively used to study infant cognition. In this variant, hypotheses are formed about infant "knowledge." For instance, in the visual *drawbridge paradigm* (Baillargeon, 1987a), the hypothesis is tested that infants "know" that solid bodies cannot occupy the same location in space. Infants are habituated to a visual stimulus in which a wooden panel (the "drawbridge") rotates from horizontal back to horizontal 180 degrees toward and away from the infant (Figure 8.11). Then one out of two new stimuli is presented. One is "novel": It differs from the habituation stimulus more than the other new stimulus. In the novel condition, a wooden block is visible to the infant, and the drawbridge is lifted until it occludes and touches that block, which leads to less rotation, about 120 degrees. This stimulus "respects" the infant's knowledge that solids cannot move through each other because the drawbridge's motion is stopped by the block, or so it appears. It is the "possible" stimulus in light of the infant's hypothesized knowledge. The other "familiar" condition is similar to the habituation stimulus, in that the drawbridge makes the same 180-degree rotation away and toward the infant. This condition is "impossible" because it violates the infant's knowledge about solids: Initially, a block is visible that must lie in the path of the drawbridge, but the drawbridge continues motion through the block (in reality, the block is dropped down through an opening in the table).

The logic of these types of paradigms is this: Infants are expected to dishabituate more to the novel than to the familiar stimulus. If they look longer at the familiar but impossible stimulus, then this is evidence that there is something surprising about that stimulus—the violation of infant expectation derived from their knowledge. Looking longer at impossible than at possible stimuli is thus considered evidence in support of the hypothesis regarding infant knowledge that the impossible stimulus challenges.

DFT suggests that the pattern of looking in this kind of paradigm can be accounted for without any reference to knowledge or violated expectation. In DFT, the metrics of perceptual experience alone can account for the pattern of results. To illustrate the idea, we analyzed the metrics of the drawbridge stimulus (see Figure 8.11). In this paradigm, the block is a new element that appears in both test stimuli, but not in the habituation stimulus. So the presence of the block alone leads to some dishabituation through increased input strength. Now the spatial structure of the stimulus is changed by the block: The drawbridge moves through space covering a particular area that overlaps with the block. This overlap is more extensive when the drawbridge moves

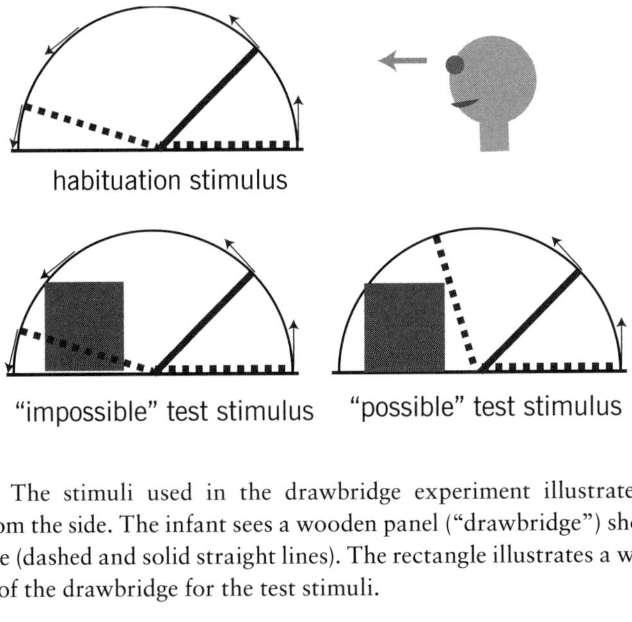

FIGURE 8.11. The stimuli used in the drawbridge experiment illustrated by seeing the arrangement from the side. The infant sees a wooden panel ("drawbridge") shown here at three moments in time (dashed and solid straight lines). The rectangle illustrates a wooden block that sits in the path of the drawbridge for the test stimuli.

through the spatial location occupied by the block than when the drawbridge stops short of that location. That difference accounts for why there is stronger perceptual input for the "impossible" stimulus, leading to more looking.

Similar analyses can be made for many of the classical habituation scenarios that are hypothesized to probe infant knowledge. The DFT model also makes specific predictions that support the account (Schöner & Thelen, 2006). In particular, the temporal order in which test stimuli are presented is shown to matter: When the familiar–impossible is presented first on the test, its stronger overlap with the habituation stimulus makes that activation start from a higher level of activation, leading to more dishabituation than when the novel–possible stimulus is presented first. In experiments, this interaction between the order of presentation and the preference for the impossible is routinely observed. In fact, in the original experiment (Baillargeon, 1987a), the looking advantage of the impossible stimulus comes entirely from this order of presentation.

Similarly, individual differences are understood in the DFT account (Schöner & Thelen, 2006). Infants with a higher initial level of activation will have a long initial look, leading to faster and deeper habituation and a larger difference between the two test stimuli. Again, in the original experiment, the difference between the two test stimuli came entirely, statistically speaking, from the fast habituators (Baillargeon, 1987a).

This DFT account is only a first sketch of all that is entailed in understanding infant perception and visual habituation. More recently, Perone and Spencer (2013) have substantially extended this account into a comprehensive theory of visual habituation and visual cognition. In their model, the control that infants exert over their own visual stimulation is modeled in more depth than was done in Schöner and Thelen (2006) by adding a dynamic fixation system constrained by what is understood from work on adult visual fixation (Kopecz, 1995). More importantly, this work links the familiarity-to-novelty transition to the emergence of working memory. The (spatial) precision hypothesis, shown in earlier work to account for the increase in the capacity of working memory in infancy (Perone, Simmering, & Spencer, 2011) is thus being confirmed as a pervasive principle of development.

Conclusion

The two worked examples have underscored how far we have moved from the self-organization metaphor that was based on analogies with physics to neurally mechanistic accounts of infants' and children's behavior and cognition. The strong neural emphasis in this transition from metaphor to theory may have come as a surprise to some readers. But it is consistent with the need to ground the concepts of self-organization in empirically valid mechanisms: that is, the patterns of neural interaction. The critical question is if this neural grounding preserves key features of the metaphor. Does the neural theory address the notion of the emergence of cognitive and motor competences under favorable environmental conditions? Does the theory articulate how these competences may emerge from experience and account for individual routes of development?

Emergence is, in fact, an inherent property of DFT. Cognitive competences, such as the ability to stabilize decisions and keep them in working memory, are captured by attractor states of the neural dynamics: stable peaks of activation within neural populations. The stability of these states emerges from the confluence of different contributions. For instance, the neural dynamics of an older infant may stabilize a peak that keeps a reaching goal in working memory just based on its more strongly developed neural interaction. A younger infant may achieve the same with weaker neural interaction if he or she receives additional broad input, for instance, from an environment with rich perceptual structure (Schöner & Dineva, 2007). Thus, whether a hidden object is "out of sight, out of mind" for an infant depends not just on age, but on environmental conditions. The neural dynamics of younger infants make stronger demands on the environment to achieve the cognitive capacity that older infants achieve more easily. Perone and colleagues (2011) have provided a similar analysis of how the capacity of visual working memory emerges depending on the task settings. The conceptual implications are lucidly discussed in Simmering and Perone (2013), who review 50 studies of visual working memory, highlighting how the confluence of different factors impacts on how infants perform in different circumstances.

Individual differences are captured through differences in the neural dynamics formalized as different parameter values. This accounts for covariation of behavioral signatures of the dynamics. For instance, in the DFT account of infant habituation (Schöner & Thelen, 2006), looking patterns during the habituation phase are predictive of the amount of dishabituation on test. Perone and Spencer (2013) have pushed the account of DFT for individual differences to a new level. They show how individual differences may emerge from learning processes. In simulations of long looking episodes in a visual habituation task, fluctuations in the looking behavior early during a trial may lead the system to build visual memories more efficiently, leading to characteristic signatures of a more mature memory system.

This chapter has illustrated the transition from metaphor to neural theory based on elementary forms of cognition, close to cognition's sensory–motor origin. Social interaction is, of course, central to development. Can the neural theory be put to work on language and social interaction? The answer is yes, although moving to higher cognition is one of the frontiers of dynamical systems thinking. In a trailblazing study, Samuelson, Smith, Perry, and Spencer (2011) show how the reference problem that is key to how children learn words while interacting with parents may be solved by activation fields over visual space. Their DFT model accounts for experimental data on word learning and leads to new, tested predictions, suggesting that the neural theory is as productive as the metaphor was.

The neural grounding of cognitive development is a strategy that dynamical systems thinking shares with connectionist and neuroconstructivist approaches to development. The convergence of these currents of theoretical thinking was the topic of a conference and book (Spencer, Thomas, & McClelland, 2009). The joint frontier of these approaches was clearly recognized as the problem of understanding autonomous development. How do infants and children shape their own behavior to provide the stimulation and experience through which they learn and develop? And how may the reproducible pattern of development emerge through autonomous learning?

ACKNOWLEDGMENTS

I thank Conny Jing Feng and Melissa Clearfield for providing access to their unpublished data, and for extensive discussion and collaboration. I also thank John Spencer, Anne Schutte, Vanessa Simmering, and Sammy Perone for their longstanding collaboration. The late Esther Thelen introduced me to development and strongly shaped my view in a marvelous collaboration. I was lucky enough to find in John Spencer a partner to move that view beyond metaphor.

REFERENCES

Baillargeon, R. (1987a). Object permanence in 3.5 and 4.5 month-old infants. *Developmental Psychology, 23*, 655–664.

Baillargeon, R. (1987b). Young infants' reasoning about the physical and spatial characteristics of a hidden object. *Cognitive Development, 2*, 179–200.

Bastian, A., Schöner, G., & Riehle, A. (2003). Preshaping and continuous evolution of motor cortical representations during movement preparation. *European Journal of Neuroscience, 18*, 2047–2058.

Blumberg, M. (2010). *Freaks of nature: What anomalies tell us about development and evolution.* New York: Oxford University Press.

Cohen, L. (1972). Attention-getting and attention-holding processes of infant visual preferences. *Child Development, 43*, 869–879.

Colombo, J. (2001). The development of visual attention in infancy. *Annual Reviews of Psychology, 52*, 337–367.

Colombo, J., Frick, J. E., & Gorman, S. A. (1997). Sensitization during visual habituation sequences: Procedural effects and individual differences. *Journal of Experimental Child Psychology, 67*, 223–235.

Colombo, J., & Mitchell, D. (2009). Infant visual habituation. *Neurobiology of Learning and Memory, 92*, 225–234.

Elman, J. L., Bates, E. A., Johnson, M. H., Karmiloff-Smith, A., Parisi, D., & Plunkett, K. (1997). *Rethinking innateness: A connectionist perspective on development.* Cambridge, MA: MIT Press.

Erlhagen, W., Bastian, A., Jancke, D., Riehle, A., & Schöner, G. (1999). The distribution of neuronal population activation (DPA) as a tool to study interaction and integration in cortical representations. *Journal of Neuroscience Methods, 94*, 53–66.

Erlhagen, W., & Schöner, G. (2002). Dynamic field theory of movement preparation. *Psychological Review, 109*, 545–572.

Fodor, J. (1981). Fixation of belief and concept acquisition. In M. Piatelli-Palmanni (Ed.), *Language and learning: The debate between Piaget and Chomsky* (pp. 143–149). London: Routledge.

Fox Keller, E. (2002). *Making sense of life: Explaining biological development with models, metaphors, and machines.* Cambridge, MA: Harvard University Press.

Fuster, J. M. (1995). *Memory in the cerebral cortex: An empirical approach to neural networks in the human and nonhuman primate.* Cambridge, MA: MIT Press.

Georgopoulos, A. P., Schwartz, A. B., & Kettner, R. E. (1986). Neural population coding of movement direction. *Science, 233*, 1416–1419.

Gierer, A., & Meinhardt, H. (1972). A theory of biological pattern formation. *Kybernetik, 12*, 30–39.

Gottlieb, G. (2001). A developmental psychobiological systems view: Early formulation and current status. In S. Oyama, P. S. Griffiths, & R. D. Gray (Eds.), *Cycles of contingency: Developmental systems and evolution* (pp. 41–54). Cambridge, MA: MIT Press.

Griffiths, P. E., & Gray, R. D. (2005). Discussion: Three ways to misunderstand developmental systems theory. *Biology and Philosophy, 20*, 417–425.

Groh, J. M., Born, R. T., & Newsome, W. T. (1997). How is a sensory map read out?: Effects of microstimulation in visual area MT on saccades and smooth pursuit eye movements. *Journal of Neuroscience, 17*, 4312–4330.

Haken, H. (1983). *Synergetics: An introduction* (3rd ed.). Berlin: Springer Verlag.

Harrison, L. G. (2011). *The shaping of life.* Cambridge, UK: Cambridge University Press.

Hock, H. S., Schöner, G., & Hochstein, S. (1996). Perceptual stability and the selective adaptation of perceived and unperceived motion directions. *Vision Research, 36*, 3311–3323.

James, W. (1890). *The principles of psychology.* New York: Holt.

Kaplan, P., & Werner, J. (1986). Habituation, response to novelty, and dishabituation: Tests of a dual-process theory. *Journal of Experimental Child Psychology, 42*, 199–217.

Kopecz, K. (1995). Saccadic motor planning by integrating visual information and preinformation on neural dynamic fields. *Biological Cybernetics, 73*, 49–60.

Kriegeskorte, N., & Kreima, G. (Eds.). (2012). *Visual population codes: Toward a common multivariate framework for cell recording and functional imaging.* Cambridge, MA: MIT Press.

Lee, C., Rohrer, W., & Sparks, D. (1988). Population coding of saccadic eye movements by neurons in the superior colliculus. *Nature, 332*, 357–360.

Lipscomb, L., Swanson, J., & West, A. (2004). Scaffolding. In M. Orey (Ed.), *Emerging perspectives on learning, teaching, and technology.* Retrieved March 18, 2013, from *http://projects.coe.uga.edu/epltt.*

Molenaar, P. C. M. (1986). On the impossibility of acquiring more powerful structures: A neglected alternative. *Human Development, 29*, 245–251.

Munakata, Y., McClelland, J. L., Johnson, M. H., & Siegler, R. S. (1997). Rethinking infant knowledge: Toward an adaptive process account of successes and failures in object permanence tasks. *Psychological Review, 104*, 686–719.

Murray, J. D. (2002). *Mathematical biology II* (3rd ed.). New York: Springer Verlag.

Newell, K. M., Liu, Y.-T., & Mayer-Kress, G. (2009). Timescales of change in connectionist and dynamical systems approaches to learning and development. In J. P. Spencer, M. S. C. Thomas, & J. L. McClelland (Eds.), *Toward a unified theory of development* (pp. 119–138). New York: Oxford University Press.

Perone, S., Simmering, V. R., & Spencer, J. P. (2011). Stronger neural dynamics capture changes in infants' visual working memory capacity over development. *Developmental Science, 14*, 1379–1392.

Perone, S., & Spencer, J. P. (2013). Autonomy in action: Linking the act of looking to memory formation in infancy via dynamic neural fields. *Cognitive Science, 37*, 1–60.

Piaget, J. (1954). *The construction of reality in the child.* New York: Basic Books.

Quartz, S. R. (1993). Neural networks, nativism, and the plausibility of constructivism. *Cognition, 48*, 223–242.

Roder, B. J., Bushnell, E. W., & Sasseville, A. M. (2000). Infants' preferences for familiarity and novelty during the course of visual processing. *Infancy, 1*, 491–507.

Samuelson, L. K., Smith, L. B., Perry, L. K., & Spencer, J. P. (2011). Grounding word learning in space. *PLoS One, 6*(12), e28095.

Schöner, G. (2008). Dynamical systems approaches to cognition. In R. Sun (Ed.), *Cambridge handbook of computational cognitive modeling* (pp. 101–126). Cambridge, UK: Cambridge University Press.

Schöner, G., & Dineva, E. (2007). Dynamic instabilities as mechanisms for emergence. *Developmental Science, 10*, 69–74.

Schöner, G., & Kelso, J. A. S. (1988). Dynamic pattern generation in behavioral and neural systems. *Science, 239*, 1513–1520.

Schöner, G., & Thelen, E. (2006). Using dynamic field theory to rethink infant habituation. *Psychological Review, 113*, 273–299.

Schutte, A. R., & Spencer, J. P. (2009). Tests of the dynamic field theory and the spatial precision hypothesis: Capturing a qualitative developmental transition in spatial working memory. *Journal of Experimental Psychology: Human Perception and Performance, 35*, 1698–1725.

Schutte, A. R., Spencer, J. P., & Schöner, G. (2003). Testing the dynamic field theory: Working memory for locations becomes more spatially precise over development. *Child Development, 74*, 1393–1417.

Simmering, V. R., & Perone, S. (2012). Working memory capacity as a dynamic process. *Frontiers in Psychology, 3*, 567.

Simmering, V. R., Schutte, A. R., & Spencer, J. P. (2008). Generalizing the dynamic field theory of spatial cognition across real and developmental timescales. *Brain Research, 1202*, 68–86.

Smith, L. B., Thelen, E., Titzer, R., & McLin, D. (1999). Knowing in the context of acting: The task dynamics of the A-not-B error. *Psychological Review, 106*, 235–260.

Spencer, J. P., Clearfield, M., Corbetta, C., Ulrich, B., Buchanan, P., & Schöner, G. (2006). Moving toward a grand theory of development: In memory of Esther Thelen. *Child Development, 77*, 1521–1538.

Spencer, J. P., Smith, L. B., & Thelen, E. (2001). Tests of a dynamic systems account of the A-not-B error: The influence of prior experience on the spatial memory abilities of 2-year-olds. *Child Development, 72*, 1327–1346.

Spencer, J. P., Thomas, M. S. C., & McClelland, J. L. (Eds.). (2009). *Toward a unified theory of development*. New York: Oxford University Press.

Thelen, E., Schöner, G., Scheier, C., & Smith, L. (2001). The dynamics of embodiment: A field theory of infant perseverative reaching. *Brain and Behavioral Sciences, 24*, 1–33.

Thelen, E., & Smith, L. B. (1994). *A dynamic systems approach to the development of cognition and action*. Cambridge, MA: MIT Press.

Turing, A. M. (1952). The chemical basis of morphogenesis. *Philosophical Transactions of the Royal Society of London, B237*, 37–72.

Usher, M., & McCelland, J. L. (2001). On the time course of perceptual choice: The leaky competing accumulator model. *Psychological Review, 108*, 550–592.

Waddington, C. H. (1953). *The epigenetics of birds*. Cambridge, UK: Cambridge University Press.

Wellman, H. M., Cross, D., & Bartsch, K. (1986). Infant search and object permanence: A meta analysis of the A-not-B error. *Monographs of the Society for Research in Child Development* (No. 214), *51*, 1–67.

Yin, H. H., & Knowlton, B. J. (2006). The role of the basal ganglia in habit formation. *Nature Reviews Neuroscience, 7*, 464–476.

Young, M. P., & Yamane, S. (1992). Sparse population coding of faces in the inferotemporal cortex. *Science, 256*, 1327–1331.

PART V

DYNAMICS OF DEVELOPMENT

CHAPTER 9

Estimating the Technology of Cognitive and Noncognitive Skill Formation

The Linear Case

Flavio Cunha and James Heckman

The importance of cognitive skills in explaining socioeconomic success is now firmly established. An emerging body of empirical research documents the importance of noncognitive skills for predicting wages, schooling, and participation in risky behaviors.[1] Borghans, Duckworth, Heckman, and ter Weel (2008), Almlund, Duckworth, Heckman, and Kautz (2011), and Heckman, Stixrud, and Urzua (2006) demonstrate that cognitive and noncognitive skills are equally important in explaining a variety of aspects of social and economic life in the sense that movements from the bottom to the top of the noncognitive and cognitive distributions have comparable effects on many outcomes.

There is a substantial body of empirical research on the determinants and evolution of cognitive test scores and their growth.[2] There is no previous research on the determinants and evolution of noncognitive skills. This chapter identifies and estimates models of the stechnology of skill formation. Building on the theoretical analyses of Cunha and Heckman (2007, 2009), Cunha, Heckman, and Schennach (2010), and Cunha, Heckman, Lochner, and Masterov (2006), we estimate the joint evolution of cognitive and noncognitive skills over the life cycle of children.

We model the self-productivity of skills as well as their dynamic complementarity. Our technology formalizes the notion that noncognitive skills foster acquisition of cognitive skills by making children more adventuresome and open to learning.[3] It also formalizes the notion that cognitive skills can promote the formation of noncognitive skills. With our estimated technology, it is possible to define and quantify the importance of critical and sensitive periods in the life cycle of child development, and to determine at which ages inputs are most effective in shaping the evolution of child skills.

Psychologists who study child development have long advocated the importance of understanding the formation of noncognitive skills for interpreting the effects of early childhood intervention programs (see Raver & Zigler, 1997; Zigler & Butterfield, 1968). Heckman, Pinto, and Savelyev (2013) note that the Perry Preschool Program did not raise IQ but did promote success among its participants in a variety of aspects pertinent to social and economic life. The main channel was through noncognitive skill formation. Our analysis of noncognitive skills, their role in shaping cognitive skills, our investigation of the role of cognitive skills in shaping noncognitive skills, and our determination of the effectiveness of parental inputs on the formation of both types of skills over the life cycle provide a framework for developing unified treatment of the early intervention and family influence literatures.

The conventional approach to estimating cognitive production functions is best exemplified by the research of Todd and Wolpin (2003, 2007). A central problem with the production function approach is accounting for the endogeneity of inputs. Another problem is the wealth of candidate parental input measures available in many datasets. The confluence of these two problems—endogeneity and the multiplicity of input measures—places great demands on standard instrumental variable (IV) and fixed-effect procedures, such as those used by Todd and Wolpin. It is common in studies of cognitive production functions for analysts to have more inputs than instruments. Indices of inputs are used to circumvent this problem and reduce the parental input data to more manageable dimensions. The constructed indices often have an ad hoc quality about them and may be poor proxies for the true combination of inputs that enter the technology.

Our approach to the identification of the technology of skill formation bypasses these problems. We estimate a dynamic factor model that exploits cross-equation restrictions (covariance restrictions in linear systems) to secure identification using a version of dynamic state–space models (Shumway & Stoffer, 1982; Watson & Engle, 1983). The idea underlying our approach is to model cognitive and noncognitive skills, as well as parental investments, as low-dimensional latent variables. Building on the analyses of Jöreskog and Goldberger (1975), Jöreskog, Sörbom, and Magidson (1979), Bollen (1989), and Carneiro, Hansen, and Heckman (2003), we use a variety of measurements related to skills and investments to proxy latent skills and investments. With enough measurements relative to the number of latent skills and investments, we can identify the latent state–space dynamics generating the evolution of skills through cross-equation restrictions. When instruments are required, they are internally justified by the model of Cunha and Heckman (2007). We economize on the instruments required to secure identification, which are often scarce. We solve the problem of the multiplicity of measures of parental investments by using all of them as proxies for low-dimensional latent investments. Instead of creating an arbitrary index of parental inputs, we estimate an index that best predicts latent skill dynamics.

We also address a recurring problem in the literature on cognitive production functions. Studies in this tradition typically use a test score as a measure of output (e.g., see Hanushek, 2003). Yet test scores are arbitrarily normalized. Any monotonic transformation of a test score is also a valid test score. Value added—the change in test scores over stages (or grades)—is not invariant to monotonic transformations.

We solve the problem of defining a scale for output by anchoring our test scores using the adult earnings of the child, which have a well-defined cardinal scale. Other anchors such as high school graduation, college enrollment, and the like could also be used. Thus, we anchor the scale of the latent factors that generate test scores by determining how the factors predict adult outcomes.[4] This approach sets the scale of the test scores and factors in an interpretable metric.

Applying our methodology to Children of the National Longitudinal Study of Youth (CNLSY) data, we find that:

1. Both cognitive and noncognitive skills change over the life cycle of the child.
2. Parental inputs affect the formation of both noncognitive skills and cognitive skills. Direct measures of mothers' ability affect the formation of cognitive skills but not noncognitive skills.
3. Parental inputs appear to affect cognitive skill formation more strongly at earlier ages. They affect noncognitive skill formation more strongly at later ages. Those ages for which parental inputs have higher marginal productivity, holding all inputs constant, are called *sensitive periods.* The sensitive periods for cognitive skills occur earlier in the life cycle of the child than do sensitive periods for noncognitive skills. Our evidence is consistent with the evidence presented in Carneiro and Heckman (2003) that noncognitive skills are more malleable at later ages than cognitive skills. See also the evidence in Heckman (2007), in Borghans and colleagues (2008), and in Almlund and colleagues (2011).
4. Measurement error in inputs is substantial, and correcting for measurement error greatly affects our estimates.

The plan of this chapter is as follows. The first section briefly summarizes our previous research on models of skill formation. The next section presents our analysis of identification using dynamic factor models. The third section discusses our empirical findings, and the final section offers our conclusions. Appendix 9.1 develops an economic model of the measurement system as a derived demand for inputs system. Appendix 9.2 presents our likelihood function.

A Model of Cognitive and Noncognitive Skill Formation

Cunha and Heckman (2007, 2009) and Cunha and colleagues (2010) analyze multiperiod models of childhood skill formation followed by a period of adulthood.[5] They extend the model of Becker and Tomes (1986), who assume that childhood lasts one period and that investment inputs at different stages of the life cycle of a child are perfect substitutes and are equally productive. Becker and Tomes do not distinguish cognitive from noncognitive skills. Cunha and Heckman (2007, 2009) and Cunha and colleagues (2010) analyze models with two kinds of skills—θ^C and θ^N—where θ^C is cognitive skill and θ^N is noncognitive skill.

Let $\theta^I_{k,t}$ denote parental investments in child skill k in period t, $k \in \{C, N\}$, and $t \in \{1,\ldots,T\}$, where T is the number of periods of childhood. Let h be the level of human capital as the child starts adulthood, which depends on both θ^C_{T+1} and θ^N_{T+1}. The parents fully control the investment in the child. A better model would incorporate investment decisions of the child as influenced by the parent through the process of preference formation, and through parental incentives for influencing child behavior. We leave the development of that model for another occasion.

Assume that each agent is born with initial conditions $\theta_1' = \left(\theta_1^C, \theta_1^N\right)$. Family environmental and genetic factors may influence these initial conditions (see Levitt, 2003; Olds, 2002). At each stage t let $\theta_t' = \left(\theta_t^C, \theta_t^N\right)$ denote the 1×2 vector of skill or ability stocks. The technology of production of skill k in period t is

$$\theta^k_{t+1} = f^k_t\left(\theta_t, \theta^I_{k,t}\right) \quad \textbf{(9.1)}$$

for $k \in \{C, N\}$ and $t \in \{1,\ldots,T\}$.[6] In this model, stocks of both skills and abilities produce next-period skills and influence the productivity of investments. Cognitive skills can promote the formation of noncognitive skills, and vice versa, because θ_t is an argument of Equation 9.1. Cunha and Heckman (2007) summarize the evidence in economics and psychology about the interaction between cognitive and noncognitive skills in the production of human capital.

Adult human capital h is a combination of period $T + 1$ skills accumulated by the end of childhood:

$$h = g\left(\theta^C_{T+1}, \theta^N_{T+1}\right) \quad \textbf{(9.2)}$$

The function g is assumed to be continuously differentiable and increasing in θ^C_{T+1} and θ^N_{T+1}. This specification of human capital assumes that there is no comparative advantage in the labor market or in other areas of social performance.[7]

Early stocks of abilities and skills promote later skill acquisition by making later investment more productive. Students with greater early cognitive and noncognitive abilities are more efficient in later learning of both cognitive and noncognitive skills. The evidence from the early intervention literature suggests that the enriched early environments of the Abecedarian, Perry, and Child–Parent Center programs promote greater efficiency in learning in schools and reduce problem behaviors (see Almlund et al., 2011; Blau & Currie, 2006; Borghans et al., 2008; Cunha & Heckman, 2007; Cunha et al., 2006).

Equation 9.1 is sufficiently rich to capture the evidence on learning in rodents and macaque monkeys documented by Meaney (2001) and Cameron (2004), respectively. See Knudsen, Heckman, Cameron, and Shonkoff (2006) for a review of the literature. Emotionally nurturing early environments producing motivation and self-discipline create preconditions for later cognitive learning. More emotionally secure young animals explore their environments more actively and learn more quickly. This is an instance of complementarity among skills in producing skills.

Using Equation 9.1, Cunha and Heckman (2007) define critical and sensitive periods for investment. At some ages, and for certain skills, parental investment may be more productive than in other periods. Such periods, as noted previously, are *sensitive periods*. If investment is productive only in a single period, it is a *critical period* for that investment.

Cunha and Heckman (2007) discuss the role of complementarity in investments. If early investments are complementary with later investments, then low early investments, associated with disadvantaged childhoods, make later investments less productive. High early investments have a multiplier effect in making later investments more productive. If investment inputs are not perfect substitutes but are instead complements, government investment in the early years for disadvantaged children promotes investment in the later years. This is dynamic complementarity

Cunha and Heckman (2007) show that there is no tradeoff between equity and efficiency in early childhood investments. Government policies to promote early accumulation of human capital should be targeted to the children of poor families. However, the optimal later-period interventions for a child from a disadvantaged environment depend critically on the nature of the technology of skill production. If early and late investments are perfect complements, on efficiency grounds a low early investment should be followed up by low later investments.

If inputs are perfect substitutes, later interventions can, in principle, eliminate initial skill deficits. At a sufficiently high level of later-period investment, it is technically possible to offset low initial investments. However, it may not be cost effective to do so. Cunha and Heckman (2007, 2009) and Cunha and colleagues (2010) identify exact conditions for no investment to be an efficient outcome in this case. Under those conditions, it would be more efficient to give children bonds that earn interest, rather than invest in their human capital in order to raise their incomes.

The key to understanding optimal investment in children is to understand the technology and market environment in which agents operate. This chapter focuses on identifying and estimating the technology of skill formation, which is a vital ingredient for designing skill formation policies and evaluating their performance.

Identifying the Technology Using Dynamic Factor Models

Identifying and estimating Equation 9.1 is a challenging task. Both the inputs and outputs can only be proxied, and measurement error is likely to be a serious problem. In addition, the inputs are endogenous because parents choose them.

General nonlinear specifications of Equation 9.1 raise additional problems regarding measurement error in latent variables in nonlinear systems (see Schennach, 2004). This chapter estimates linear specifications of Equation 9.1. A more general nonlinear, nonparametric analysis requires addressing additional econometric and computational considerations, which are addressed in Cunha and colleagues (2010).

Identifying a Linear Technology

Using a linear specification, we can identify critical and sensitive periods for inputs. We can also identify cross-effects, as well as self-productivity of the stocks of skills. If we find little evidence of self-productivity, sensitive or critical periods, or cross-effects in a simpler setting, it is unlikely that a more general nonlinear model will overturn these results. Identifying a linear technology raises many challenges, which we address in this chapter.

There is a large body of research that estimates the determinants of the evolution of cognitive skills. Todd and Wolpin (2003) survey this literature. To our knowledge, there is no previous research on estimating the evolution of noncognitive skills.

The empirical analysis reported in Todd and Wolpin (2007) represents the state of the art in modeling the determinants of the evolution of cognitive skills.[8] In their paper, they use a scalar measure of cognitive ability $\left(\theta_{t+1}^C\right)$ in period $t + 1$ that depends on period t cognitive ability $\left(\theta_t^C\right)$ and investment. We denote investment by θ_t^I in this and remaining sections, rather than $\theta_{k,t}^I$, as in the preceding section. This notation reflects the fact that we cannot empirically distinguish between investment in cognitive skills and investment in noncognitive skills. Todd and Wolpin assume a linear-in-parameters technology:

$$\theta_{t+1}^C = a_t\theta_t^C + b_t\theta_t^I + \eta_t \tag{9.3}$$

where η_t represents unobserved input. They allow inputs to have different effects at different stages of the child's life cycle. They use the components of the "home score" measure to proxy parental investment.[9] We use a version of the inputs into the home score as well, but in a different way than they do.

Todd and Wolpin (2003, 2007) discuss problems arising from endogenous inputs $\left(\theta_t^C, \theta_t^I\right)$ that depend on unobservable η_t. In their 2005 paper, they use IV methods coupled with fixed-effect methods.[10] Reliance on IV methods is problematic because of the ever-present controversy about the validity of exclusion restrictions. As stressed by Todd and Wolpin, fixed-effect methods require very special assumptions about the nature of the unobservables, their persistence over time, and the structure of agent decision rules.[11] The CNLSY data used by Todd and Wolpin (2007) and this chapter have a multiplicity of investment measures subsumed in a home score measure that combines many diverse parental input measures into a score that weights all components equally.[12] As we note below, use of arbitrary aggregates calls into question the validity of IV estimation strategies for inputs.

Todd and Wolpin (2007) and the large literature they cite use a cognitive test score as a measure of output. This choice imparts a certain arbitrariness to their analysis. Test scores are arbitrarily normed. Any monotonic function of a test score is a perfectly good alternative test score. A test score is only a relative rank. Although Todd and Wolpin use raw scores and others use ranks (e.g., see Carneiro & Heckman, 2003; Carneiro, Heckman, & Masterov, 2005), none of these measures are intrinsically satisfactory because there is no meaningful cardinal scale for test scores.

We address this problem in this chapter by using adult outcomes to anchor the scale of the test score. Cunha and colleagues (2010) address this problem in a more general way for arbitrary monotonically increasing transformations of the factors. In this chapter, we develop an interpretable scale for θ_t^C, θ_t^N that is robust to all affine transformations of the units in which factors $\left(\theta_t^C, \theta_t^N\right)$ are measured. For example, using adult earnings Y as the anchor, we write

$$\ln Y = \mu + \delta^C \theta_{T+1}^C + \delta^N \theta_{T+1}^N + \varepsilon \quad \textbf{(9.4)}$$

where the scales of θ_{T+1}^C and θ_{T+1}^N are unknown. For any affine transformation of θ_{T+1}^k, corresponding to different units of measuring the factors, the value of δ^k and the intercept adjust and we can uniquely identify the left-hand side of

$$\frac{\partial \ln Y}{\partial \theta_t^I} = \delta^k \left(\frac{\partial \theta_{T+1}^k}{\partial \theta_t^I} \right) \text{ for } k \in \{C, N\};\ t \in \{1, \ldots, T\} \quad \textbf{(9.5)}$$

for any scale of θ_t^I. Thus, although the scale of δ^k is not uniquely determined, nor is the scale of θ_{T+1}^k, the scale of $\delta^k \theta_{T+1}^k$ is uniquely determined by its effect on log earnings, and we can define the effects of all inputs on $\ln Y$ relative to their effects on earnings.

The scale for measuring investment θ_t^I is also arbitrary. We report results for alternative normalizations of the units of investment. Natural scales are in dollars or log dollars. Using elasticities,

$$\left(\frac{\partial \ln Y}{\partial \theta_t^I} \right) \theta_t^I = \left(\delta^k \frac{\partial \theta_{T+1}^k}{\partial \theta_t^I} \right) \theta_t^I$$

produces parameters that are invariant to linear transformations of the units in which investment is measured. This approach generalizes to multiple factors and multiple anchors, and we apply it in this chapter. We now develop our empirical approach to identifying and estimating the technology of skill formation.

Estimating the Technology of Production of Cognitive and Noncognitive Skills

Our analysis departs from that of Todd and Wolpin (2007) in six ways:

1. We analyze the evolution of both cognitive and noncognitive outcomes using the equation system

$$\begin{pmatrix} \theta_{t+1}^N \\ \theta_{t+1}^C \end{pmatrix} = \begin{pmatrix} \gamma_1^N & \gamma_2^N \\ \gamma_1^C & \gamma_2^C \end{pmatrix} \begin{pmatrix} \theta_t^N \\ \theta_t^C \end{pmatrix} + \begin{pmatrix} \gamma_3^N \\ \gamma_3^C \end{pmatrix} \theta_t^I + \begin{pmatrix} \eta_t^N \\ \eta_t^C \end{pmatrix} \quad \textbf{(9.6)}$$

where θ_t^I can be a vector. We define $\eta_t' = \left(\eta_t^N, \eta_t^C\right)$.

2. We determine how stocks of cognitive and noncognitive skills at date t affect the stocks at date $t + 1$, examining both self-productivity (the effects of θ_t^N on θ_{t+1}^N, and θ_t^C on θ_{t+1}^C) and cross productivity (the effects of θ_t^C on θ_{t+1}^N and the effects of θ_t^N on θ_{t+1}^C) at each stage of the life cycle.
3. We develop a dynamic factor model where we proxy $\theta_t' = \left(\theta_t^N, \theta_t^C, \theta_t^I\right)$ by vectors of measurements on skills and investments that can include test scores as well as outcome measures.[13] In our analysis, test scores and parental inputs are indicators of the latent skills and latent investments. We account for measurement errors in output and input variables. We find substantial measurement errors in the proxies for parental investment and in the proxies for cognitive and noncognitive skills.
4. Instead of imposing a particular index of parental input based on components of the home score, we estimate an index that best fits the data.
5. Instead of relying solely on exclusion restrictions to generate instruments to correct for measurement error in the proxies for θ_t and for endogeneity, we use covariance restrictions that exploit a feature of our data: that there are many more measurements on θ_{t+1} and θ_t than there are latent factors. This feature allows us to secure identification from cross-equation restrictions using multiple indicator–multiple cause (MIMIC) (Jöreskog & Goldberger, 1975) and linear structural relationship (LISREL) (Jöreskog et al., 1979) models.[14] When instruments are needed, they arise from the internal logic of the model developed in Cunha and colleagues (2006) and Cunha and Heckman (2007), using methods developed by Madansky (1964) and Pudney (1982).
6. Instead of relying on test scores as measures of output and change in output due to parental investments, we anchor the scale of the test scores using adult outcome measures: log earnings and the probability of high school graduation. We thus estimate the effect of parental investments on the adult earnings of the child and on the probability of high school graduation.

A Model for the Measurements

We assume access to measurement systems that can be represented by a dynamic factor structure:

$$Y_{j,t}^k = \mu_{j,t}^k + \alpha_{j,t}^k \theta_t^k + \varepsilon_{j,t}^k, \text{ for } j \in \{1,\ldots,m_t^k\},\ k \in \{C, N, I\}$$

where m_t^k is the number of measurements on cognitive skills, noncognitive skills, and investments in period t; and where θ_t^k is a dynamic factor for component k, $k \in \{C, N, I\}$. $\text{Var}\left(\varepsilon_{j,t}^k\right) = \sigma_{k,j,t}^2$. We account for latent initial conditions of the process, $\left(\theta_1^C, \theta_1^N\right)$, which correspond to endowment of abilities. Because we have multiple measurements of abilities in the first period of our data, we can identify the distribution of the latent initial conditions. We also identify the distribution of each $\theta_t = \left(\theta_t^C, \theta_t^N, \theta_t^I\right)$, as well as the dependence across θ_t and $\theta_{t'}$, $t \neq t'$. The $\mu_{j,t}^k$ and the $\alpha_{j,t}^k$ can depend on regressors, which we keep implicit.

As above, let θ_t^C denote the stock of cognitive skill of the agent in period t. We do not observe θ_t^C directly. Instead, we observe a vector of measurements, such as test scores, $Y_{j,t}^C$, for $j \in \{1,2,\ldots,m_t^C\}$. Assume that

$$Y_{j,t}^C = \mu_{j,t}^C + \alpha_{j,t}^C \theta_t^C + \varepsilon_{j,t}^C \text{ for } j \in \{1,2,\ldots,m_t^C\} \tag{9.7}$$

and set $\alpha_{1,t}^C = 1$ for all t. Some normalization is needed to set the scale of the factors. The $\mu_{j,t}^C$ may depend on regressors.

We have a similar equation for noncognitive skills at age t, relating θ_t^N to proxies for it:

$$Y_{j,t}^N = \mu_{j,t}^N + \alpha_{j,t}^N \theta_t^N + \varepsilon_{j,t}^N \text{ for } j \in \{1,\ldots,m_t^N\} \tag{9.8}$$

and we normalize $\alpha_{1,t}^N = 1$ for all t. Finally, we model the measurement equations for investments, θ_t^I:

$$Y_{j,t}^I = \mu_{j,t}^I + \alpha_{j,t}^I \theta_t^I + \varepsilon_{j,t}^I \text{ for } j \in \{1,\ldots,m_t^I\} \tag{9.9}$$

and the factor loading $\alpha_{1,t}^I = 1$. The ε's are measurement errors that account for the fallibility of our measures of latent skills and investments.[15] Again, the $\mu_{j,t}^I$ and $\alpha_{j,t}^I$ may depend on the regressors, which we keep implicit.

We analyze a linear law of motion for skills:

$$\theta_{t+1}^k = \gamma_0^k + \gamma_{1,t}^k \theta_t^N + \gamma_{2,t}^k \theta_t^C + \gamma_{3,t}^k \theta_t^I + \eta_t^k \tag{9.10}$$
$$\text{for } k \in \{C,N\} \text{ and } t \in \{1,\ldots,T\}$$

where the error term η_t^k is independent across agents and over time for the same agents, but η_t^C and η_t^N are freely correlated. We assume that the η_t^k, $k \in \{C,N\}$ are independent of $\left(\theta_1^C, \theta_1^N\right)$. Below, we show how to relax the independence assumption and allow for unobserved inputs. Thus the γ_0^k and the $\gamma_{l,t}^k$, $l = 1, 2, 3$, may depend on regressors, which we keep implicit.

We allow the components of θ_t to be freely correlated for any t and with any vector $\theta_{t'}$, $t' \neq t$, and we can identify this dependence. We assume that any variables in the $\mu_{j,t}^k$ are independent of θ_t, $\varepsilon_{j,t}^k$, and η_t^k for $k \in \{C,N,I\}$ and $t \in \{1,\ldots,T\}$. We now establish conditions under which the technology parameters are identified.

Semiparametric Identification

The goal of the analysis is to recover the joint distribution of $\left\{\theta_t^C, \theta_t^N, \theta_t^I\right\}_{t=1}^T$, the distributions of $\left\{\eta_t^k\right\}_{t=1}^T$ and $\left\{\varepsilon_{j,t}^k\right\}_{j=1}^{m_t^k}$ nonparametrically, as well as the parameters $\left\{\alpha_{j,t}^k\right\}_{j=1}^{m_t^k}, \left\{\gamma_{j,t}^k\right\}_{j=1}^3$ for $k \in \{C,N\}$ and for $t \in \{1,\ldots,T\}$. Identification of the means of the measurements is straightforward under our assumptions.[16]

Classical Measurement Error for the Case of Two Measurements per Latent Factor: $m_t^C = m_t^N = m_t^I = 2$

We make the following assumptions about the $\varepsilon_{j,t}^k$:

> **Assumption 1.** $\varepsilon_{j,t}^k$ *is mean zero and independent across agents and over time for* $t \in \{1,\ldots,T\}$; $j \in \{1,2\}$; *and* $k \in \{C,N,I\}$.
>
> **Assumption 2.** $\varepsilon_{j,t}^k$ *is mean zero and independent of* $\left(\theta_\tau^C, \theta_\tau^N, \theta_\tau^I\right)$ *for all* $t, \tau \in \{1,\ldots,T\}$; $j \in \{1,2\}$; *and* $k \in \{C,N,I\}$.
>
> **Assumption 3.** $\varepsilon_{j,t}^k$ *is mean zero and independent from* $\varepsilon_{i,t}^l$ *for* $i,j \in \{1,2\}$ *and* $i \neq j$ *for* $k = l$; *otherwise* $\varepsilon_{j,t}^k$ *is mean zero and independent from* $\varepsilon_{i,t}^l$ *for* $i,j \in \{1,2\}$; $k \neq l$, $k,l \in \{C,N,I\}$; *and* $t \in \{1,\ldots,T\}$.

IDENTIFICATION OF THE FACTOR LOADINGS

Since we observe $\left\{\left[Y_{j,t}^k\right]_{j=1}^2\right\}_{t=1}^T$ for every person, we can compute $\text{cov}\left(Y_{1,t}^k, Y_{2,\tau}^l\right)$ from the data for all t,τ and k,l pairs, where $t,\tau \in \{1,\ldots,T\}$; $k,l \in \{C,N,I\}$. Consider, for example, measurements on cognitive skills. Recall that $\alpha_{1,t}^C = 1$. We know the left-hand side of each of the following equations:

$$\text{cov}\left(Y_{1,t}^C, Y_{1,t+1}^C\right) = \text{cov}\left(\theta_t^C, \theta_{t+1}^C\right) \tag{9.11}$$

$$\text{cov}\left(Y_{2,t}^C, Y_{1,t+1}^C\right) = \alpha_{2,t}^C \text{cov}\left(\theta_t^C, \theta_{t+1}^C\right) \tag{9.12}$$

$$\text{cov}\left(Y_{1,t}^C, Y_{2,t+1}^C\right) = \alpha_{2,t+1}^C \text{cov}\left(\theta_t^C, \theta_{t+1}^C\right) \tag{9.13}$$

We can identify $\alpha_{2,t}^C$ by taking the ratio of Equation 9.12 to Equation 9.11 and $\alpha_{2,t+1}^C$ from the ratio of Equation 9.13 to Equation 9.11. Proceeding in the same fashion, we can identify $\alpha_{j,t}^k$ for $t \in \{1,\ldots,T\}$ and $j \in \{1,2\}$, up to the normalization $\alpha_{1,t}^k = 1$, $k \in \{C,N,I\}$. We assume that $\alpha_{2,t}^k \neq 0$ for $k \in \{C,N,I\}$ and $t \in \{1,\ldots,T\}$. If $\alpha_{2,t}^k = 0$, we would violate the condition that states that there are exactly $m_t^k = 2$ valid measurements for the factor θ_t^k.

IDENTIFICATION OF THE JOINT DISTRIBUTION OF $\left\{\left(\theta_t^C, \theta_t^N, \theta_t^I\right)\right\}_{t=1}^T$

Once the parameters $\alpha_{1,t}^k$ and $\alpha_{2,t}^k$ are identified (up to the normalization $\alpha_{1,t}^k = 1$), we can rewrite Equations 9.7, 9.8, and 9.9 as

$$\frac{Y_{j,t}^k}{\alpha_{j,t}^k} = \frac{\mu_{j,t}^k}{\alpha_{j,t}^k} + \theta_t^k + \frac{\varepsilon_{j,t}^k}{\alpha_{j,t}^k},\ j \in \{1,2\} \text{ for } \alpha_{j,t}^k \neq 0,\ k \in \{C,N,I\};\ t \in \{1,\ldots,T\}$$ [17]

Now define

$$Y_j = \left\{ \left(\frac{Y_{j,t}^C}{\alpha_{j,t}^C}, \frac{Y_{j,t}^N}{\alpha_{j,t}^N}, \frac{Y_{j,t}^I}{\alpha_{j,t}^I} \right) \right\}_{t=1}^{T} \quad \text{for } j = 1,2$$

Similarly, define

$$\varepsilon_j = \left\{ \left(\frac{\varepsilon_{j,t}^C}{\alpha_{j,t}^C}, \frac{\varepsilon_{j,t}^N}{\alpha_{j,t}^N}, \frac{\varepsilon_{j,t}^I}{\alpha_{j,t}^I} \right) \right\}_{t=1}^{T} \quad \text{for } j = 1,2$$

and

$$\mu_j = \left\{ \left(\frac{\mu_{j,t}^C}{\alpha_{j,t}^C}, \frac{\mu_{j,t}^N}{\alpha_{j,t}^N}, \frac{\mu_{j,t}^I}{\alpha_{j,t}^I} \right) \right\}_{t=1}^{T} \quad \text{for } j = 1,2$$

Let θ denote the vector of all factors in all time periods:

$$\theta = \left\{ \left(\theta_t^C, \theta_t^N, \theta_t^I \right) \right\}_{t=1}^{T}$$

We rewrite the measurement equations as

$$Y_1 = \mu_1 + \theta + \varepsilon_1$$
$$Y_2 = \mu_2 + \theta + \varepsilon_2$$

Under the assumption that measurement error is classical, we can apply Kotlarski's (1967) theorem and identify the joint distribution of θ as well as the distributions of ε_1 and ε_2. Since $\alpha_{j,t}^k$ is identified, it is possible to recover the distribution of $\varepsilon_{j,t}^k$ for $j \in \{1,2,\ldots,m_t^k\}$; $k \in \{C,N,I\}$ and $t \in \{1,2,\ldots,T\}$.

EXAMPLE 1. Suppose that $\theta \sim N(0,\Sigma)$, $\varepsilon_{j,t}^k \sim N\left(0,\sigma_{k,j,t}^2\right)$. We observe the vectors Y_1 and Y_2, μ_1 and μ_2 are identified, and the Y_1 and Y_2 can be adjusted accordingly. As previously established, we can identify the factor loadings $\alpha_{j,t}^k$ by taking the ratio of covariances such as Equation 9.12 to Equation 9.11. To identify the distribution of the factors, we need to identify the variance–covariance matrix Σ. We can compute the variance of the factor θ_t^k from the covariance between $Y_{1,t}^k$ and $Y_{2,t}^k$:

$$\operatorname{cov}\left(Y_{1,t}^k, Y_{2,t}^k\right) = \alpha_{2,t}^k \operatorname{var}\left(\theta_t^k\right) \text{ for } k \in \{C,N,I\}$$

Recall that $\alpha_{2,t}^k$ is identified and the covariance on the left-hand side can be formed from the data. The covariance of any two elements of θ can be computed from the corresponding moments:

$$\operatorname{cov}\left(Y_{1,t}^{k}, Y_{1,\tau}^{l}\right) = \operatorname{cov}\left(\theta_{t}^{k}, \theta_{\tau}^{l}\right) \text{ for } k,l \in \{C,N,I\} \text{ and } t,\tau \in \{1,\ldots,T\} \quad \textbf{(9.14)}$$

and

$$\operatorname{cov}\left(Y_{j,t}^{k}, Y_{k,\tau}^{l}\right) = \alpha_{j,t}^{k}\alpha_{k,\tau}^{l}\operatorname{cov}\left(\theta_{t}^{k}, \theta_{\tau}^{l}\right) \quad \textbf{(9.15)}$$

where the coefficients $\alpha_{j,t}^{k}$, $\alpha_{k,\tau}^{l}$ are known by the previous argument. Since we know $\operatorname{var}\left(Y_{j,t}^{k}\right), \left(\alpha_{j,t}^{k}\right)$, and $\operatorname{var}\left(\theta_{j,t}^{k}\right)$, we can identify $\sigma_{k,j,t}^{2}$ from these ingredients:

$$\operatorname{var}\left(Y_{j,t}^{k}\right) - \left(\alpha_{j,t}^{k}\right)^{2}\operatorname{var}\left(\theta_{j,t}^{k}\right) = \sigma_{k,j,t}^{2},\ k \in \{C,N,I\},\ t \in \{1,\ldots,T\}$$ [18]

IDENTIFICATION OF THE TECHNOLOGY PARAMETERS ASSUMING INDEPENDENCE OF η

Assume that η_t^k is independent of $\left(\theta_t^C, \theta_t^N, \theta_t^I\right)$. Consider, for example, the law of motion for noncognitive skills:

$$\theta_{t+1}^{N} = \gamma_{0}^{N} + \gamma_{1,t}^{N}\theta_{t}^{N} + \gamma_{2,t}^{N}\theta_{t}^{C} + \gamma_{3,t}^{N}\theta_{t}^{I} + \eta_{t}^{N} \text{ for } t \in \{1,\ldots,T\} \quad \textbf{(9.16)}$$

Assume that η_t^N is serially independent but possibly correlated with η_t^C. Define

$$\tilde{Y}_{1,t+1}^{N} = Y_{1,t+1}^{N} - \mu_{1,t+1}^{N}$$
$$\tilde{Y}_{1,t}^{N} = Y_{1,t}^{N} - \mu_{1,t}^{N}$$
$$\tilde{Y}_{1,t}^{C} = Y_{1,t}^{C} - \mu_{1,t}^{C}$$
$$\tilde{Y}_{t}^{I} = Y_{t}^{I} - \mu_{t}^{I}$$

We substitute these measurement equations $\tilde{Y}_{1,t+1}^{N}, \tilde{Y}_{1,t}^{N}, \tilde{Y}_{1,t}^{C}, \tilde{Y}_{t}^{I}$ as proxies for θ_{t+1}^{N}, θ_{t}^{N}, θ_{t}^{C}, θ_{t}^{I}, respectively:

$$\begin{aligned}\tilde{Y}_{1,t+1}^{N} &= \gamma_{0}^{N} + \gamma_{1}^{N}\tilde{Y}_{1,t}^{N} + \gamma_{2}^{N}\tilde{Y}_{1,t}^{C} + \gamma_{3}^{N}\tilde{Y}_{1,t}^{I} \\ &\quad + \left(\varepsilon_{1,t+1}^{N} - \gamma_{1,t}^{N}\varepsilon_{1,t}^{N} - \gamma_{2,t}^{N}\varepsilon_{1,t}^{C} - \gamma_{3,t}^{N}\varepsilon_{1,t}^{I} + \eta_{t}^{N}\right)\end{aligned} \quad \textbf{(9.17)}$$

If we estimate Equation 9.17 by least squares, we do not obtain consistent estimators of γ_k^N for $k \in \{1,2,3\}$ because the regressors $\tilde{Y}_{1,t}^{N}, \tilde{Y}_{1,t}^{C}, \tilde{Y}_{1,t}^{I}$ are correlated with the error term ω_{t+1}, where

$$\omega_{t+1} = \varepsilon_{1,t+1}^{N} - \gamma_{1,t}^{N}\varepsilon_{1,t}^{N} - \gamma_{2,t}^{N}\varepsilon_{1,t}^{C} - \gamma_{3,t}^{N}\varepsilon_{1,t}^{I} + \eta_{t}^{N}$$

However, we can instrument $\tilde{Y}_{1,t}^{N}, \tilde{Y}_{1,t}^{C}, \tilde{Y}_{1,t}^{I}$, using $Y_{2,t}^{N}, Y_{2,t}^{C}, Y_{2,t}^{I}$ as instruments by applying two-stage least squares to recover the parameters γ_k^N for $k = 1,2,3$. (See Madansky, 1964, or Pudney, 1982, for the precise conditions on the factor loadings.) The suggested instruments are also independent of η_t^N because of the assumed lack of serial correlation in η_t^N. We can repeat the argument for different time periods. In this way, we can identify stage-specific technologies for each stage of the child's life cycle. We can perform a parallel analysis for the cognitive skill equation.

Nonclassical Measurement Error

We can replace Assumption 3 with the following assumption and still obtain full identification of the model.

> **Assumption 4.** *$\varepsilon_{1,t}^k$ is independent of $\varepsilon_{j,\tau}^l$ for $j \in \{2,\ldots,m_t^k\}$; $k,l \in \{C,N,I\}$ and $t, \tau \in \{1,2,\ldots,T\}$, $m_t^k \geq 2$. $\varepsilon_{1,t}^k$ is independent of $\varepsilon_{1,\tau}^k$, for $t \neq \tau$. Otherwise the $\varepsilon_{j,\tau}^k$, for $j \in \{2,\ldots,m_t^k\}$; $k,l \in \{C,N,I\}$ and $t,\tau \in \{1,2,\ldots,T\}$ can be arbitrarily dependent.*

The proof of identification is as follows. Let $Y_{j,t}^k = \alpha_{j,t}^k \theta_t^k + \varepsilon_{j,t}^k$, for $j \in \{1,\ldots,m_t^k\}$; $t \in \{1,\ldots,T\}$ and $k \in \{C,N,I\}$. Normalize $\alpha_{1,t}^k = 1$ for all $k \in \{C,N,I\}$ and $t \in \{1,\ldots,T\}$. Within a k system, for a fixed t, we can compute $\text{cov}\left(Y_{j,t}^k, Y_{1,t}^k\right) = \alpha_{j,t}^k \text{var}\left(\theta_t^k\right)$, $j \in \{1,\ldots,m_t^k\}$. For temporally adjacent systems, we can compute

$$\begin{aligned} \text{cov}\left(Y_{1,t-1}^k, Y_{1,t}^k\right) &= \text{cov}\left(\theta_{t-1}^k, \theta_t^k\right), \\ \text{cov}\left(Y_{1,t-1}^k, Y_{j,t}^k\right) &= \alpha_{j,t}^k \text{cov}\left(\theta_{t-1}^k, \theta_t^k\right),\ j \in \{2,\ldots,m_t^k\} \end{aligned} \tag{9.18}$$

Hence we can identify $\alpha_{j,t}^k$, $j \in \{1,\ldots,m_t^k\}$; $t \in \{1,\ldots,T\}$; and $k \in \{C,N,I\}$ and thus $\text{var}\left(\theta_t^k\right)$, $t \in \{1,\ldots,T\}$; $k \in \{C,N,I\}$. With these ingredients in hand, we can identify $\text{var}\left(\varepsilon_{j,t}^k\right)$, $t \in \{1,\ldots,T\}$, as well as

$$\text{cov}\left(\varepsilon_{j,t}^k, \varepsilon_{j',t}^k\right) = \text{cov}\left(Y_{j,t}^k, Y_{j',t}^k\right) - \alpha_{j,t}^k \alpha_{j',t}^k \text{var}\left(\theta_t^k\right)$$

since we know every ingredient on the right-hand side of the preceding equation. By a similar argument, we can identify

$$\text{cov}\left(\varepsilon_{j,t}^k, \varepsilon_{j',\tau}^l\right) = \text{cov}\left(Y_{j,t}^k, Y_{j',\tau}^l\right) - \alpha_{j,t}^k \alpha_{j',\tau}^l \text{cov}\left(\theta_t^k, \theta_\tau^l\right) \tag{9.19}$$

We can rewrite the measurement equations as a system:

$$\frac{Y_{j,t}^k}{\alpha_{j,t}^k} = \frac{\mu_{j,t}^k}{\alpha_{j,t}^k} + \theta_t^k + \frac{\varepsilon_{j,t}^k}{\alpha_{j,t}^k},\ j \in \{1,\ldots,m_t^k\};\ t \in \{1,\ldots,T\};\ k \in \{C,N,I\}$$

Applying Schennach (2004), we can identify the joint distribution of $(\theta_1^C,\ldots,\theta_T^C, \theta_1^N\ldots,\theta_T^N,\theta_1^I,\ldots,\theta_T^I)$ as well as the joint distribution of $\{\varepsilon_{j,t}^k\}$, $j \in \{1,\ldots,m_t^k\}$; $t \in \{1,\ldots,T\}$ and $k \in \{C,N,I\}$ using multivariate deconvolution.

EXAMPLE 2. Assume access to three measurements for cognitive, noncognitive, and investment factors, respectively. Suppose that $\theta = \left(\theta^C, \theta^N, \theta^I\right) \sim N(0,\Sigma), \varepsilon_{1,t}^k \sim N\left(0, \sigma_{k,1,t}^2\right)$, and are independent of $(\varepsilon_2, \varepsilon_3)$ where

$$\varepsilon_j = \left\{ \left(\frac{\varepsilon_{j,t}^C}{\alpha_{j,t}^C}, \frac{\varepsilon_{j,t}^N}{\alpha_{j,t}^N}, \frac{\varepsilon_{j,t}^I}{\alpha_{j,t}^I} \right) \right\}_{t=1}^T$$

for $j = 1, 2, 3$, and $\varepsilon_{j,t}^k$ and $\varepsilon_{j,t}^k$ are independent for $t \neq \tau$ but $(\varepsilon_2, \varepsilon_3) \sim N(0,\Omega)$, where Ω need not be diagonal. As discussed above, we identify var(θ_t^k), $k \in \{C, N, I\}$. Again, any element of the variance–covariance matrix Σ is obtained from Equation 9.14. Furthermore, any element of the matrix Ω can be obtained from Equation 9.19. Finally, we can identify $\sigma_{k,j,t}^2$ from var($Y_{j,t}^k$).

For this more general measurement-error system, we can identify stage-specific technologies using the same proof structure as was used for the case with classical measurement error.

The Identification of the Technology with Correlated Omitted Inputs

It is unrealistic to assume that omitted inputs are serially independent. Fortunately, we can relax this assumption. Assume now that η_t^k is *not* independent of $\theta_t' = (\theta_t^C, \theta_t^N, \theta_t^I)$. Consider a model in which η_t^k can be decomposed into two parts:

$$\eta_t^N = \gamma_4^N \lambda + \nu_t^N \text{ and } \eta_t^C = \gamma_4^C \lambda + \nu_t^C$$

so that the equations of motion can be written as

$$\theta_{t+1}^N = \gamma_0^N + \gamma_1^N \theta_t^N + \gamma_2^N \theta_t^C + \gamma_3^N \theta_t^I + \gamma_4^N \lambda + \nu_t^N \tag{9.20}$$

and

$$\theta_{t+1}^C = \gamma_0^C + \gamma_1^C \theta_t^N + \gamma_2^C \theta_t^C + \gamma_3^C \theta_t^I + \gamma_4^C \lambda + \nu_t^C \tag{9.21}$$

In this section we normalize $\gamma_4^N = 1$. The term λ is a time-invariant input permitted to be freely correlated with θ_t. We allow λ to have a different impact on cognitive and noncognitive skill accumulation. Let $\nu_t = (\nu_t^N, \nu_t^C)$. We make the following assumption.

Assumption 5. *The error term ν_t is independent of $\theta_t, \lambda, \nu_\tau$ conditional on regressors for any $\tau \neq t$.*

Under this assumption, we can identify both a stage-invariant technology and a stage-varying technology. We first analyze the stage-invariant case. Consider, for example, the law of motion for noncognitive skills. For any periods t, $t + 1$ we can compute the difference

$$\begin{aligned}\theta_{t+1}^N - \theta_t^N &= \gamma_1^N\left(\theta_t^N - \theta_{t-1}^N\right) + \gamma_2^N\left(\theta_t^C - \theta_{t-1}^C\right)\\ &+ \gamma_3^N\left(\theta_t^I - \theta_{t-1}^I\right) + \nu_t^N - \nu_{t-1}^N,\ t \geq 2\end{aligned} \tag{9.22}$$ [19]

where $t = 1$ is assumed to be the initial period. We use the measurement equations to proxy the unobserved θ's:

$$\begin{aligned}\tilde{Y}_{1,t+1}^N - \tilde{Y}_{1,t}^N &= \gamma_1^N\left(\tilde{Y}_{1,t}^N - \tilde{Y}_{1,t-1}^N\right) + \gamma_2^N\left(\tilde{Y}_{1,t}^C - \tilde{Y}_{1,t-1}^C\right)\\ &+ \gamma_3^N\left(\tilde{Y}_{1,t}^I - \tilde{Y}_{1,t-1}^I\right) + \nu_t^N - \nu_{t-1}^N\\ &+ \left\{\left(\varepsilon_{1,t+1}^N - \varepsilon_{1,t}^N\right) - \gamma_1^N\left(\varepsilon_{1,t}^N - \varepsilon_{1,t-1}^N\right)\right.\\ &\left. - \gamma_2^N\left(\varepsilon_{1,t}^C - \varepsilon_{1,t-1}^C\right) - \gamma_3^N\left(\varepsilon_{1,t}^I - \varepsilon_{1,t-1}^I\right)\right\},\ t \geq 2\end{aligned} \tag{9.23}$$

where again $t = 1$ is assumed to be the initial period. Ordinary least squares (OLS) applied to Equation 9.23 does not produce consistent estimates of γ_1^N, γ_2^N, and γ_3^N because the regressors $(\tilde{Y}_{1,t}^k - \tilde{Y}_{1,t-1}^k)$ are correlated with the error term ω, where

$$\omega = \left(\varepsilon_{1,t+1}^N - \varepsilon_{1,t}^N\right) - \gamma_1^N\left(\varepsilon_{1,t}^N - \varepsilon_{1,t-1}^N\right) - \gamma_2^N\left(\varepsilon_{1,t}^C - \varepsilon_{1,t-1}^C\right) - \gamma_3^N\left(\varepsilon_{1,t}^I - \varepsilon_{1,t-1}^I\right)$$

However, we can instrument $(\tilde{Y}_{1,t}^k - \tilde{Y}_{1,t-1}^k)$ using $\left\{\left(Y_{j,t-1}^k - Y_{j,t-2}^k\right)\right\}_{j=2}^{m_t^k}$ as instruments. These instruments are valid because of the generalization of investment equation (9) in Cunha and Heckman (2007, 2009) to a T period model.[20] Using a two-stage least squares regression with these instruments allows us to recover the parameters γ_1^N, γ_2^N, and γ_3^N. We can identify γ_0^N if we assume that $E(\lambda) = 0$. Following a parallel argument, we can identify γ_0^N, γ_1^N, γ_2^N, and γ_3^N using the data on the evolution of cognitive test scores.

Next, define

$$\psi_{t+1}^k = \theta_{t+1}^k - \left(\gamma_0^k + \gamma_1^k\theta_t^N + \gamma_2^k\theta_t^C + \gamma_3^k\theta_t^I\right)$$

From the measurement equations, we know the joint distribution of $\left(\theta_{t+1}^k, \theta_t^N, \theta_t^C, \theta_t^I\right)$ for $k \in \{C, N\}$. We have established how to obtain the parameter values γ_0^N, γ_1^N, γ_2^N, and γ_3^N. Consequently, we know the distribution of ψ_t^k for $k \in \{C, N\}$ and $t \in \{1, \ldots, T\}$. We have $2T$ equations:

$$\begin{aligned}
\psi_T^N &= \lambda + \nu_T^N & \psi_T^C &= \gamma_4^C \lambda + \nu_T^C \\
\psi_{T-1}^N &= \lambda + \nu_{T-1}^N & \psi_{T-1}^C &= \gamma_4^C \lambda + \nu_{T-1}^C \\
&\vdots & &\vdots \\
\psi_1^N &= \lambda + \nu_1^N & \psi_1^C &= \gamma_4^C \lambda + \nu_1^C
\end{aligned}$$

Under Assumption 5 we can apply Kotlarski's theorem to this system and obtain the distribution of λ and ν_t for any t. Note that we can identify the parameter γ_4^C from the covariance:

$$\operatorname{cov}\left(\psi_t^N, \psi_\tau^C\right) = \gamma_4^C \operatorname{var}(\lambda)$$

for any t, $\tau \in \{1,\ldots,T\}$ since the variance of λ is known. This approach solves the problem raised by correlated omitted inputs for stage-invariant technologies.

For the stage-varying case, a similar but subtler argument applies. Recall that the first period of life is $t = 1$. In place of Equation 9.22, we can write

$$\begin{aligned}
\theta_{t+1}^N - \theta_t^N &= \gamma_{0,t}^N - \gamma_{0,t-1}^N + \gamma_{1,t}^N \theta_t^N - \gamma_{1,t-1}^N \theta_{t-1}^N + \gamma_{1,t}^N \theta_t^C \\
&\quad - \gamma_{2,t-1}^N \theta_{t-1}^C + \gamma_{3,t}^N \theta_t^I - \gamma_{3,t-1}^N \theta_{t-1}^I + \nu_t^N - \nu_{t-1}^N
\end{aligned} \tag{9.24}$$

Using the measurement equations to proxy the θ's, we obtain

$$\begin{aligned}
\tilde{Y}_{1,t+1}^N - \tilde{Y}_{1,t}^N &= \gamma_{0,t}^N - \gamma_{0,t-1}^N + \gamma_{1,t}^N \tilde{Y}_{1,t}^N - \gamma_{1,t-1}^N \tilde{Y}_{1,t-1}^N + \gamma_{2,t}^N \tilde{Y}_{1,t}^C - \gamma_{2,t-1}^N \tilde{Y}_{1,t-1}^C \\
&\quad + \gamma_{3,t}^N \tilde{Y}_{1,t}^I - \gamma_{3,t-1}^N \tilde{Y}_{1,t-1}^I + \nu_t^N - \nu_{t-1}^N + \Big\{\left(\varepsilon_{1,t+1}^N - \varepsilon_{1,t}^N\right) \\
&\quad - \left(\gamma_{1,t}^N \varepsilon_{1,t}^N - \gamma_{1,t-1}^N \varepsilon_{1,t-1}^N\right) - \left(\gamma_{2,t}^N \varepsilon_{1,t}^C - \gamma_{2,t-1}^N \varepsilon_{1,t-1}^C\right) \\
&\quad - \left(\gamma_{3,t}^N \varepsilon_{1,t}^I - \gamma_{3,t-1}^N \varepsilon_{1,t-1}^I\right)\Big\}, \ t \geq 2
\end{aligned}$$

We can instrument $\tilde{Y}_{1,t}^k$, $\tilde{Y}_{1,t-1}^k$, $k \in \{C,N,I\}$, using $\left\{Y_{j,t-l}^k\right\}_{j=2}^{m_t^k}$, $k \in \{C,N,I\}$ and $l \geq 2$, as instruments. The validity of the instruments is based on the generalization of investment equation (9) in Cunha and Heckman (2007), discussed in our analysis of stage-invariant technologies. Thus we can identify the coefficients of Equation 9.24 except for the intercepts. We can identify relative intercepts $(\gamma_{0,t}^N - \gamma_{0,t-1}^N)$, $t \in \{2,\ldots,T\}$. With these intercepts in hand, we can identify the remaining parameters by the preceding proof, provided we have enough proxies for each factor in each period.[21]

Anchoring the Factors in the Metric of Earnings

We can set the scale of the factors by estimating their effects on log earnings for children when they become adults. Let Y be adult earnings. We write

$$\ln Y = \mu_T + \delta_N \theta_T^N + \delta_C \theta_T^C + \varepsilon \tag{9.25}$$

where ε is not correlated with θ_t or $\varepsilon^k_{j,T}$. Define

$$D = \begin{pmatrix} \delta_N & 0 \\ 0 & \delta_C \end{pmatrix}$$

Assume $\delta_N \neq 0$ and $\delta_C \neq 0$.[22] For any given normalization of the test scores, we can transform the θ_t to an earnings metric by multiplying Equation 9.6 by D:

$$D\theta_{t+1} = \left(DAD^{-1}\right)\left(D\theta_t\right) + \left(DB\right)\theta^I_t + \left(D\eta_t\right) \quad \textbf{(9.26)}$$

and work with $D\theta_{t+1}$ and $D\theta_t$ in place of θ_{t+1} and θ_t. The cross terms in $\left(DAD^{-1}\right)$ are affected by this change of units, but not the self-productivity terms. The relative magnitude of θ^I_t on the outcomes can be affected by this change in scale. We can use other anchors besides earnings. We report results from two anchors in this chapter: (1) log earnings and (2) the probability of graduating from high school. For the latter, we use a linear probability model.

Estimating the Technology of Skill Formation

We use a sample of 1,053 white males from the Children of the National Longitudinal Survey of Youth, 1979 (CNLSY/79; Center for Human Resource Research, 2004) dataset. Starting in 1986, the children of the NLSY/79 female respondents have been assessed every 2 years. The assessments measure cognitive ability, temperament, motor and social development, behavior problems, and self-confidence of the children as well as their home environment. Data were collected via direct assessment and maternal report during home visits at every biannual wave. Table 9.1 presents summary statistics of the measures of skill and investment used in our study.

The measures of quality of a child's home environment that are included in the CNLSY/79 survey are the components of the Home Observation Measurement of the Environment—Short Form (HOME-SF). They are a subset of the measures used to construct the HOME scale designed by Bradley and Caldwell (1980, 1984) to assess the emotional support and cognitive stimulation children receive through their home environment, planned events, and family surroundings. These measurements have been used extensively as inputs to explain child outcomes (e.g., see Todd & Wolpin, 2007).[23] Our empirical study uses measurements on the following parental investments: the number of books available to the child, a dummy variable indicating whether the child has a musical instrument, a dummy variable indicating whether the family receives a daily newspaper, a dummy variable indicating whether the child receives special lessons, a variable indicating how often the child goes to museums, and a variable indicating how often the child goes to the theater. We also report results from some specifications that use family income as a proxy for parental inputs, but none of our empirical conclusions rely on this particular measure.

TABLE 9.1. Summary Dynamic Measurements, White Children, CNLSY/1979

	Ages 6 and 7			Ages 8 and 9			Ages 10 and 11			Ages 12 and 13		
	Obser-vations	Mean	Standard error	Obser-vations	Mean	Standard error	Obser-vations	Mean	Standard error	Obser-vations	Mean	Standard error
PIAT Math[a]	753	−1.0376	0.5110	799	0.0423	0.6205	787	0.7851	0.6101	690	1.2451	0.5783
PIAT Reading Recognition[a]	751	−1.0654	0.4303	795	−0.0932	0.6543	783	0.6179	0.7334	688	1.1442	0.7852
Antisocial Score[a]	753	0.0732	0.9774	801	−0.0843	1.0641	787	−0.0841	1.0990	717	−0.0658	1.0119
Anxious Score[a]	778	0.1596	1.0016	813	−0.0539	1.0187	813	−0.0753	1.0771	730	−0.0664	1.0561
Headstrong Score[a]	780	0.0192	0.9882	813	−0.2127	1.0000	812	−0.2146	1.0416	729	−0.2123	1.0572
Hyperactive Score[a]	780	−0.0907	0.9673	815	−0.1213	1.0148	813	−0.0983	0.9902	729	−0.0349	0.9910
Conflict Score[a]	779	0.0177	0.9977	815	−0.0057	0.9935	814	−0.0441	1.0304	731	−0.0472	1.0420
Number of Books[b]	629	3.9173	0.3562	821	3.9220	0.3104	676	3.6746	0.6422	730	3.6315	0.6768
Musical Instrument[c]	628	0.4650	0.4992	821	0.4896	0.5002	674	0.5504	0.4978	728	0.5907	0.4921
Newspaper[c]	629	0.5326	0.4993	821	0.5043	0.5003	674	0.4985	0.5004	728	0.5000	0.5003
Child has special lessons[c]	627	0.5470	0.4982	820	0.7049	0.4564	672	0.7247	0.4470	727	0.7717	0.4200
Child goes to museums[d]	628	2.2596	0.9095	821	2.3082	0.8286	672	2.2604	0.8239	729	2.2195	0.8178
Child goes to theater[d]	630	1.8111	0.8312	820	1.8012	0.7532	674	1.8309	0.8000	728	1.8475	0.7920

Natural logarithm of family income[e]	865	10.4915	1.3647	936	10.4494	1.5689	881	10.5454	1.3168	795	10.6169	1.1877
Mother's highest grade completed[f]	1,053	12.9620	2.2015	1,053	12.9620	2.2015	1,053	12.9620	2.2015	1,053	12.9620	2.2015
Mother's Arithmetic Reasoning Score[g]	776	0.6050	1.0132	776	0.6050	1.0132	776	0.6050	1.0132	776	0.6050	1.0132
Mother's Word Knowledge Score[g]	776	0.5894	0.7666	776	0.5894	0.7666	776	0.5894	0.7666	776	0.5894	0.7666
Mother's Paragraph Composition Score[g]	776	0.5464	0.7311	776	0.5464	0.7311	776	0.5464	0.7311	776	0.5464	0.7311
Mother's Numerical Operations Score[g]	776	0.4945	0.8189	776	0.4945	0.8189	776	0.4945	0.8189	776	0.4945	0.8189
Mother's Coding Speed Score[g]	776	0.4554	0.8084	776	0.4554	0.8084	776	0.4554	0.8084	776	0.4554	0.8084
Mother's Mathematical Knowledge Score[g]	776	0.5297	1.0259	776	0.5297	1.0259	776	0.5297	1.0259	776	0.5297	1.0259

[a]The variables are standardized with mean 0 and variance 1 across the entire CNLSY/79 sample.
[b]The variable takes the value *1* if the child has no books, *2* if the child has 1 or 2 books, *3* if the child has 3–9 books, and *4* if the child has 10 or more books.
[c]For example, for musical instrument, the variable takes value *1* if the child has a musical instrument at home and *0* otherwise. Other variables are defined accordingly.
[d]For example, for "museums," the variable takes the value *1* if the child never went to the museum in the last calendar year, *2* if the child went to the museum once or twice in the last calendar year, *3* if the child went to the museum several times in the past calendar year, *4* if the child went to the museum about once a month in the last calendar year, and *5* if the child went to a museum once a week in the last calendar year.
[e]Family income is CPI adjusted. Base year is 2000.
[f]Mother's highest grade completed by age 28.
[g]Components of the ASVAB Battery. The variables are standardized with mean 0 and variance 1 across the entire CNLSY/79 sample.

As measurements of noncognitive skills we use components of the Behavior Problem Index (BPI), created by Peterson and Zill (1986) and designed to measure the frequency, range, and type of childhood behavior problems for children age 4 and older, although in our empirical analysis we only use children ages 6–13. The Behavior Problem score is based on responses from the mothers to 28 questions about specific behaviors that children age 4 and over may have exhibited in the previous 3 months. Three response categories are used in the questionnaire: *often true*, *sometimes true*, and *not true*. In our empirical analysis we use the following subscores of the BPI: (1) antisocial, (2) anxious/depressed, (3) headstrong, (4) hyperactive, and (5) peer problems. We standardize these variables so that among other characteristics, a child who scores low on the antisocial subscore is a child who often cheats or tells lies, is cruel or mean to others, and does not feel sorry for misbehaving. A child who displays a low score on the anxious/depressed measurement is a child who experiences sudden changes in mood, feels no one loves him or her, is fearful, or feels worthless or inferior. A child with low scores on the headstrong measurement is tense, nervous, argues too much, and is disobedient at home. Children will score low on the hyperactivity subscale if they have difficulty concentrating, act without thinking, and are restless or overly active. Finally, children will be assigned a low score on the peer problem subscale if they have problems getting along with other children, are not liked by others, and are not involved with others.

For measurements of cognitive skills we use the Peabody Individual Achievement Test (PIAT), which is a wide-ranging measure of academic achievement of children age 5 and over. It is commonly used in research on child development. Todd and Wolpin (2007) use the raw PIAT test score as their measure of cognitive outcomes. The CNLSY/79 includes two subtests from the full PIAT battery: PIAT Mathematics and PIAT Reading Recognition.[24] The PIAT Mathematics test measures a child's attainment in mathematics as taught in mainstream education. It consists of 84 multiple-choice items of increasing difficulty. It begins with basic skills such as recognizing numerals and progresses to measuring advanced concepts in geometry and trigonometry. The PIAT Reading Recognition subtest measures word recognition and pronunciation ability. Children read a word silently, then say it aloud. The test contains 84 items, each with four options, which increase in difficulty from preschool to high school levels. Skills assessed include the ability to match letters, name names, and read single words aloud.

Our dynamic factor models allow us to exploit the wealth of measures available in these data and enable us to solve several problems. First, there are many proxies for parental investments in children's cognitive and noncognitive development. Even if all parents provided responses to all of the measures of family input, we would still face the problem of selecting which variables to use and how to find enough instruments for so many endogenous variables. Applying the dynamic factor model, we let the data tell us the best combination of family input measures to use in predicting the levels and growth in the test scores instead of relying on an arbitrary index. Measured inputs that are not very informative on family investment decisions will have estimated factor loadings that are close to zero. Covariance restrictions in our model substitute for the missing instruments to secure identification.

Second, our models have the additional advantage of helping us solve the problem of missing data. It often happens that mothers do not provide responses to all items of the HOME-SF score. Similarly, some children may take the PIAT Reading Recognition exam, but not the PIAT Mathematics test. Another missing data problem that arises is that the mothers may provide information about whether the child has peer problems or not, but may refuse to issue statements regarding the child's hyperactivity level. For such cases, some researchers drop the observations for the parents who do not respond to certain items, or do not analyze the items that are not responded to by many parents, even though these items may be very informative. With our setup, we do not need to drop the parents or entire items in our analysis. Assuming that the data are missing randomly, we integrate out the missing items from the sample likelihood. Appendix 9.2 presents our sample likelihood. We now present and discuss our empirical results using the CNLSY data.

Empirical Results

We first present our estimates of an age-invariant version of the technology where we assume no critical and sensitive periods. We report estimates of a model with critical and sensitive periods in the subsection "Evidence of Sensitive Periods of Investment in Skills," below.

Estimates of Time-Invariant Technology Parameters

Using the CNLSY data, we estimate the simplest version of the model that imposes the restriction that the coefficients on the technology equations do not vary over periods of the child's life cycle, there are no omitted inputs correlated with θ_t, and the measurement error is classical. In Table 9.2 we report results in the scale of standardized test scores. We normalize the scale of the investment factor θ_t^I on different measures. Columns (1) and (4) show the estimated noncognitive and cognitive skill technologies, respectively, when we normalize the investment factor on family income. Columns (2) and (5) show the estimated parameters when we normalize the investment factor on "trips to the museum." Finally, in columns (3) and (6) we show the results when we normalize the factor loading in "trips to the theater." The estimated technology is robust to different normalization assumptions.[25]

Table 9.2 shows the estimated parameter values and their standard errors. From this table, we see that (1) both cognitive and noncognitive skills show strong persistence over time; (2) noncognitive skills in one period affect the accumulation of next-period cognitive skills, but cognitive skills in one period do not affect the accumulation of next-period noncognitive skills; (3) the estimated parental investment factor affects noncognitive skills slightly more strongly than cognitive skills, but the differences are not statistically significant; (4) the mother's ability affects the child's cognitive ability but not noncognitive ability; and (5) the mother's education plays no role in affecting the evolution of ability after controlling for parental investments and mother's ability. We contrast the OLS estimates of this model (presented in Table 9.16) with our

TABLE 9.2. Unanchored Technology Equations, White Males, CNLSY/79

	Noncognitive skill $\left(\theta_{t+1}^{N}\right)$			Cognitive skill $\left(\theta_{t+1}^{C}\right)$		
Independent variable	(1)	(2)	(3)	(4)	(5)	(6)
Lagged noncognitive skill $\left(\theta_t^N\right)$	0.884 (0.021)	0.884 (0.021)	0.884 (0.021)	0.028 (0.013)	0.028 (0.013)	0.028 (0.013)
Lagged cognitive skill $\left(\theta_t^C\right)$	0.003 (0.013)	0.003 (0.012)	0.003 (0.013)	0.977 (0.038)	0.977 (0.038)	0.977 (0.038)
Parental investment $\left(\theta_t^I\right)$	0.072 (0.020)	0.078 (0.021)	0.080 (0.024)	0.064 (0.013)	0.069 (0.014)	0.071 (0.015)
Mother's education, S	0.004 (0.008)	0.004 (0.008)	0.004 (0.008)	0.003 (0.010)	0.003 (0.010)	0.003 (0.010)
Mother's cognitive skill, A	−0.006 (0.006)	−0.006 (0.006)	−0.006 (0.006)	0.025 (0.009)	0.025 (0.009)	0.025 (0.009)

Note. Measurement error is classical; absence of omitted inputs is correlated with θ_t. Let $\theta_t' = \left(\theta_t^N, \theta_t^C, \theta_t^I\right)$ denote the noncognitive, cognitive, and investment dynamic factors, respectively. Let S denote mother's education and A denote mother's cognitive ability. The technology equations is

$$\theta_{t+1}^k = \gamma_1^k \theta_t^N + \gamma_2^k \theta_t^C + \gamma_3^k \theta_t^I + \psi_1^k S + \psi_2^k A + \eta_t^k$$

In this table we show the estimated parameter values and standard errors (in parentheses) of $\gamma_1^k, \gamma_2^k, \gamma_3^k, \psi_1^k$, and ψ_2^k in columns (1) through (6). In columns (1) and (4), the parental investment factor is normalized in the log-family income equation. In columns (2) and (5), the parental investment factor is normalized on trips to the museum. In columns (3) and (6), we normalize the parental investment factor on trips to the theater.

measurement-error corrected versions in the subsection "Estimating the Components of the Home Investment Dynamic Factor," below.

The dynamic factors are statistically dependent. Table 9.3 shows the evolution of the correlation patterns across the dynamic factors. The correlation between cognitive and noncognitive skills is .18 at ages 6 and 7 and grows to around .28 at ages 12 and 13. There is a strong contemporaneous correlation among noncognitive skills and the home investment. The correlation starts off at .40 at ages 6 and 7 and grows to .55 by ages 12 and 13. The same pattern is true for the correlation between cognitive skills and home investments. The correlation between these two variables goes from .38 at ages 6 and 7 to .61 at ages 12 and 13.

Allowing for Nonclassical Measurement Error

We check the robustness of our findings by relaxing the assumption that the error terms in the measurement equations are classical. We allow the measurement errors (except for the first measurement) to be freely correlated and estimate their dependence. Table 9.4 shows the estimated technologies for noncognitive and cognitive

TABLE 9.3. Contemporaneous Correlation Matrices, White Males, CNLSY/79

	Noncognitive	Cognitive	Investments
	Period 1—Children ages 6 and 7		
Noncognitive	1.0000	0.1892	0.3426
Cognitive	0.1892	1.0000	0.2921
Investments	0.3426	0.2921	1.0000
	Period 2—Children ages 8 and 9		
Noncognitive	1.0000	0.2334	0.4065
Cognitive	0.2334	1.0000	0.3835
Investments	0.4065	0.3835	1.0000
	Period 3—Children ages 10 and 11		
Noncognitive	1.0000	0.2643	0.4785
Cognitive	0.2643	1.0000	0.4892
Investments	0.4785	0.4892	1.0000
	Period 4—Children ages 12 and 13		
Noncognitive	1.0000	0.2845	0.5511
Cognitive	0.2845	1.0000	0.6111
Investments	0.5511	0.6111	1.0000

Note. Measurement error is classical; absence of omitted inputs is correlated with θ_t.

TABLE 9.4. Unanchored Technology Equations, White Males, CNLSY/79

Independent variable	Noncognitive skill $\left(\theta_{t+1}^{N}\right)$	Cognitive skill $\left(\theta_{t+1}^{C}\right)$
Lagged noncognitive skill $\left(\theta_t^{N}\right)$	0.8672 (0.024)	0.0264 (0.011)
Lagged cognitive skill $\left(\theta_t^{C}\right)$	0.0045 (0.014)	0.9739 (0.038)
Parental investment $\left(\theta_t^{I}\right)$	0.0801 (0.018)	0.0647 (0.012)
Maternal education, S	0.0041 (0.008)	0.0026 (0.010)
Maternal cognitive skill, A	−0.0092 (0.006)	0.0252 (0.009)

Note. Measurement error is nonclassical; absence of omitted inputs is correlated with θ_t. Let $\theta_t' = \left(\theta_t^{N}, \theta_t^{C}, \theta_t^{I}\right)$ denote the noncognitive, cognitive, and investment dynamic factors, respectively. Let S denote mother's education and A denote mother's cognitive ability. The technology equations are

$$\theta_{t+1}^{k} = \gamma_1^k \theta_t^{N} + \gamma_2^k \theta_t^{C} + \gamma_3^k \theta_t^{I} + \psi_1^k S + \psi_2^k A + \eta_t^k$$

In this table we show the estimated parameter values and standard errors (in parentheses) of $\gamma_1^k, \gamma_2^k, \gamma_3^k, \psi_1^k$, and ψ_2^k for noncognitive (k = N) and cognitive (k = C) skills. Investment is normalized in family income.

skills estimated under these more general conditions.[26] The main conclusions based on Table 9.2 are robust to the assumption that measurement error is classical.[27] In Table 9.5 we show the estimated contemporaneous correlation across the measurement errors in our measures of noncognitive skills. Most of the correlations across the error terms are low. In fact, no correlation in any period exceeds, in absolute value, .2, and most are well below it.

Table 9.6 reports the contemporaneous correlation of the error terms in the measurement equations for investment. We assume that the error term in family income is independent of the remaining error terms. Virtually all correlations are well below .04 in absolute value. The only exceptions are the correlations between "trips to the museum" and "trips to the theater" in periods 1 and 2. In sum, these findings suggest that the assumption that the measurement error is classical is not at odds with the data we analyze, and allowing for correlation in errors does not change the main conclusions obtained from the simpler technology assuming classical measurement error.

TABLE 9.5. Contemporaneous Correlation Matrices in Measurement Error, Measurements for Noncognitive Skills, White Males, CNLSY/79

	Anti-Social	Anxious	Headstrong	Hyperactive	Peer Conflict
		Period 1—Children ages 6 and 7			
Anti-Social	1.0000	0.0000	0.0000	0.0000	0.0000
Anxious	0.0000	1.0000	−0.0054	−0.0083	0.0479
Headstrong	0.0000	−0.0054	1.0000	0.0193	−0.1113
Hyperactive	0.0000	−0.0083	0.0193	1.0000	−0.1721
Peer Conflict	0.0000	0.0479	−0.1113	−0.1721	1.0000
		Period 2—Children ages 8 and 9			
Anti-Social	1.0000	0.0000	0.0000	0.0000	0.0000
Anxious	0.0000	1.0000	−0.0023	−0.0020	0.0117
Headstrong	0.0000	−0.0023	1.0000	0.0328	−0.1941
Hyperactive	0.0000	−0.0020	0.0328	1.0000	−0.1652
Peer Conflict	0.0000	0.0117	−0.1941	−0.1652	1.0000
		Period 3—Children ages 10 and 11			
Anti-Social	1.0000	0.0000	0.0000	0.0000	0.0000
Anxious	0.0000	1.0000	0.0196	0.0001	−0.0007
Headstrong	0.0000	0.0196	1.0000	0.0067	−0.0312
Hyperactive	0.0000	0.0001	0.0067	1.0000	−0.0002
Peer Conflict	0.0000	−0.0007	−0.0312	−0.0002	1.0000
		Period 4—Children ages 12 and 13			
Anti-Social	1.0000	0.0000	0.0000	0.0000	0.0000
Anxious	0.0000	1.0000	−0.0797	−0.1495	−0.0105
Headstrong	0.0000	−0.0797	1.0000	0.0692	0.0049
Hyperactive	0.0000	−0.1495	0.0692	1.0000	0.0092
Peer Conflict	0.0000	−0.0105	0.0049	0.0092	1.0000

TABLE 9.6. Contemporaneous Correlation Matrices in Measurement Error, Measurements for Parental Investments, White Males, CNLSY/79

	Income	Books	Musical	Newspaper	Lessons	Museum	Theater
			Period 1—Children ages 6 and 7				
Family Income	1.0000	0.0000	0.0000	0.0000	0.0000	0.0000	0.0000
Number of Books	0.0000	1.0000	−0.0044	0.0050	−0.0029	−0.0269	−0.0678
Musical Instruments	0.0000	−0.0044	1.0000	−0.0047	0.0027	0.0257	0.0647
Newspaper Subscriptions	0.0000	0.0050	−0.0047	1.0000	−0.0031	−0.0290	−0.0731
Number of Special Lessons	0.0000	−0.0029	0.0027	−0.0031	1.0000	0.0168	0.0423
Trips to Museum	0.0000	−0.0269	0.0257	−0.0290	0.0168	1.0000	0.3960
Trips to Theater	0.0000	−0.0678	0.0647	−0.0731	0.0423	0.3960	1.0000
			Period 2—Children ages 8 and 9				
Family Income	1.0000	0.0000	0.0000	0.0000	0.0000	0.0000	0.0000
Number of Books	0.0000	1.0000	−0.0008	0.0052	0.0018	−0.0160	−0.0484
Musical Instruments	0.0000	−0.0008	1.0000	−0.0019	−0.0006	0.0058	0.0175
Newspaper Subscriptions	0.0000	0.0052	−0.0019	1.0000	0.0039	−0.0355	−0.1076
Number of Special Lessons	0.0000	0.0018	−0.0006	0.0039	1.0000	−0.0121	−0.0366
Trips to Museum	0.0000	−0.0160	0.0058	−0.0355	−0.0121	1.0000	0.3291
Trips to Theater	0.0000	−0.0484	0.0175	−0.1076	−0.0366	0.3291	1.0000
			Period 3—Children ages 10 and 11				
Family Income	1.0000	0.0000	0.0000	0.0000	0.0000	0.0000	0.0000
Number of Books	0.0000	1.0000	−0.0001	−0.0001	−0.0001	0.0052	0.0007
Musical Instruments	0.0000	−0.0001	1.0000	0.0002	0.0003	−0.0137	−0.0017
Newspaper Subscriptions	0.0000	−0.0001	0.0002	1.0000	0.0002	−0.0083	−0.0010
Number of Special Lessons	0.0000	−0.0001	0.0003	0.0002	1.0000	−0.0130	−0.0016
Trips to Museum	0.0000	0.0052	−0.0137	−0.0083	−0.0130	1.0000	0.0693
Trips to Theater	0.0000	0.0007	−0.0017	−0.0010	−0.0016	0.0693	1.0000
			Period 4—Children ages 12 and 13				
Family Income	1.0000	0.0000	0.0000	0.0000	0.0000	0.0000	0.0000
Number of Books	0.0000	1.0000	0.0003	−0.0007	0.0000	0.0017	0.0158
Musical Instruments	0.0000	0.0003	1.0000	−0.0006	0.0000	0.0016	0.0151
Newspaper Subscriptions	0.0000	−0.0007	−0.0006	1.0000	0.0000	−0.0034	−0.0313
Number of Special Lessons	0.0000	0.0000	0.0000	0.0000	1.0000	0.0001	0.0010
Trips to Museum	0.0000	0.0017	0.0016	−0.0034	0.0001	1.0000	0.0803
Trips to Theater	0.0000	0.0158	0.0151	−0.0313	0.0010	0.0803	1.0000

Allowing for Correlated Omitted Inputs

We next investigate the assumption that the error term in the technology equations η_t is independent of the vector θ_t, by allowing for the presence of a time-invariant omitted input λ, as discussed in the subsection "Identification of the Technology with Correlated Inputs."[28] The results, displayed in Table 9.7, are consistent with the results shown in Table 9.2. Accounting for correlated omitted inputs does not reverse any major conclusion. Note that for purposes of identification, we normalize the coefficient on λ in the cognitive technology equation to one, $\gamma_4^C = 1$, and we estimate the coefficient on the noncognitive technology equation, $\gamma_4^N = 0.2835$.

Anchoring Our Estimates of the Factor Scale Using Adult Outcomes

Table 9.8 reports estimates of the time-invariant technology that uses the earnings data for persons ages 23–28 to anchor the output of the production function in a log dollar metric.[29] We initially assume that η_t is serially uncorrelated and that measurement error is classical. We relax these assumptions below, when we report estimates of more general specifications. Our fitted earnings function is linear in age and depends on the final level of the factors θ_{T+1}^C and θ_{T+1}^N. The coefficient on cognitive skills in the log earnings equations is estimated to be 0.052 (standard error is 0.0109). For noncognitive skills, we estimate a loading of 0.14 (with a standard error of 0.054). These estimates are consistent with estimates reported in Heckman and colleagues (2006). From Equation 9.26, it is clear that anchoring does not affect the estimates of

TABLE 9.7. Unanchored Technology Equations, White Males, CNLSY/79

Independent variable	Noncognitive skill $\left(\theta_{t+1}^N\right)$	Cognitive skill $\left(\theta_{t+1}^C\right)$
Lagged noncognitive skill $\left(\theta_t^N\right)$	0.8848 (0.021)	0.0276 (0.013)
Lagged cognitive skill $\left(\theta_t^C\right)$	0.0022 (0.013)	0.9891 (0.039)
Parental investment $\left(\theta_t^I\right)$	0.0797 (0.020)	0.0844 (0.017)
Omitted correlated inputs, λ	0.2835 (0.134)	1.0000 (normalized)

Note. Measurement error is classical, allows for omitted input λ correlated with θ_t. Let $\theta_t' = \left(\theta_t^N, \theta_t^C, \theta_t^I\right)$ denote the noncognitive, cognitive, and investment dynamic factors, respectively. Let λ denote omitted inputs that are potentially correlated with θ_t. The technology equations are

$$\theta_{t+1}^k = \gamma_1^k \theta_t^N + \gamma_2^k \theta_t^C + \gamma_3^k \theta_t^I + \gamma_4^k \lambda + v_t^k$$

In this table we show the estimated parameter values and standard errors (in parentheses) of $\gamma_1^k, \gamma_2^k, \gamma_3^k$, and γ_4^k. Note that for identification purposes, we normalize $\gamma_4^C = 1$. Investment is normalized on family income.

TABLE 9.8. Anchored Technology Equations, Anchoring on Log Earnings and Graduation from High School, White Males, CNLSY/79

	Noncognitive skill $\left(\theta_{t+1}^{N}\right)$		Cognitive skill $\left(\theta_{t+1}^{C}\right)$	
Independent variable	(1)	(2)	(3)	(4)
Lagged noncognitive skill $\left(\theta_t^N\right)$	0.8844 (0.0210)	0.8843 (0.0210)	0.0100 (0.0046)	0.0687 (0.0319)
Lagged cognitive skill $\left(\theta_t^C\right)$	0.0084 (0.0364)	0.0012 (0.0053)	0.9777 (0.0380)	0.9771 (0.0380)
Parental investment $\left(\theta_t^I\right)$	0.0101 (0.0028)	0.0079 (0.0022)	0.0032 (0.0007)	0.0173 (0.0035)
Maternal education, S	0.0006 (0.0011)	0.0004 (0.0009)	0.0002 (0.0005)	0.0008 (0.0027)
Maternal cognitive skill, A	−0.0008 (0.0008)	−0.0007 (0.0007)	0.0013 (0.0005)	0.0068 (0.0024)

Note. Measurement error is nonclassical; no omitted inputs are correlated with θ_t. Let $\theta_t' = \left(\theta_t^N, \theta_t^C, \theta_t^I\right)$ denote the noncognitive, cognitive, and investment dynamic factors, respectively. Let S denote mother's education and A denote mother's cognitive ability. The technology equations are

$$\theta_{t+1}^k = \gamma_1^k \theta_t^N + \gamma_2^k \theta_t^C + \gamma_3^k \theta_t^I + \psi_1^k S + \psi_2^k A + \eta_t^k$$

In this table we show the estimated parameter values and standard errors (in parentheses) of $\gamma_1^k, \gamma_2^k, \gamma_3^k, \psi_1^k$, and ψ_2^k in columns (1) through (4). In columns (1) and (3), we anchor the skill factors in log earnings of the child when adult. In columns (2) and (4), we anchor the skill factors in the probability of graduating from high school using a linear probability model. The investment factor is normalized in family income.

self-productivity but can affect the estimates of cross-productivity. It can also affect the magnitude of the estimated effect of θ_T^I on outcomes.

Columns (1) and (3) in Table 9.8 transform the estimates in Table 9.2 by D into a log earnings metric. The two cross-effects are ordered in the same direction as in the model reported in Table 9.2 where we use the metric of test scores. The effect of noncognitive skills on cognitive skills is precisely estimated.

One problem that might arise in using log earnings as an anchor for this sample is that log earnings are observed for the children who are born to very young mothers, making it a very selected sample. To check the robustness of these conclusions with regard to the log earnings anchor, we also use high school graduation for a person at least 19 years old to anchor the parameters of the technology equations. We model the probability of high school graduation as a linear probability equation. It is interesting to note that in the metric of the probability of graduating from high school, the estimated parental investment factor affects *cognitive* skills more strongly than noncognitive skills. This is because cognitive skills receive higher weight in the high school graduation equation than in the log earnings equation. The relative strength of these effects is reversed across the two metrics. The choice of a metric is not innocuous.

Evidence of Sensitive Periods of Investment in Skills

We now report evidence on critical and sensitive periods. Our analysis in this section presents conditions under which we can identify the parameters of the technology when they are allowed to vary over stages of the life cycle. We can identify whether or not there are sensitive periods in the development of skills provided that we normalize the investment factor on an input that is used at all stages of the child's life cycle. Results for an unanchored stage-specific technology, not correcting for nonclassical measurement error and serially correlated omitted inputs, are presented in Table 9.9. Using several alternative measures, including family income, trips to the museum, and trips to the theater, we estimate the same qualitative ordering on the sensitivity of parental investments at different stages of the life cycle.[30] Using a likelihood ratio test, we test and reject the hypothesis that the parameters describing the technologies are invariant over stages of the life cycle.[31]

Although we use test scores as a measure of output, transformation of output units by D will not affect our inference about sensitive periods because D is time invariant. When we allow the coefficients of the technology to vary over time, we find evidence of sensitive periods for parental investment in both cognitive and noncognitive skills. Sensitive periods for parental investments in cognitive skills occur at

TABLE 9.9. Unanchored Stage-Specific Technology Equations, White Males, CNLSY/79

	Noncognitive skill $\left(\theta_{t+1}^N\right)$			Cognitive skill $\left(\theta_{t+1}^C\right)$		
Independent variable	Stage 1	Stage 2	Stage 3	Stage 1	Stage 2	Stage 3
Lagged noncognitive skill $\left(\theta_t^N\right)$	0.9849	0.9383	0.7570	0.0605	0.0212	0.0014
	(0.014)	(0.015)	(0.010)	(0.012)	(0.008)	(0.008)
Lagged cognitive skill $\left(\theta_t^C\right)$	0.0508	−0.0415	0.0412	0.9197	0.8845	0.9099
	(0.043)	(0.041)	(0.041)	(0.023)	(0.021)	(0.019)
Parental investment $\left(\theta_t^I\right)$	0.0533	0.1067	0.0457	0.1125	0.0364	0.0379
	(0.013)	(0.022)	(0.019)	(0.032)	(0.014)	(0.014)
Maternal education, S	0.0034	−0.0028	0.0138	0.0050	0.0131	0.0021
	(0.007)	(0.007)	(0.008)	(0.010)	(0.012)	(0.014)
Maternal cognitive skill, A	0.0007	−0.0077	−0.0134	0.0506	0.0044	0.0194
	(0.001)	(0.001)	(0.002)	(0.013)	(0.008)	(0.007)

Note. Measurement error is classical; no omitted inputs are correlated with θ_t. Let $\theta_t' = \left(\theta_t^N, \theta_t^C, \theta_t^I\right)$ denote the noncognitive, cognitive, and investment dynamic factors, respectively. Let S denote mother's education and A denote mother's cognitive ability. The technology equations are

$$\theta_{t+1}^k = \gamma_{1,t}^k \theta_t^N + \gamma_{2,t}^k \theta_t^C + \gamma_{3,t}^k \theta_t^I + \psi_{1,t}^k S + \psi_{2,t}^k A + \eta_t^k$$

In this table we show the estimated parameter values and standard errors (in parentheses) of $\gamma_{1,t}^k, \gamma_{2,t}^k, \gamma_{3,t}^k, \psi_{1,t}^k$, and $\psi_{2,t}^k$. Stage 1 is the transition from ages 6–7 to ages 8–9. Stage 2 refers to the transition from ages 8–9 to 10–11. Stage 3 is the transition from ages 10–11 to 12–13.

earlier ages than sensitive periods for parental investments in noncognitive skills. The coefficient on investments in the technology for cognitive skills for the transition from period 1 to period 2 (ages 6 and 7 to ages 8 and 9) is around 0.11 (with a standard error of 0.032). For the transition from period 2 to period 3 (ages 8 and 9 to 10 and 11) the corresponding coefficient decreases rather sharply to 0.0364 (with a standard error of 0.014). For the final transition (ages 10 and 11 to ages 12 and 13), the estimate is about the same, 0.0379, with a standard error of 0.014. The difference between the early coefficient and the two later coefficients is statistically significant. This finding is consistent with periods 1 and 2 being sensitive periods for cognitive skills.[32]

For noncognitive skills in period 1, the coefficient on investments is only 0.0533, with a standard error of 0.013. Then it increases to 0.1067 in period 2. It decreases to 0.0457 in the final transition. This evidence suggests that sensitive periods for the development of noncognitive skills occur at later ages in comparison to sensitive periods for cognitive skills.[33]

For the sake of completeness, in Table 9.10 we show the estimated technologies for cognitive and noncognitive skills when we allow the error term in the measurement equations for noncognitive skills and investments to be correlated. Again, the estimates in Tables 9.9 and 9.10 are very similar, suggesting that the assumption of independence across measurement errors does not substantially affect our estimates.

TABLE 9.10. Unanchored Stage-Specific Technology Equations, White Males, CNLSY/79

	Noncognitive skill $\left(\theta_{t+1}^{N}\right)$			Cognitive skill $\left(\theta_{t+1}^{C}\right)$		
Independent variable	Stage 1	Stage 2	Stage 3	Stage 1	Stage 2	Stage 3
Lagged noncognitive skill $\left(\theta_t^N\right)$	0.9884 (0.016)	0.9427 (0.018)	0.7568 (0.012)	0.0597 (0.012)	0.0211 (0.008)	0.0014 (0.008)
Lagged cognitive skill $\left(\theta_t^C\right)$	0.0497 (0.046)	−0.0463 (0.042)	0.0418 (0.042)	0.9192 (0.024)	0.8846 (0.023)	0.9101 (0.021)
Parental investment $\left(\theta_t^I\right)$	0.0532 (0.013)	0.1002 (0.022)	0.0435 (0.019)	0.1116 (0.033)	0.0367 (0.002)	0.0378 (0.001)
Maternal education, *S*	0.0032 (0.007)	−0.0029 (0.007)	0.0138 (0.008)	−0.0050 (0.010)	0.0131 (0.010)	0.0021 (0.010)
Maternal cognitive skill, *A*	−0.0008 (0.001)	−0.0062 (0.001)	−0.0119 (0.002)	0.0510 (0.021)	0.0045 (0.010)	0.0194 (0.004)

Note. Measurement error is nonclassical; no omitted inputs are correlated with θ_t. Let $\theta_t' = \left(\theta_t^N, \theta_t^C, \theta_t^I\right)$ denote the noncognitive, cognitive, and investment dynamic factors, respectively. Let *S* denote mother's education and *A* denote mother's cognitive ability. The technology equations are

$$\theta_{t+1}^k = \gamma_{1,t}^k \theta_t^N + \gamma_{2,t}^k \theta_t^C + \gamma_{3,t}^k \theta_t^I + \psi_{1,t}^k S + \psi_{2,t}^k A + \eta_t^k$$

In this table we show the estimated parameter values and standard errors (in parentheses) of $\gamma_{1,t}^k, \gamma_{2,t}^k, \gamma_{3,t}^k, \psi_{1,t}^k$, and $\psi_{2,t}^k$. Stage 1 is the transition from ages 6–7 to ages 8–9. Stage 2 refers to the transition from ages 8–9 to 10–11. Stage 3 is the transition from ages 10–11 to 12–13.

TABLE 9.11. Anchored Stage-Specific Technology Equations, White Males, CNLSY/79

	Noncognitive skill $\left(\theta_{t+1}^{N}\right)$			Cognitive skill $\left(\theta_{t+1}^{C}\right)$		
Independent variable	Stage 1	Stage 2	Stage 3	Stage 1	Stage 2	Stage 3
Lagged noncognitive skill $\left(\theta_t^N\right)$	0.9849 (0.014)	0.9383 (0.015)	0.7570 (0.010)	0.0216 (0.004)	0.0076 (0.003)	0.0005 (0.003)
Lagged cognitive skill $\left(\theta_t^C\right)$	0.1442 (0.120)	−0.1259 (0.115)	0.1171 (0.115)	0.9197 (0.023)	0.8845 (0.021)	0.9099 (0.019)
Parental investment $\left(\theta_t^I\right)$	0.0075 (0.002)	0.0149 (0.003)	0.0064 (0.003)	0.0056 (0.002)	0.0018 (0.001)	0.0019 (0.001)
Maternal education, S	0.0005 (0.001)	−0.0004 (0.001)	0.0019 (0.001)	−0.0003 (0.001)	0.0007 (0.001)	0.0001 (0.001)
Maternal cognitive skill, A	0.0001 (0.000)	−0.0011 (0.000)	−0.0019 (0.000)	0.0025 (0.001)	0.0002 (0.000)	0.0010 (0.000)

Note. Anchor is log earnings of the child between ages 23 and 28. Measurement error is classical; no omitted inputs are correlated with θ_t. Let $\theta_t' = \left(\theta_t^N, \theta_t^C, \theta_t^I\right)$ denote the noncognitive, cognitive, and investment dynamic factors, respectively. Let S denote mother's education and A denote mother's cognitive ability. The technology equations are

$$\theta_{t+1}^k = \gamma_{1,t}^k \theta_t^N + \gamma_{2,t}^k \theta_t^C + \gamma_{3,t}^k \theta_t^I + \psi_{1,t}^k S + \psi_{2,t}^k A + \eta_t^k$$

In this table we show the estimated parameter values and standard errors (in parentheses) of $\gamma_{1,t}^k, \gamma_{2,t}^k, \gamma_{3,t}^k, \psi_{1,t}^k$, and $\psi_{2,t}^k$. Stage 1 is the transition from ages 6–7 to ages 8–9. Stage 2 refers to the transition from ages 8–9 to 10–11. Stage 3 is the transition from ages 10–11 to 12–13.

Table 9.11 shows that the qualitative evidence on sensitive periods reported in Table 9.9 is robust to anchoring. Period 1 is the sensitive period for cognitive skills; period 2 is the sensitive period for noncognitive skills in all of these specifications. The effects of parental investment on noncognitive skill remain strong at all stages, and are stronger and more precisely determined than in the case where we impose a stage-invariant technology.[34]

Estimating the Components of the Home Investment Dynamic Factor

The CNLSY/1979 reports an aggregate HOME score by taking a simple mean of the variables presented in Table 9.12, which assigns each component of the score the same weight. For expositional purposes we call these ad hoc weights.

The logic of the factor model speaks against uniform weighting. From Equation 9.9, it follows that different components of the HOME score, $H_{i,t}$, weight latent θ_T^I differently. A uniformly weighted average of the mean adjusted components of the scores for person i is

$$\tilde{H}_{i,t} = \frac{1}{m_t^I} \sum_{j=1}^{m_t^I} \left(Y_{i,j,t}^I - \mu_{j,t}^I\right) = \left(\frac{1}{m_t^I} \sum_{j=1}^{m_t^I} \alpha_{j,t}^I\right) \theta_{i,t}^I + \frac{1}{m_t^I} \sum_{j=1}^{m_t^I} \varepsilon_{i,j,t}^I \tag{9.27}$$

TABLE 9.12. The Weights in the Construction of the Investment Factor

	Estimated weights[a]	Ad hoc weights[b]	Share of total residual variance due to factors[c]	Share of total residual variance due to uniqueness[d]
		Ages 6 and 7		
Log family income	0.0787	—	0.1188	0.8812
Number of books	0.0919	0.1667	0.1359	0.8641
Musical instrument	0.0917	0.1667	0.1358	0.8642
Newspaper	0.1083	0.1667	0.1564	0.8436
Child has special lessons	0.2251	0.1667	0.2783	0.7217
Child goes to museums	0.2019	0.1667	0.2569	0.7431
Child goes to theater	0.2025	0.1667	0.2575	0.7425
		Ages 8 and 9		
Log family income	0.0646	—	0.0686	0.9314
Number of books	0.0987	0.1667	0.1011	0.8989
Musical instrument	0.1338	0.1667	0.1323	0.8677
Newspaper	0.0828	0.1667	0.0862	0.9138
Child has special lessons	0.1990	0.1667	0.1848	0.8152
Child goes to museums	0.1912	0.1667	0.1789	0.8211
Child goes to theater	0.2299	0.1667	0.2076	0.7924
		Ages 10 and 11		
Log family income	0.0721	—	0.0537	0.9463
Number of books	0.1310	0.1667	0.0934	0.9066
Musical instrument	0.1566	0.1667	0.1097	0.8903
Newspaper	0.0973	0.1667	0.0711	0.9289
Child has special lessons	0.1386	0.1667	0.0983	0.9017
Child goes to museums	0.1785	0.1667	0.1232	0.8768
Child goes to theater	0.2260	0.1667	0.1510	0.8490
		Ages 12 and 13		
Log family income	0.0862	—	0.0349	0.9651
Number of books	0.1314	0.1667	0.0523	0.9477
Musical instrument	0.1109	0.1667	0.0445	0.9555
Newspaper	0.0968	0.1667	0.0390	0.9610
Child has special lessons	0.1036	0.1667	0.0417	0.9583
Child goes to museums	0.1890	0.1667	0.0735	0.9265
Child goes to theater	0.2821	0.1667	0.1059	0.8941

[a]See text derivation. We assume mutually uncorrelated measurement errors.

[b]Ad hoc weighting is uniform weighting. If there are m_t^I measures, each measure has weight $\frac{1}{m_t^I}$.

[c]$\text{var}(\tilde{Y}_{k,t}^I) = (\alpha_{k,t}^I)^2 \text{var}(\theta_t^I) + \text{var}(\varepsilon_{k,t}^I)$. The share of the variance due to the factor is $(\alpha_{k,t}^I)^2 \text{var}(\theta_t^I) / \text{var}(\tilde{Y}_{k,t}^I)$.

[d]$\text{var}(\varepsilon_{k,t}^I) / \text{var}(\tilde{Y}_{k,t}^I)$.

There is no guarantee that the term in front of $\theta^I_{i,t}$ in Equation 9.27 is 1. Thus the mean-adjusted HOME scores may be biased for $\theta^I_{i,t}$ for each person even if the $\varepsilon^I_{i,j,t}$ are mutually independent and m^I_t gets large so that the second term converges to zero.

The measurement error of the standard mean adjusted home score $\tilde{H}_{i,t}$ for $\theta^I_{i,t}$ for person i is thus

$$\tilde{H}_{i,t} - \theta^I_{i,t} = \left(\frac{1}{m^I_t} \sum_{j=1}^{m^I_t} \alpha^I_{j,t} - 1 \right) \theta^I_{i,t} + \frac{1}{m^I_t} \sum_{j=1}^{m^I_t} \varepsilon^I_{i,j,t}$$

Unless the term in parentheses on the right-hand side equals zero, the measurement error is correlated with the true score. Instrumenting $\tilde{H}_{i,t}$ by a variable $Z_{i,t}$ correlated with $\theta^I_{i,t}$, but uncorrelated with $\varepsilon^I_{i,j,t}$ and $j \in \{1,\ldots,m^I_t\}$, will not produce consistent estimates of the skill technology. Trivially, it would produce consistent estimates of the technology parameter for $\theta^I_{i,t}$ divided by $\frac{1}{m^I_t} \sum_{j=1}^{m^I_t} \alpha^I_{j,t}$.[35]

Rewriting Equation 9.9 and removing the means $\left(\tilde{Y}^I_{j,t} = Y^I_{j,t} - \mu^I_{j,t}\right)$, we obtain

$$\frac{\tilde{Y}^I_{i,j,t}}{\alpha^I_{j,t}} = \theta^I_{i,t} + \frac{\varepsilon^I_{i,j,t}}{\alpha^I_{j,t}}$$

An unweighted average of the inverse-factor-weighted mean-adjusted scores is unbiased for θ^I_t for each person. The minimum variance unbiased combination of the inverse-factor loading weighted $\tilde{Y}^I_{i,j,t}$ in the case of uncorrelated $\frac{\varepsilon^I_{i,j,t}}{\alpha^I_{j,t}}$ assigns weight (dropping the i subscript to simplify the notation)

$$\omega_{j,t} = \frac{\left(\alpha^I_{j,t}\right)^2}{\operatorname{var}\left(\varepsilon^I_{j,t}\right)} \left[\sum_{k=1}^{m^I_t} \frac{\left(\alpha^I_{k,t}\right)^2}{\operatorname{var}\left(\varepsilon^I_{k,t}\right)} \right]^{-1}$$

to $\frac{\tilde{Y}^I_{j,t}}{\alpha^I_{j,t}}$, where $\sum_{j=1}^{m^I_{k,t}} \omega_{j,t} = 1$.

Arraying $\frac{\tilde{Y}^I_{j,t}}{\alpha^I_{j,t}}$ into a vector,

$$\tilde{Y}^I_t = \left(\frac{\tilde{Y}^I_{1,t}}{\alpha^I_{1,t}}, \ldots, \frac{\tilde{Y}^I_{m^I_t,t}}{\alpha^I_{m^I_t,t}} \right)$$

and the $\frac{\varepsilon^I_{i,j,t}}{\alpha^I_{j,t}}$ into a vector,

$$\varepsilon^I_t = \left(\frac{\varepsilon^I_{1,t}}{\alpha^I_{1,t}}, \ldots, \frac{\varepsilon^I_{m^I_t,t}}{\alpha^I_{m^I_t,t}} \right)$$

and defining $V_t = E\left[\left(\varepsilon_t^I\right)' \varepsilon_t^I\right]$, where in this expression E denotes expectation, we can produce optimal weights ω_t as the solution to

$$\min \omega_t' V_t \omega_t \text{ subject to } \iota' \omega_t = 1$$

where t is a 1 by m_t^I vector of ones. The solution in the general case is

$$\omega_t = \frac{1}{\left(\iota' V_t^{-1} \iota\right)} \left(V_t^{-1} \iota\right)$$

which specializes to the weight previously given when V_t is diagonal. These weights are optimal normalized home scores in the sense that they produce a minimum variance unbiased estimator of θ_t^I that will produce less bias for the true coefficient of θ_t^I in a least squares regression using θ_t^I as a regressor.[36]

The importance of these weights depends on the importance of the measurement error in the components of these scores. For example, consider the number of books available to the child. This variable is correlated with parental inputs because parents who invest more in the development of their children will tend to spend more resources on books. The number of books is unlikely to be a perfect indicator of total parental input. Our method allows for imperfect proxies. Under our method, the number of books a child has at age t (R_t) is modeled as $R_t = \alpha_{R,t}^I \theta_t^I + \varepsilon_{R,t}^I$ so that $\text{var}(R_t) = (\alpha_{R,t}^I)^2 \text{var}(\theta_t^I) + \text{var}(\varepsilon_{R,t}^I)$ because of the independence between θ_t^I and $\varepsilon_{R,t}^I$. We can decompose the total unobserved variance into two terms: one that is due to the parental input, the other that is orthogonal to it. The latter arises from measurement error. The relative importance of the two sources of error can be computed as

$$s_{I,R,t} = \frac{\left(\alpha_{R,t}^I\right)^2 \text{var}\left(\theta_t^I\right)}{\left(\alpha_{R,t}^I\right)^2 \text{var}\left(\theta_t^I\right) + \text{var}\left(\varepsilon_{R,t}^I\right)}$$

and

$$s_{I,\varepsilon_R,t} = \frac{\text{var}\left(\varepsilon_{R,t}^I\right)}{\left(\alpha_{R,t}^I\right)^2 \text{var}\left(\theta_t^I\right) + \text{var}\left(\varepsilon_{R,t}^I\right)}$$

Table 9.12 reports that $s_{I,R,1} = 0.1359$ (for the first stage, corresponding to ages 6 and 7), and $s_{I,\varepsilon_R,t} = 0.8641$. Most of the unobservable variance in "the number of books a child has" is actually not informative on the unobserved parental input θ_t^I. We report the same measures for the other input variables in Table 9.12. Over stages of the life cycle, all of the input measures tend to become relatively more error laden as proxies for θ_t^I.

Table 9.12 also displays the estimated optimal weights $\omega_{j,t}$ for each measurement j at each period t. The weights are far from uniform across inputs, as is assumed in constructing the traditional home score. Note further that the weights change over the life cycle, reflecting the differential importance of measurement-error variance at different ages. The change in the error variance reflects in part the change in $\alpha^I_{j,t}$ with t. Our estimates show that whether the child has special lessons has high weight early on (ages 6 and 7 to 8 and 9), but the weight declines considerably in the later periods (ages 10 and 11 to 12 and 13). The variable that indicates the number of books at home, on the other hand, exhibits the opposite behavior. It starts small in early ages, but becomes more important at later ages. It is interesting to note that variables that describe the number of books at home and whether the family takes a newspaper, although informative about home investments, receive lower weight in our method than other components of the home score. The optimal weighting differs greatly from the uniform weighting traditionally used in constructing home scores.

In our sample, the covariance between the measurement error $(\tilde{H}_{i,t} - \theta^I_{i,t})$ and the true score θ^I_t is relatively weak (see Table 9.13). This happens because $\frac{1}{m^I_t}\sum_{k=1}^{m^I_t}\alpha^I_{t,k}$ is close to 1. See Table 9.14 for the factor loadings and normalized factor loadings for noncognitive, cognitive, and parental investment (HOME score) components. The fact that $\frac{1}{m^I_t}\sum_{k=1}^{m^I_t}\alpha^I_{t,k} = 1$ implies that in our sample, standard IV methods designed to protect against classical measurement error in the standard HOME score are likely to be effective.

Table 9.15 displays the reliability in the test scores for cognitive and noncognitive skills in a manner comparable to the estimates of the share of measurement error for the components of the HOME score in Table 9.12. The components of both cognitive and noncognitive tests are measured with substantial error. Simple unweighted averages of the components of these tests are biased for θ^C_t and θ^N_t for each person. We

TABLE 9.13. Covariance between Measurement Error and the Dynamic Factors, White Males, CNLSY/79

	Period 1	Period 2	Period 3	Period 4
$\text{cov}\left(\frac{1}{m^N_t}\sum_{k=1}^{m^N_t}\left(\alpha^I_{t,k}-1\right)\theta^I_t + \frac{1}{m^N_t}\sum_{k=1}^{m^N_t}\varepsilon^I_{t,k}, \theta^I_t\right)$	−0.0271	−0.0593	−0.0498	−0.0099
$\text{cov}\left(\frac{1}{m^C_t}\sum_{k=1}^{m^C_t}\left(\alpha^I_{t,k}-1\right)\theta^I_t + \frac{1}{m^C_t}\sum_{k=1}^{m^C_t}\varepsilon^I_{t,k}, \theta^I_t\right)$	0.0113	0.0346	0.0421	0.0441
$\text{cov}\left(\frac{1}{m^I_t}\sum_{k=1}^{m^I_t}\left(\alpha^I_{t,k}-1\right)\theta^I_t + \frac{1}{m^I_t}\sum_{k=1}^{m^I_t}\varepsilon^I_{t,k}, \theta^I_t\right)$	0.0237	0.0216	0.0066	0.0029

TABLE 9.14. Estimated Factor Loadings and Standard Errors, White Males, CNLSY/79

	Period 1	Period 2	Period 3	Period 4
	Noncognitive Skills (Normalization: Anti-Social Score)			
Anxiety Score	0.9006	0.8910	0.9364	1.0122
	(0.0231)	(0.0236)	(0.0233)	(0.0234)
Headstrong Score	1.0671	0.9692	0.9590	1.1071
	(0.0366)	(0.0368)	(0.0368)	(0.0372)
Hyperactivity Score	1.0028	0.8980	0.8673	0.9208
	(0.0329)	(0.0331)	(0.0332)	(0.0337)
Peer Conflict Score	0.7647	0.7252	0.7974	0.8472
	(0.0188)	(0.0182)	(0.0189)	(0.0194)
	Cognitive Skills (Normalization: PIAT Math Score)			
Reading Recognition Score	1.2995	1.4878	1.6533	1.8307
	(0.0244)	(0.0274)	(0.0322)	(0.0411)
	Parental Investments (Normalization: Family Income)			
Number of Books	0.2710	0.2514	0.6244	0.6797
	(0.0098)	(0.0097)	(0.0163)	(0.0167)
Number of Musical Instruments	0.3908	0.4346	0.5180	0.4622
	(0.0103)	(0.0151)	(0.0158)	(0.0153)
Newspaper Subscriptions	0.4191	0.3559	0.4318	0.4430
	(0.0311)	(0.0308)	(0.0317)	(0.0318)
Special Lessons	0.5546	0.4678	0.4487	0.3899
	(0.0373)	(0.0351)	(0.0350)	(0.0346)
Trips to the Museum	0.9874	0.8619	0.9247	0.9384
	(0.0412)	(0.0427)	(0.0433)	(0.0434)
Trips to the Theater	0.8895	0.8113	0.9764	1.0578
	(0.0277)	(0.0257)	(0.0299)	(0.0331)

display the proportion of the variance due to measurement error in each of these scores for each test in the second column. The share of measurement error is roughly stable across ages but fluctuates for some components (e.g., hyperactivity).

Our evidence of substantial measurement error in all of the measures of inputs and outputs suggests that simple OLS estimates of the technology of skill formation are likely to be considerably biased. Table 9.16 presents an OLS version of the model with estimates reported in columns (1) and (4) of Table 9.2 that use income as the investment anchor.[37] The contrast between the estimates reported in the first and fourth columns of Table 9.2 and the least squares estimates in Table 9.16 is striking. Generally, OLS coefficients are downward-biased, showing much smaller self-productivity, cross-productivity, and investment productivity effects. The estimated effect of the HOME score on the Math score is perverse.

TABLE 9.15. Fraction of Total Variance Explained by Skill Factor versus Uniqueness, White Males, CNLSY/79

	Share of total variance explained by factor	Share of total variance explained by uniqueness
	Ages 6–7	
	Noncognitive measurements	
Anti-Social Score	0.5321	0.4679
Anxiety Score	0.3673	0.6327
Headstrong Score	0.6289	0.3711
Hyperactivity Score	0.5500	0.4500
Peer Conflict Score	0.2073	0.7927
	Cognitive measurements	
PIAT Math	0.3512	0.6488
PIAT Reading Recognition	0.9473	0.0527
	Ages 8–9	
	Noncognitive measurements	
Anti-Social Score	0.5409	0.4591
Anxiety Score	0.3983	0.6017
Headstrong Score	0.5620	0.4380
Hyperactivity Score	0.4371	0.5629
Peer Conflict Score	0.2005	0.7995
	Cognitive measurements	
PIAT Math	0.3938	0.6062
PIAT Reading Recognition	0.9119	0.0881
	Ages 10–11	
	Noncognitive measurements	
Anti-Social Score	0.5266	0.4734
Anxiety Score	0.4460	0.5540
Headstrong Score	0.5368	0.4632
Hyperactivity Score	0.4286	0.5714
Peer Conflict Score	0.2738	0.7262
	Cognitive measurements	
PIAT Math	0.3835	0.6165
PIAT Reading Recognition	0.9181	0.0819
	Ages 12–13	
	Noncognitive measurements	
Anti-Social Score	0.5040	0.4960
Anxiety Score	0.4803	0.5197
Headstrong Score	0.6324	0.3676
Hyperactivity Score	0.4064	0.5936
Peer Conflict Score	0.2613	0.7387
	Cognitive measurements	
PIAT Math	0.3561	0.6439
PIAT Reading Recognition	0.9149	0.0851

Note. For $\tilde{Y}^k_{j,t} = \alpha^k_{j,t}\theta^k_t + \varepsilon^k_{j,t}$, $(\alpha^k_{j,t})^2 \mathrm{var}(\theta^k_t) / \mathrm{var}(\tilde{Y}^k_{j,t})$ is share of variance explained by factor. $\mathrm{var}(\varepsilon^k_{j,t}) / \mathrm{var}(\tilde{Y}^k_{j,t})$ is share of variance explained by uniqueness.

TABLE 9.16. OLS Estimation of the Technology Equations, White Males, CNLSY/79

Independent variable	Anti-Social score ($t + 1$)	PIAT Math ($t + 1$)
Anti-Social Score, t	0.6431 (0.0165)	0.0333 (0.0096)
PIAT Math, t	0.0933 (0.0317)	0.5909 (0.0184)
HOME Score, t	0.0147 (0.0059)	−0.0137 (0.0034)
Maternal Education	0.0358 (0.0091)	0.0208 (0.0053)
Maternal ASVAB Arithmetics	−0.0254 (0.0190)	0.0658 (0.0110)

Note. Measurement error is classical; absence of omitted inputs is correlated with θ_t.

Conclusion

This chapter identifies and estimates a model of investment in child cognitive and noncognitive skills using dynamic factor models. The model is based on the analysis of Cunha and Heckman (2007, 2009) and Cunha and colleagues (2006).

Our empirical methodology accounts for the proxy nature of the measurements of parental investments and outcomes and for the endogeneity of inputs. It allows us to utilize the large number of potentially endogenous proxy variables available in our dataset without exhausting the available instruments. Our instruments are justified by the model of Cunha and Heckman (2007). To avoid the arbitrariness that arises in using test scores to measure the output of parental investments, we anchor estimated effects of investment in the metric of adult earnings and in the metric of the probability of high school graduation. The choice of the metric affects our conclusions about the relative productivity of parental investment on cognitive and noncognitive skills. We report results for alternative normalizations of the scale of parental investment and generally find agreement among alternative specifications.

We reach the following major conclusions.

1. We find high levels of self-productivity in the production of cognitive and noncognitive skills.
2. We find evidence of sensitive periods for parental investments in both types of skills, with the sensitive period for cognitive skill investments occurring earlier in the life cycle than the sensitive period for investments in noncognitive skills.

3. We also find substantial evidence of measurement error in the home input proxies and corollary evidence of attenuation bias in the OLS estimates of the technology of skill formation.
4. The estimated relative effect of parental input on cognitive and noncognitive skills depends on the metric in which we measure input.

Different adult outcomes are affected differently by cognitive and noncognitive skills. Sensitive periods occur at different stages for cognitive and noncognitive skills. Therefore, different stages of the child's life cycle are sensitive periods for investment to achieve different adult outcomes.

To show this, we simulate the effect of a 10% increase in investment at different stages of the life cycle of the child on log earnings at age 23 (Table 9.17A) and on high school graduation (Table 9.17B). These estimates include the cross-effects of each skill on the other, self-productivity, and direct investment effects.[38]

For the log earnings outcome, the strongest effect is for investment in period 2. This operates primarily through its effect on noncognitive skills, which then percolate into the next period and raise both cognitive and noncognitive skills. The strongest effect of investment on earnings operating through effects on cognitive skills is in

TABLE 9.17A. The Percentage Impact on Log Earnings at Age 23 of an Exogenous Increase by 10% in Investments at Different Periods, White Males, CNLSY/79

Total percentage impact on earnings	Percentage impact on log earnings exclusively through cognitive skills	Percentage impact on log earnings exclusively through noncognitive skills
	Period 1	
0.2487 (0.0302)	0.1247 (0.0151)	0.1240 (0.0150)
	Period 2	
0.3065 (0.0358)	0.0445 (0.0052)	0.2620 (0.0306)
	Period 3	
0.2090 (0.0230)	0.0540 (0.0059)	0.1550 (0.0170)

Note. Let $\tilde{Y}^I_{j,t}$ denote the jth measurement on the parental investment dynamic factor θ^I_t with the mean removed. We obtain the predicted parental investment $\hat{\theta}^I_t$ by applying the weights reported in Table 9.12 and measurements in the following way:

$$\hat{\theta}^I_t = \sum_{j=1}^{m^I_t} \omega_{j,t} \tilde{Y}^I_{j,t}$$

We simulate the model and obtain the adult level of cognitive and noncognitive skills. Using the anchoring equation, we predict baseline log earnings, log E. We then perform a counterfactual simulation. We investigate the level of adult skills if investments at different periods were increased by 10%, and we check the impact on log earnings, log E_τ, where E_τ is the counterfactual earnings if investment in period τ were 10% higher, $\tau = 1, 2, 3$. In this table, we report the percentage change in earnings, that is, log E_τ − log E.

TABLE 9.17B. The Percentage Impact on the Probability of Graduating from High School of an Exogenous Increase by 10% in Investments at Different Periods, White Males, CNLSY/79

Total percentage impact	Percentage impact through cognitive skills	Percentage impact exclusively through noncognitive skills
	Period 1	
0.6441	0.5480	0.0961
(0.0789)	(0.0672)	(0.0118)
	Period 2	
0.3980	0.1951	0.2029
(0.0466)	(0.0229)	(0.0238)
	Period 3	
0.3565	0.2366	0.1198
(0.0389)	(0.0258)	(0.0131)

Note. Let $\tilde{Y}_{j,t}^{I}$ denote the jth measurement on the parental investment dynamic factor θ_t^I with the mean removed. We obtain the predicted parental investment $\hat{\theta}_t^I$ by applying the weights reported in Table 9.12 and measurements in the following way:

$$\hat{\theta}_t^I = \sum_{j=1}^{m_t^I} \omega_{j,t} \tilde{Y}_{j,t}^{I}$$

We simulate the model and obtain the adult level of cognitive and noncognitive skills. Using the anchoring equation, we predict the probability of graduating from high school, p. We then perform a counterfactual simulation, investigate the level of adult skills if investments at different periods were increased by 10%, and check the impact on the probability of graduating from high school, p_τ, where p_τ is the counterfactual graduation probability if investment in period τ were 10% higher. In this table, we report the percentage change in probability of graduating, that is, $\log p_\tau - \log p$. Standard errors are given in parentheses.

Stage 1. Even at Stage 1, however, the effects of investment on cognitive skills and noncognitive skills is equally strong.

For high school graduation, the strongest effect of investment comes in period 1, and it operates primarily through its effects on cognitive skills. Even though period 1 is important, the effects of investment in later periods are substantial.

Missing from this chapter is an estimate of the key substitution parameters that determine the cost of later remediation relative to early investment that does not assume perfect substitutability as imposed here. To recover these crucial parameters requires a more general specification of the technology and more advanced econometric methods. This problem is addressed in Cunha and colleagues (2010), who also present a more general nonlinear approach to anchoring the test scores in an outcome measure.

ACKNOWLEDGMENTS

This research was supported in part by the American Bar Foundation; the J. B. & M. K. Pritzker Family Foundation; the Susan Thompson Buffett Foundation; the Geary Institute; University

College Dublin, Ireland; Grant Nos. R37 HD065072 and R01 HD054702 from the National Institute of Child Health and Human Development; a European Research Council grant hosted by University College Dublin (No. DEVHEALTH 269874); a grant from the Institute for New Economic Thinking; and an anonymous funder. The views expressed in this chapter are those of the authors and not necessarily those of the funders or commentators mentioned here. An early draft of this chapter was presented at a conference at the Minneapolis Federal Reserve, October 2003. We received helpful comments from Robert Pollak at a seminar at Washington University, February 2004; from Susanne Schennach, who is a coauthor of a more general paper; and from Petra Todd and Kenneth Wolpin. We also thank the editor Peter Molenaar for comments on the final draft. Molenaar (2003) contains closely related analyses of psychometric state–space models.

NOTES

1. See Bowles, Gintis, and Osborne (2001), Heckman and Rubinstein (2001), and Heckman et al. (2006).

2. Todd and Wolpin (2003) survey the educational production function literature, as well as the child development literature.

3. Cameron (2004) reports evidence for such effects in her experimental studies of macaque monkeys, and Meaney (2001) reports similar results for rodents. See the evidence in Knudsen et al. (2006) and the evidence summarized in Cunha and Heckman (2007).

4. Cawley, Heckman, and Vytlacil (1999) anchor test scores in earnings outcomes. We substantially extend their analysis by allowing for investment at different life cycle stages to affect the evolution of test scores.

5. See also their Web appendix, where more general models of skill formation are analyzed.

6. We assume that f_t^k is twice continuously differentiable, increasing, and concave in $\theta_{k,t}^I$. Twice continuous differentiability is only a convenience.

7. Thus we rule out one potentially important avenue of compensation: that agents can specialize in tasks that do not require the skills in which they are deficient. Borghans, ter Weel, and Weinberg (2007) discuss evidence against this assumption. Cunha et al. (2006) present a more general task function that captures the notion that different tasks require different combinations of skills and abilities. If we assume that the output (reward) in adult task j is $g_j\left(\theta_{T+1}^C, \theta_{T+1}^N, \eta\right)$, where η is a person-specific parameter and there are J distinct tasks, we can define $g_{\hat{j}}\left(\theta_{T+1}^C, \theta_{T+1}^N, \eta\right) = \arg\max_{j} \left\{g_j\left(\theta_{T+1}^C, \theta_{T+1}^N, \eta\right)\right\}_{j=1}^{J}$ and capture the operation of comparative advantage in the labor market.

8. Todd and Wolpin (2007) discuss a paper by Fryer and Levitt (2004) that uses inappropriate static methods to estimate a dynamic model of investment. Fryer and Levitt assume that parental inputs do not cumulate. Alternatively, they assume 100% depreciation of investment in each period. They also do not account for endogeneity of inputs or measurement error in inputs, which we find to be substantial.

9. This measure originates in the work of Bradley and Caldwell (1980, 1984) and is discussed further in the section "Estimating the Technology of Skill Formation."

10. See Hsiao (1986), Baltagi (1995), and Arellano (2003) for descriptions of these procedures.

11. Fixed-effect methods do not easily generalize to the nonlinear frameworks that are suggested by our analysis of the technology of skill formation. See, however, the analysis of Altonji and Matzkin (2005) for one approach to fixed effects in nonlinear systems.

12. There are many other papers that use this score. For example, see Baydar and Brooks-Gunn (1991) and the papers cited by Todd and Wolpin (2007).

13. In this and later sections, θ_t includes the investment factor, whereas in the section "A Model of Cognitive and Noncognitive Skill Formation" it only includes stocks of skills at date t.

14. See Carneiro et al. (2003) and Hansen, Heckman, and Mullen (2004) for some recent extensions.

15. Measurement Equations 9.7, 9.8, and 9.9 can be interpreted as output-constant demand equations arising from the following two-stage maximization problem. Families use inputs $X_{j,t}$ with prices $P_{j,t}$, $j \in \{1,\ldots,m_t^I\}$ to produce family investment $\theta_t^I = \phi_t^I\left(X_{1,t},\ldots,X_{m^I,t}\right)$. For the problem of minimizing the cost of achieving a given output, one can derive demand functions $X_{j,t} = h_{j,t}\left(P_{1,t},\ldots,P_{m_t^I,t},\theta_t^I\right)$ under general conditions. Specifications in Equations 9.7–9.9 are consistent with Cobb–Douglas and Leontief technologies, when θ_t^I is measured in logs. Prices appear in the intercepts. These technologies impose restrictions on the factor loadings of the inputs. See Appendix 9.1, which develops this point further.

16. Obviously, we cannot separately identify the mean of the factor, $E\left(\theta_t^k\right)$, and the intercepts $\mu_{j,t}^k$. It is necessary either to normalize the intercept in one equation $\mu_{j,t}^k = 0$ and identify $E\left(\theta_t^k\right)$ or to normalize $E\left(\theta_t^k\right) = 0$ and identify all of the intercepts $\mu_{j,t}^k$.

17. The same remark applies as in note 16. We cannot separately identify the mean of the factor, $E\left(\theta_t^k\right)$, and the intercepts $\mu_{j,t}^k$. It is necessary either to normalize the intercept in one equation $\mu_{j,t}^k$ and identify $E\left(\theta_t^k\right)$ or to normalize $E\left(\theta_t^k\right) = 0$ and identify all of the intercepts $\mu_{j,t}^k$.

18. Cunha et al. (2010) consider identification under weaker conditions.

19. We work only with first differences and not levels and hence do not model the initial conditions.

20. See their Web appendix.

21. Identification of the distribution of v_1 follows from the following observation. We know the distribution of $\varepsilon_{1,t+1}^N$, $\varepsilon_{1,t}^N$ (and $(1+\gamma_{1,t}^N)\varepsilon_{1,t}^N$), $\gamma_{1,t-1}^N\varepsilon_{1,t-1}^N$, $\gamma_{2,t}^N\varepsilon_{1,t}^C$, $\gamma_{2,t-1}^N\varepsilon_{1,t-1}^C$, $\gamma_{3,t}^N\varepsilon_{1,t}^I$, and $\gamma_{3,t-1}^N\varepsilon_{1,t-1}^I$ for $t \geq 2$ from the measurement system and from the IV estimation of the equation just below Equation 9.24. From the residuals of the error term for successive stages, we can use deconvolution to isolate the distribution of $v_{t+1}^N - v_t^N$, $t \geq 1$. By Kotlarski's theorem we can identify the distributions of v_t^N, $t = 1,\ldots,T$.

22. Note that we can identify the loadings δ_N and δ_C from

$$\operatorname{cov}\left(\ln Y, Y_{1,T}^N\right) = \delta_N \operatorname{var}\left(\theta_T^N\right) + \delta_C \operatorname{cov}\left(\theta_T^N, \theta_T^C\right)$$

and

$$\operatorname{cov}\left(\ln Y, Y_{1,T}^C\right) = \delta_N \operatorname{cov}\left(\theta_T^N, \theta_T^C\right) + \delta_C \operatorname{var}\left(\theta_T^C\right)$$

which gives us two linearly independent equations in two unknowns (δ_N, δ_C). The solution is

$$\begin{pmatrix} \delta_N \\ \delta_C \end{pmatrix} = \frac{1}{\mathrm{var}\left(\theta_T^N\right)\mathrm{var}\left(\theta_T^C\right) - \mathrm{cov}\left(\theta_T^N, \theta_T^C\right)^2} \begin{bmatrix} \mathrm{var}\left(\theta_T^C\right) & -\mathrm{cov}\left(\theta_T^N, \theta_T^C\right) \\ -\mathrm{cov}\left(\theta_T^N, \theta_T^C\right) & \mathrm{var}\left(\theta_T^N\right) \end{bmatrix}$$

23. As discussed in Linver, Brooks-Gunn, and Cabrera (2004), some of these items are not useful because they do not vary much among families (i.e., more than 90–95% of all families make the same response).

24. We do not use the PIAT Reading Comprehension battery since it is not administered to the children who score low in the PIAT Reading Recognition.

25. The magnitude of the estimated parental investment effect clearly depends on the scale in which investments are measured.

26. We use family income to normalize investment.

27. As discussed in the subsection "Nonclassical Measurement Error," to generalize our results to allow for nonclassical measurement error, we need to assume that the error term in one of the measures is independent of all measurement errors. For the measurements for noncognitive skills, we impose this assumption on the error term in the antisocial score equation.

28. We normalize investment on family income.

29. Investment is normalized on family income.

30. In the text we report the results for the normalization of investment relative to family income. In our Web appendix, we report estimates of alternative normalizations using "trips to the theater" and "trips to the museum."

31. Under the restricted model, we estimate 277 parameters, and the value of the log likelihood at the maximum is –53,877. Under the unrestricted model, we estimate 305 parameters, and the log likelihood attains the maximum value of –53,800. The statistic $\Lambda = -2(\ln L_R - \ln L_U)$, where R denotes restricted and U denotes unrestricted, is distributed as chi-square with 28 (= 305 – 277) degrees of freedom. We find that Λ is 155, significantly above the critical value of 41.337 at a 5% significance level.

32. For the coefficients on cognitive skills, the lower bound for the t statistic for the hypothesis $\gamma_{I,2}^C = \gamma_{I,1}^C$ is 2.73. For the hypothesis $\gamma_{I,2}^C = \gamma_{I,3}^C$ it is 3.43.

33. For the coefficients of investments on noncognitive skills, the lower bound for the t statistic for the hypothesis $\gamma_{I,2}^N = \gamma_{I,1}^N$ is 2.16, and for the hypothesis $\gamma_{I,2}^N = \gamma_{I,3}^N$ it is 2.34.

34. When we anchor on high school graduation instead of log earnings, we find that parental effects on the cognitive factor are stronger than on the noncognitive factor. This is also found in the stage-invariant technology.

35. The proof is straightforward. Divide both sides of Equation 9.27 by $\frac{1}{m_t^I}\sum_{j=1}^{m_t^I} \alpha_{j,t}^I$, substitute into Equation 9.6, and apply standard IV.

36. These weights are period by period. An alternative approach, used in psychometrics, would use all periods in a Kalman filter estimator of the factors (see Dolan & Molenaar, 1991).

37. We get similar results for other anchors.

38. Allowing for stage-specific coefficients in Equation 9.6, we obtain the effect of investment in a period k stages before the terminal period on adult abilities as

$$\theta_{T+1} = \left(\prod_{j=0}^{k} A_{T-j} \right) B_{T-k} \theta_{T-k}^{I}$$

The effects of variations in components of θ_{T-k}^{I} operating through cognitive and noncognitive skills are reported in Tables 9.17A and 9.17B. (The top element of B_{T-k} corresponds to the noncognitive effect of investment in the period; the bottom element corresponds to the cognitive effect of investment in the period.)

REFERENCES

Almlund, M., Duckworth, A., Heckman, J. J., & Kautz, T. (2011). Personality psychology and economics. In E. A. Hanushek, S. Machin, & L. Wössmann (Eds.), *Handbook of the economics of education* (Vol. 4, pp. 1–181). Amsterdam: Elsevier.

Altonji, J. G., & Matzkin, R. L. (2005). Cross section and panel data estimators for nonseparable models with endogenous regressors. *Econometrica, 73*(4), 1053–1102.

Arellano, M. (2003). *Panel data econometrics*. New York: Oxford University Press.

Baltagi, B. H. (1995). *Econometric analysis of panel data*. New York: Wiley.

Baydar, N., & Brooks-Gunn, J. (1991). Effectos of maternal employment and child care arrangements in infancy on preschoolers' cognitive and behavioral outcomes: Evidence from the Children of the NLSY. *Developmental Psychology, 27*, 918–931.

Becker, G. S., & Tomes, N. (1986). Human capital and the rise and fall of families. *Journal of Labor Economics, 4*(3, Pt. 2), S1–S39.

Blau, D., & Currie, J. (2006). Preschool, daycare, and afterschool care: Who's minding the kids? In E. Hanushek & F. Welch (Eds.), *Handbooks in economics: Vol. 2. Handbook of the economics of education* (pp. 1163–1278). Amsterdam: North-Holland.

Bollen, K. A. (1989). *Structural equations with latent variables*. New York: Wiley.

Borghans, L., Duckworth, A. L., Heckman, J. J., & ter Weel, B. (2008). The economics and psychology of personality and motivation. *Journal of Human Resources, 43*(4), 972–1059.

Borghans, L., ter Weel, B., & Weinberg, B. A. (2007). *Interpersonal styles and labor market outcomes* (Working Paper No. 12846). Cambridge, MA: National Bureau of Economic Research.

Bowles, S., Gintis, H., & Osborne, M. (2001). The determinants of earnings: A behavioral approach. *Journal of Economic Literature, 39*(4), 1137–1176.

Bradley, R. H., & Caldwell, B. M. (1980). The relation of home environment, cognitive competence, and IQ among males and females. *Child Development, 51*(4), 1140–1148.

Bradley, R. H., & Caldwell, B. M. (1984). The relation of infants' home environments to achievement test performance in first grade: A follow-up study. *Child Development, 55*(3), 803–809.

Cameron, J. (2004). *Evidence for an early sensitive period for the development of brain systems underlying social affiliative behavior.* Unpublished manuscript, Oregon National Primate Research Center, Beaverton, OR.

Carneiro, P., Hansen, K., & Heckman, J. J. (2003). Estimating distributions of treatment effects with an application to the returns to schooling and measurement of the effects of uncertainty on college choice. Lawrence R. Klein Lecture. *International Economic Review, 44*(2), 361–422.

Carneiro, P., & Heckman, J. J. (2003). Human capital policy. In J. J. Heckman, A. B. Krueger, & B. M. Friedman (Eds.), *Inequality in America: What role for human capital policies?* (pp. 77–239). Cambridge, MA: MIT Press.

Carneiro, P., Heckman, J. J., & Masterov, D. V. (2005). Labor market discrimination and racial differences in pre-market factors. *Journal of Law and Economics, 48*(1), 1–39.

Cawley, J., Heckman, J. J., & Vytlacil, E. J. (1999). On policies to reward the value added by educators. *Review of Economics and Statistics, 81*(4), 720–727.

Cawley, J., Heckman, J. J., & Vytlacil, E. J. (2001). Three observations on wages and measured cognitive ability. *Labour Economics, 8*(4), 419–442.

Center for Human Resource Research. (Ed.). (2004). *NLSY79 child and young adult data user's guide.* Columbus: Ohio State University.

Cunha, F., & Heckman, J. J. (2007). The technology of skill formation. *American Economic Review, 97*(2), 31–47.

Cunha, F., & Heckman, J. J. (2009). The economics and psychology of inequality and human development. *Journal of the European Economic Association, 7*(2–3), 320–364.

Cunha, F., Heckman, J. J., Lochner, L. J., & Masterov, D. V. (2006). Interpreting the evidence on life cycle skill formation. In E. A. Hanushek & F. Welch (Eds.), *Handbook of the economics of education* (pp. 697–812). Amsterdam: North-Holland.

Cunha, F., Heckman, J. J., & Schennach, S. M. (2010). Estimating the technology of cognitive and noncognitive skill formation. *Econometrica, 78*(3), 883–931.

Dolan, C. V., & Molenaar, P. C. M. (1991). A note on the calculation of latent trajectories in the quasi Markov simplex model by means of regression and the discrete Kalman filter. *Kwantitatieve Methoden, 38*, 29–44.

Durbin, J., & Koopman, S. J. (2001). *Time series analysis by state space methods.* New York: Oxford University Press.

Fryer, R., & Levitt, S. (2004). Understanding the black–white test score gap in the first two years of school. *Review of Economics and Statistics, 86*(2), 447–464.

Hansen, K. T., Heckman, J. J., & Mullen, K. J. (2004). The effect of schooling and ability on achievement test scores. *Journal of Econometrics, 121*(1–2), 39–98.

Hanushek, E. (2003). The failure of input-based schooling policies. *Economic Journal, 113*(485), F64–F98.

Harvey, A. C. (1989). *Forecasting, structural time series models and the Kalman filter.* New York: Cambridge University Press.

Heckman, J. J. (2007). The technology and neuroscience of capacity formation. *Proceedings of the National Academy of Sciences, 104*(3), 13250–13255.

Heckman, J. J., Pinto, R., & Savelyev, P. (2013). Understanding the mechanisms through which an influential early childhood program boosted adult outcomes. *American Economic Review, 103*(6), 2052–2086.

Heckman, J. J., & Rubinstein, Y. (2001). The importance of noncognitive skills: Lessons from the GED testing program. *American Economic Review, 91*(2), 145–149.

Heckman, J. J., Stixrud, J., & Urzua, S. (2006). The effects of cognitive and noncognitive abilities on labor market outcomes and social behavior. *Journal of Labor Economics, 24*(3), 411–482.

Herrnstein, R. J., & Murray, C. A. (1994). *The bell curve: Intelligence and class structure in American life.* New York: Free Press.

Hsiao, C. (1986). *Analysis of panel data*. New York: Cambridge University Press.

Jöreskog, K. G., & Goldberger, A. S. (1975). Estimation of a model with multiple indicators and multiple causes of a single latent variable. *Journal of the American Statistical Association, 70*(351), 631–639.

Jöreskog, K. G., Sörbom, D., & Magidson, J. (1979). *Advances in factor analysis and structural equation models*. Cambridge, MA: Abt Books.

Knudsen, E. I., Heckman, J. J., Cameron, J., & Shonkoff, J. P. (2006). Economic, neurobiological, and behavioral perspectives on building America's future workforce. *Proceedings of the National Academy of Sciences, 103*(27), 10155–10162.

Kotlarski, I. I. (1967). On characterizing the gamma and normal distribution. *Pacific Journal of Mathematics, 20*, 69–76.

Levitt, P. (2003). Structural and functional maturation of the developing primate brain. *Journal of Pediatrics, 143*(4, Suppl.), S35–S45.

Linver, M. R., Brooks-Gunn, J., & Cabrera, N. (2004). The Home Observation for Measurement of the Environment (HOME) inventory: The derivation of conceptually designed subscales. *Parenting: Science and Practice, 4*(2/3), 99–114.

Madansky, A. (1964). Instrumental variables in factor analysis. *Psychometrika, 29*(2), 105–113.

Meaney, M. J. (2001). Maternal care, gene expression, and the transmission of individual differences in stress reactivity across generations. *Annual Review of Neuroscience, 24*(1), 1161–1192.

Molenaar, P. C. M. (2003). State space techniques in structural equation modeling: Transformation of latent variables in and out of latent variable models. Available online at *www.hhdev.psu.edu/dsg/staff*.

Murnane, R. J., Willett, J. B., & Levy, F. (1995). The growing importance of cognitive skills in wage determination. *Review of Economics and Statistics, 77*(2), 251–266.

Olds, D. L. (2002). Prenatal and infancy home visiting by nurses: From randomized trials to community replication. *Prevention Science, 3*(2), 153–172.

Peterson, J. L., & Zill, N. (1986). Marital disruption, parent–child relationships, and behavior problems in children. *Journal of Marriage and the Family, 48*(2), 295–307.

Pudney, S. E. (1982). Estimating latent variable systems when specification is uncertain: Generalized component analysis and the eliminant method. *Journal of the American Statistical Association, 77*(380), 883–889.

Raver, C. C., & Zigler, E. F. (1997). Social competence: An untapped dimension in evaluating Head Start's success. *Early Childhood Research Quarterly, 12*(4), 363–385.

Schennach, S. M. (2004). Estimation of nonlinear models with measurement error. *Econometrica, 72*(1), 33–75.

Shumway, R. H., & Stoffer, D. S. (1982). An approach to time series smoothing and forecasting using the EM algorithm. *Journal of Time Series Analysis, 3*(3), 253–264.

Todd, P. E., & Wolpin, K. I. (2003). On the specification and estimation of the production function for cognitive achievement. *Economic Journal, 113*, F3–F33.

Todd, P. E., & Wolpin, K. I. (2007). *The production of cognitive achievement in children: Home, school and racial test score gaps*. Unpublished manuscript.

Watson, M. W., & Engle, R. F. (1983). Alternative algorithms for the estimation of dynamic factor, MIMIC and varying coefficient regression models. *Journal of Econometrics, 23*(3), 385–400.

Zigler, E. F., & Butterfield, E. C. (1968). Motivational aspects of changes in IQ test performance of culturally deprived nursery school children. *Child Development, 39*(1), 1–14.

APPENDIX 9.1. Interpretation of the Measurement Equations as Derived Demands

Write a production function

$$\theta_t^I = \varphi_t^I\left(X_{1,t},\ldots,X_{m_t^I,t}\right)$$

which is the output of the investment sector. Let $X_t = \left(X_{1,t},\ldots,X_{m_t^I,t}\right)'$. $P_t = \left(P_{1,t},\ldots,P_{m_t^I,t}\right)$ is the price vector.

The problem of the family is to minimize costs,

$$\min\left[P_t X_t + \lambda\left(\theta_t^I - \varphi_t^I\left(X_t\right)\right)\right]$$

The first-order condition for this problem is

$$P_{i,t} - \lambda\frac{\partial\varphi_t^I\left(X_{i,t}\right)}{\partial X_{i,t}} = 0,\ i \in \{1,\ldots,m_t^I\}$$

for an interior solution. We can derive input demands as a function of prices and output levels

$$X_{i,t} = h_{j,t}\left(P_t,\theta_t^I\right),\ j \in \{1,\ldots,m_t^I\}$$

which implicitly define the measurement equations.

For the Cobb–Douglas case, the technology is

$$\theta_t^I = A_t\prod_{i=1}^{m_t^I} X_{i,t}^{\alpha_i}$$

The input demand function for input *i* is

$$\ln X_{i,t} = \frac{1}{\sum_{j=1}^{m_t^I}\alpha_j}\ln\theta_t^I - \frac{\ln A_t}{\sum_{j=1}^{m_t^I}\alpha_j} - \sum_{j=1}^{m_t^I}\ln\left(\frac{\alpha_j P_i}{\alpha_i P_j}\right),\ i \in \{1,\ldots,m_t^I\}$$

Accounting for measurement error,

$$Y_{i,t}^I = \ln X_{i,t} + \varepsilon_{i,t}^I$$

In the Cobb–Douglas case, all inputs (measurements) have the same factor loading on $\ln\theta_t^I$. Only the intercepts, which depend on the share parameters and the prices, are different. In the Cobb–Douglas case, one would use logs of the factors.*

In the Leontief case,

$$\theta_t^I = \min\left\{\frac{X_{1,t}}{\alpha_1},\ldots,\frac{X_{m_t^I,t}}{\alpha_{m_t^I}}\right\}$$

The input demand equation is

$$X_{i,t} = \alpha_i\theta_t^I,\ i \in \{1,\ldots,m_t^I\}$$

* One can always write $\tilde{\theta}_t^I = \log\theta_t^I$ and work with $\tilde{\theta}_t^I$ everywhere.

and in logs,

$$\ln X_{i,t} = \ln \alpha_i + \ln \theta_t^I$$

so

$$Y_{i,t}^I = \ln X_{i,t} + \varepsilon_{i,t}^I,\ i \in \{1,\ldots,m_t^I\}$$

Thus all factor loadings on $\ln \theta_t^I$ are unity.

A generalized Leontief function writes

$$\theta_t^I = \min\left\{\frac{X_{1,t}^{\tau_1}}{\alpha_1},\ldots,\frac{X_{m_t^I,t}^{\tau_{m_t^I}}}{\alpha_{m_t^I}}\right\}$$

Thus,

$$\ln X_{i,t} = \frac{1}{\tau_i}\ln \alpha_i + \frac{1}{\tau_i}\ln \theta_t^I,\ i \in \{1,\ldots,m_t^I\}$$

In this case, the factor loadings are input-specific.

APPENDIX 9.2. Sample Likelihood for the Basic Estimation Strategy

We derive the likelihood and describe the basic estimation strategy for the model with classical measurement error and without serially correlated η_t. The likelihood for the more general models we estimate follows from a straightforward modification of the analysis in this appendix. In period t, let $m_t = m_t^N + m_t^C + m_t^I$, where m_t^N is the number of measurements on the noncognitive factor, and m_t^C and m_t^I are defined accordingly for the cognitive and investment factors. Here we explicitly allow for the number of measurements to be period specific. Let Y_t denote the $(m_t \times 1)$ vector

$$Y_t' = \left(Y_{1,t}^N,\ldots,Y_{m_t^N,t}^N, Y_{1,t}^C,\ldots,Y_{m_t^C,t}^C, Y_{1,t}^I,\ldots,Y_{m_t^I,t}^I\right)$$

In each period t, let $\theta_t' = \left(\theta_t^N, \theta_t^C, \theta_t^I\right)$. We use α_t to denote the $(m_t \times 3)$ matrix containing the factor loadings:

$$\alpha_t = \begin{bmatrix} 1 & 0 & 0 \\ \vdots & \vdots & \vdots \\ \alpha_{m_t^N,t}^N & 0 & 0 \\ 0 & 1 & 0 \\ \vdots & \vdots & \vdots \\ 0 & \alpha_{m_t^C,t}^C & 0 \\ 0 & 0 & 1 \\ \vdots & \vdots & \vdots \\ 0 & 0 & \alpha_{m_t^I,t}^I \end{bmatrix}$$

Let ε_t denote the $(m_t \times 1)$ vector of uniquenesses and $K_t = \text{var}\left(\varepsilon_t\right)$, where K_t is an $(m_t \times m_t)$ matrix. With this notation, we can write the observation equations in period t as:

$$Y_t = \alpha_t \theta_t + \varepsilon_t \tag{9B.1}$$

Recall that we use S and A to denote the mother's education and cognitive ability, respectively. Let G_t be a (3×3) matrix of coefficients. Let $\psi_{1,t}$ and $\psi_{2,t}$ denote (3×1) vectors. The G_t matrix and the vectors $\psi_{1,t}$ and $\psi_{2,t}$ contain the technology parameters for both the cognitive and noncognitive factors:

$$\theta_{t+1} = G_t \theta_t + \psi_{1,t} S + \psi_{2,t} A + \eta_t$$

where η_t is a (3×1) vector of error terms in the technology equations. Define $Q_t = \text{var}(\eta_t)$.

We assume that $\theta_1 | S, A \sim N(a_1, P_1)$. In the text, we establish the conditions for identification of a_1 and P_1. We also assume that $\varepsilon_t \sim N(0, K_t)$ and $\eta_t \sim N(0, Q_t)$. Then, given the normality assumption, together with linearity, it follows that $Y_1 \sim N(\mu_1, F_1)$, where

$$\mu_1 = \alpha_1 a_1 \text{ and } F_1 = \alpha_1 P_1 \alpha_1' + K_1$$

Normality is not required for identification, but it facilitates computation. To proceed in the normal case, we apply the Kalman filtering procedure (for details on the derivations, see, e.g., Durbin & Koopman, 2001; Harvey, 1989). If we define $Y^t = \left(Y_1, \ldots, Y_t\right), a_{t+1} = E\left(\theta_{t+1} \middle| S, A, Y^t\right)$, and $P_{t+1} = \text{var}\left(\theta_{t+1} \middle| S, A, Y^t\right)$, it is straightforward to establish that

$$a_{t+1} = G_t a_t + G_t P_t \alpha_t' \left(\alpha_t P_t \alpha_t' + K_t\right)^{-1} (Y_t - \alpha_t a_t) + \psi_{1,t} S + \psi_{2,t} A$$

and

$$P_{t+1} = G_t P_t G_t' - G_t P_t \alpha_t' \left(\alpha_t P_t \alpha_t' + K_t\right)^{-1} \alpha_t P_t G_t' + Q_t$$

Consequently, using Equation 9.28 we obtain $Y_{t+1} \middle| S, A, Y^t \sim N\left(\mu_t, F_t\right)$, where

$$\mu_t = \alpha_t a_t \text{ and } F_t = \alpha_t P_t \alpha_t' + K_t$$

Assuming that we observe mother's schooling, S, and mother's cognitive ability, A, we can decompose the contribution of individual i to the likelihood as

$$f\left(Y_{i,T}, Y_{i,T-1}, \ldots, Y_{i,1} \middle| S_i, A\right) = f\left(Y_{i,1} \middle| S_i, A\right) \prod_{t=2}^{T} f\left(Y_{i,t} \middle| S_i, A, Y_i^{t-1}\right)$$

where Y_i^{t-1} is the history of Y_i up to time period t–1. In general, we observe S but not A. However, we have shown that we can identify the distribution of A if we have a set of cognitive test scores for the mother, M. Consequently, we can integrate A out:

$$f\left(Y_{i,T}, Y_{i,T-1}, \ldots, Y_{i,1} \middle| S_i\right) = \int f\left(Y_{i,1} \middle| S_i, A\right) \prod f\left(Y_{i,t} \middle| S_i, A, Y_i^{t-1}\right) f_A(A) \, dA$$

Assuming that observations are i.i.d. over children, the likelihood of the sample is

$$\prod_{i=1}^{n} f\left(Y_{i,T}, Y_{i,T-1}, \ldots, Y_{i,1} \middle| S_i\right) = \prod_{i=1}^{n} \int f\left(Y_{i,1} \middle| S_i, A\right) \prod_{t=2}^{T} f\left(Y_{i,t} \middle| S_i, A, Y_i^{t-1}\right) f_A(A)\, dA$$

Missing data can be integrated out, and so all cases can be used even in the presence of missing data. Extensions to the other cases are straightforward and are deleted for the sake of brevity.

CHAPTER 10

Dynamics of Development

A Complex Systems Approach

Han L. J. van der Maas, Kees Jan Kan, Abe Hofman, and Maartje E. J. Raijmakers

Complex systems are everywhere. They are studied in fields such as mathematics, physics, chemistry, biology, and economy. Figure 10.1 gives an impression of the diverse topics and applications in complex systems research. The psychological system is probably the most complex system that we can study because it involves the brain's neural network but also the various social and societal networks in which it is imbedded. In addition, it is a developmental system. Its development in a newborn with only basic reflexes to an adult with the ability to lecture, reason, organize, and many more things, is even astonishing from a complex systems view.

Clearly, the study of human development is very challenging because of (1) its complexity, (2) the many (ethical) restrictions on research methods, and (3) its paradoxical status. With regard to the latter, it would be problematic to conclude that humans are incapable of scientific study. One consequence of these severe challenges is that psychology is still largely characterized by informal verbal descriptions, and most research is based on unconnected verbal mini-theories. In our view, one route to progress in psychological science is in studying and copying ways to model and investigate complex systems in the natural sciences.

In this chapter we follow this route by providing three examples of modeling and investigating complex systems. First, we present a new model for general intelligence based on a mathematical model for ecological networks. Second, we discuss ways to study phase transitions in psychological systems. Third, we introduce a completely new approach to collect high-frequency data on children's development, which is a necessity for studying complex systems. With these three examples we hope to demonstrate the viability of the complex systems approach to the study of human development.

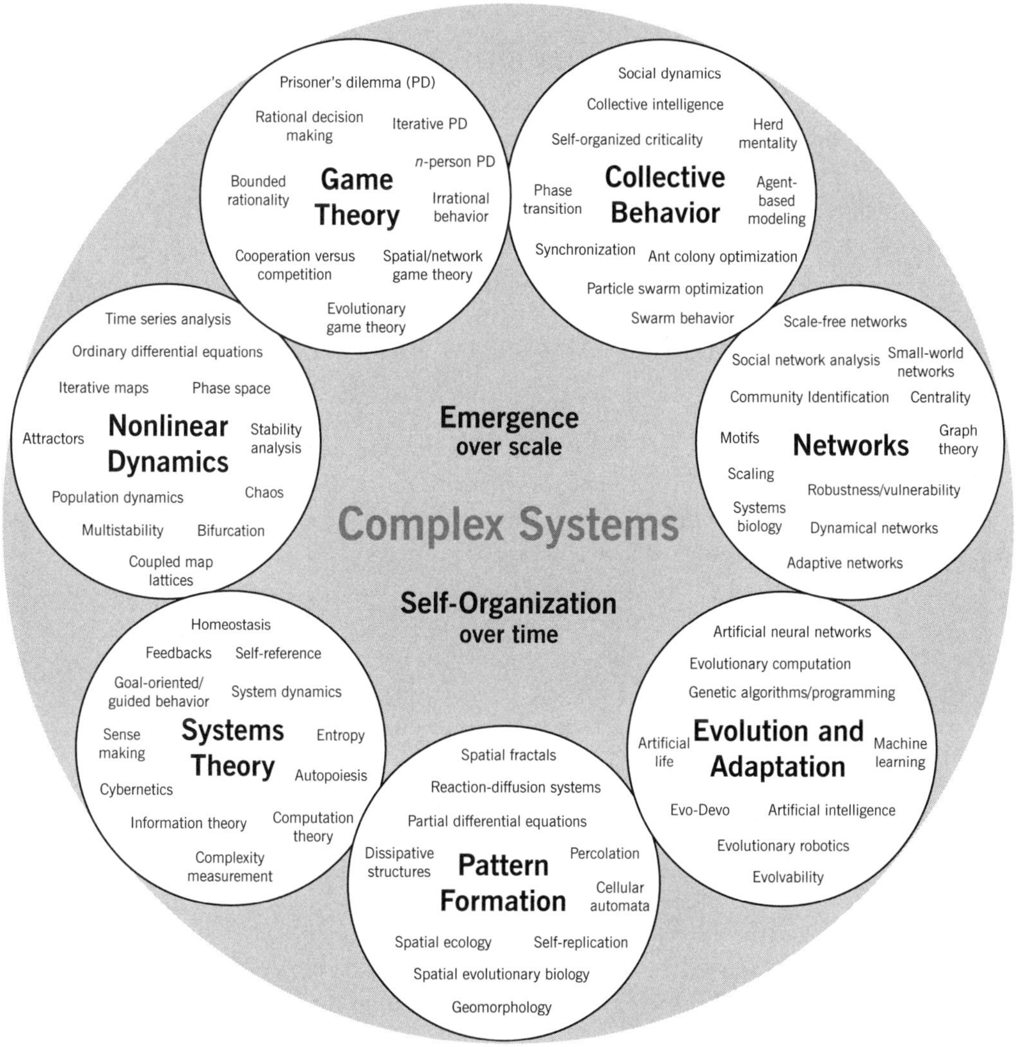

FIGURE 10.1. The field of complex system research. Created by Hiroki Sayama, DSc, Collective Dynamics of Complex Systems (CoCo) Research Group at Binghamton University, State University of New York.

Modeling Complex Systems: The Case of General Intelligence

One of the most replicated findings in differential psychology is the positive manifold of correlations between scores on all kinds of cognitive tests in unrestricted samples of human subjects (Jensen, 1998). Simply put, if you score well on one type of cognitive test, you will probably also score well on any other cognitive test. If these correlations are subjected to factor analysis, we usually arrive at some sort of factor model (e.g., hierarchical) with one main factor at the apex, Spearman's *g* factor. The impact of Spearman's (1904, 1927) factor-analytic work on practical and theoretical work in psychology can hardly be overstated. Psychologists routinely propose new latent psychological constructs, develop ability tests or questionnaires, and analyze the data according to this *g*-factor protocol (cf. Cramer, Waldorp, van der Maas, & Borsboom, 2010).

Although we do not have objections to the statistical work, we challenge the theoretical account of the *g* factor. First, the *g*-factor model of intelligence lacks any developmental component (Ackerman & Lohman, 2003). Are we born with *g*? Does *g* itself develop? Standard *g* theory does not provide answers, but a variant of this theory, investment theory, might. Cattell's (1971) investment theory distinguishes fluid (*gf*) and crystallized intelligence (*gc*). During development *gf* is invested to result in *gc*. Apart from difficulties with the definitions of *gf* and *gc* (Kan, Kievit, Dolan, & van der Maas, 2011), this theory is still rather vague on development. Apparently, we should not take the metaphor of investment too literally, since *gf* does not decrease in this process (a good example of the problems of verbally stated theories in psychology). But more importantly, it is now unclear what the status of *gf* is.

Second, apart from its developmental role, we need to know how *g* relates to brain variables and how *g* is connected to cognitive processing. Many have attempted to identify *g* with measurable variables (e.g., speed of nerve conductance, reaction time, glucose metabolism in the brain). These studies have produced interesting results, but have not revealed the single underlying cause of the *g* factor (Ackerman, Beier, & Boyle, 2005; Luciano et al., 2005).

Sampling Theory

Interestingly, there are alternative mechanisms for general intelligence that can produce exactly the same correlational data as the *g*-factor model. The first is called *sampling theory*, proposed by Thompson in 1914 (Bartholomew, Deary, & Lawn, 2009; Thorndike, 1927). It says that solving cognitive tasks requires many lower-order uncorrelated (neural) processes or modules. It is assumed that tasks will call upon overlapping samples of modules, causing positive correlations between the test scores. Therefore, complex tasks, involving many modules, will load strongest on the *g* factor. Theoretically, if we were able to devise unidimensional cognitive tests for each module, the positive manifold would disappear. In that sense, general intelligence is a measurement artifact.

Sampling theory is largely ignored in the intelligence literature but seems to be at least partly correct. Psychologists are unable to create strictly unidimensional tests (Lumsden, 1976), and sampling from the same underlying processes seems inevitable. In addition, this mechanism of sampling may also play a role in the relation between genes and intelligence. The so-called *watershed model* of Cannon and Keller (2005) describes how specific genes play a role in "upstream," narrowly defined endophenotypes, which in turn play a role in a number of more upstream cognitive abilities. In our opinion, besides the genetic and measurement parts of the explanation of general intelligence, a second, developmental mechanism is relevant.

Mutualism in Networks of Cognitive Abilities

van der Maas and colleagues (2006) proposed another mechanism to explain the positive manifold of correlations. As in sampling theory, it is assumed that the cognitive system consists of many basic, lower-order cognitive processes that are initially uncorrelated. The development of each process has two parts: an autonomous part, best described by a simple logistic growth equation, and a mutualistic part. The mutualistic part is based on the assumption that the growth of each ability is influenced by other abilities, and that all abilities are organized in a network with mainly positive interactions. An example would be the growth of working memory and arithmetic ability. An increase in working memory probably allows for more complicated mental calculations, and vice versa; daily training in arithmetic in schools might increase general working memory capacity (Siegler & Alibali, 2005). Other examples are syntactic and semantic bootstrapping (Fisher, Hall, Rakowitz, & Gleitman, 1994; Pinker, 1994).

van der Maas and colleagues (2006) applied a well-studied mathematical model for mutualistic networks that has been developed for ecological systems (the Lotka–Volterra mutualism model; May, 1973; Murray, 2002):

$$\frac{dx_i}{dt} = a_i x_i (1 - x_i / K_i) + a_i \sum_{\substack{j=1 \\ j \neq i}}^{W} M_{ij} x_j x_i / K_i \quad \text{for } i = 1 \ldots W \tag{10.1}$$

The x_i represents W cognitive processes. Parameters a_i are growth parameters, influencing the steepness of the logistic growth function associated with each x_i. K_i represents the limited resources of the logistic growth processes. The matrix $\boldsymbol{M}$ encompasses the strength, M_{ij}, of the interactions between the basic processes, $\boldsymbol{x}$. Parameters $\boldsymbol{x}_0$, $\boldsymbol{a}$, and $\boldsymbol{K}$ are random parameters, whereas $\boldsymbol{M}$ is a population parameter, meaning that we assume that the interaction strengths are equal over subjects. In simulations, we sample uncorrelated values for $\boldsymbol{x}_0$, $\boldsymbol{a}$, and $\boldsymbol{K}$, choose $\boldsymbol{M}$, and compute the development of $\boldsymbol{x}$ over time using an LSODA (a solver for first-order ordinary differential equations). This equation gives a developmental pattern as shown in Figure 10.2.

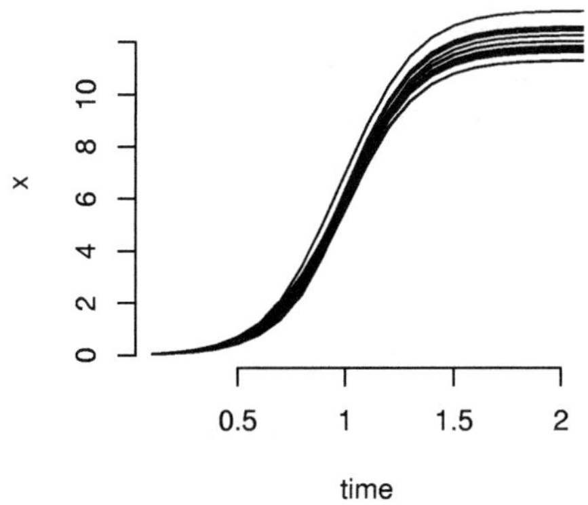

FIGURE 10.2. The development of a set of cognitive abilities in the mutualism model.

We repeat this procedure for N subjects and take the values of x at some point in time (e.g., $t = 2$) and subject these values to correlation and factor analysis. van der Maas and colleagues (2006) have shown that this results in typical positive manifold correlation matrices, which yield one dominant factor as in the *g* model. If all connections have the same positive value, we obtain the standard one-factor model. If the connections are sampled from a normal distribution with positive mean, we obtain data that fit with a hierarchical factor model.

If we assume the simplest additive structure for the effects of genes and environment on the K parameter (resources parameter), the heritability of the dominant factor obtained by factor analysis shows an increase with age, as found in empirical data (Bartels, Rietveld, van Baal, & Boomsma, 2002). Also, the low predictability of intelligence from early childhood performance and the differentiation effects in intelligence are easily explained in the mutualism model.

The Cattell–Jensen Effect

One key effect we could not easily explain is the so-called *Jensen effect* for heritability. This effect pertains to the correlations between the vector of heritabilities and the vector of *g* loadings of cognitive tests (i.e., the method of correlated vectors). These correlations are typically larger than .5, which is taken to suggest a high influence of the genetic component of *g* (Jensen, 1998). It appears that the mechanism of mutualism leads to such a correlation only if we introduce weak positive correlations between the genetic parts of resources K. However, introducing correlations between model parameters weakens the idea of the mutualism model.

This issue was reason to further study this Jensen effect. In Kan, Wicherts, Dolan, and van der Maas (in press), we analyzed the results of 23 twin studies with regard to the heritability coefficients and *g* loadings (see Figure 10.3). In most cases we indeed found a positive correlation between these parameters. However, detailed analysis of these correlations revealed an unexpected pattern. It appears that the most heritable and most *g*-loaded tests are the typical *gc* subtests. These crystallized or culture-loaded tests are tests such as vocabulary, spelling, and arithmetic. When culture load is reduced, fluid tests such as picture completion and digit span show lower heritability coefficients and *g* loadings. In these analyses we control for differences in reliability.

From a standard *g* theoretical point of view this effect, which we call the Cattell–Jensen effect, is hard to explain. Jensen himself, for instance, did not expect much of these cultural tests:

> Information tests consisting of questions like "Who was the first President of the United States?," "Whose picture is on a penny?," and so on, make poor test items mainly for two reasons: (a) they do not get at complex mental process, and (b) they cannot be steeply graded in difficulty level without introducing items of information to which there is a relatively low probability of exposure, in which case social status and educational differences become practically impossible to avoid. The same holds true for vocabulary tests. . . . The difficulty levels differ only because of frequency of

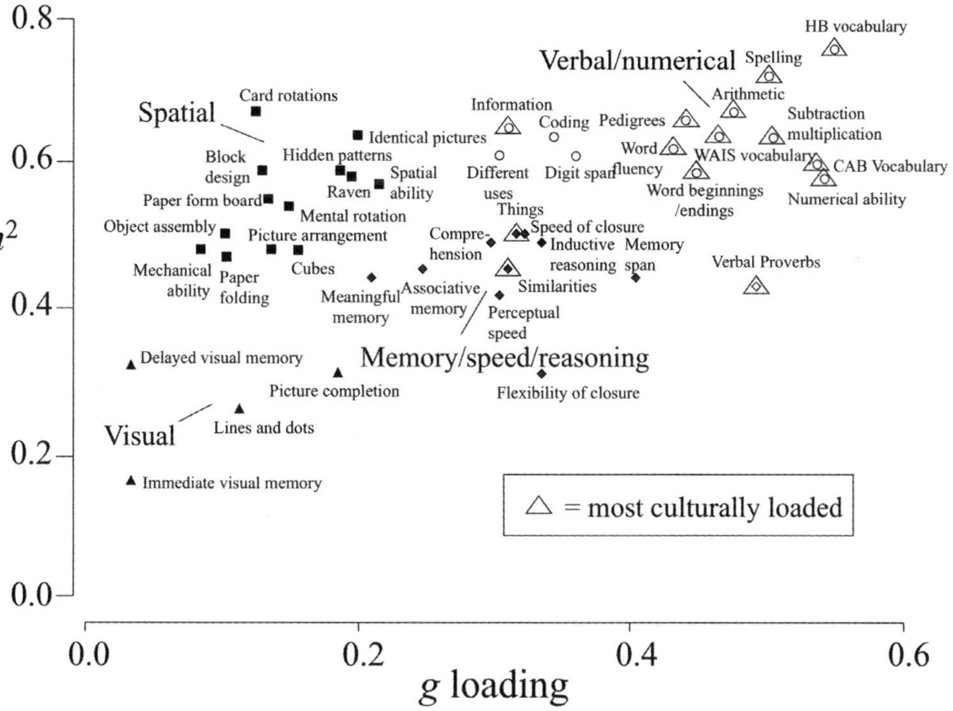

FIGURE 10.3. The relations among *g* loading, heritability coefficients (h^2), and cultural load in the Minnesota Twin Study. Culture-loaded tests are the most *g* loaded and show the highest heritability coefficients. Adapted from Kan, Wicherts, Dolan, and van der Maas (in press).

> exposure. Such items based on information and vocabulary are rightly regarded as more culturally loaded than items which vary in difficulty because of the complexity of the mental processes involved. (1973, p. 184)

Yet precisely these information tests are responsible for the Jensen effect.

Explaining the Cattell–Jensen Effect

Since the Cattell–Jensen effect concerns a differentiation in cultural (crystallized) and noncultural (fluid) ability tests, we have to think of an extension of the mutualism model that allows for these abilities. In Kan (2011, Chap. 7) such a mutualism model is proposed. In this simulation model there are eight fluid and eight crystallized abilities and an external environment. The mutualistic weights are specified such that:

- Fluid → fluid interactions are mildly positive.
- Fluid → crystallized interactions are strong.

- Crystallized → fluid interactions are mildly positive.
- Crystallized → crystallized interactions are mildly positive.
- Effect of environment on crystallized and fluid abilities is (equally) high.
- Environment is only influenced (strongly) by crystallized abilities.

As in van der Maas and colleagues (2006), we used a standard additive genetic model for the resources *K*, with small intercorrelations between the genetic parts of *K*. The latter effect represents genetic sampling. Kan shows that this scenario, as well as some related alternative scenarios, lead to the Cattell–Jensen effect. The main mechanism in this explanation is a gene–environment correlation.

Kan (2011) summarizes our view of general intelligence in Figure 10.4. Correlations between fluid abilities result from mutual beneficial interactions between basic cognitive processes during development. Crystallized abilities are the result of these basic cognitive processes. The development of crystallized abilities has a beneficial effect on the development of fluid abilities. Some individuals turn out to be more intelligent because they possess higher levels of cognitive abilities. These individuals will be more likely to end up in cognitive environments conducive to the further development

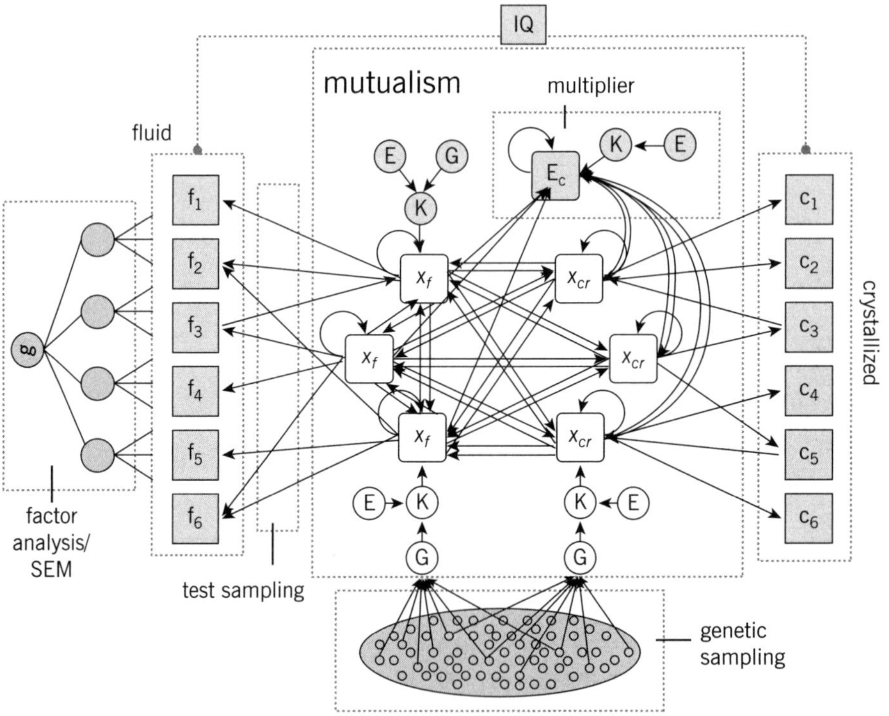

FIGURE 10.4. The explanation of general intelligence based on the combined effects of developmental mutualism, environmental multipliers, genetic sampling, and test sampling. The factor *g* is, like IQ, nothing more than an index of general intelligence, not the common cause.

of crystallized abilities (which is, in turn, beneficial to the further development of fluid abilities). The growth of the cognitive abilities will be constrained by genetically and environmentally influenced limited capacities. These capacities are possibly weakly intercorrelated due to genetic sampling. Sampling on the test level may also be present. According to this model general intelligence, as the outcome of factor analysis, is an index of cognitive functioning.

Analyzing Complex Systems: Phase Transitions

Although the mutualism model explains a wide variety of phenomena in intelligence research, its dynamics are rather dull. Normally, we will see curves like those in Figure 10.2, where all abilities will raise to some stable value. The dynamics depend on the matrix of mutualistic weights, which are all positive in the basic mutualism model. If some are also negative, implying competitive relations, the behavior becomes more irregular. These irregularities could be chaotic but could also result from alternative stable states in the system. Such irregularities have been demonstrated in similar dynamic models by van Geert (1991, 1994).

The occurrence of alternative stable states in complex systems is especially relevant when in developmental systems (Waddington, 1966). We find it notable that the equilibration theory of Piaget (1964/1997) does fit very well in the modern complex systems view of development. Piagetian concepts such as *disequilibrium, stages, transitions*, and *reorganization* are all terms that have well-defined meanings in current dynamical theory (Molenaar & Raijmakers, 2000).

Interestingly, this correspondence goes beyond theory. It has been extremely difficult to test Piaget's theory empirically (Flavell & Wohlwill, 1969). Take, for instance, the idea of transitions between developmental stages in the development of the conservation ability. The developmental researchers in the 20th century could never reach consensus on what a transition was exactly and how it could be detected (Brainerd, 1978; Wohlwill, 1973). But in complex systems theory, especially catastrophe theory or bifurcation theory, phase transitions are well defined mathematically. Moreover, there are many new methods with which to investigate phase transitions empirically.

The Cusp Catastrophe

In fact, there are many different types of phase transitions, among other things, depending on the types of equilibria or attractors involved. The attractors can be simple point attractors but also strange attractors associated with complex chaotic behavior. In most cases, the crucial point is that the current attractor loses its stability and the system moves (often abruptly) to a new state.

A prototypical example of a phase transition is the cusp catastrophe, which models the transition between two point attractors as a function of two control variables. An illustrative example is shown in Figure 10.5. The cusp model is the most applied phased transition model in catastrophe theory.

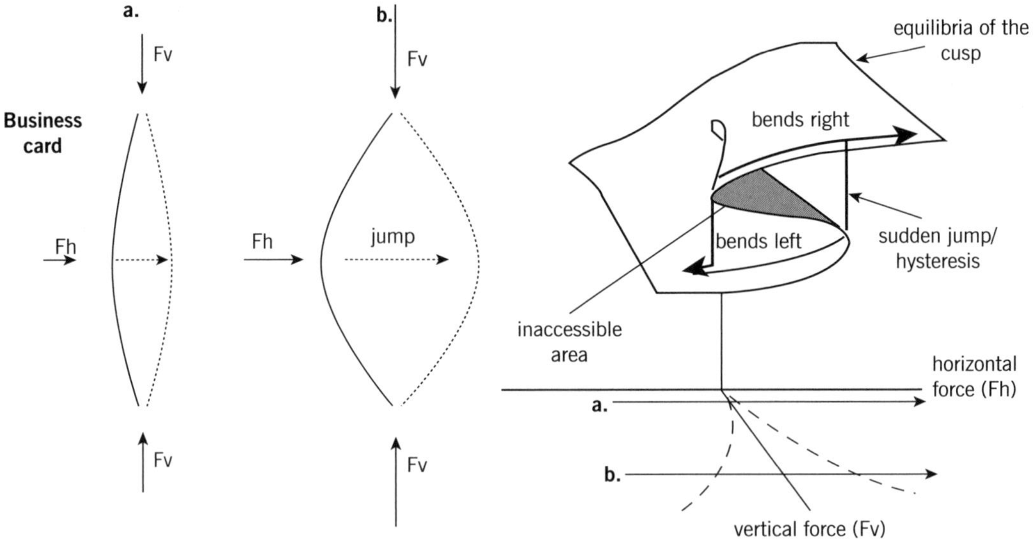

FIGURE 10.5. The cusp catastrophe can be illustrated with a simple business card. Vertical force (Fv) can be exerted by holding the card between the thumb and index finger of one hand. By pressing with the other index finger (Fh), the card bends from left to right. For high Fv this change will be sudden or catastrophic. The dynamics of this system are described by the cusp catastrophe on the right, in which Fv and Fh are the control variables (the splitting and normal variable, respectively). The surface represents the bending of the card. Note that for high Fv, the middle position (straight) card is instable, something that can be verified easily.

To explain the usefulness of this model in developmental psychology, we discuss the example of Jansen and van der Maas (2001). They studied the transition from rule I to rule II on the balance scale test of proportional reasoning. In this test children have to predict the movement of a balance scale on which weights are placed at varying distances from the fulcrum. Rule I users, typically between 4 and 8 years of age, completely ignore the distance information. Rule II users, somewhat older, also ignore distance when the number of weights left and right differs, but they do use distance information when the number of weights left and right is equal. Such items, with equal weights and unequal distances, are called *distance items*. Higher, more advanced rules also incorporate distance in other items. To distinguish these rules other item types are required, but to distinguish between rules I and II, a test with only distance items suffices.

An important first indication of a catastrophic switch from rule I to rule II is the bimodal distribution of sum scores on sets of distance items (van der Maas & Molenaar, 1992). Jansen and van der Maas (2001) show bimodality for several sample datasets. Further evidence for the phase transition hypothesis can be gathered by testing for the presence of other indicators or flags of catastrophic change (Gilmore, 1981). The most appealing flag, hysteresis, the phenomenon that jumps up and down is delayed (see Figure 10.5). Hysteresis can be proven by testing where jumps take place when slowly increasing and decreasing the normal control variable. It can be shown, for instance, that in disturbance-free conditions, water freezes at −4°C, whereas ice melts at 0°C.

Jansen and van der Maas (2001) applied such a design to balance scale learning. They used a series of items for which the distance difference on distance items was systematically varied. The *distance difference* refers to the difference in distance between the equal number of weights left and right on the balance scale. Figure 10.6 displays the item set and an answer pattern associated with hysteresis.

Jansen and van der Maas (2001) found a statistically significant number of hysteresis patterns in the responses of 314 children. They tested for hysteresis in sets of items administered in alternative orders and controlled for alternative explanations by using item sets in which other stimulus characteristics were varied. The combined evidence for bimodality and hysteresis makes a strong case for a real developmental transition. Another recent example of this empirical approach is described in Dutilh, Wagenmakers, Visser, and van der Maas (2010).

The Link to Categorical Latent Structure Modeling

A logical next step in testing the cusp catastrophe model would be to fit the cusp model to data consisting of measurements of the behavioral variable and both the control variables. Based on the work of Cobb (1980) we developed an R package that does exactly this (Grasman, van der Maas, & Wagenmakers, 2009). We refer to this paper for an extended explanation of this technique. An application of Cobb's method can be found in Ploeger, van der Maas, and Hartelman (2002).

Here we discuss an alternative line of work in which we make a connection between complex systems theory and a set of statistical techniques known as *categorical latent structure modeling*. This type of latent structure modeling is used when

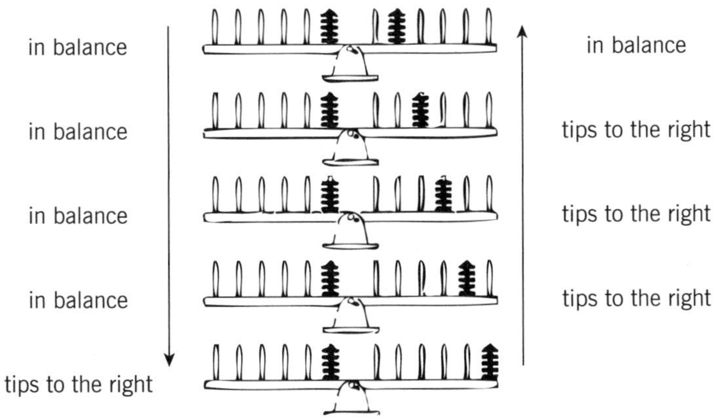

FIGURE 10.6. Balance scale items used to detect hysteresis in the transition from rule I to rule II in proportional reasoning. The administration order of items is depicted with the arrows. Answers left and right display a typical hysteresis pattern.

individual differences are categorical or discrete. We think complex systems theory sheds a new light on the continuing debate in psychology concerning the dichotomy of categorical/discrete versus dimensional/continuous (e.g., De Boeck, Wilson, & Acton, 2005).

A good example to explain our point concerns attitudes, such as people's ideas on abortion. The question is whether the individual differences (pro-life vs. pro-choice) are categorical or continuous. Is pro-life (or pro-choice) a typology or a dimension? A simple cusp model of this attitude has two control variables. The normal variable ("Fh" in Figure 10.5) relates to political and especially religious beliefs of the person, since religious beliefs are more associated with the pro-life point of view. The splitting variable (vertical pressure; "Fv" in Figure 10.5) can best be interpreted as involvement in the issue. Within the group of highly involved subjects we expect a strong discrete typology of pro-life and pro-choice positions. In this group change is hard and always sudden. When involvement is low, however, we expect a continuum of positions that can change more easily. It is reasonable to expect a typology in the United States where involvement in the abortion issue is high, and a continuum in the Netherlands, where abortion is not a big issue. Thus, whether a variable is discrete or continuous might depend on a third variable ("Fv," or involvement) in a continuous way.

Another point to make is that, in general, both categorical/discrete and dimensional/continuous patterns occur in complex systems. However, there is a hierarchy here. Phase transitions or bifurcations demarcate qualitatively different dynamical regimens in systems. These regimens are the types or categories. Quantitative continuous variation may exist within regimens. Thus, we have to look for typologies first and then, within types, we may apply standard quantitative analyses. A clear example is the butterfly's life cycle. We have to distinguish between the different stages (egg, caterpillar, pupa, adult) before doing quantitative analysis on, say, length and weight.

This point has implications for the way we analyze developmental data. We should first look for clusters or classes with techniques such as latent class analysis, finite mixture modeling, or cluster analysis. Within these classes or clusters we can do item response modeling, factor analysis, and so on. Ideally, we do both analyses in one step in latent structure models in which dimensional analyses are nested within classes. One such technique is mixture item response theory (IRT) modeling (Rost, 1990) for discrete data. Here we present an example for continuous data using mixture factor models.

Dolan and van der Maas (1998) take a usual conservation task as example. In this so-called conservation anticipation task, water is poured from a filled glass into an empty glass with different dimensions (smaller or wider glass; see Figure 10.7). Children have to indicate how high the water level will rise in the empty glass. This level is measured in millimeters. Nonconservers are expected to predict a level equal to the level in the filled glass, since they ignore the differences in width of the two glasses. Conservers do note this difference and estimate a different level. In Dolan and van der Maas these predicted levels are modeled with a finite mixture model with two normal distributions as components, representing the conservers and nonconservers. Over sets of items nonconservers make small unsystematic errors in aligning the level

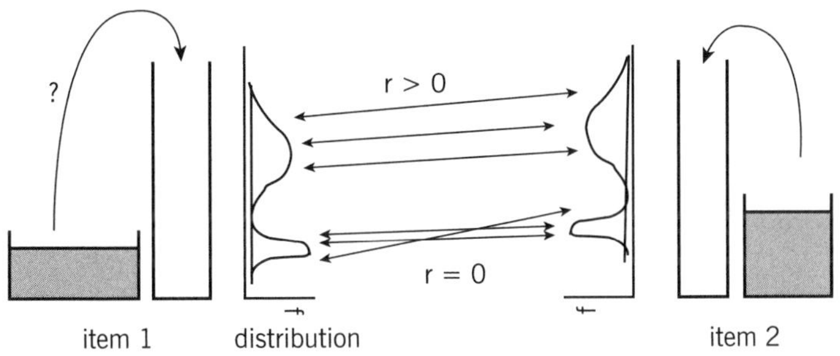

FIGURE 10.7. An example of a mixture factor modeling. Children have to indicate the level of water in the empty glass after pouring from the filled glass. The distribution of indicated levels is bimodal (a mixture of two normal distributions). Within each mode we can do standard factor analysis. In the nonconserver mode correlations are zero, but in the conserver mode a one-factor model describes the choice of levels over items.

to the level of the filled glass. The correlations between items within this group are roughly zero. Conservers that overestimate (underestimate) on one item tend to overestimate (underestimate) the level on other items too. This bias can be described with a one-factor model. The complete model (mixture with factor models) can be estimated simultaneously (e.g., Lubke & Muthén, 2005). Note that this statistical analysis follows the complex system idea of continuous traits within qualitative types or regimens.

Measuring Complex Systems: Math Garden

The last topic we discuss concerns data collection for the study of complex developmental systems. Again we seek inspiration in other disciplines. A quick assessment of complex systems research in other fields shows us that researchers in these fields put much effort into collecting high-quality, high-frequency data. An extreme case is the study of stock markets. The data in this field are collected with amazing sampling rates (e.g., nanosecond trading). But also in ecosystem research, climate research, biology, physiology, and so on, time-intensive data collection is essential. In longitudinal studies in developmental or educational research data are quite different, usually with sample rates of one or two per year. We asked ourselves if it were possible to collect daily data on many children over long periods.

In an effort to acquire such data, we first tried to analyze the exercise books and tests used in standard educational methods. However, this task turned out to be impractical, tedious, and ultimately unsatisfactory. In 2007 we developed a Web-based adaptive training and monitoring system for primary education to collect time-intensive data. This system is called *Rekentuin* in the Netherlands, which means Math Garden (see *www.mathsgarden.com* for an English version).

In this system children play different cognitive games based on either scholastic abilities or more abstract reasoning skills. Games consist of items that should be answered within a limited amount of time (Klinkenberg, Straatmeier, & van der Maas, 2011). Currently, most children use Math Garden as an additional learning tool, but as development of the system continues, future versions may replace, and expand upon, traditional exercise books. By framing exercises in arithmetic and other topics as games and including direct feedback, Math Garden provides an engaging and rewarding platform that greatly enhances children's motivation to train their abilities. One indication of the popularity of Math Garden is the large proportion of item responses (about 25%) that are generated after school. The immediate feedback provided to the children frees the teachers from correcting the children's exercise books. Teachers are also provided with detailed information about the progress and ability of children, including the errors they make. They can use this information, which is accumulated over the time that the child spends on the system, to optimize (individual) instruction.

Furthermore, Math Garden provides researchers with an invaluable dataset. The high-frequency data of a large number of children makes it possible to investigate fundamental questions about the dynamics of cognitive development. These data stem from a subgroup of children who visit Math Garden almost daily and play for extended periods. The data provided by these children are rich in quantity and in dynamics, as the following graph of a child solving the arithmetic problem "21 divided by 3" illustrates (Figure 10.8). The child starts by clicking the question mark (don't know answer) and occasional incorrect answers, followed by a sequence of correct responses that first increases and later decreases in response time. This sequence clearly indicates different phases of learning.

Psychometrics of the Math Garden

The basis of Math Garden is an extension of classic computerized adaptive testing (CAT) methods. CAT is a testing method based on IRT, which contains a variety of item response models. Generally, these models assume one-dimensionality and conditional independence. The method used by Math Garden relies on the simplest item response model, the 1-PL or Rasch model.

In CAT, the order of presentation of items depends on responses to previous items provided by the testee (Wainer, 2000): If the immediately preceding response is correct

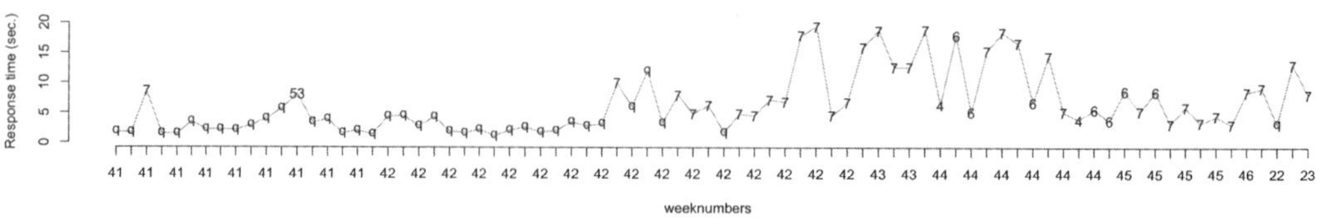

FIGURE 10.8. The x-axis displays the week of playing, the y-axis, the response time. The answer "q" means that the child selected the "I don't know" answer option. Initially, the child quickly selects this answer. Halfway (week 42), the child learns the correct answer and starts to respond with highly fluctuating response times. This is followed by a period with quick and generally correct responses.

(incorrect), a more (less) difficult item is presented next. The advantage of using CAT is that abilities, such as arithmetic ability, can be estimated using fewer items than in standard tests. Currently, CAT is primarily used for testing, but in Math Garden it is used for testing and training at the same time. Therefore Math Garden uses an extended CAT technique based on two crucial innovations described here.

First, Math Garden uses a new self-organizing system that incorporates an "on the fly" Elo estimation algorithm (Klinkenberg et al., 2011), which originated in chess competitions (Elo, 1978). Elo estimation provides a self-organizing testing system in which both the ability estimates of children and the difficulty estimates of items are continually updated in real time, based on the responses of the children. The reliability of the Elo estimation system is well analyzed analytically and in simulations (Batchelder & Bershad, 1979; Glickman, 2001). The most prominent advantage of this system is that it does not require pretested items, as in normal CAT. This requirement of pretested items makes standard CAT very expensive and applicable only in large-scale educational testing applications.

The second crucial innovation is the use of both accuracy and response time when updating ability and difficulty estimates by means of a new scoring rule (Klinkenberg et al., 2011; Maris & van der Maas, 2012). This new rule elicits important additional information about the ability of the child and renders the whole computerized procedure more game-like in practice. In Math Garden, items usually have a time limit of 20 seconds. In the scoring rule that is applied, the score equals the remaining time (RT; 20 seconds minus RT) in case of a correct response, but equals –1 times the response time if the response is incorrect. As a consequence, guessing is risky, and if a child has no clue about the answer, he or she can best refrain from responding, which provides a score of zero. This scoring rule is presented visually, such that even very young children can understand it. This new scoring rule has two important advantages. First, it solves the notorious speed–accuracy tradeoff problem (Wickelgren, 1977) since subjects now know how speed and accuracy are weighted in the scoring. Second, Maris and van der Maas (2012) have shown that under certain mild statistical assumptions, this scoring rule implies a standard two-parameter IRT model, in which discrimination is a linear function of the time limit of an item. Therefore, already a lot is known about the model properties; for example, about the marginal and conditional distributions of the model estimates.

Math Garden is very successful. Currently (July, 2013), it is used by more than 800 schools and many private families. In total over 100,000 Dutch children use Math Garden. We collected more than 200 million responses in 4 years.

Conclusions

In this chapter we intended to reveal some new research fields in developmental psychology based on the idea that the developing psychological system is a typical complex system. In fact, it may be the most complex system studied in science. Recognition of this fact helps us find new ways to model, analyze, and measure developmental systems. We did not attempt to give an overview of all new options for research. Instead we gave examples from our own work in each of these areas (modeling, analysis, measurement). For each of these examples we expect a bright future. The network modeling that we applied to intelligence has great potential in other psychological fields—for instance, in the modeling of psychological disorders (Cramer et al., 2010). Phase transition research has a great promise, especially since much better tools have become available (Grasman et al., 2009). Finally, we expect many results from our Math Garden system. Analyzing the invaluable dataset provided by this system is challenging, but the richness of data is amazing.

REFERENCES

Ackerman, P. L., Beier, M. E., & Boyle, M. O. (2005). Working memory and intelligence: The same or different constructs? *Psychological Bulletin, 131*, 30–60.

Ackerman, P. L., & Lohman, D. F. (2003). Education and g. In H. Nyborg (Ed.), *The scientific study of general intelligence* (pp. 275–292). Amsterdam: Pergamon Press.

Bartels, M., Rietveld, J. H., van Baal, G. C. M., & Boomsma, D. I. (2002). Genetic and environmental influences on the development of intelligence. *Behavior Genetics, 32*, 236–249.

Bartholomew, D. J., Deary, I. J., & Lawn, M. (2009). A new lease of life for Thomson's bonds model of intelligence. *Psychological Review, 116*, 567–579.

Batchelder, W. H., & Bershad, N. J. (1979). The statistical analysis of a Thurstonian model for rating chess players. *Journal of Mathematical Psychology, 19*, 39–60.

Brainerd, C. J. (1978). The stage question in cognitive-developmental theory. *Behavioral and Brain Sciences, 2*, 173–213.

Cannon, T. D., & Keller, M. C. (2005). Endophenotypes in genetic analyses of mental disorders. *Annual Review of Clinical Psychology, 2*, 267–290.

Cattell, R. B. (1971). *Abilities: Their structure, growth, and action.* Oxford, UK: Houghton Mifflin.

Cobb, L. (1980). Estimation theory for the cusp catastrophe model. *Proceedings of the Section on Survey Research Methods*, pp. 772–776.

Cramer, A. O. J., Waldorp, L. J., van der Maas, H. L. J., & Borsboom, D. (2010). Comorbidity: A network perspective. *Behavioral and Brain Sciences, 33*, 137–150.

De Boeck, P., Wilson, M., & Acton, G. (2005). A conceptual and psychometric framework for distinguishing categories and dimensions. *Psychological Review, 112*(1), 129–158.

Dolan, C. V., & van der Maas, H. L. J. (1998). Fitting multivariate normal mixtures subject to structural equation modeling. *Psychometrika, 63*, 227–253.

Dutilh, G., Wagenmakers, E.-J., Visser, I., & van der Maas, H. L. J. (2010). A phase transition model for the speed–accuracy trade-off in response time experiments. *Cognitive Science*, 34, 211–250.

Elo, A. (1978). *The rating of chessplayers, past and present.* Georgetown, CT: Arco.

Fisher, C., Hall, D. J., Rakowitz, S., & Gleitman, L. (1994). When it is better to receive than to give: Syntactic and conceptual constraints on vocabulary growth. In L. Gleitman & B. Landua (Eds.), *The acquisition of the lexicon* (pp. 333–375). Cambridge, MA: MIT Press.

Flavell, J. H., & Wohlwill, J. F. (1969). Formal and functional aspects of cognitive development. In D. Elkind & J. H. Flavell (Eds.), *Studies in cognitive development: Essay in honor of Jean Piaget* (pp. 67–120). New York: Oxford University Press.

Gilmore, R. (1981). *Catastrophe theory for scientists and engineers.* New York: Wiley.

Glickman, M. E. (2001). Dynamic paired comparison models with stochastic variances. *Journal of Applied Statistics, 28,* 673–689.

Grasman, R. P. P. P., van der Maas, H. L. J., & Wagenmakers, E.-J. (2009). Fitting the cusp catastrophe in R: A cusp-package primer. *Journal of Statistical Software, 32,* 1–28.

Jansen, B. R. J., & van der Maas, H. L. J. (2001). Evidence for the phase transition from rule I to rule II on the balance scale task. *Developmental Review, 21,* 450–494.

Jensen, A. R. (1973). *Educability and group differences.* New York: Harper & Row.

Jensen, A. R. (1998). *The g factor: The science of mental ability.* Westport, CT: Praeger/Greenwood Press.

Kan, K.-J. (2011). *The nature of nurture: The role of gene–environment interplay in development.* Unpublished doctoral dissertation, University of Amsterdam.

Kan, K.-J., Kievit, R. A., Dolan, C. V., & van der Maas, H. L. J. (2011). On the interpretation of the CHC factor Gc. *Intelligence, 39*(5), 292–302.

Kan, K.-J., Wicherts, J. M., Dolan, C. V., & van der Maas, H. L. J. (in press). On the nature and nurture of intelligence and specific cognitive abilities: The more heritable, the more culture-dependent? *Psychological Science.*

Klinkenberg, S., Straatemeier, M., & van der Maas, H. L. J. (2011). Computer adaptive practice of maths ability using a new item response model for on the fly ability and difficulty estimation. *Computers and Education, 57,* 1813–1824.

Lubke, G. H., & Muthén, B. O. (2005). Investigating population heterogeneity with factor mixture models. *Psychological Methods, 10,* 21–39.

Luciano, M., Posthuma, D., Wright, M. J., de Geus, E. J. C., Smith, G. A., & Geffen, G. M. (2005). Perceptual speed does not cause intelligence, and intelligence does not cause perceptual speed. *Biological Psychology, 70,* 1–8.

Lumsden, J. (1976). Test theory. *Annual Review of Psychology, 27,* 251–280.

Maris, G., & van der Maas, H. L. J. (2012). Speed-accuracy response models: Scoring rules based on response time and accuracy. *Psychometrika, 77*(4), 615–633.

May, R. M. (1973). *Stability and complexity in model ecosystems.* Princeton, NJ: Princeton University Press.

Molenaar, P. C. M., & Raijmakers, M. E. J. (2000). A causal interpretation of Piaget's theory of cognitive development: Reflections on the relationship between epigenesis and nonlinear dynamics. *New Ideas in Psychology, 18,* 41–55.

Murray, J. D. (2002). *Mathematical biology: I. An introduction* (3rd ed.). Berlin: Springer Verlag.

Piaget, J. (1997) Development and learning. In M. Gauvain & M. Cole (Eds.), *Readings on the development of children* (2nd ed., pp. 19–28). New York: Freeman. (Original work published 1964)

Pinker, S. (1994). How could a child use verb syntax to learn verb semantics? In L. Gleitman & B. Landua (Eds.), *The acquisition of the lexicon* (pp. 377–410). Cambridge, MA: MIT Press.

Ploeger, A., van der Maas, H. L. J., & Hartelman, P. A. I. (2002). Stochastic catastrophe analysis of switches in the perception of apparent motion. *Psychonomic Bulletin and Review, 9*(1), 26–42.

Rost, J. (1990). Rasch models in latent classes: An integration of two approaches to item analysis. *Applied Psychological Measurement, 14,* 271–282.

Siegler, R. S., & Alibali, M. W. (2005). *Children's thinking* (4th ed.). Upper Saddle River, NJ: Prentice Hall.

Spearman, C. (1904). General intelligence, objectively determined and measured. *American Journal of Psychology, 15*, 201–293.

Spearman, C. (1927). *The abilities of man: Their nature and measurement.* New York: Macmillan.

Thorndike, R. L. (1927). *The measurement of intelligence.* New York: Teachers College.

van der Maas, H. L. J., Dolan, C. V., Grasman, R. P. P. P., Wicherts, J. M., Huizenga, H. M., & Raijmakers, M. E. J. (2006). A dynamical model of general intelligence: The positive manifold of intelligence by mutualism. *Psychological Review, 113*, 842–861.

van der Maas, H. L. J., & Molenaar, P. C. M. (1993). Stagewise cognitive development: An application of catastrophe theory. *Psychological Review, 99*(3), 395–417.

van Geert, P. (1991). A dynamic systems model of cognitive and language growth. *Psychological Review, 98*, 3–53.

van Geert, P. (1994). *Dynamic systems of development: Change between complexity and chaos.* New York: Harvester.

Waddington, C. H. (1966). *New patterns in genetics and development.* New York: Columbia University Press.

Wainer, H. (Ed.). (2000). *Computerized adaptive testing: A primer* (2nd ed.). Mahwah, NJ: Erlbaum.

Wickelgren, W. A. (1977). Speed–accuracy trade-off and information processing dynamics. *Acta Psychologica, 41*, 67–85.

Wohlwill, J. F. (1973). *The study of behavioral development.* New York: Academic Press.

CHAPTER 11

Dynamic Development of Brain and Behavior

Kurt W. Fischer and Paul van Geert

How does the growth of neural systems in the brain relate to children's psychological development? The study of neuroscience is moving beyond speculation to discovering how brain and behavior connect—how development of brain functioning relates to actions, thoughts, and emotions. This knowledge provides possibilities for moving beyond global correlations to real breakthroughs in understanding. Analysis of patterns of growth can illuminate how brain functions and behaviors develop through common developmental mechanisms and produce similar growth curves.

We propose that two characteristics are especially important for analyzing and explaining the developmental mechanisms underlying brain–behavior relations. First, many growers are connected, with important variations in the types of connections; growers powerfully influence each other's growth. Second, growers commonly move through periods of rapid change or developmental discontinuity, in which new capacities or forms of activity emerge, creating the transformations of childhood and adolescence. Advances in methods for studying development make it possible to analyze the processes of development of brain and behavior.

Relations between Brain and Behavior in Development

Brain activity, myelination, brain mass, synaptic density, dendritic branching, and pruning of neurons and synapses all change systematically with age during childhood (Benes, 1994; Conel, 1939–1963; Diamond, Werker, & Lalonde, 1994; Fischer, 2008; Fischer & Bidell, 2006; Huttenlocher, 1994; Thatcher, 1994; Thatcher, Lyon, Rumsey, & Krasnegor, 1996; Yakovlev & Lecours, 1967). Many aspects of children's behavior show developmental change and are necessarily correlated globally.

Progress in research on analyzing development has set the stage for better research about relations between brain and behavior in development. For example, large areas of the cerebral cortex in rhesus monkeys show synaptic growth and pruning in parallel cycles that relate to broad cognitive changes (Changeux & Danchin, 1976; Fischer, 1987; Goldman-Rakic, 1987; John, 1977; Rakic, Bourgeois, Eckenhoff, Zecevic, & Goldman-Rakic, 1986). Similarly, cortical networks and connections among them in people, indexed by brain electrical activity, demonstrate developmental patterns that relate to cognitive and language development (Fischer & Bidell, 2006; Grossberg, Mingolla, & Fazi, 2009; Hudspeth & Pribram, 1992; John, 1977; Steenbeek, Jansen, & van Geert, 2012; Thatcher, 1994; Thatcher et al., 1996; van der Molen & Molenaar, 1994). Moreover, measures of patterns of brain activity in infants indicate the possibility of prediction of specific types of learning difficulties (Gabrieli, 2009).

Similarly for behavior, specific growth functions for cognitive development (e.g., understanding classes of concepts) show spurts and upper limits in individual children at specific ages, tested by provision of instruction and practice for the children (Fischer & Bidell, 2006; Halford & Boyle, 1985; Kitchener, Lynch, Fischer, & Wood, 1993; Stein, Dawson, & Fischer, 2010). Different knowledge domains show parallel development, including synchronies beyond global correlations with age. Strong discontinuities in growth occur for specific domains at specific ages (Eichorn & Bayley, 1962; Fischer & Bullock, 1984; McCall, 1983 ; Ruhland & van Geert, 1998; Steenbeek et al., 2012; van der Maas & Molenaar, 1992; van Geert, 1991). Individual understanding varies systematically within a range that changes across domains (Brown & Reeve, 1987; Fischer & Bidell, 2006; Fischer & Lamborn, 1989; Granott, 2002; Vygotsky, 1978).

We propose a framework for analyzing brain–behavior relations in development based on the hypothesis that major developmental changes involve coordination of brain–behavior components into higher-order control systems, which we call dynamic skills (Bullock & Grossberg, 1988; Fischer, 1980; Fischer & Bidell, 2006; Grossberg, 1988; Mascolo & Fischer, 2010). Skills comprise multiple elements following the principles of dynamic systems (Smith & Thelen, 1993; Stein et al., 2010; van Geert, 1991, 1998). Before coordination, connections of these elements are mostly weak. With development of coordination, connections become strong and shape the growth functions.

Neural networks support the skills following principles of parallel, distributed, resonant processing (Bullock & Grossberg, 1988; Grossberg et al., 2009). Each network is composed of multiple connected parallel components, interacting dynamically and regulated through feedback. Important components include not only neurons but also sense organs, body parts, and environmental contexts. For reaching for an object, for example, a person coordinates the eyes, the structure of the arm and hand, and the nature of the object into a dynamic system for grasping. Sometimes top-down information from higher network levels matches bottom-up information from world or body, and then the network can produce behavior at a high degree of complexity—beyond the usual (Fischer, Bullock, Rotenberg, & Raya, 1993). In other words, the network functions at different levels based on the configuration of component inputs. In this way contexts have strong effects on action, thought, and emotion.

When these parallel distributed networks grow, they move in regular cycles from competition among comparable networks to coordination of networks. With the cycles each new system starts the coordination process anew, marked by a surge in growth for a potential component that can coordinate into a complex system. General cycles of network growth are evident in brain function and behavior. The cycles lead to alternation between periods of rapid growth and periods of slower growth, with discontinuity (sudden change) being a primary index of a cycle. Figure 11.1 provides an example of a curve for relative power in the alpha band of the electroencephalogram (EEG) (Matousek & Petersén, 1973). Besides the general upward growth, note the regular cycles of spurts followed by slow growth or even plateau and decrease.

The general shape of this curve (spurts and plateaus) is common for psychological and brain development. Figure 11.2 shows growth in adulthood of a complex kind of reasoning, called *reflective judgment* by John Dewey (Dewey, 1933; Kitchener & Fischer, 1990; Kitchener et al., 1993). Reflective judgment addresses knowledge dilemmas with no simple answer. For example, do chemical additives to food keep people healthy by preventing food poisoning, or do they harm people by producing diseases such as cancer? Prior research has identified a sequence of seven stages in coordination of viewpoint, argument, and evidence in thorny issues like this one about reflective judgment (Kitchener, King, Wood, & Davison, 1989).

As Figure 11.2 demonstrates, adolescents and young adults showed rapid jumps in their understanding of each new kind of reflective judgment that they constructed. This spurt-and-plateau pattern occurs most commonly for people's optimal performance, the most sophisticated skills that they control for a specific topic (Kitchener et al., 1989).

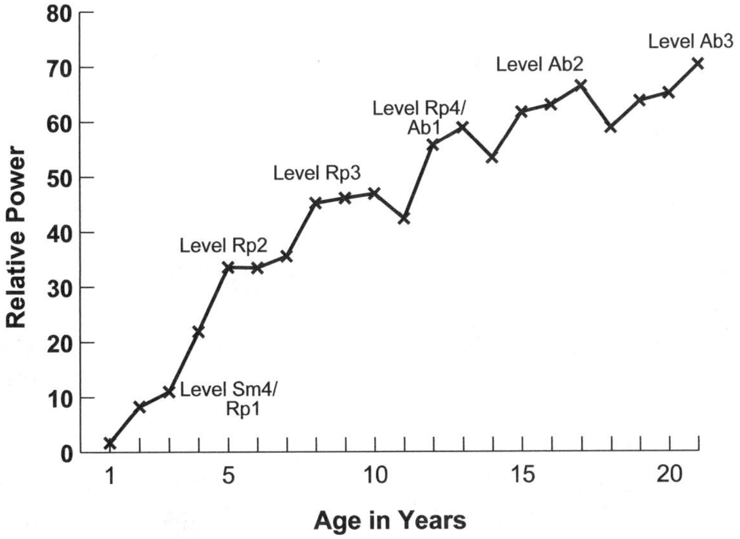

FIGURE 11.1. Development of relative power in alpha EEG in the occipitoparietal area in Swedish children and adolescents. Sm, sensorimotor; Rp, representational; Ab, abstract.

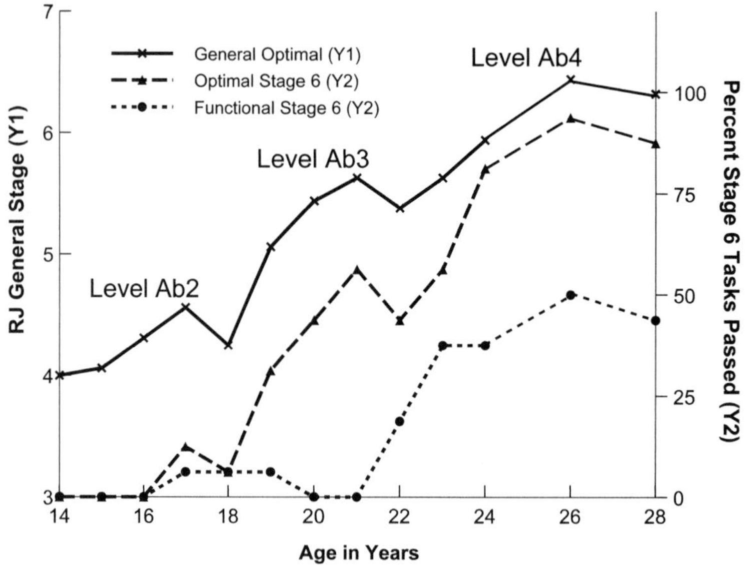

FIGURE 11.2. Development of reflective judgment: Optimal and functional levels.

Figures 11.1 and 11.2 demonstrate the sorts of parallels that often occur for discontinuities in brain–behavior development (Fischer & Bidell, 2006). For example, spurts appear at approximately 15 and 20 years for both reflective judgment and EEG energy. Spurts and drops in growth can occur for many different reasons, however, so that interpreting them can be difficult (van Geert, 1991). For example, a person's activities can vary abruptly as a result of changes in context or emotional state (Smith & Thelen, 1993).

Despite the challenge of these vagaries, the systematic complexity of skills provides a tool for analyzing growth functions for brain and behavior. Researchers can search for links between discontinuities in growth functions, such as those in Figures 11.1 and 11.2.

Competition and Coordination

Earlier in the history of brain modeling, competition was often considered the central process in brain development. But competition and coordination work together, interacting dynamically in brain and behavioral development. The large-scale changes that occur in brain and cognitive development involve the coordination of competing elements to form control systems of components that collaborate and support each other as well as compete. Brain functioning involves coordination of components into networks, as well as competition between components (Immordino-Yang et al., 2012; Poldrack, Halchenko, & Hanson, 2009).

Competition was especially emphasized in the classic analysis of neurological growth, and it certainly is an important part of brain development (Changeux & Danchin, 1976). Neurons are pruned away when they receive little input and so are not active; the ones that are active are not pruned away and continue to function. This competition seems to account for many of the effects of specific experience, as cells that fire are sustained. For example, a young child who sees through only one eye sustains connections to that eye and loses them to the eye that does not see (Hubel & Wiesel, 1977; Singer, 1995).

However, the focus on competition has led to neglect of the role of coordination in development. Recent work highlighting the centrality of neural networks in learning and development has helped to correct this bias in interpretation. One purpose of the framework we are presenting is to correct this neglect. To explain the importance of the shift from competition to coordination, we use three lenses to analyze dynamic developmental processes. First, research on children's cognitive development shows that as children build new understandings, they move from competition to coordination. Second, models of the dynamics among growers show a shift from competition to coordination, including the creation of jumps in growth functions such as those in Figures 11.1 and 11.2. Third, research on cortical functioning shows some of the ways that coordination is mediated by the brain (Grossberg, 1988; Grossberg et al., 2009; Spencer, Thomas, & McClelland, 2009).

In developmental advances, coordination can become dominant. For an infant to find an object hidden under a cloth (Piaget's *object permanence*), he or she needs to hold in mind where the object last disappeared. With that information he or she can reach for the object where it disappeared, and so search under the correct cloth. Piaget (1936/1952) proposed that infants coordinate action systems as the foundation for finding the hidden object. Subsequent theorists have elaborated models of coordination in which large increases in the sophistication of behavior require the coordination of multiple components of action or representation into new, control systems (Case, 1992; Fischer, 1980; Fischer & Bidell, 2006; Halford & Boyle, 1985).

This coordination can be seen directly in the ways that children act as they develop new understandings. First they demonstrate a new understanding by co-occurrence of two competing behaviors. They begin to move toward coordination when they can sustain those behaviors independently within a single control system. For example, in learning new concepts for arithmetic, children often use two contradictory numerical procedures. They show one in their gestures, and they describe a different one with their words (Goldin-Meadow, Nusbaum, Garber, & Church, 1993). Gradually, they combine the two procedures to create a new, coordinated understanding.

Development of negative and positive emotional reactions demonstrates a similar emerging coordination. Children who are coping with social conflicts often think of the other child as being either nice or mean (not both). In fact, people are frequently both nice *and* mean, not just one or the other, and children gradually integrate the opponent concepts (Fischer & Ayoub, 1994; Fischer et al., 1990; Harter & Buddin, 1987). Typically children react initially in one way or the other (a strong form of competition), but then they alternate between positive and negative. Eventually they coordinate positive and negative, and then go on to build more complex coordinations. For example, in constructive criticism, someone criticizes and thus is mean, but the intention is to act nice by helping the friend improve (Fischer & Lamborn, 1989).

We have proposed that this pattern of alternation or co-occurrence of competing actions and then coordination has been observed so often in diverse tasks that it is a general rule (Fischer, 1980; Fischer & Bidell, 2006; Granott, 2002): First, a person develops the ability to produce two actions separately, and then he or she becomes able to sustain both of them in parallel. Finally, he or she uses the co-occurrence to coordinate the two actions and form a higher-level control system. In this way the previously separate skills become mutually supporting.

Development of Coordination between Growers

Dynamic modeling of development can specify how competition and coordination interact and predict the shapes of growth. van Geert (1991, 1998) describes the general nature of dynamic growth equations, and Fischer and Bidell (2006) and Thatcher (1994; Thatcher et al., 1996) show how they apply to brain development. The simple model for three hierarchical growers, A, B, and C, in Figure 11.3 illustrates the general nature of cognitive and brain growth curves (Fischer & Rose, 1994).

The model starts with three growers, A, B, and C, each at its own level or step, called Steps a, b, and c. At first the growers are separate, but eventually they coordinate

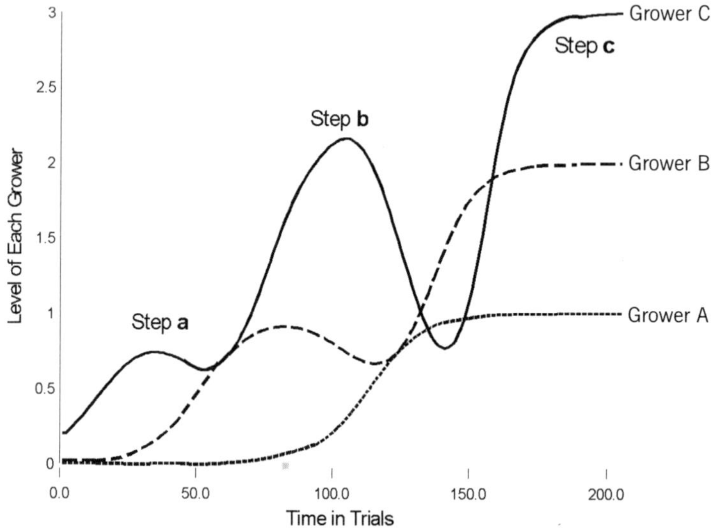

FIGURE 11.3. Development of individual Growers A, B, and C at Steps a, b, and c. As a new grower begins to spurt, it induces a spurt in the prior grower. Also, competition between growers produces a drop before each spurt in a grower.

and support each other. Each grower builds on components from the previous step. For example, Grower B builds on coordination of Grower A with at least one other grower at Step a.

At the start the three growers begin with zero-order levels. Then, Grower A begins to grow, competing with other growers at Step a, but unaffected by Grower B because its level remains low. Next, Grower A permits growth to start in B, which coordinates A with parallel growers at Step a. Later Growers B and C undergo a similar process, with Grower A serving as precursor for Grower B. Similarly Grower B serves as precursor for Grower C, so that when Grower C develops, it subsumes Grower B as a component. In this way the three growers become related hierarchically. The model can become more complex with additional connected growers for each step (e.g., Growers A1, A2, A3, A4, and A5 at Step a) as well as more than three steps.

Growth equation analysis makes it possible to specify how competition and coordination interact and to predict the form of growth curves. Figure 11.3 shows a straightforward example of this interaction for three growers, A, B, and C, producing growth curves similar to those in Figures 11.1 and 11.2, as illustrated in the combined scores in Figure 11.3.

Development routinely involves hierarchical relations similar to these, such as development of Dewey's reflective judgment and development of nice and mean (positive and negative) interactions. Likewise, neural network development is characterized by this kind of hierarchical property, in which later networks develop from and subsume earlier networks (Fischer & Bidell, 2006).

In a study to understand arithmetic operations, we tested arithmetic concepts, which produced growth curves similar to those in Figure 11.3 (Fischer, Hand, & Russell, 1984). In this case, Grower A specified explanation of addition problems such as 3 + 11 = 14. Grower B involved the abstract concept of addition, such as when a child explains that addition combines smaller numbers to produce a larger one, applying the definition to a specific problem such as 3 + 11 = 14. Grower A, the understanding of simple addition calculations, serves as a precursor for Grower B, the abstract understanding of addition. As Grower B emerges, it subsumes A because the abstract meaning of addition includes specific calculations of addition problems.

In the same way, Grower C includes relating the operation of addition to that of subtraction. For instance, a child might explain that addition and subtraction are opposites because addition puts together individual numbers to create a larger one, whereas subtraction takes away a number to get a smaller one. He or she can also indicate how the addition–subtraction relation applies to the problems 3 + 11 = 14 and 14 – 3 = 11. Grower B, comprehending the operation of addition, serves as a precursor for Grower C, which connects the operations of addition and subtraction. Grower C similarly subsumes Grower B as a component.

The model starts with the growth equation for one grower such as A, which is not yet linked to Grower B (van Geert, 1991). The level that the grower will eventually reach derives from (1) its previous level, (2) its growth rate, and (3) the limit on its growth (its highest possible level), called its *carrying capacity*. This equation produces the common S-shaped growth curve (the basic growth function for living things) as well as other more complex (and sometimes chaotic) patterns.

$$L_{A_{t+f}} = L_{A_t} + L_{A_t} R_A - \frac{L^2_{A_t} R_A}{K_A} \quad \textbf{(11.1)}$$

where $L_{A_{t+f}}$ is the level for Grower A; subscript t marks the initial assessment time and subscript $t + f$ denotes an assessment at a later time; R_A is the growth rate for Grower A; and K_A is the carrying capacity for Grower A, the system's limit on its growth.

A similar equation applies for Grower B, which here is simplified through grouping the common factor L_{B_t}.

$$L_{B_{t+f}} = L_{B_t}(1 + R_B - \frac{L_{B_t} R_B}{K_B}) \quad \textbf{(11.2)}$$

To specify the model for hierarchical development of three growers connected by both competition and support, we make the following assumptions: Grower B begins to grow when A reaches a mimimum specified level (precursor). After A reaches it, B starts to grow, both competing with and supporting A (C indicates competition, and S indicates support). In the same manner, Grower B grows only when B reaches a minimum precursor level. After Grower C starts to grow, it both supports and competes with Grower B.

These equations represent the specification of competition and support processes for Growers A and B and Growers B and C, respectively:

$$L_{B_{t+2}} = L_{B_{t+1}}(1 + R_B - \frac{L_{B_{t+1}} R_B}{K_B} - C_{C\to B}(L_{C_{t+1}} - L_{C_t}) + S_{C\to B} L_{C_{t+1}}) \quad \textbf{(11.3)}$$

$$L_{A_{t+2}} = L_{A_{t+1}}(1 + R_A - \frac{L_{A_{t+1}} R_A}{K_A} - C_{B\to A}(L_{B_{t+1}} - L_{B_t}) + S_{B\to A} L_{B_{t+1}}) \quad \textbf{(11.4)}$$

where $C_{B\to A}$ indicates the competition of Grower B with Grower A; $C_{C\to B}$ indicates the competition of Grower C with Grower B; $S_{B\to A}$ indicates the support of Grower B by Grower A; and $S_{C\to B}$ indicates the support of Grower C by Grower B.

Note that competition and support can be based either on the change in a grower between two times of assessment—that is, $L_{t+1} - L_t$, the difference between levels—or on L_{t+1}, the level at a given time.

In this model, competition depends on the difference, but support depends on the level. The reason for the dependence of competition on the difference between levels is that large change in Grower B or C is likely to disrupt Grower A or B during the process of change. For example, when a student changes his or her conception of the relation of addition to subtraction (Grower C), this will produce a temporary disruption in the concept of addition alone (Grower B). We do not assume that the level itself will have a competitive effect because conceptions of the relation of addition to subtraction do not themselves interfere with conceptions of addition alone. Of course, different models can be tested by simply specifying them and doing the calculations on a computer.

In contrast, support depends on the level at $t + 1$, not the difference between levels. Growers B and C involve coordination of components with Growers A and B, respectively. We expect a higher level of understanding of the later grower to support understanding of the earlier one. For example, we expect that when students have a better understanding of how addition and subtraction relate, this will also facilitate their understanding of addition alone.

This hierarchical support has powerful effects. First, Grower B changes the level of Grower A, directly increasing its carrying capacity. Likewise Grower C changes the level of Grower B, raising its carrying capacity. This change leads, in turn, to a further change in Grower A. The sharp growth in Grower B leads to a spurt in Grower A; and similar changes happen for Growers C and B.

In our research on children's development, we do in fact find these effects for hierarchical growers (Fischer & Bidell, 2006; Fischer & Bullock, 1984; Kitchener et al., 1993). Evidence suggests that when a grower at a later level spurts, the performance level for the prior grower also surges to something like the ceiling of its scale. This effect of support seems to move down successive steps in the hierarchy, as shown in Figure 11.3. That is, a spurt in a later grower in a hierarchy does appear to increase the level and carrying capacity of the earlier growers that it supports.

However, in other domains of development the relation may go the other way round. In studies of early language development, for instance (Bassano & van Geert, 2007), we found that particular grammatical skills tend to first emerge and then disappear due to the likely competition with new skills that build upon the older ones. For instance, children begin with simple undifferentiated, so-called *holophrastic* or one-word sentences, and then start to combine words in simple two- to three-word combinations. These relatively simple combinatorial principles gradually disappear because of a competition with grammatical skills based on syntactic rules of the language.

Cortical Mediation of Coordination

A basic function of the prefrontal cortex is to hold information "online" about prior activity while other activities occur. This capacity is essential for coordinating two activities (as described above), moving them from independence or competition to co-occurrence and then finally coordination. In the object search task, Goldman-Rakic and her colleagues have demonstrated that specific columns of cells in the prefrontal cortex of the rhesus monkey hold key object location information online. These cells fire during the period when the information needs to be held and stop firing after the information has been used to retrieve the object (Diamond et al., 1994; Goldman-Rakic, 1987).

In a similar manner, the prefrontal cortex may be essential for many, if not most, major cognitive-developmental advances, as suggested by Thatcher's (1994; Thatcher et al., 1996) research on development of coherence in the EEG. High coherence suggests that two cortical areas may show synchronization in the EEG, which is often an index of connectivity between areas. Neural networks show synchronization when they are connected. Thatcher found that the prefrontal cortex was involved in more than 90% of coherence patterns showing systematic development in the first 20 years of life. The prefrontal cortex was the most prominent brain area, apparently regulating connections between the front and back of the brain. (There can be issues of bias by volume conduction.)

This specialization of the prefrontal cortex for sustaining information online is fundamental for the coordination of components, including the succession of large-scale developmental changes that were originally outlined by Piaget (1983). The prefrontal cortex probably provides the foundation for these large developmental reorganizations. For example, the prefrontal region would seem to be fundamental to the emergence of a new unit of activity, such as representation and complex language at about 2 years, playing a major role in making it available for coordination (Ruhland & van Geert, 1998).

Coordination of components into control systems is essential to development of both brain and behavior. Skills and networks both compete and coordinate, in part through the mediation of the prefrontal cortex. Models of these connections help explain the shapes of brain and behavioral development, and provide tools for analyzing the ways that brain and behavior work together in development.

The sketch of developmental levels that we have provided shows that each level requires a control system to coordinate components, and each produces a cluster of discontinuities in behavioral and cognitive growth. The simple brain-growth hypothesis is that each level is grounded in a broad-based spurt in brain growth that creates a new type of neural network and thus a new control system. At each level the key capacity is coordination of components to form a more complex control system for prior skills. The most important cortical region for that capacity seems to be the prefrontal area, which is specialized for holding information online from diverse cortical regions while other activities occur (Case, 1992). We propose that this coordination is the fundamental capacity that makes cognitive and emotional development possible.

The Dynamic Nature of Skills

Dynamic skills are defined as higher-order control systems based on the coordination of specific brain and behavior components. Coordination of these control systems is based primarily on processes of growth and dynamic coupling. The dynamic coupling consists either of a correlated increase in the levels of the components (correlated because the components positively affect the growth of other components) or of a correlated increase in one component and a decrease in another—a process that can be referred to by the term *pruning*. A dynamic skill can thus be viewed as an attractor state of a developing system. That is, given the supportive or competitive relations between brain and behavioral components of a developing system—for instance, a specific child in his or her environmental and educational niche—particular self-sustaining control systems will emerge based on experience that correspond with the person's developing dynamic skills.

Dynamic skills are typically complex phenomena in the sense that they depend on the real-time coordination of many components distributed across both brain components and the person's actual environment. For instance, a child's writing skill requires a coordination of components on various levels, one of which is the level of goals (e.g., a young child's goal may be to "write a word," whereas an adult's goal may be to pass a particular message to someone who is currently absent). Another level of coordination is the actual means and objects present in the situation where the goal must be accomplished, and these means may range from pencils to laptops or from clay tablets to chisels and granite stones. The neural networks that serve these goals and means are typically distributed and must coordinate perceptual, motor, and planning processes and memory. (See Thelen & Smith, 1994, for examples of how dynamic skills are emergent from perception–action couplings that dynamically incorporate components of the context as constituents.)

Dynamic skills are typically defined by this combination of goals, material means, and abilities that are represented in the form of bodily properties, in particular, the neural networks that form essential parts of the bodily substrate of a specific ability. In other words, dynamic skills are embodied phenomena (Overton, 2013, and Chapter 2, this volume). Dynamic skills are also coordinated control systems that a person can learn and develop, for which the person can develop a particular level of skill ranging from virtual absence to exceptional performance. A person in an environment must invest effort to specifically control these functions. Dynamic skills are also typically correlated, in the sense that a person who has attained a particular level in a specific skill is likely to have an associated level in related skills. For instance, an Olympic gold medal winner in the 50-meter freestyle swimming competition is likely to also be good in 200-meter freestyle swimming. This within-person correlation between dynamic skills is traditionally explained by treating a particular performance as a sum of constituents that are common to a wide variety of performances. This approach has been especially prevalent in research on intelligence, where the *g* factor (general intelligence) is traditionally seen as a separate, person-specific intelligence component that makes a particular contribution to any intelligent act (Horn & Hofer, 1992). Of course, there are also components specific to the contents such as leg use, verbal skills, and so on. In contrast with this componential approach, van der Maas and colleagues (2006, 2011) have argued that the *g* factor is not a separate component, but is instead an emergent property of coordinations (support and competition relations) among a wide variety of domain-specific intellectual skills.

In educational contexts, dynamic skills are typically emergent from the interplay between a particular child's abilities and an immediate, real-time educational context, which consists of an educator interacting with the child and/or specific educational material, including texts and schemes. This process of intertwining child–educator activities results from dynamic coordination, which leads to self-sustaining patterns that are characteristic of this particular child, educator, and activity. These patterns often change in discontinuous ways. For example, Steenbeek, Jansen, and van Geert (2012) observed a teacher and a child interacting about math assignments in a special education setting over a 2-year period. They focused on matches or mismatches between the actions of one participant and the reactions of the other. For this child, the long-term process passed through different patterns of action and reaction, which showed successive changes that were relatively discontinuous and dynamic for particular (individual) math learning. The long-term results of these instruction encounters seemed to be strongly determined by matches and mismatches between the child's and the educator's goals or concerns. These goals constitute implicit or unconscious drives that function as control parameters of the participants' activities (Steenbeek & van Geert, 2013). For instance, the teacher sometimes had a strong mastery-oriented focus, emphasizing the autonomous thinking and effort of the child, whereas the child sometimes had a strong performance-oriented concern focusing on reducing personal effort. Learning often involves teacher and student negotiating a match between these kinds of goals or concerns.

Since these concerns emerge and function in the actual interplay between the student and the teacher in a concrete math lesson, they can only be measured—if that is the right word—by means of detailed observations of the concrete student–teacher interaction. Key questions are: What is being assessed? How can we observe a covert property such as a person's concerns? A test is a distinct context of emergence, which at best relates probabilistically with what happens in the observed math interaction. From a dynamic perspective, assessment of the student's and teacher's concerns must always involve coordinating various perspectives by means of observations and tests. The real thing is always dynamic and multimodal, based on multiple perspectives. For instance, prior research demonstrates that children cannot be characterized by a single level of some particular measure of development. Instead they should be characterized by a range varying from functional to optimal levels of performance that emerge under different conditions (Fischer & Ayoub, 1994; Fischer, Bernstein, & Immordino-Yang, 2007; Fischer & Bidell, 2006; Vygotsky, 1978). Assessment requires multiple dimensions in dynamic interaction, not just one or two.

Another example of long-term coordination in social situations comes from research on adaptation of parent and child to the child's developing language use (Bassano et al., 2011; van Dijk et al., 2012). Child and parent jointly adapt to the child's changing skills in using grammatical constituents such as determiners or properties (e.g., mean length of utterance). These growth patterns show not only moment-to-moment fluctuations but also long-term adaptation in both children and adults. Assessments should capture the ranges and fluctuations—the dynamics of growth—and not assume stability that is ultimately incorrect. People act dynamically, varying their responses across contexts. They do not respond in means or medians or averages.

Models of Dynamic Networks of Brain and Cognitive Development

Broadly, development of brain and behavior are processes in a dynamic network of components. The properties of the components change over time as a function of how the components relate to one another, and these relations change over time. To understand a dynamic network, we return to the example of development of coordination among growers. We start with three growers: Grower A represents a child's skill for adding, Grower B represents the child's abstract concept of addition, and Grower C represents the child's understanding of the relation between addition and subtraction. These growers involve two types of relations. The first is the potential precursor relationship; for example, a certain level of the skill of adding numbers is necessary for the child to develop an abstract concept of addition. The second is the supportive or competitive relation between one grower and another; for example, rapid growth in one grower may temporarily reduce growth in another.

Consider the network representation of precursor relations. A network can typically be represented as a graph, which consists of nodes (the components A, B, and C, also called the *vertices* of the graph) and lines that connect one node to another (the precursor relations between components, also called the *edges* of the graph). In this particular graph, the relations are typically asymmetrical: A is a precursor to B, and not the other way around. The relation is a directed edge and is represented by an arrow. Since component C represents the child's understanding of the relation between addition and subtraction, we must, by implication, extend our graph with two other nodes: One is the child's skill at doing subtraction problems, and the other is the child's abstract understanding of what subtraction is. The complete precursor graph for this small network is represented in Figure 11.4.

A comparable type of graph can be drawn for the other type of relation, the supportive or competitive relation between two nodes. We know that there is a supportive relation from A (addition skill) to B (understanding what addition means). There is also a supportive relation from B to A, in that increasing understanding of the principle of addition leads to better addition skill. In this case we need a directed relation (arrow) from A to B, and a directed relation from B to A. They are different arrows because of the possibility that the nature of the relation from one to the other differs from the nature of the relation from the other to the first. The meaning of these arrows can

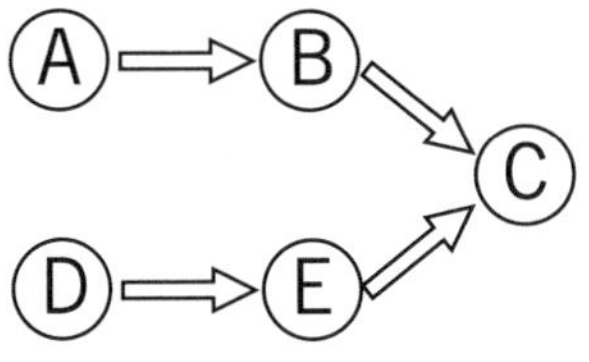

FIGURE 11.4. A theoretical graph representation of the precursor relationships described in the text.

be expressed in the precise mathematical form of Equations 11.1–11.4 above. These equations define relations between a set of growers in terms of the influence a grower has on itself (the growth part of the equation) and on other growers (the competition–support part of the equation).

The precursor graph can be represented by a matrix, which in this case consists of a 5 × 5 array, mapping the precursor relations among the components A, B, C, D, and E (Figure 11.5). This matrix is but a small part of a big matrix that describes all possible components for precursor relations, including verbal components such as understanding of the concepts of *more* and *less*, emotional and motivational components to perform and understand, and so forth. Such a network of relations becomes intractable, in that the complexity of the child's cognitive, verbal, emotional, and social system prevents describing each and every precursor relation. However, the good news is that major properties of such complex networks can be inferred from a set of general characteristics. One such general characteristic of a complex network is the average density or sparsity of the connections. As can be seen in Figure 11.5, the number of actual precursor relations is but a small subset of the number of possible relations in the network. Networks in which the relations are sparse are called small-world networks (Albert & Barabási, 2002; Fagiolo, 2007; Newman, 2010; Watts & Strogatz, 1998). These networks have characteristic properties relating to forms of self-organization, discontinuous change, and the like. A key question, addressed later, concerns the general properties of dynamic developmental networks that correspond with what is known about general patterns of development.

Network models can be applied to any set of dynamically connected components and can be used to explain the phenomena that emerge in such networks (for general accounts, see Albert & Barabási, 2002; Barabási, 2005, 2009; Newman, 2010; Watts & Stogatz, 1998). This type of network has been widely used in the study of ecological systems (e.g., Anderson & Jensen, 2005) such as food webs (Kingsland, 1995), diseases (Barabási, Gulbahce, & Loscalzo, 2011), representations of cells (Barabási & Oltvai, 2004), and biological systems in general (Mason & Verwoerd, 2007). Application of network models to mega-networks such as the World Wide Web has boosted the application of such models to a wide variety of fields (Kingsland, 1995; Newman, 2003, 2010). General network models of connected variables have been used to model the development of intelligence—in particular, the emergence of the *g* factor (van der Maas et al.,

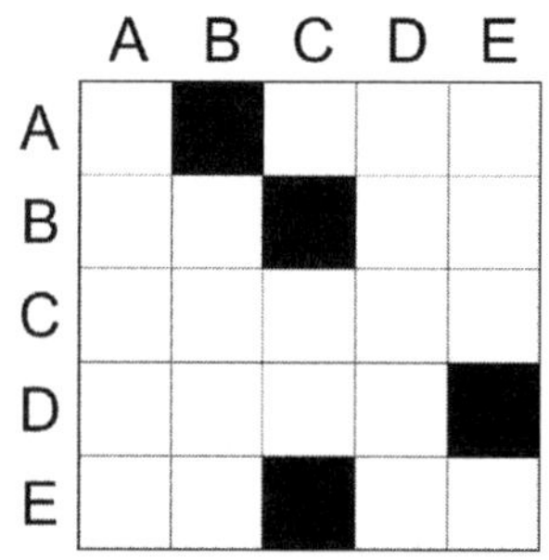

FIGURE 11.5. A matrix representation of the precursor relations in Figure 11.4.

2006, 2011), cognition (McIntosh, 2000), language (Solé, Corominas-Murtra, Valverde, & Steels, 2010), the emergence and sustenance of psychopathological conditions such as depression (Cramer, Waldorp, van der Maas, & Borsboom, 2010), and psychological disorders (Borsboom, Cramer, Schmittmann, Epskamp, & Waldorp, 2011). In psychology, network models usually pertain to social relations and are often referred to as *agent models* (e.g., Borgatti, Mehra, Brass, & Labianca, 2009; Elliott & Kiel, 2004; Smith & Conrey, 2007). Other examples are connectionist or neural net models that focus on computational relations between input and output connections. (For a comparison with general dynamic systems models, see Spencer, Thomas, & McClelland, 2009.)

Particularly fascinating is the recent work on complex network measures of brain connectivity, describing the brain as a complex dynamic network (see Figure 11.6) (Bullmore & Sporns, 2009; Rubinov & Sporns, 2010; Sporns, 2011; Sporns, Tononi, & Edelman, 2000).

Nodes and links in the brain network are determined on the basis of brain mapping, anatomical isolation, and measures of connectivity. General properties of brain networks, such as small-world connectivity, create dynamic properties such as clustering and patterns of coordination (described earlier), the emergence of modularity, and brain plasticity (see Immordino-Yang, 2007). Other general features, such as the

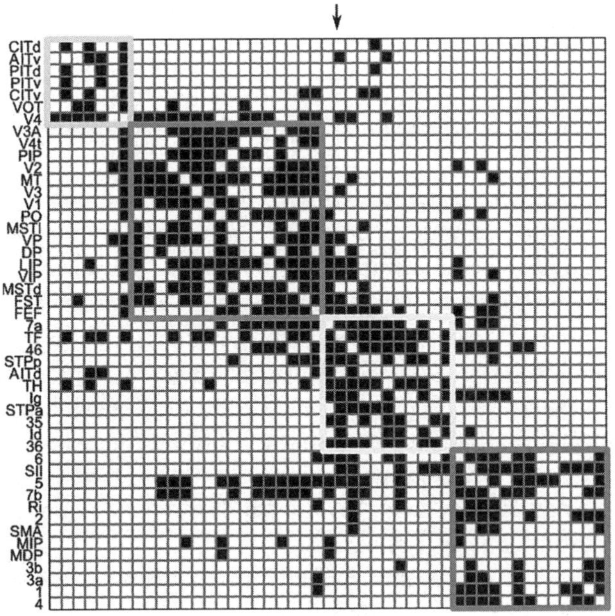

FIGURE 11.6. Structure of anatomical connectivity in the macaque cortex, with reordered nodes to show the modular structure. Rows and columns represent the functional areas of the brain (codes referring to the areas appear on the vertical axis). Each black square corresponds with a specific, direct connection between two areas. For instance, the black square on the first row and second column represents a direct connection between the areas CITd and AITv. Areas form clusters of strong interconnection (groups of black squares in the matrix). From Rubinov and Sporns (2010). Copyright 2010 by Elsevier. Reprinted by permission.

distribution of the degree of connectivity of a component with other components, predict dynamic properties such as the resilience of the brain network to damage, and the sensitivity to damage occurring in highly central nodes (those with a high degree of connectivity to other ones). In addition to studying the effects of general properties of connectivity, research should include study of effects of specific forms of connectivity, such as the predator–prey connectivity described by Thatcher (1994; Thatcher et al., 1996). These specific forms of connectivity tend to repeat across the brain and are technically referred to as *motifs* (e.g., the asymmetric competition–support relation between two components that is typical of a predator–prey relation; Alon, 2007; Milo et al., 2002; Newman, 2003).

We have suggested a number of common motifs in dynamic growth models where the network consists of brain and behavioral components (including context-related). Typical motifs include (1) combinations of unilinear precursors with symmetric support relations; and (2) combinations of unilinear precursors with asymmetric support and competitive relations, which we find, for example, in language development (Bassano & van Geert, 2007; van Geert, 1991, 1994). These motifs can explain a host of dynamic developmental phenomena such as stepwise growth, temporary regressions, and inverted U-shaped patterns (see the section "Development of Coordination between Growers," for examples).

Applications to Talent

In more recent work, some of the authors have explored the possibilities of studying the dynamics generated by general properties of developmental networks, such as sparse connection and small-world relations (i.e., the number of direct connections between any two nodes in a network is small, but most of the nodes of the network can be reached from another node by means of a relatively small number of intermediate steps; van Geert, Steenbeek, & van Dijk, 2012). This work arose from an attempt to provide a general explanation for the emergence of talent, which we define as the potential for exceptional performance in specific dynamic skills. It turns out that a general "minimalist" network model can provide a natural explanation of the emergence of talent and exceptional performance, in addition to providing explanations for nonlinear dynamical growth phenomena such as coordination, stepwise increase, and temporary regressions. The reason why such a general model can provide these explanations is that the general principles of connections among components in a developmental network subsume the special cases.

Existing models of talent development and giftedness make a distinction between internal factors and external factors. Internal factors can be domain-specific, such as the person's interest in a particular domain or his or her domain-specific persistence in pursuing a goal (e.g., Gagné, 2004; Lubinski & Benbow, 2000, 2006; Perleth & Heller, 2007; Sternberg, 2003; Ziegler, 2005). Internal factors can also be domain-general, as with spatial ability or excellence in science and technology (Lubinski, 2010; Wai, Lubinski, & Benbow, 2009), general intelligence, perceptual speed, or working memory (e.g., Sternberg, 2003; Vandervert, Schimpf, & Liu, 2007). The so-called *generalist genes* are also examples of domain-general components (see Plomin, Kovas, & Haworth, 2007).

Although the relations between a particular talent or performance domain and a specific additional factor can be supportive (as when spatial ability supports the growth of scientific reasoning), relations may also be competitive or negative, as when a factor counteracts the growth of a skill. For instance, a person's dyslexia (reading difficulties) may negatively affect his or her science learning, since science heavily depends on text (in addition to many other things). However, the factor that exerts a negative influence on a particular talent or performance domain in one person may have no effect or even a positive one in another person. Dyslexia, for instance, has been shown to be related to global visual–spatial ability, and may thus positively affect a person's science learning, as when a scientific discipline relies heavily on visual–spatial ability and less on reading of texts (Schneps, Rose, & Fischer, 2007; von Károlyi, 2001; von Károlyi, Winner, Gray, & Sherman, 2003; Winner et al., 2001). A similar type of idiosyncratic relation exists between various performance domains such as mathematics, technology, or art, and aspects of autism such as problems with theory of mind, hypersystematizing, hyperattention, and sensory hypersensitivity (Baron-Cohen, Ashwin, Ashwin, Tavassoli, & Chakrabarti, 2009; Baron-Cohen, Wheelwright, Burtenshaw, & Hobson, 2007; Happé & Vital, 2009). In some autistic individuals these hyper characteristics may hamper learning, whereas in others, for whom the conditions are "just right," the exceptional attention to detail may lead to exceptional levels of performance in particular domains.

Prevailing models of talent and giftedness also emphasize the importance of external factors, such as specific educational experiences that facilitate developing particular talents. Also relevant are general effects, such as a supportive family climate, which help a child pursue his or her own goals (e.g., Bloom, 1985). Examples of important components are the amount of appropriate challenges in the environment (Bleske-Rechek, Lubinski, & Benbow, 2004; Lubinski & Benbow, 2006; Wai, Lubinski, Benbow, & Steiger, 2010), the individual's feeling of success and satisfaction (Lubinski & Benbow, 2000, 2006), and the individual's perception of his or her own ability and relation with the environment (Asbury, Almeida, Hibel, Harlaar, & Plomin, 2008; Chamorro-Premuzic, Harlaar, Greven, & Plomin, 2010; Greven, Harlaar, Kovas, Chamorro-Premuzic, & Plomin, 2009; Oliver, Pike, & Plomin, 2008).

These relations are again highly idiosyncratic and characteristic of a particular person's profile (Achter, Benbow, & Lubinski, 1997; Robertson, Smeets, Lubinski, & Benbow, 2010; Webb, Lubinski, & Benbow, 2002; for a general account, see Molenaar, 2004, and Molenaar & Campbell, 2009). They should be treated as nodes in a dynamic network, affecting the growth of other nodes (growers, components, . . .) and not as independent variables in the traditional sense of the word.

Graph theoretically, a particular talent or domain of performance or expertise can be represented by a small network, consisting of a node corresponding with the talent or skill at issue and other nodes consisting of factors that positively or negatively affect the talent, such as particular interests, domain-specific persistence, general factors such as spatial ability, and external conditions such as particular educational contexts that nourish (or suppress) the talent. The connections between these nodes are directed, in that a relation runs from one node to another (e.g., from domain-specific persistence to domain-specific skill). Connections may be symmetric or asymmetric; for example, domain-specific persistence positively affects domain-specific skill, which may eventually affect persistence, such as when children think that their increasing mastery in a particular domain eliminates the necessity for effortful practice. Relations may be direct or indirect, for example, when a particularly good math teacher positively affects a student's math knowledge, which positively affects the student's learning of physics, for which the student has developed a particular interest.

Each person possesses a particular performance or talent profile—that is, a number of performance domains in which he or she is particularly strong or highly interested. A person's profile can thus be represented as a network of smaller subnetworks that represent the specific talent or performance domains (see Figure 11.5). It is highly likely that such subnetworks are also interconnected, but with a degree of connection that is considerably lower than the degree of connection within the subnetwork. Subnetworks can be connected via a domain-general node, such as spatial ability. For example, if mathematical understanding positively affects spatial ability, and spatial ability is in turn related to physics understanding (Figures 11.7a and 11.7c), the multiple connecting hubs in Figure 11.7c might correspond with the generalist genes discussed earlier. Subnetworks can also be connected via domain-specific components; for example, the network of nodes relating to a person's skill or talent in physics can be related to a person's skill or talent for math because the math knowledge positively affects the understanding of physics (Figure 11.7b).

Specific performance domains are likely to be characterized by a denser pattern of connections than the connections that occur across domains. This leads to a modular structure that is the result of self-organization of coordination arising from the interactions among components. If the modular structures are interconnected via general nodes that represent general abilities such as working memory, trait motivation, or general intelligence, the network probably also classifies as a so-called *scale-free network*, where the types of relation between subcomponents resemble the types of relation between components and higher-order structures (e.g., Barabási, 2009).

Finally, the network is dynamic. This means that the values of the nodes (the levels) change as a consequence of interactions with other nodes, and nodes may appear or disappear over developmental time, as with particular forms of educational support such as academic teaching or professional experiences relating to work satisfaction (Lubinski & Benbow, 2000, 2006).

The general network model of performance growth takes the form of a so-called *neutral generative model* (Frank, 2009). Neutral generative models are theoretically simple in that they confine themselves to specifying a set of randomly distributed, and in this case, sparsely connected, relations between the variables, which can vary from negative to positive connections, with an average connection strength which is equal to zero. (These are special cases of the general model.)

This general dynamic network model makes qualitative predictions regarding development that are supported by existing data. One prediction is that if one focuses

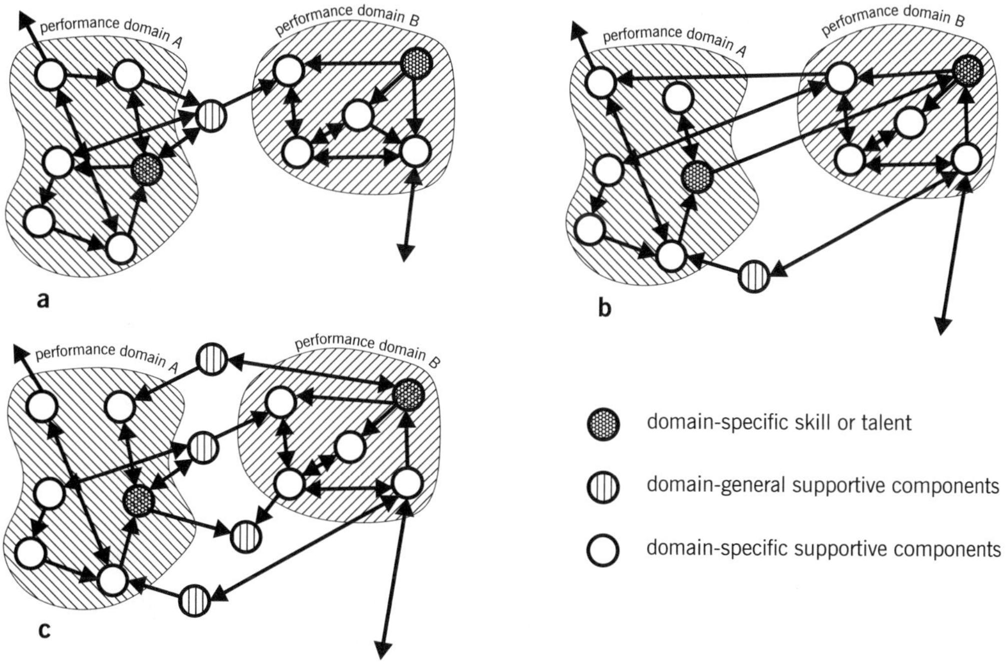

FIGURE 11.7. Graph-theoretical representation of two domain-specific skills, A and B (e.g., technological reasoning skill and scientific reasoning skill), represented as networks of interconnected domain-specific and domain-overarching (general) components. Figures 11.a, 11.b, and 11.c represent three possible networks, and thus three possible individuals. In the first network, representing person a, domains are relatively isolated and connected via a single domain-general "hub" component (e.g., mathematical knowledge, functioning as a connecting component between a person's technological reasoning skill and scientific reasoning skill). In the second person, b, the domain-specific skills are characterized by connections between domain-specific components; that is, there is a direct relationship between the person's technological skill and science reasoning skill. In the third network, the two performance domains are connected via a number of domain-general support components, such as general mathematical knowledge, the person's domain-general conscientiousness, the quality of science and technology teaching the person receives, etc.

on individual developmental processes, nonlinear trajectories are the rule rather than the exception. The network automatically generates a host of nonlinear growth patterns, such as stepwise growth and temporary regressions, inverted U-shaped growth, and so on. Development will take the form of a developmental web, with "stages" or "levels" involving coordinations between relatively rapid changes in clusters of connected variables (Fischer & Bidell, 2006; Stein et al., 2010). Phenomena such as a so-called *autistic savant syndrome* are specific cases of the general principle of network connections, which occur if connections tend to occur in relatively isolated units, and so-called *hubs* are lacking in the network (Heaton & Wallace, 2004; Howlin, Goode, Hutton, & Rutter, 2009; Miller, 1999, 2005; Treffert, 2009). The model also shows how different types of relations between the same personal and educational factors, including motivation, may lead to either successful or unsuccessful learning trajectories (Steenbeek & van Geert, 2013).

The network model likewise predicts the somewhat counterintuitive but well-corroborated finding that heritability increases during the life span (Bergen, Gardner, & Kendler, 2007; Davis, Haworth, & Plomin, 2009; Deary, Johnson, & Hoolihan, 2009; Haworth et al., 2010; Hoekstra, Bartels, & Boomsma, 2007; Johnson, 2010). Regarding exceptional performance, the model predicts that early indicators of talent are weak, despite the occurrence of child prodigies and individual cases of early exceptionality. The model also predicts that in different individuals, a similar form of talent may emerge at different ages and that the features and characteristics of a particular talent in a person can change during the person's life span. That is, talent is not necessarily stable: It can change or even disappear during a person's lifetime (the empirical evidence for these predictions has been described by Simonton; e.g., see Simonton, 1999, 2001, for overview articles).

A very interesting prediction of the network model is that life-span developmental processes will lead to the development of domain-specific abilities, that is, dynamic skills that are strongly log-normally distributed in the population (right-skewed distribution). At first sight, this prediction might seem incompatible with what is sometimes taken as an iron rule in the behavioral sciences: that abilities are symmetrically and normally distributed. However, if we look at the distribution of the products of domain-specific skills (i.e., of domain-specific forms of expertise that occur in professional skills, artistic skills, scientific skills, sports skills, etc.), there is overwhelming evidence that these products are asymmetrically distributed and that they are not symmetrical or Gaussian (for an overview, see Huber, 2000; van Geert et al., 2012; see Figure 11.8 for two typical examples).

Clearly, the distribution of the products of a person's abilities is not the same as the distribution of the levels of such abilities. Nevertheless, the application of simple, empirically corroborated product models to the dynamic network shows that the empirically obtained product distributions require strongly right-skewed distributions of the underlying ability. Think of the number of publications by scientists, objective indicators of their scientific importance such as the number of citations, the number of patents registered by inventors, the number of goals made by soccer players, and so on.

Conclusion: The Virtues of Dynamic Network Models

In conclusion, a general dynamic network model specifying development as a change in a network of brain and behavioral components, including educational and environmental influences, generates a wide variety of qualitative predictions that are supported by the available evidence. Dynamic systems models commonly relate to data in this manner. Instead of highly specific and precise quantitative predictions of specific phenomena, they predict many qualitative phenomena that are subsumed under a small set of general dynamic principles. Instead of providing specific, precise estimates

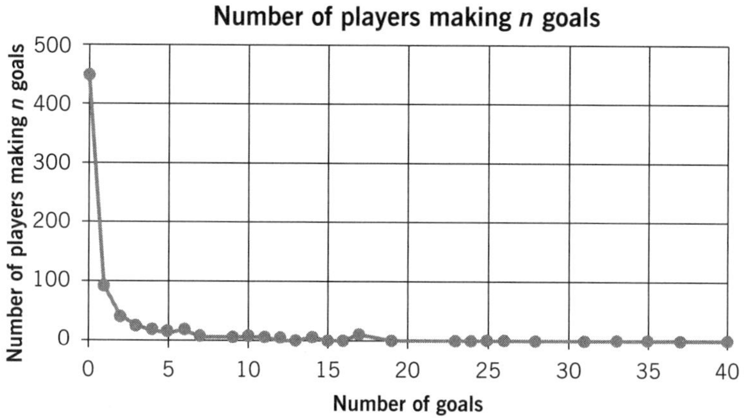

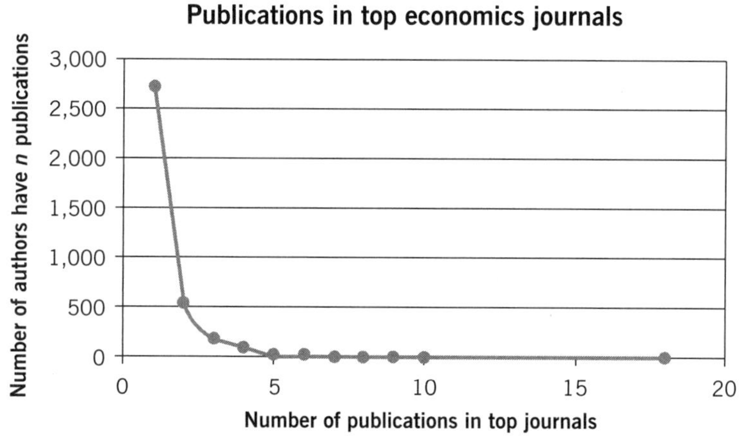

FIGURE 11.8. Typical distributions of skill-related products: the number of goals made by a Dutch soccer player during between-country matches (*top*), and the number of publications by a first author in top-ranked economics journals (*bottom*). The distributions are anything but normal or symmetric. These extremely right-skewed distributions appear in virtually all product distributions, ranging from science to arts and sports. They do not fit normal distributions.

of parameter values, their contributions involve predicting a wide variety of qualitative phenomena related to a small set of general dynamic principles. In the majority of cases, alternative models that typically make precise parameter explanations and quantitative fits to data do not make these kinds of predictions, which is not to say that parametric models are incapable of making qualitative predictions (several chapters in Molenaar & Newell, 2010, describe parametric dynamic systems models yielding such qualitative predictions). In this way, dynamic systems models portray a wide range of phenomena in approximate terms and thus sketch the kinds of phenomena that these models can explain.

REFERENCES

Achter, J. A., Benbow, C. P., & Lubinski, D. (1997). Rethinking multipotentiality among the intellectually gifted: A critical review and recommendations. *Gifted Child Quarterly, 41*(1), 5–15.

Albert, R., & Barabási, A. L. (2002). Statistical mechanics of complex networks. *Reviews of Modern Physics, 74*, 47–97.

Alon, U. (2007). Network motifs: Theory and experimental approaches. *Nature Reviews Genetics, 8*(6), 450–461.

Anderson, P. E., & Jensen, H. J. (2005). Network properties, species abundance and evolution in a model of evolutionary ecology. *Journal of Theoretical Biology, 232*(4), 551–558.

Asbury, K., Almeida, D., Hibel, J., Harlaar, N., & Plomin, R. (2008). Clones in the classroom: A daily diary study of the nonshared environmental relationship between monozygotic twin differences in school experience and achievement. *Twin Research and Human Genetics, 11*(6), 586–595.

Barabási, A.-L. (2005). Taming complexity. *Nature Physics, 1*, 68–70.

Barabási, A.-L. (2009). Scale-free networks: A decade and beyond. *Science, 325*, 412–413.

Barabási, A.-L., Gulbahce, N., & Loscalzo, J. (2011). Network medicine: A network-based approach to human disease. *Nature Reviews Genetics, 12*(1), 56–68.

Barabási, A.-L., & Oltvai, Z. N. (2004). Network biology: Understanding the cell's functional organization. *Nature Reviews Genetics, 5*(2), 101–113.

Baron-Cohen, S., Ashwin, E., Ashwin, C., Tavassoli, T., & Chakrabarti, B. (2009). Talent in autism: Hyper-systemizing, hyper-attention to detail and sensory hypersensitivity. *Philosophical Transactions of the Royal Society of London: Series B, Biological Sciences, 364*, 1377–1383.

Baron-Cohen, S., Wheelwright, S., Burtenshaw, A., & Hobson, E. (2007). Mathematical talent is linked to autism. *Human Nature, 18*(2), 125–131.

Bassano, D., Maillochon, I., Korecky-Kroll, K., van Dijk, M., Laaha, S., Dressler, W. U., et al. (2011). A comparative and dynamic approach to the development of determiner use in three children acquiring different languages. *First Language, 31*(3), 253–279.

Bassano, D., & van Geert, P. (2007). Modeling continuity and discontinuity in utterance length: A quantitative approach to changes, transitions and intra-individual variability in early grammatical development. *Developmental Science, 10*(5), 588–612.

Benes, F. (1994). Development of the corticolimbic system. In G. Dawson & K. W. Fischer (Eds.), *Human behavior and the developing brain* (pp. 176–206). New York: Guilford Press.

Bergen, S. E., Gardner, C. O., & Kendler, K. S. (2007). Age-related changes in heritability of behavioral phenotypes over adolescence and young adulthood: A meta-analysis. *Twin Research and Human Genetics, 10*(3), 423–433.

Bleske-Rechek, A., Lubinski, D., & Benbow, C. P. (2004). Meeting the educational needs of special populations: Advanced placement's role in developing exceptional human capital. *Psychological Science, 15*(4), 217–224.

Bloom, B. S. (1985). *Developing talent in young people.* New York: Ballantine Books.

Borgatti, S. P., Mehra, A., Brass, D. J., & Labianca, G. (2009). Network analysis in the social sciences. *Science, 323*, 892–895.

Borsboom, D., Cramer, A. O. J., Schmittmann, V. D., Epskamp, S., & Waldorp, L. J. (2011). The small world of psychopathology. *PloS One, 6*(11), e27407.

Brown, A. L., & Reeve, R. (1987). Bandwidths of competence: The role of supportive contexts in learning and development. In L. S. Liben (Ed.), *Development and learning: Conflict or congruence?* (pp. 173–223). Hillsdale, NJ: Erlbaum.

Bullmore, E., & Sporns, O. (2009). Complex brain networks: Graph theoretical analysis of structural and functional systems. *Nature Reviews Neuroscience, 10*(3), 186–198.

Bullock, D., & Grossberg, S. (1988). Neural dynamics of planned arm movements: Emergent invariants and speed–accuracy properties during trajectory formation. *Psychological Review, 95*, 49–90.

Case, R. (1992). The role of the frontal lobes in the regulation of human development. *Brain and Cognition, 20*, 51–73.

Chamorro-Premuzic, T., Harlaar, N., Greven, C. U., & Plomin, R. (2010). More than just IQ: A longitudinal examination of self-perceived abilities as predictors of academic performance in a large sample of UK twins. *Intelligence, 38*(4), 385–392.

Changeux, J.-P., & Danchin, A. (1976). Selective stabilisation of developing synapses as a mechanism for the specificiation of neuronal networks. *Nature, 264*, 705–712.

Conel, J. L. (1939–1963). *The postnatal development of the human cerebral cortex* (7 vols.). Cambridge, MA: Harvard University Press.

Cramer, A. O. J., Waldorp, L. J., van der Maas, H. L. J., & Borsboom, D. (2010). Comorbidity: A network perspective. *Behavioral and Brain Sciences, 33*(2–3), 137–150.

Davis, O. S. P., Haworth, C. M. A., & Plomin, R. (2009). Dramatic increase in heritability of cognitive development from early to middle childhood: An 8-year longitudinal study of 8,700 pairs of twins. *Psychological Science, 20*(10), 1301–1308.

Deary, I. J., Johnson, W., & Houlihan, L. (2009). Genetic foundations of human intelligence. *Human Genetics, 126*, 215–232.

Dewey, J. (1933). *How we think: A restatement of the relation of reflective thinking to the educative process.* Lexington, MA: Heath.

Diamond, A., Werker, J. F., & Lalonde, C. (1994). Toward understanding commonalities in the development of object search, detour navigation, categorization, and speech perception. In G. Dawson & K. W. Fischer (Eds.), *Human behavior and the developing brain* (pp. 380–426). New York: Guilford Press.

Eichorn, D., & Bayley, N. (1962). Growth in head circumference from birth through young adulthood. *Child Development, 33*, 257–271.

Elliott, E., & Kiel, L. D. (2004). Agent-based modeling in the social and behavioral sciences. *Nonlinear Dynamics, Psychology, and Life Sciences, 8*(2), 121–130.

Fagiolo, G. (2007). Clustering in complex directed networks. *Physical Review E, 76*(2), 1–8.

Fischer, K. W. (1980). A theory of cognitive development: The control and construction of hierarchies of skills. *Psychological Review, 87*, 477–531.

Fischer, K. W. (1987). Relations between brain and cognitive development. *Child Development, 57*, 623–632.

Fischer, K. W. (2008). Dynamic cycles of cognitive and brain development: Measuring growth in mind, brain, and education. In A. M. Battro, K. W. Fischer, & P. Léna (Eds.), *The educated brain* (pp. 127–150). Cambridge, UK: Cambridge University Press.

Fischer, K. W., & Ayoub, C. (1994). Affective splitting and dissociation in normal and maltreated children: Developmental pathways for self in relationships. In D. Cicchetti & S. L. Toth (Eds.), *Disorders and dysfunctions of the self* (Vol. 5, pp. 149–222). Rochester, NY: Rochester University Press.

Fischer, K. W., Bernstein, J. H., & Immordino-Yang, M. H. (Eds.). (2007). *Mind, brain, and education in reading disorders.* Cambridge, UK: Cambridge University Press.

Fischer, K. W., & Bidell, T. R. (2006). Dynamic development of action and thought. In W.

Damon & R. M. Lerner (Eds.), *Theoretical models of human development: Handbook of child psychology* (6th ed., Vol. 1, pp. 313–399). New York: Wiley.

Fischer, K. W., & Bullock, D. (1984). Cognitive development in school-age children: Conclusions and new directions. In W. A. Collins (Ed.), *Development during middle childhood: The years from six to twelve* (pp. 70–146). Washington, DC: National Academy Press.

Fischer, K. W., Bullock, D., Rotenberg, E. J., & Raya, P. (1993). The dynamics of competence: How context contributes directly to skill. In R. Wozniak & K. Fischer (Eds.), *Development in context: Acting and thinking in specific environments* (pp. 93–117). Hillsdale, NJ: Erlbaum.

Fischer, K. W., Hand, H. H., & Russell, S. L. (1984). The development of abstractions in adolescence and adulthood. In M. Commons, F. A. Richards, & C. Armon (Eds.), *Beyond formal operations* (pp. 43–73). New York: Praeger.

Fischer, K. W., & Lamborn, S. (1989). Mechanisms of variation in developmental levels: Cognitive and emotional transitions during adolescence. In A. de Ribaupierre (Ed.), *Transition mechanisms in child development* (pp. 33–67). New York: Cambridge University Press.

Fischer, K. W., & Rose, S. P. (1994). Dynamic development of coordination of components in brain and behavior: A framework for theory and research. In G. Dawson & K. W. Fischer (Eds.), *Human behavior and the developing brain* (pp. 3–66). New York: Guilford Press.

Fischer, K. W., Shaver, P., & Carnochan (1990). How emotions develop and how they organise development. *Cognition and Emotion, 4*, 81–127.

Frank, S. A. (2009). The common patterns of nature. *Journal of Evolutionary Biology, 22*(8), 1563–1585.

Gabrieli, J. D. E. (2009). Dyslexia: A new synergy between education and cognitive neuroscience. *Science, 5938*, 280–283.

Gagné, F. (2004). Transforming gifts into talents: The DMGT as a developmental theory. *High Ability Studies, 15*(2), 119–147.

Goldin-Meadow, S., Nusbaum, H., Garber, P., & Church, R. B. (1993). Transitions in learning: Evidence for simultaneously activated rules. *Journal of Experimental Psychology: Human Perception and Performance, 19*, 92–107.

Goldman-Rakic, P. S. (1987). Connectionist theory and the biological basis of cognitive development. *Child Development, 58*, 601–622.

Granott, N. (2002). How microdevelopment creates macrodevelopment: Reiterated sequences, backward transitions, and the zone of current development. In N. Granott & J. Parziale (Eds.), *Microdevelopment: Transition processes in development and learning* (pp. 213–242). Cambridge, UK: Cambridge University Press.

Greven, C. U., Harlaar, N., Kovas, Y., Chamorro-Premuzic, T., & Plomin, R. (2009). School achievement is predicted by self-perceived abilities—but for genetic rather than environmental reasons. *Psychological Science, 20*(6), 753–763.

Grossberg, S. (1988). Nonlinear neural networks: Principles, mechanisms, and architectures. *Neural Networks, 1*, 17–61.

Grossberg, S., Mingolla, E., & Fazi, A. (2009). View-invariant object category learning, recognition, and search: How spatial and object attention are coordinated using surface-based attentional shrouds. *Cognitive Psychology, 58*(1), 1–48.

Halford, G. S., & Boyle, F. M. (1985). Do young children understand conservation of number? *Child Development, 56*, 165–176.

Happé, F., & Vital, P. (2009). What aspects of autism predispose to talent? *Philosophical*

Transactions of the Royal Society of London: Series B, Biological Sciences, 364, 1369–1375.

Harter, S., & Buddin, B. (1987). Children's understanding of the simultaneity of two emotions: A five-stage developmental sequence. *Developmental Psychology, 23*, 388–399.

Haworth, C. M., Wright, M. J., Martin, N. W., Martin, N. G., Boomsma, D. I., Bartels, M., et al. (2009). A twin study of the genetics of high cognitive ability selected from 11,000 twin pairs in six studies from four countries. *Behavior Genetics, 39*(4), 359–370.

Heaton, P., & Wallace, G. L. (2004). Annotation: The savant syndrome. *Journal of Child Psychology and Psychiatry, and Allied Disciplines, 45*(5), 899–911.

Hoekstra, R., Bartels, M., & Boomsma, D. (2007). Longitudinal genetic study of verbal and nonverbal IQ from early childhood to young adulthood. *Learning and Individual Differences, 17*(2), 97–114.

Horn, J. L., & Hofer, S. M. (1992). Major abilities and development in the adult period. In R. J. Sternberg & C. A. Berg (Eds.), *Intellectual development* (pp. 44–99). Cambridge, UK: Cambridge University Press.

Howlin, P., Goode, S., Hutton, J., & Rutter, M. (2009). Savant skills in autism: Psychometric approaches and parental reports. *Philosophical Transactions of the Royal Society of London: Series B, Biological Sciences, 364*, 1359–1367.

Hubel, D. H., & Wiesel, T. N. (1977). Functional architecture of macaque monkey visual cortex. *Proceedings of the Royal Society of London: Series B, Biological Sciences, 193*, 1–59.

Huber, J. C. (2000). A statistical analysis of special cases of creativity. *Journal of Creative Behavior, 34*(3), 203–225.

Hudspeth, W. J., & Pribram, K. H. (1992). Psychophysiological indices of cerebral maturation. *International Journal of Psychophysiology, 12*, 19–29.

Huttenlocher, P. (1994). Synaptogenesis in human cerebral cortex. In G. Dawson & K. W. Fischer (Eds.), *Human behavior and the developing brain* (pp. 137–152). New York: Guilford Press.

Immordino-Yang, M. H. (2007). A tale of two cases: Lessons for education from the study of two boys living with half their brains. *Mind, Brain, and Education, 1*(2), 66–83.

Immordino-Yang, M. H., Christodoulou, J. A., & Singh, V. (2012). Rest is not idleness: Implications of the brain's default mode for human development and education. *Perspectives in Psychological Science, 7*(4), 352–364.

John, E. R. (1977). *Functional neuroscience: Vol. 2. Neurometrics*. Hillsdale, NJ: Erlbaum.

Johnson, W. (2010). Understanding the genetics of intelligence: Can height help? Can corn oil? *Current Directions in Psychological Science, 19*(3), 177–182.

Kingsland, S. E. (1995). *Modeling nature: Episodes in the history of population ecology* (2nd ed.). Chicago: University of Chicago Press.

Kitchener, K. S., & Fischer, K. W. (1990). A skill approach to the development of reflective thinking. *Developmental Perspectives on Teaching and Learning Thinking Skills, 21*(4), 48–62.

Kitchener, K. S., King, P. M., Wood, P. K., & Davison, M. L. (1989). Sequentiality and consistency in the development of reflective judgment: A six year longitudinal study. *Journal of Applied Developmental Psychology, 10*, 73–95.

Kitchener, K. S., Lynch, C. L., Fischer, K. W., & Wood, P. K. (1993). Developmental range of reflective judgment: The effect of contextual support and practice on developmental stage. *Developmental Psychology, 29*, 893–906.

Lubinski, D. (2010). Spatial ability and STEM: A sleeping giant for talent identification and development. *Personality and Individual Differences, 49*(4), 344–351.

Lubinski, D., & Benbow, C. P. (2000). States of excellence. *American Psychologist, 55*(1), 137–150.

Lubinski, D., & Benbow, C. P. (2006). Study of mathematically precocious youth after 35 years: Uncovering antecedents for the development of math–science expertise. *Perspectives on Psychological Science, 1*(4), 316–345.

Mascolo, M. F., & Fischer, K. W. (2010). The dynamic development of thinking, feeling, and acting over the lifespan. In R. M. Lerner (Ed.-in-Chief) & W. F. Overton (Vol. Ed.), *Handbook of life-span development: Vol. 1. Biology, cognition, and methods* (pp. 149–194). Hoboken, NJ: Wiley.

Mason, O., & Verwoerd, M. (2007). Graph theory and networks in biology. *IET Systems Biology, 1*(2), 89–119.

Matousek, M., & Petersén, I. (1973). Frequency analysis of the EEG in normal children and adolescents. In P. Kelleway & I. Petersén (Eds.), *Automation of clinical electroencephalography* (pp. 75–102). New York: Raven Press.

McCall, R. B. (1983). Exploring developmental transitions in mental performance. In K. W. Fischer (Ed.), Levels and transitions in children's development. *New Directions for Child Development* (Vol. 21, pp. 65–80). San Francisco: Jossey-Bass.

Mclntosh, R. (2000). Towards a network theory of cognition. *Neural Networks, 13*(8–9), 861–870.

Miller, L. K. (1999). The savant syndrome: Intellectual impairment and exceptional skill. *Psychological Bulletin, 125*(1), 31–46.

Miller, L. K. (2005). What the savant syndrome can tell us about the nature and nurture of talent. *Journal for the Education of the Gifted, 28*(3–4), 361–373.

Milo, R., Shen-Orr, S., Itzkovitz, S., Kashtan, N., Chklovskii, D., & Alon, U. (2002). Network motifs: Simple building blocks of complex networks. *Science, 298*, 824–827.

Molenaar, P. C. M. (2004). A manifesto on psychology as idiographic science: Bringing the person back into scientific psychology, this time forever. *Measurement, 2*, 201–218.

Molenaar, P. C. M., & Campbell, C. G. (2009). The new person-specific paradigm in psychology. *Current Directions in Psychological Science, 18*(2), 112–117.

Molenaar, P. C. M., & Newell, K. M. (Eds.). (2010). *Individual pathways of change: Statistical models for analyzing learning and development.* Washington, DC: American Psychological Association.

Newman, M. E. J. (2003). The structure and function of complex networks. *SIAM Review, 45*(2), 167–256.

Newman, M. E. J. (2010). *Networks: An introduction.* Oxford, UK: Oxford University Press.

Oliver, B. R., Pike, A., & Plomin, R. (2008). Nonshared environmental influences on teacher-reported behaviour problems: Monozygotic twin differences in perceptions of the classroom. *Journal of Child Psychology and Psychiatry, and Allied Disciplines, 49*(6), 646–653.

Overton, W. F. (2006). Developmental psychology: Philosophy, concepts, methodology. In W. Damon & R. M. Lerner (Eds.), *Handbook of child psychology* (6th ed., Vol. 1, pp. 20–88). New York: Wiley.

Overton, W. F. (2013). A new paradigm for developmental science: Relationism and relational-developmental systems. *Applied Developmental Science, 17*(2), 94–107.

Perleth, C., & Heller, K. (2007). Adapting conceptual models for cross-cultural applications. *Educational Research and Evaluation, 13*(6), 539–554.

Piaget, J. (1952). *The origins of intelligence in children* (M. Cook, Trans.). New York: International Universities Press. (Original work published 1936)
Piaget, J. (1983). Piaget's theory. In W. Kessen (Ed.), *Handbook of child psychology: History, theory, and methods* (4th ed., Vol. 1, pp. 103–126). New York: Wiley.
Plomin, R., Kovas, Y., & Haworth, C. M. (2007). Generalist genes: Genetic links between brain, mind, and education. *Mind, Brain, and Education, 1*(1), 11–19.
Poldrack, R. A., Halchenko, Y. O., & Hanson, S. J. (2009). Decoding the large-scale structure of brain function by classifying mental states across individuals. *Psychological Science, 20*(11), 1364–1372.
Rakic, P., Bourgeois, J.-P., Eckenhoff, M. F., Zecevic, N., & Goldman-Rakic, P. (1986). Concurrent overproduction of synapses in diverse regions of the primate cerebral cortex. *Science, 232*, 232–235.
Robertson, K. F., Smeets, S., Lubinski, D., & Benbow, C. P. (2010). Beyond the threshold hypothesis: Even among the gifted and top math/science graduate students, cognitive abilities, vocational interests, and lifestyle preferences matter for career choice, performance, and persistence. *Current Directions in Psychological Science, 19*(6), 346–351.
Rubinov, M., & Sporns, O. (2010). Complex network measures of brain connectivity: Uses and interpretations. *NeuroImage, 52*(3), 1059–1069.
Ruhland, R., & van Geert, P. (1998). Jumping into syntax: Transitions in the development of closed class words. *British Journal of Developmental Psychology, 16*(Pt. 1), 65–95.
Schneps, M. H., Rose, L. T., & Fischer, K. W. (2007). Visual learning and the brain: Implications for dyslexia. *Mind, Brain, and Education, 1*(3), 128–139.
Simonton, D. K. (1999). Talent and its development: An emergenic and epigenetic model. *Psychological Review, 106*(3), 435–457.
Simonton, D. K. (2001). Talent development as a multidimensional, multiplicative, and dynamic process. *Current Directions in Psychological Science, 10*(2), 39–43.
Singer, W. (1995). Development and plasticity of cortical processing architectures. *Science, 270*, 758–764.
Smith, E. R., & Conrey, F. R. (2007). Agent-based modeling: A new approach for theory building in social psychology. *Personality and Social Psychology Review, 11*(1), 1–18.
Smith, L. B., & Thelen, E. (Eds.). (1993). *Dynamical systems in development: Applications.* Cambridge, MA: MIT Press.
Solé, R. V., Corominas-Murtra, B., Valverde, S., & Steels, L. U. C. (2010). Language networks: Their structure, function, and evolution. *Complexity, 15*(6), 20–26.
Spencer, J. P., Thomas, M. S. C., & McClelland, J. (2009). *Toward a new grand theory of development?: Connectionism and dynamic systems theory reconsidered.* Oxford, UK: Oxford University Press.
Sporns, O. (2011). *Networks of the brain.* Cambridge, MA: MIT Press.
Sporns, O., Tononi, G., & Edelman, G. M. (2000). Connectivity and complexity: The relationship between neuroanatomy and brain dynamics. *Neural Networks, 13*(8–9), 909–922.
Steenbeek, H., Jansen, L., & van Geert, P. (2012). Scaffolding dynamics and the emergence of problematic learning trajectories. *Learning and Individual Differences, 22*(1), 64–75.
Steenbeek, H. W., & van Geert, P. L. C. (2013). The emergence of learning–teaching trajectories in education: A complex dynamic systems approach. *Nonlinear Dynamics, Psychology, and Life Sciences, 17*, 233–267.
Stein, Z., Dawson, T., & Fischer, K. W. (2010). Redesigning testing: Operationalizing the new science of learning. In M. S. Khine & I. M. Saleh (Eds.), *New science of learning: Cognition, computers, and collaboration in education* (pp. 207–224). New York: Springer.

Sternberg, R. J. (2003). WICS as a model of giftedness. *High Ability Studies, 14*(2), 109–137.

Thatcher, R. W. (1994). Cyclic cortical reorganization: Origins of human cognitive development. In G. Dawson & K. W. Fischer (Eds.), *Human behavior and the developing brain* (pp. 232–266). New York: Guilford Press.

Thatcher, R. W., Lyon, G. R., Rumsey, J., & Krasnegor, N. (Eds.). (1996). *Developmental neuroimaging: Mapping the development of brain and behavior.* New York: Academic Press.

Thelen, E., & Smith, L. B. (1994). *A dynamic systems approach to the development of perception and action.* Cambridge, MA: MIT Press.

Treffert, D. (2009). The savant syndrome: An extraordinary condition—a synopsis: Past, present, future. *Philosophical Transactions of the Royal Society of London: Series B, Biological Sciences, 364*, 1351–1357.

van der Maas, H. J., Dolan, C., Grasman, R. P., Wicherts, J., Huizenga, H., & Raijmakers, M. J. (2006). A dynamical model of general intelligence: The positive manifold of intelligence by mutualism. *Psychological Review, 113*(4), 842–861.

van der Maas, H. J., Molenaar, D., Maris, G., Kievit, R.A., & Borsboom, D. (2011). Cognitive psychology meets psychometric theory: On the relation between process models for decision making and latent variable models for individual differences. *Psychological Review, 118*(2), 339–356.

van der Maas, H. J., & Molenaar, P. (1992). A catastrophe-theoretical approach to cognitive development. *Psychological Review, 99*, 395–417.

van der Molen, M. W., & Molenaar, P. C. M. (1994). Cognitive psychophysiology: A window to cognitive development and brain maturation. In G. Dawson & K. W. Fischer (Eds.), *Human behavior and the developing brain* (pp. 456–490). New York: Guilford Press.

Vandervert, L. R., Schimpf, P. H., & Liu, H. (2007). How working memory and the cerebellum collaborate to produce creativity and innovation. *Creativity Research Journal, 19*(1), 1–18.

van Dijk, M. W. G., van Geert, P. L. C., Bassano, D., Maillochon, I., Korecky-Kroll, K., Laaha, S., et al. (2012). Dynamic adaptation in child–adult language interaction. *Language Learning, 63*(2), 243–270.

van Geert, P. L. C. (1991). A dynamic systems model of cognitive and language growth. *Psychological Review, 98*, 3–53.

van Geert, P. L. C. (1994). *Dynamic systems of development: Change between complexity and chaos.* New York: Harvester.

van Geert, P. L. C. (1998). A dynamic systems model of basic developmental mechanisms: Piaget, Vygotsky, and beyond. *Psychological Review, 105*, 634–677.

van Geert, P. L. C., Steenbeek, H. W., & van Dijk, M. W. G. (2012). *A dynamic systems and network model of talent and excellence across the lifespan.* Manuscript submitted for publication.

von Károlyi, C. (2001). Visual–spatial strength in dyslexia: Rapid discrimination of impossible figures. *Journal of Learning Disabilities, 34*(4), 380–391.

von Károlyi, C., Winner, E., Gray, W., & Sherman, G. F. (2003). Dyslexia linked to talent: Global visual–spatial ability. *Brain and Language, 85*(3), 427–431.

Vygotsky, L. (1978). *Mind in society: The development of higher psychological processes* (M. Cole & V. John-Steiner, Eds.; S. Scribner & E. Souberman, Trans.). Cambridge, MA: Harvard University Press.

Wai, J., Lubinski, D., & Benbow, C. P. (2009). Spatial ability for STEM domains: Aligning over 50 years of cumulative psychological knowledge solidifies its importance. *Journal of Educational Psychology, 101*(4), 817–835.

Wai, J., Lubinski, D., Benbow, C. P., & Steiger, J. H. (2010). Accomplishment in science, technology, engineering, and mathematics (STEM) and its relation to STEM educational dose: A 25-year longitudinal study. *Journal of Educational Psychology, 102*, 860–871.

Watts, D. J., & Strogatz, S. H. (1998). Collective dynamics of "small-world" networks. *Nature, 393*, 440–442.

Webb, R. M., Lubinski, D., & Benbow, C. P. (2002). Mathematically facile adolescents with math–science aspirations: New perspectives on their educational and vocational development. *Journal of Educational Psychology, 94*(4), 785–794.

Winner, E., von Károlyi, C., Malinsky, D., French, L., Seliger, C., Ross, E., et al. (2001). Dyslexia and visual–spatial talents: Compensation versus deficit model. *Brain and Language, 76*(2), 81–110.

Yakovlev, P. I., & Lecours, A. R. (1967). The myelogenetic cycles of regional maturation of the brain. In A. Minkowsky (Ed.), *Regional development of the brain in early life* (pp. 3–70). Oxford, UK: Blackwell.

Ziegler, A. (2005). The actiotope model of giftedness. In R. J. Sternberg & J. E. Davidson (Eds.), *Conceptions of giftedness* (2nd ed., pp. 411–436). New York: Cambridge University Press.

CHAPTER 12

Dynamics of Motor Learning and Development across the Life Span

Karl M. Newell and Yeou-Teh Liu

The central feature that supports inferences about motor learning and development is the change in movement and action over time. In spite of the common denominator of change, it is the case that the subdomains of motor learning and motor development have, over the years, largely considered the phenomena and theory of movement independently. This division has spawned and sustained the idea of motor learning and motor development as separate fields of study, with all the manifestations that such a view entails.

The traditions in these fields of motor learning and motor development have focused investigation on different categories of action, different age groups of subjects, and different theories of change in movement and action (Newell & van Emmerik, 1990). It is no wonder that they became different fields of study where the twain shall never meet. In motor learning, the emphasis has been on the incremental change in movement outcome of young adults as reflected in the investigation of learning curves (Newell, Mayer-Kress, & Liu, 2006; Newell & Rosenbloom, 1981). In motor development, the emphasis has been on new movement forms as manifest in the emergence of the fundamental movement sequence of infants and young children (Thelen & Smith, 1994; Vereijken & Adolf, 1999). And, the recent emphasis on the loss of motor performance and complexity with aging and disease (Lipsitz & Goldberger, 1992; Vaillancourt & Newell, 2002) has provided additional considerations regarding the question and nature of change in movement and action over the life span.

A central implication of these diverse trends and emphases in motor learning and development is that there are *multiple* processes underlying the change in movement and action over the life span. Thus, the decomposition of the change processes in movement into *one* of the life-span-change-over-time categories of learning, development, and adaptation is fraught with challenges. Similarly, the decomposition of change into the more immediate timescale processes of warm-up, fatigue, attentional deficits, and so on, that influence movement and outcome only adds to the theoretical and experimental challenges of revealing the processes of change across the life span. The traditions of learning theory viewed the change in movement and outcome apparent in learning, transfer, and retention as reflecting primarily the single process of forming a representation for the movement and action (Adams, 1987). This single memorial process created the formation of persistent features of learning as reflected in the change in movement and outcome over practice and time (Hilgard & Bower, 1975).

In our view, a consideration of movement in action across the life span necessarily integrates the multiple emphases to the behavioral change of the relevant change-related subdomains (adaptation, development, learning, aging) because there is common ground in the phenomenological properties of the change that occurs and in the processes that organize the change through the life span. Thus, among other considerations, motor development needs to accommodate, in addition to the emergence of new movement forms, the task-relevant change in movement outcome. And, motor learning research needs to consider, in addition to the change in movement outcome, the emergence of new movement forms in motor skill acquisition.

The introduction of the self-organization construct to research on movement in action (Kugler, Kelso, & Turvey, 1980, 1982), together with the theory and methods of the dynamical systems approach to motor control (Haken, Kelso, & Bunz, 1985; Kelso, 1995), has provided the background for a general framework to consider the question of change in motor learning (Fowler & Turvey, 1978; Newell, Kugler, van Emmerik, & McDonald, 1989; Schöner, 1989; Zanone & Kelso, 1992) and motor development (Thelen & Smith, 1994; Vereijken & Adolph, 1999), together with experimental and analytical strategies to reveal the nature of the processes that support and drive the observed change (Savelsbergh, van der Maas, & van Geert, 1999). In this chapter, we elaborate on one aspect of this emerging framework for motor learning and development based on our epigenetic landscape model, which is constructed as a low-dimensional approximation of a high-dimensional (or infinitely dimensional) complex dynamical system (Mayer-Kress, Liu, & Newell, 2006; Newell, Liu, & Mayer-Kress, 2001, 2003, 2009).

The focus here is the identification of a common set of dynamical principles in the individual pathways of change in movement and outcome across the life span (Molenaar & Newell, 2010). The approach and developmental problem fit squarely within the theoretical framework and focus of this book: namely, developmental systems theory (Ford & Lerner, 1992; Gunar & Thelen, 1989; Spencer, Thomas, & McClelland, 2009; Thelen & Smith, 1994).

Understanding Change in Movement and Action

Movement is a necessary but not sufficient condition for action. *Movement* refers to the motion of the body and limbs produced as a consequence of the spatial and temporal patterns of muscular contractions. Movement can also occur, however, independently of the performer's efforts through the creation of an imbalance of forces acting externally on the body. Thus, actions encompass movement and are distinguished by the goal to which they are directed or by specifying certain criteria to which the performer complies in what he or she does (e.g., sitting, walking). Furthermore, actions are characterized by the intention on the part of the performer to realize a certain goal (Lewis, 1946)—a criterion that rules out, for example, passive movement as part of an action. The robustness of the distinction between action and movement is challenged, however, by the properties of that class of closed skills (Poulton, 1957), wherein the movement pattern is the goal of the task, such as in gymnastics and high-board diving. The open and closed skills notion refers, respectively, to the relative degree of predictability of the environment.

That actions are identified by the goal to which they are directed leads naturally to the index of the outcome score of action being the primary consideration of change in learning. In the acquisition of ontogenetic movement activities, the emphasis has been on the incremental continuous change in performance outcome as reflected in learning curves (Newell & Rosenbloom, 1981). And, in phylogenetic activities, measures of discontinuous change and the emergence of the new movement forms have been primary with even little emphasis on the change in an outcome score in motor development. The distinction between outcome and movement indices to measure the change in motor learning and development has, however, become increasingly less clear-cut through efforts to integrate them both in contemporary studies of movement and action.

This trend is due to the growing theoretical influence of dynamical approaches to the study of motor learning and development and the technical gains in the recording and analysis of movement dynamics, including that of movement forms. On the other hand, these developments in the analysis of movement in action have brought with them an embarrassing array of riches in the sense that there is now the new challenge of rationalizing *which* properties of movement are relevant to the theory of change in motor learning and development. This challenge is only magnified when one also considers the potential for analysis of change in subsystems at other levels of analysis of movement during action, such as in brain activity (e.g., electroencephalogram [EEG], functional magnetic resonance imaging [fMRI]), muscle activity (electromyography [EMG]), and so on. Here we focus our analysis of change in motor learning and development on the behavioral macroscopic movement and outcome properties of action, though there is currently vigorous investigation of brain–behavior relations in movement.

Table 12.1 shows the primacy of three behavioral emphases in the analysis of change in motor learning and development (adapted from Saltzman & Kelso, 1987). One is the outcome of action that has been emphasized by the traditions in psychology and learning theory, in particular (Newell & Rosenbloom, 1981). Another is the movement kinematics (linear and angular displacements, velocities, and accelerations) and kinetics (forces that cause movement) of the movement trajectories of individual joint and body segment degrees of freedom that have been emphasized by the traditions in the biomechanics of movement (Winter, 2009). And, finally, there is the more contemporary interdisciplinary paradigm of coordination dynamics that emphasizes,

TABLE 12.1. Categories of Task Dynamics

Dynamics of change of movement in action
• Task outcome (psychology)
• Individual degrees of freedom (biomechanics)
• Coordination of the degrees of freedom (coordination dynamics)

in motor learning and development, the change in the relations of the movement of body segments to themselves and to the environment (Jirsa & Kelso, 2004; Kelso, 1995; Turvey, 1990).

The integration of these three categories of emphasis in movement and action is critical for advancing our understanding of motor learning and motor development. This interpretation is particularly relevant because there is not a direct mapping between the movement of individual degrees of freedom, their coordination, and the outcome in action. As a consequence, a variety of movements can be generated to complete any one act (Bernstein, 1967) and, by the same token, a variety of movements can be identified as a particular act (Mischel, 1969). Bernstein (1967) first articulated the problem directly by referring to motor skill acquisition as the mastery of the redundant degrees of freedom. *Degrees of freedom* is generally interpreted in a variety of contexts and domains as the parts of a system that are free to vary.

Expressed in a related way (Edelman & Gally, 2001), the degeneracy of the movement system in action is a (for some, *the*) central problem in motor control and the understanding of the change question in motor learning and development. It is the degeneracy of the system that provides the flexibility and adaptability of functional equivalence in movement and action across the life span—that is, the capacity of individuals to realize the same action goal with solutions of different qualitative and/or quantitative properties of movement patterns. The large number of degrees of freedom, per se, of complex systems is therefore not necessarily the central problem, as has generally been interpreted from Bernstein's (1967, 1996) writings, in that they contribute to the degeneracy of the system and its contextual adaptability and flexibility (Latash, 2012; Lipsitz & Goldberger, 1992; Mayer-Kress et al., 2006).

Continuity and Discontinuity of Change in Learning and Development

Change in all of the contextual subdomains of human learning and development, including motor learning and development, has generally been characterized under the encompassing measurement categories of *continuous* and *discontinuous* (Emde & Harmon, 1984). Motor learning research has given more emphasis to continuous change in movement outcome as expressed in the form of learning curves (Newell & Rosenbloom, 1981), whereas motor development research has emphasized discontinuous change through the emergence of new movement forms (Thelen & Smith, 1994). Fitting formal functions to the change in performance outcome over practice time has a history in motor learning research (Lane, 1987; Newell & Rosenbloom, 1981; Snoddy, 1926), but a mathematical account of the discontinuous change in motor development has been much more difficult to capture (Savelsbergh et al., 1999; van Geert, Savelsbergh, & van der Maas, 1999).

The descriptive and explanatory challenge of formulating a common set of principles to encompass the change processes in motor learning and development has been made more difficult by the varied definitions of qualitative change, including the often-used notion of *stage* in learning and development literatures. In motor learning and development *stage* has been used to refer to a particular qualitative state of the system in regard to a single behavior (e.g., standing or walking), a group of behaviors (e.g., sensorimotor) (Piaget, 1952), or the categorical state of a collection of processes evident in the performance of the movement skill (Fitts, 1964). Brainerd (1978) proposed that it is most useful to reserve the notion of stage for an age-related broad collective of behaviors (but see critique by Molenaar, 1986). The focus here on the transitions between movement forms is not interpreted within the construct of stages of learning and/or development.

Figure 12.1a reproduces a schematic from the classic study of infant motor development by Shirley (1931). It captures the relatively invariant order to which the fundamental movement patterns emerge in infancy. This sequence is traditionally based on categorical observation strategies of infant development and reflects discontinuous qualitative change in the dynamics of the basic movement patterns of human action (Goldfield, 1995). The individual variation in the timing of the onset and order of the new qualitative movement forms has also been a focus of developmental analysis. The observation of a sequence to the onset of movement forms in the fetus (Prechtl, 1986) has broadened and enriched our understanding of the early changes in the qualitative and quantitative properties of movement in development.

Figure 12.1b shows a schematic from Muchisky, Gershkoff-Stowe, Cole, and Thelen (1996) that is an elaboration of Waddington's (1957) epigenetic landscape for development. The schematic captures the dimensions of developmental time, the capacity for physical activities (as identified on the axis of the figure by collective variables), and their stability (depth of well). The figure presents an image as metaphor of the continuities and discontinuities of the movement forms of development. It nicely captures the flow of the qualitative and quantitative change in the dynamics of development and the emergence and even dissolution of movement activity patterns (collective variables).

The schematic of Figure 12.1b could be extended in developmental time to include the acquisition of the ontogenetic movement activities beyond infancy and to show the dynamics of change (including those of behavioral decline) across the life span to death. Nevertheless, there are several limitations to the logic and veracity of the developmental schematic as presented (Newell et al., 2003, 2009). A major problem is that the figure shows more than a single activity as stable at any developmental time point. It is not possible for all activities or even smaller subsets of the collective variables to be stable at the same time. Thus, the schematic represents the *potential* of a collective variable to be stable at a developmental time point rather than the actual dynamics of movement that are realized at any moment in time.

(a)

(b)

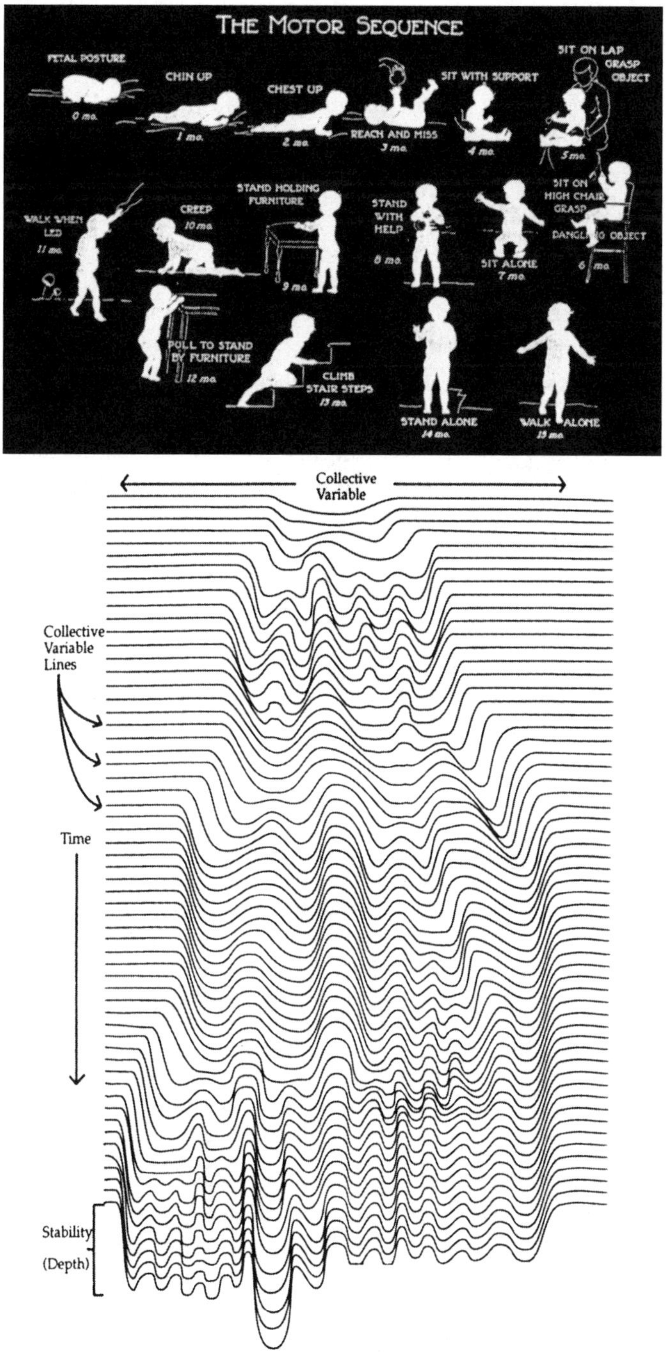

FIGURE 12.1. (a) The infant motor development sequence from Shirley (1931). (b) Multiple timescales of development. This figure depicts an elaborated form of Waddington's landscape. From Muchisky, Gershkoff-Stowe, Cole, and Thelen (1996). Copyright by Elsevier. Adapted by permission.

The acquisition of the so-called *ontogenetic physical activities* in later segments of the life span can also lead to qualitative change in the movement form of an action. Bernstein (1967) outlined a three-stage model of motor skill acquisition that captures the order to these changes in movement form. In Stage 1 the learner freezes the available degrees of freedom to a minimum. In Stage 2 there is a release of the ban on the degrees of freedom so that movements of the body segments are gradually introduced into the movement solution. And, finally, in Stage 3 there is the exploitation of the reactive forces that arise from movement in action. These sequential changes lead to a progression in the movement form of the individual and the performance outcome of action as a function of practice.

Nevertheless, although there is evidence for elements of each of the Bernstein (1967) stages in motor skill acquisition (e.g., Schneider, Zernicke, Schmidt, & Hart, 1989; Vereijken, van Emmerik, Whiting, & Newell, 1992), no general principles have emerged from the pathways of change evident in the learning of new movement patterns (Newell & McDonald, 1994; Newell et al., 2001). Furthermore, and consistent with the principles of parsimony, these changes in the qualitative properties of movement coordination in adults learning ontogenetic activities (e.g., tennis, unicycle riding) do not seem to be fundamentally different from the coordination changes in the phylogenetic activities (e.g., sitting, walking) expressed in infancy, though of course there are differences in ages, activities, and constraints to action (Newell, 1986; Thelen, 1986). This observation invites the application of a common dynamical framework to consider the general principles of change in movement across the life span (Newell et al., 2001; Newell & van Emmerik, 1990).

Figure 12.2 shows the classic learning curves of two adult individual subjects from Bryan and Harter (1897) learning to receive and send Morse code over many weeks of practice. These figures reflect plots of the respective outcome scores—the number of letters sent or received per minute—as a function of weeks of practice. The data points are averaged over trials, but the learning curves are instructive because they show many different rates of change in the performance dynamics. They show the persistent trend of the negatively accelerating rate of the change in performance over the practice sessions, the fluctuations in performance from practice session to session, and even plateaus of no apparent change over segments of practice. Nevertheless, the averaging of data over trials masks some of the other individual rates of change that undoubtedly exist in the sequence of performances over practice time (Bahrick, Fitts, & Briggs, 1957; Liu, Mayer-Kress, & Newell, 2003; Molenaar, 2004).

As noted previously there has been a substantial effort to fit functions to motor learning data such as those of Bryan and Harter (1897). The power law has been traditionally taken as the best-fitting function to the continuous change in the performance outcome data of motor learning (Newell & Rosenbloom, 1981; Snoddy, 1926), although this interpretation has been challenged more recently with emphasis given to exponential functions (Gallistel, Fairhurst, & Balsam, 2004; Heathcote, Brown, & Mewhort, 2000; Newell et al., 2001). Much of this function fitting has been driven by the empirical demands of goodness of fit rather than the theoretical considerations of the different interpretations of mathematical functions for the processes of change. It

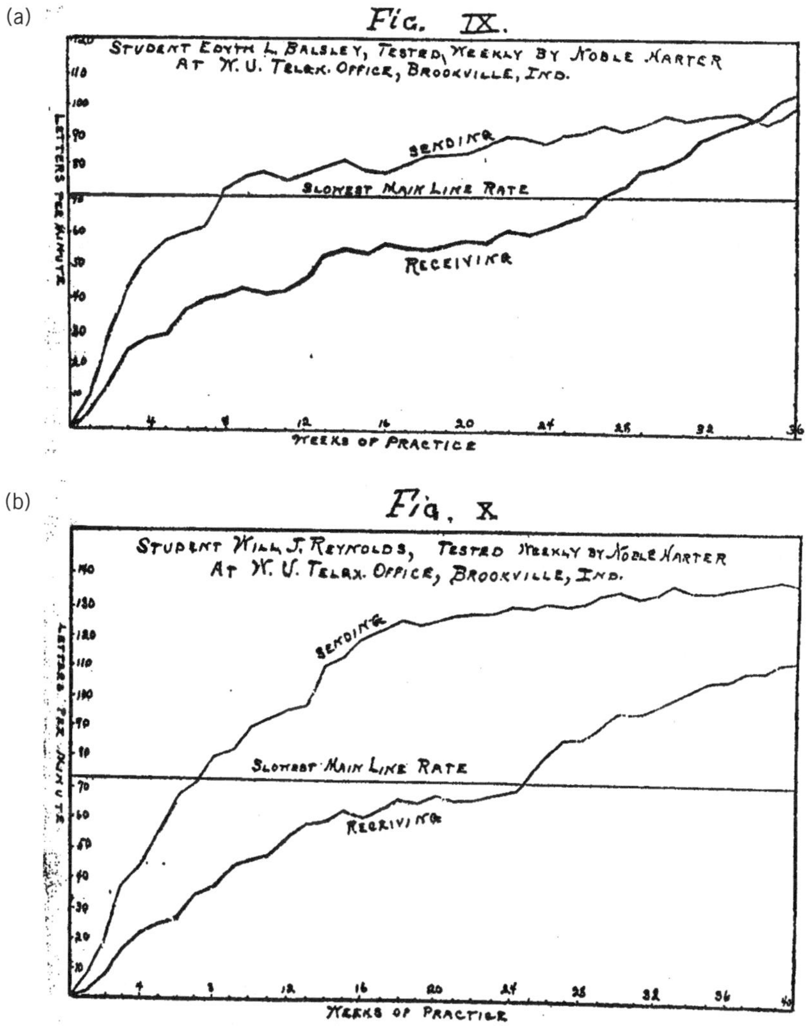

FIGURE 12.2. Data from two participants learning to receive and send Morse code. From Bryan and Harter (1897). In the public domain.

is also the case that a single process interpretation of the persistent trend in the learning curves has dominated, in spite of the multiple rates of change in performance with an inference of multiple processes to motor learning.

The fit of a function such as an exponential or power law can be influenced by the time course of the learning data collected. In contrast to Bryan and Harter's (1897) research, most motor learning studies reported are over practice periods of relatively few trials with a very limited number of days. This method narrows the possibility to observe the range of processes that tend to influence behavior and performance over the life span. One exception to this is the finding of a power law by Crossman (1959) for the change in movement time performance as a function of practice in factory workers rolling cigars over a time period of up to 7½ years and 10 million trials. Unfortunately, the Crossman experiment has each data point in the learning curve as a separate subject, which compromises the strong interpretation of learning as a power law. Nevertheless, it does seem likely that the longer the practice session, the greater the opportunity for additional processes to influence the performance dynamics (Newell et al., 2006), leaving the power law as a special and perhaps idealized case of learning that has rarely been demonstrated on single subjects learning a motor skill.

The respective continuities and discontinuities in learning and motor development have traditionally been treated independently because of the differences in the ages of the subjects, the tasks, and importantly, the theory. The infant motor development sequence provided in differential ways strong support for maturational accounts of development (e.g., Gesell, 1929), whereas the qualitative and quantitative changes in adult motor learning were framed by learning theory perspectives (Adams, 1971; Fitts, 1964; Schmidt, 1975). These traditional distinctions between the contributions of the organism and the environment to motor learning and development have lost their proprietary edges, but their influence is still pervasive in movement studies across the life span.

In contemporary accounts, the qualitative changes in motor learning and development have been taken to reflect the nonlinear processes arising from the principle of self-organization (Kugler et al., 1980, 1982; Savelsbergh et al., 1999). The application of dynamical systems theory to the movement domain has provided models and methods with which to analyze change, with the particular capacity to examine the qualitative change that characterizes movement transitions. These applications have continued to develop and become more influential in both motor learning (Newell et al., 1989; Schöner, 1989; Zanone & Kelso, 1992) and motor development fields (Goldfield, 1995; Thelen & Smith, 1994; van der Maas & Molenaar, 1992).

Toward a Dynamical Theory of Change in Movement and Action

Inferences about motor learning, motor development, aging, and also motor performance are drawn from the change in movement and action over time. An important and central feature in making these inferences is the rate or the timescale of change in outcome and movement properties. As we have noted already, the challenge of understanding this property of change is magnified because there is no single timescale of change in motor learning, development, and aging across the life span.

Indeed, the longer the period of the life span over which observations of change are to be made, the more difficult the challenge to decompose the performance dynamics because it is more likely that there are multiple timescales of influence arising from the multiple processes of learning, development, and adaptation. The traditions of motor learning researchers have tended to focus on the amount of change in the outcome score as an index of learning, with less or little emphasis on the timescale of change (cf. Schmidt & Lee, 2005). This is a convenient trend for applying analysis of variance procedures, but it does not directly approach the dynamics of change in motor learning and development. And, the emphasis on the dynamics of change needs to encompass the integrated framework of the outcome score, the motion of the individual degrees of freedom, and their coordination dynamics.

Timescales

In our epigenetic framework for the dynamics of change in movement and action, we emphasize timescales of change as a principled strategy to decompose the multiple process influences on learning, development, and aging (Newell et al., 2001, 2003, 2009). In our approach, a timescale is not merely the duration or time of an event, as it is in many definitions and interpretations. In short, we view a timescale to be more than the duration of an event in that it is an event that arises from the motions of dynamical processes. Thus, our notion of timescales is grounded in the time arising from dynamical processes.

We build from the position that there are two fundamental types of idealized dynamical systems that give rise naturally to the idea of timescales in living and nonliving systems: (1) periodic oscillations (Haken, 1983) and (2) growth–decay at a constant rate (Kaplan & Glass, 1995). In oscillatory systems the intrinsic timescale of the system is the period that is the inverse of frequency that the system requires to complete a full oscillation. In growth–decay systems the intrinsic timescale is the inverse of the growth–decay rate, which can also be expressed as doubling or half-life time.

Using a complex exponential function to describe a damped oscillation motion that consists of both types of exponential trajectory, a variable x at time t can be expressed as a complex exponential function:

$$x(t) = A\, e^{(\gamma + i\omega)t} \qquad \textbf{(12.1)}$$

Here, A is the position at time $t = 0$, the real part γ of the exponent is the growth–decay rate, and the imaginary part ω is the frequency of the oscillation. The reciprocal of the growth or decay rate is the intrinsic or characteristic timescale. We use this relation to show, in principle, the link between the two types of timescales and emphasize that we are focusing on the growth–decay rate γ for change over time processes.

The oscillatory and growth–decay processes of Equation 12.1 are fundamental to describing behavior close to a fixed point. A *fixed point* corresponds to the absence of motion in a pendulum at rest. A significant outgrowth from this dynamical perspective is that in these systems, the motion close to a fixed point can be approximated to be linear. This allows the motion of the trajectory to be characterized by the exponential growth–decay component γ of Equation 12.1.

This formal position about the rates of growth and decay processes can be elaborated to consider the processes of change in learning, development, and aging. Thus, a learning curve can be characterized by a superposition of exponential processes wherein each function has a unique exponent, given that it reflects the dynamics of a unique process (Newell et al., 2001). This leads to the projection that the multiple timescales of change in performance outcome over time are interpreted to originate from the system's trajectory on an evolving attractor landscape. The improvement of performance in this framework can be visualized and analyzed as the stabilization of behavior at a fixed-point attractor. This approach is most appropriate for modeling the learning of motor tasks wherein the participant can already produce a task-relevant movement coordination pattern (mode) that needs to be scaled appropriately to the space–time or force–time demands of the task (Newell, 1985; Newell et al., 2001). The performance curves on the learning of Morse code shown in Figure 12.2 from Bryan and Harter (1897) are candidates for this kind of modeling of landscape dynamics to a fixed-point attractor, as are most motor learning curves that have been reported in the literature.

Our theoretical and experimental framework can also analyze the transitions of movement that are evident in the emergence of the fundamental movement sequence in infancy and in the tasks that adults undertake wherein the learning of a new coordination movement pattern is required—either when it is the goal of the task or when the pattern leads to the performance outcome that is necessary (Newell et al., 2001). These transitions in the coordination mode tend to lead to more complicated dynamics of performance outcome than is typical in movement scaling tasks that are considered as moving to a fixed point. Furthermore, different bifurcations between attractor organizations and transient phenomena can lead to exponential, power law, or S-shaped learning curves. In our framework, the functions of change at the task level are products of the evolving set of dynamical subsystems at multiple levels of analysis of the organism–environment interaction, each with its own changing timescale.

We have pursued the position that a common theoretical framework with a small set of dynamical principles can realize the multiple timescales of change in motor learning, development, and aging. There has been far less experimental investigation of the movement transitions involved in motor learning and motor development than of the scaling changes of adult motor learning. Our working assumption, however, is to address the transitions of motor development with a coherent framework (Savelsbergh et al., 1999; Thelen & Smith, 1994; Veriejken & Adolph, 1999) as well as those of adults learning or relearning a new coordination mode (Newell & van Emmerik, 1990; Newell et al., 2001). This parsimony of approach to the dynamics of learning and development places emphasis on the understanding of the emerging constraints to movement and action through the life span (Newell, 1986).

Finally, it should be noted that the most common use of Equation 12.1 in the movement domain is through the imaginary part ω, which allows the determination of the frequency spectrum of oscillations of the system by the well-known spectral analysis procedures. The current focus on $1/f$ processes in behavior and physiology has provided formal linkage with, and emphasis to, the idea of multiple timescales (Bassingthwaighte, Liebovitch, & West, 1994; Schroeder, 1991; Ward, 2002) that can be revealed through the decomposition of the structure of movement variability (Davids, Bennett, & Newell, 2006). However, with our interest here in the change over time in processes of learning, development, and aging, we are more focused on the growth and decay rate, γ. The full dynamic analysis of the timescales of change in movement and action across the life span, nevertheless, requires the combined growth–decay and oscillatory components of the complex exponential function.

Timescales of Individual Data

It is important to emphasize that we focus on the timescales of change in individual data from motor learning and development studies. It is not that averaging is not required or of interest, but rather that we need to average data at the appropriate time and in the appropriate way. This point is important because averaging data tends to mask or even create timescales in the performance dynamics that do not exist in any of the data (Liu et al., 2003; Newell et al., 2001).

There are two main ways in which data in motor learning and development are averaged over subjects or over trials. Often movement data are averaged over both subjects and trials (cf. Schmidt & Lee, 2005). Compression of a dataset via averaging is justified (usually implicitly) in most cases on the grounds of simplification of the data and the averaging away of the random-like variations that are evident in single-subject trial-to-trial data. But as we note above, this analysis strategy tends to mask or create false timescales of the change in data over time. Furthermore, a consideration of the stochastic properties of the performance dynamics can be directly approached in our timescale framework to motor learning and development (Liu, Mayer-Kress, & Newell, 1999; Newell et al., 2001); they do not need to be averaged away.

In this regard it is instructive to note that most experimental reports of a power law in human learning have been based on averaged data (Lane, 1987; Newell & Rosenbloom, 1981), and that averaging across a set of exponential functions with different timescale exponents leads to the approximation of a power law (Newell et al., 2001). The inference is that the prevalence of the power law in human learning studies may be due to the artifact of averaging data (Bahrick et al., 1957; Heathcote at al., 2000; Newell et al., 2001), though there are potentially other factors that can contribute to a given function of change. The challenges of appropriately averaging data across subjects and trials that exhibit transitions *and* discontinuities of learning and development are more difficult than averaging over what appear to be continuous data.

Molenaar (2004, 2007) has shown, through the formalisms of ergodic theory, that the standard assumptions about the relations of within- and between-subject variability do not hold unless the data meet a restrictive set of assumptions around homogeneity and stationarity. Typically, the requirements of homogeneity and stationarity are not assessed in research reports, so it is fair to assume that the relevant tests have not been conducted. Given that the constructs of learning and development *presume* change over time, and that experiments are set up to test this kind of change in performance over time, it follows that the related datasets violate the assumption of stationarity. A treatise of these theoretical and technical issues around the significance of intraindividual variability is provided by Molenaar and Nesselroade (Chapter 17, this volume).

Applying the Dynamics of Learning to Scale a Motor Task Outcome to a Point Attractor

There have been many attempts to categorize motor tasks on a variety of dimensions. These categories include posture, locomotion, and manipulation; open and closed skills (Poulton, 1957); posture and movement; and ontogenetic and phylogenetic activities. However, the change in movement and outcome across these task categories tends, as noted earlier, to be either continuous and/or discontinuous. Continuous change is predominantly associated with the learning of tasks where no transition in the coordination mode is required or occurs. Discontinuous change is associated with the learning of motor tasks in which there is a transition or several transitions of the movement coordination mode. These kinds of change often occur through practice and learning on the *same* task, given that coordination, control, and skill are embedded constructs (Newell, 1985).

The study of human motor learning has been dominated by an investigation of motor tasks that do not require the acquisition of a new coordination mode (Newell, 1985; Newell et al., 2001). The traditions of physiology and psychology held that studying "simple" tasks would provide the fundamental principles that could subsequently be built on by the investigation of more "complex" tasks. The definitions of *simple* and *complex* tasks were typically not forthcoming (Wulf & Shea, 2002), but the net result was the study of motor tasks that did not require a transition in the coordination mode or indeed in the more limiting case wherein an effector was constrained to operate as a single biomechanical degree of freedom. This experimental approach led to a prevalence in the study of motor tasks such as line drawing, maze drawing, pursuit rotor, finger posture, and finger motion, together with the dominance of the associated outcome score to study the change in performance over time. The experimental investigation of children's motor learning outside of the infant motor development sequence together with that of older adults tended to follow the same traditions of motor task selection to study the age-related learning of ontogenetic movement activities.

In our dynamical systems approach to motor learning we view the tasks requiring continuous change to a task outcome to reflect motion to a fixed-point attractor (Newell et al., 2001). Given the theoretical assumptions outlined earlier about the dynamical principles of timescales, we can model the change in performance outcome as that of reflecting exponential change to a fixed point. And, based on the assumption of motor learning reflecting multiple processes of change, we can model the performance outcome as a set of superimposed characteristic timescales or exponentials on an attractor landscape.

Thus, our theoretical framework provides a rationale as to why one would fit a particular function, in this case, an exponential or a set of exponentials, to the data. This theoretical approach is not driven alone by "goodness of fit" and "the percent of variance accounted for," which have been the predominant criteria for the assessment of function fitting (Heathcote et al., 2000; Newell & Rosenbloom, 1981). We are strongly driven by the significance of theory for interpreting the qualitative properties of timescales and the degree to which they can accommodate or inform about the data in human motor learning (Thurstone, 1919).

A theory-based approach to the analysis of learning curves is also important in our view because the percent of variance differences in the function fitting of the outcome scores of motor learning tends to be very small—sometimes as low as a 1% difference between, for example, a power law and an exponential (Newell & Rosenbloom, 1981). This has also led us to use other measures of the rate of change properties of learning curves, such as the distributions of proportional change and fat tail tests, to provide converging approaches to the decomposition of the performance dynamics (Liu et al., 2003).

Example 1: Snoddy (1926) Learning Maze Drawing

In Newell, Mayer-Kress, Hong, and Liu (2009) we contrasted the fits of eight models of learning to each of four different motor learning datasets. The eight models represented what we viewed as the primary published candidate models for the motor learning of motion to a point attractor and included two versions of our two-timescale model. The models compared were (1) exponential; (2) power law; (3) Anderson, Fincham, and Douglass (AFD; 1999) power law model; (4) two-power law; (5) five-parameter two-timescale; (6) six-parameter two-timescale; (7) five-parameter two-timescale (with adaptation off during practice); and (8) six-parameter two-timescale (with adaptation off during practice). The fits of these models to the Snoddy dataset are shown in Figure 12.3. We emphasize here the contrast between the AFD model (variant of a power law) and the two-timescale model because these functions clearly fit these four motor learning datasets the best.

The original Snoddy (1926) publication provided the data for our reanalysis of the function of learning that is limited in this historical example to group-averaged data. The dependent variable was a compound property of time and space as reflecting the speed–accuracy nature of the drawing task. The subjects had 20 trials of practice a day for a total of 4 days. The Snoddy experiment is foundational on this issue of learning curves because the study is generally recognized to have been the first to fit motor learning data to a power law, albeit via log–log coordinate paper.

We modeled the data of Snoddy (1926) and the other datasets on the basis of the influence of two processes in the learning of maze drawing task. It was assumed, following the traditions of the learning domain, that one process was the slower, relatively persistent process change of learning. The second process was inferred to be the relatively faster change of warm-up; that is, the transient change at the beginning of a practice session when the system is being tuned to the task demands. Traditionally, warm-up has been interpreted as a more rapidly changing process distinct from that of learning (Adams, 1961; Nacson & Schmidt, 1971), though the modeling of the learning curves has not always held to this distinction, treating the performance data as reflecting the single process of memory strength (e.g., Anderson et al., 1999).

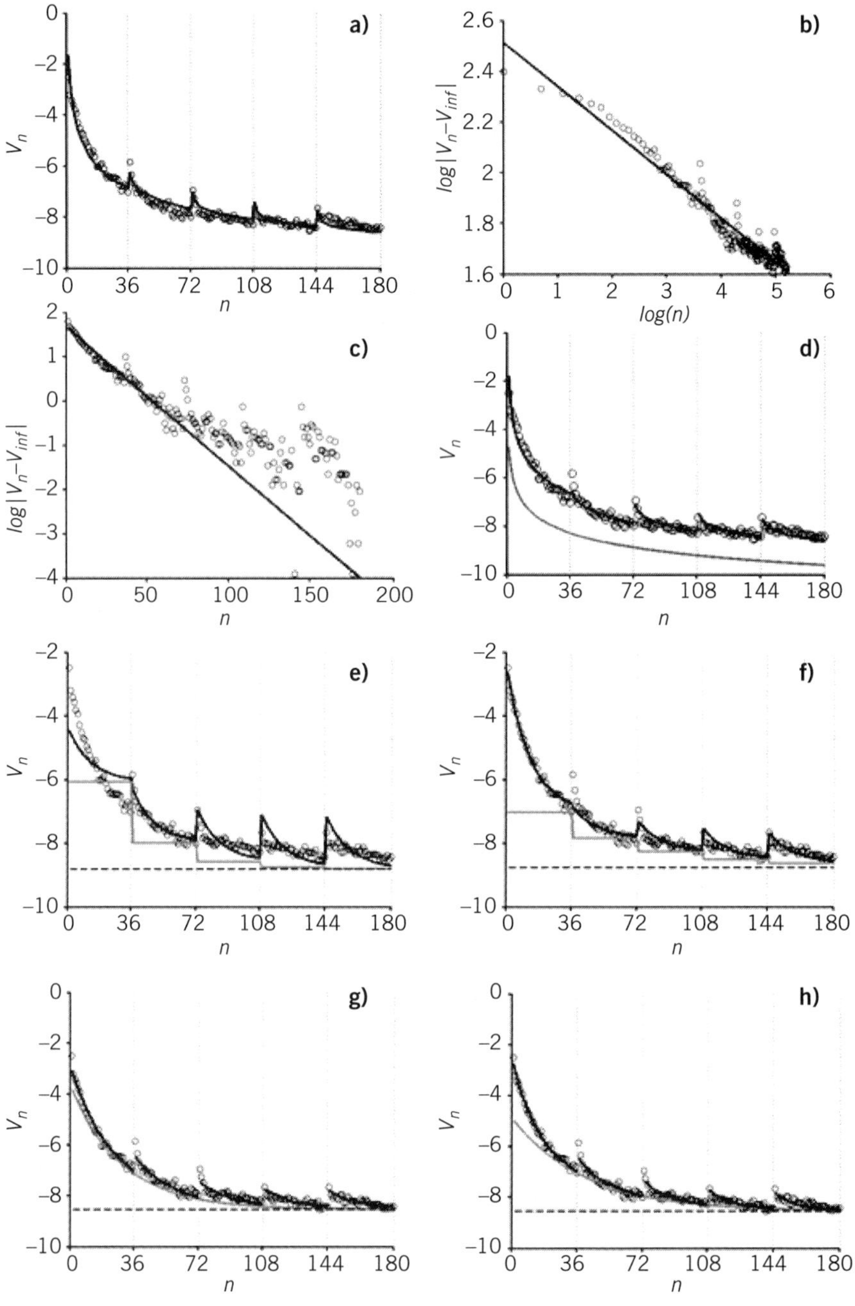

FIGURE 12.3. The eight different models fitted to the Snoddy (1926) dataset: (a) AFD; (b) log–log; (c) log–linear; (d) two-power law; (e) five-parameter two-timescale on–off model; (f) six-parameter two-timescale on–off model; (g) five-parameter two-timescale on–on model; (h) six-parameter two-timescale on–on model. Dashed lines mark V_{inf} value. The unit of V is what we call a Snoddy unit because it is a mixture of a spatial and temporal measure. From Newell, Mayer-Kress, Hong, and Liu (2009).

Previous work has suggested that the timescale of warm-up is more rapid than that of the persistent property of motor learning. Furthermore, evidence shows that the timescale of warm-up is similar from practice session to practice session, even if the initial conditions of performance are less prone to error over time (Adams, 1952, 1961). Thus in Equation 12.2 we have modeled the performance dynamics of motor learning as the superposition of two exponential functions, with different characteristic exponents that are associated with the persistent properties of learning and the transient process of warm-up.

$$V_j(n) = V_{inf} + a_s e^{-\gamma_s n} + a_j e^{-\gamma_j (n - n_{j-1})} \quad \textbf{(12.2)}$$

The model captures a performance variable $V_j(n)$ that converges to an asymptotic target value V_{inf} as time n—measured in units of trial numbers—increases. The slow timescale (s) is taken across all the data, whereas the fast timescale (j) is taken on the trials from each day. Here, j represents the jth practice session (day) and n_j represents the last trial of the session j.

The fits for the two-timescale model held up well in relation to the other models when considered with the Akaike (Burnham & Anderson, 2002) contrast, which takes into account the number of parameters in the equations. The two best-fitting models were the Anderson and colleagues (1999) power law model and the five-parameter two-timescale model that also qualitatively fit the data. The analysis further supported the hypothesis that the timescale of warm-up decrement is consistent across practice sessions, even when the level of performance from the initial conditions is different.

Figure 12.4 shows a superposition of the two exponential functions of the Snoddy (1926) data previously shown in Figure 12.3g (see Appendix in Newell et al., 2009, for a formal account of this analytical process). This is a contour representation with ellipses drawn through values of the regression function taken at the initial and final

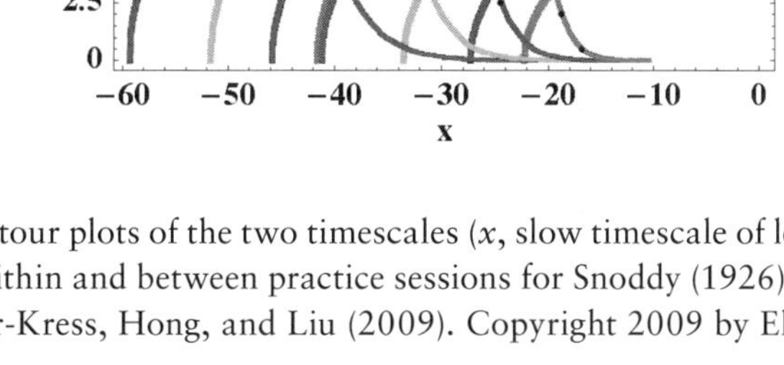

FIGURE 12.4. Contour plots of the two timescales (x, slow timescale of learning; y, fast timescale of warm-up) within and between practice sessions for Snoddy (1926) maze learning data. From Newell, Mayer-Kress, Hong, and Liu (2009). Copyright 2009 by Elsevier. Reprinted by permission.

points of each practice session in a progressive color-coded sequence corresponding to the four consecutive days of practice. Initial points whose slow components are closer to the target converge faster than other initial points on the same contour, even though their fast components are further from the target. Note too that the initial trials of the practice sessions 2–4 are further from the target point on the fast timescale of warm-up, whereas performance is closer to the target on slow timescale of learning. The timescale decomposition affords the analytical separation of the process influences on motor learning and performance.

The landscape representation of the Snoddy (1926) data, with its separation of the persistent timescale from the transient warm-up effects, also provides a strategy to reveal properties of the performance–rest practice effects on learning and performance. For example, Figure 12.4 clearly shows that the amount of change on the slow timescale of learning is greater *between* sessions than it is *during* the practice session itself. This finding emerges because the separation of the influences of the fast and slow timescales reveals the actual practice effects on the slow timescale of learning, uninfluenced by the faster timescale effects of warm-up.

The pattern of findings in Figure 12.4 shows the importance of the rest interval in the progression along the slow timescale to the fixed point and what is called *motor learning*. We have found this effect to be present in other motor learning datasets, including when individual datasets are analyzed (Newell et al., 2009). The timescale decomposition strategy of revealing the impact of multiple processes of learning can be applied to a systematic examination of the effect of time course of practice and rest on motor learning. In short, the timescale contour decomposition provides a behavioral strategy to examine consolidation effects in motor learning and retention through systematic manipulation of the performance and rest cycles and decomposing the influence of the multiple timescales of performance.

We have extended the superimposed two-timescale exponential model to include three processes (learning, warm-up, and fatigue) and an examination of individual learning data (Newell, Mayer-Kress, Hong, & Liu, 2010). In this approach, reflected in Equation 12.3, fatigue is included in the model as an exponential with a timescale that is opposite in direction to the influence of the processes of learning and warm-up. Again, Akaike tests demonstrated that the three-process model accounted for variance commensurate with the addition of parameters. This adaptive process of fatigue has a negative effect on the performance dynamics and, therefore, this process is taken into account as a reversal of performance gains along a third timescale (γ_f).

$$V_j(n) = V_{inf} + a_s e^{-\gamma_s n} + a_j e^{-\gamma_j (n-n_j-1)} + a_j e^{\gamma_f (n-n_j-1)} \qquad \textbf{(12.3)}$$

Keeping in mind that the task variable requires a minimization of the outcome score, the fatigue exponential is a growth process; hence the positive exponent of γ_f represents a movement away from better performance. When fitted to the data, different values are obtained for a_j when $j = 1$ (the first practice session) and for all $j > 1$ (all the practice sessions that are after the first one), similar to the two-timescale model. Parameter values obtained for a_f, γ_f, and γ_j were constant across all values of j.

Figure 12.5 depicts the fit with the two- and three- (bottom segments) timescale models and shows that the additional process of fatigue/inhibition is only present in one of the participants. This observation is supported by the mean square error and Akaike information criteria (AIC) comparisons. Although the addition of the third timescale reduces the *MSe* for the fit of the data from P44, the *MSe* remained unchanged for the fit of P42. Thus, the additional two parameters for the third timescale resulted in a poorer AIC score for the fit of P42, showing that the added parameters were superfluous. This is an example of how learning dynamics can clearly differ from one individual to another. Visual inspection of the data from P42 indicates that this participant continued to improve toward the later trials of the practice session. P44, on the other hand, clearly showed signs of fatigue, as evidenced by the decline in performance during the trials near the end of the session.

This analysis shows that there is agreement between the inferences made based on the dynamical approach and a qualitative visual inspection of the performance data. These individual differences in particular process contributions to the performance dynamics would have been lost in the analysis of averaged learning curves. The timescale approach provides a framework to explore individual differences in performance gains during practice, allowing inferences about different processes of adaptation and learning in the performance dynamics.

In general, as we show here, the characteristic timescale models of Equation 12.2 and Equation 12.3 can be used on individual or group-averaged learning data. We would expect there to be greater variation in the analysis of the learning data of individuals than for group-averaged data. Indeed, this difference in variation is the source of the traditions of averaging data in the study of learning and development. However, averaging data can lead to very different interpretations of the function of change over

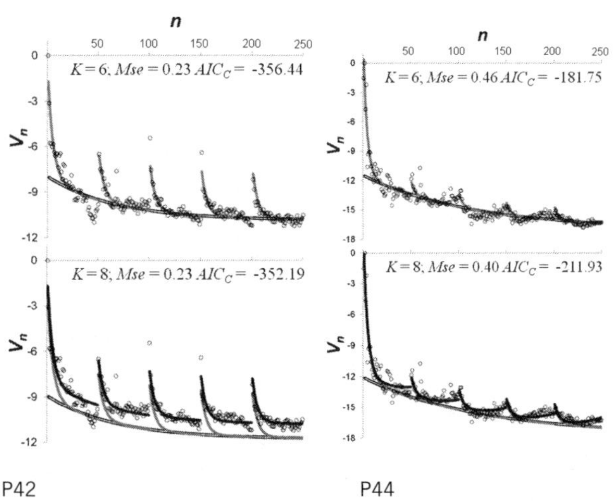

FIGURE 12.5. Data from two participants learning a maze drawing task with two- and three-timescale model fits. Data are from Newell, Mayer-Kress, Hong, and Liu (2010) on a computer graphic board version of the Snoddy (1926) task. AIC, Akaike information criterion.

time in contrast to assessments of individual data. The limits of effectively adding processes to this superimposed model have yet to be realized.

Finally, it is of interest to note that examinations of the timescales of change in scaling tasks have not been conducted across age groups, either at the children's or aging adult end of the age continuum. It is well established that older children tend to learn to perform on the respective task to a higher level than younger children (Keogh & Sugden, 1985) and that there is a tendency for decline in performance outcome on a variety of tasks with aging (Spirduso, 1995). The timescales of these age and/or developmental differences in the motor learning of scaling tasks have not, however, been investigated in a systematic way because the emphasis in age-related contrasts has been on the degree or level of performance difference rather than the rates and timescales of change in movement and action over time.

Dynamics of Learning Movement Transitions

A distinctive property of change in movement performance over time is that some tasks require, whether implicitly or explicitly, a transition(s) in the coordination mode to achieve the task goal (Newell, 1985). This kind of learning particularly takes place in the emergence of the fundamental movement patterns of infancy and the learning of, for example, sitting, standing, and walking, but also tasks like bicycle riding and juggling, where the older child or adult learner is initially unable to perform the task. In these situations, the practice-induced transition of the movement coordination mode can be represented by changes in the attractor landscape (Liu, Mayer-Kress, & Newell, 2006, 2010; Zanone & Kelso, 1992, 1994).

There are considerably fewer experimental examples of the study of transitions in the movement patterns of motor learning and development than there are of learning to rescale a movement pattern. Most studies of transitions in motor development are based on categorical judgments from natural observation of the change in behavior over time (as in the Shirley, 1931, study shown in Figure 12.1a). It is not that these categorizations of the change in behavior are not representative or useful to the study of learning and development. Clearly, they are important, if not essential. Rather, they tend to fail to provide the information about dynamics that would allow direct assessment of the timescales of change in motor learning and development. In the next section, we highlight an example of a movement transition in adult motor learning from our recent experiments (Liu, Mayer-Kress, & Newell, 2006, 2010).

Example 2: Liu et al. (2006)

We have studied a phase transition in motor learning through a task called the *roller-ball* (see Figure 12.6A). The subject holds a plastic ball that encompasses a ball inside of it. The task is to learn to manipulate the outer shell so as to preserve the rotational motion of the inner ball. The successful performance of this task, which is the maintenance of motion of the inner ball, requires the learning of a particular coordination mode: namely, the learning of a phase relation between the direction of motion of the hand on the outer shell and the direction of the motion of the axis of rotation of the inner ball. Without the learning of this coordination pattern, the ball speed tends to drop to zero within a short period of time and a "failure" trial is recorded. We anticipate that the learning dynamics of this task hold similarities to those of learning to ride a bicycle and other activities where a transition to success in performance occurs.

Figure 12.6B–D shows the learning dynamics of individuals representing each of three classes of learner that were observed in this task. Figure 12.6B shows the performance outcome of a learner who produces the transition and increases the ball acceleration beyond the threshold for "success" in the task—indicating that he or she accomplished the transition to the task-relevant coordination mode. Figure 12.6C shows the performance of a learner who increases the acceleration of the ball to some degree but does not produce the transition and a necessary level of acceleration for task success. Figure 12.6D shows a learner who failed to produce any systematic change

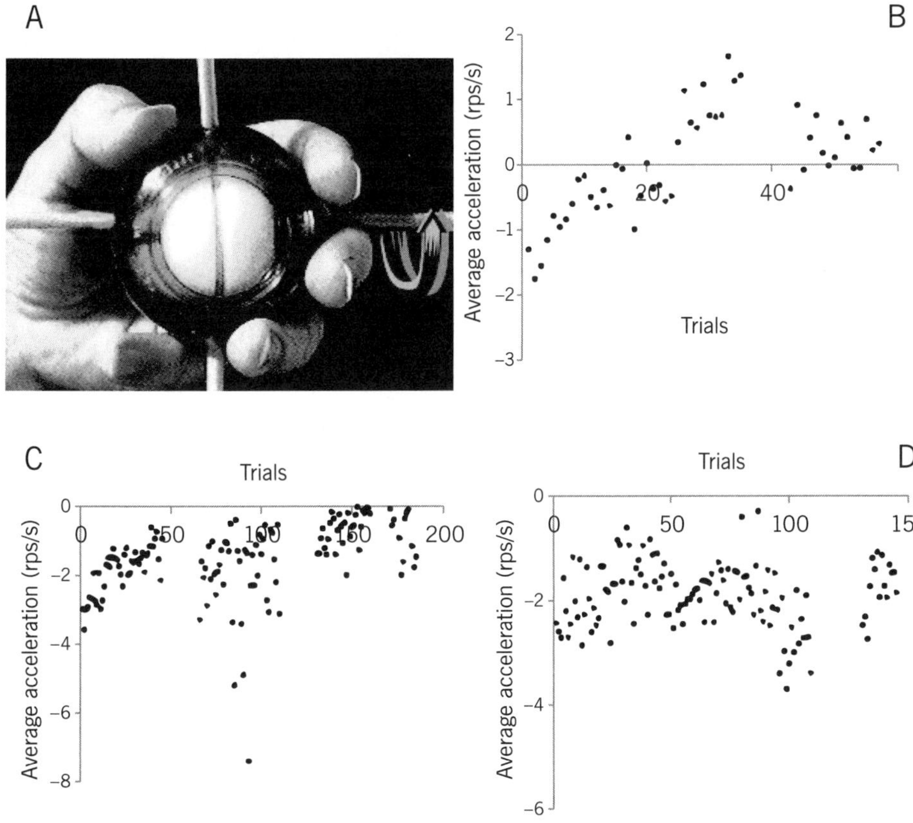

FIGURE 12.6. (A) The rollerball; (B) subject who learned the transition in the task; (C) subject who learned to enhance the scaling of the output; (D) subject who showed no systematic performance improvement over the practice sessions. Adapted from Liu, Mayer-Kress, and Newell (2006). Copyright 2006 by the American Psychological Association. Adapted by permission.

in performance over the three sessions of practice time. These three examples reveal significant individual differences in the timescales of learning the transition to the task-relevant coordination mode. It follows from the earlier discussion on individual pathways of change that averaging across these individual performance differences in learning will mask, rather than reveal, the dynamics of change in both the individuals and group.

These learning curves, particularly that of the learner in Figure 12.6D, do not reflect the change properties of those typically seen in the rescaling of a movement coordination pattern and highlighted in textbooks (Schmidt & Lee, 2005). Indeed, the data in Figure 12.6D seem to show no learning at all over the days of practice—again, a very different pattern of change from that usually shown in motor learning studies. We take this difference as evidence of the fundamental differences in learning a new coordination mode from rescaling a movement pattern that one can already produce. However, we would also propose that the pattern of practice-induced change shown in Figure 12.6D is not uncommon for learning new movement patterns of coordination across the life span.

Learning the transition dynamics of this task over a sufficient period of practice leads to the S-shaped learning curve in the performance outcome. The S-shaped learning function is not seen in its entirety in the learning curves of Figure 12.6 because each participant is at a different position of the S-shaped pathway of change. Thus, we would interpret the data to indicate that the participants are at different positions in making the transition from one stable state to another and in their progress to the emergence of a nonequilibrium phase transition (Liu et al., 2006, 2010).

Figure 12.7 shows an unpublished example from Liu and Newell of a different adult participant learning the rollerball task over 50 trials a day for 5 days. The figure clearly shows a transition between the end of day 3 and the beginning of day 4. The

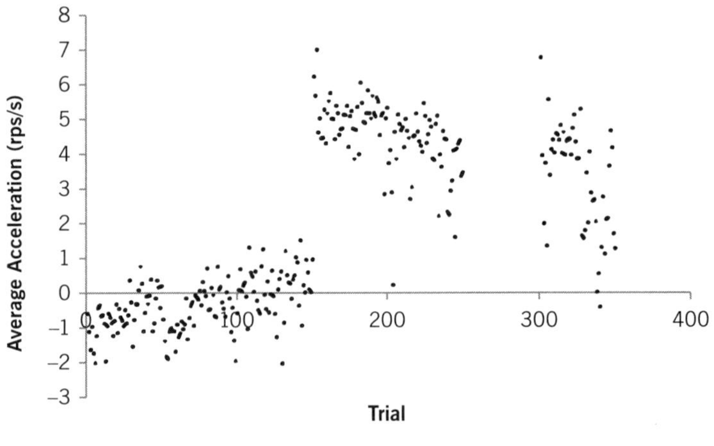

FIGURE 12.7. Data of an individual learning the rollerball task; 50 trials a day for 5 consecutive days with a 1-week interval to the last practice day. Based on unpublished data from Liu and Newell.

S-shaped learning curve is more apparent with the longer practice regimen than in the examples of Figure 12.6. The participant in Figure 12.7 did not sustain, however, the acceleration scaling of the ball at the high performance level after the transition, although the transition to the coordination mode for success was continued, even over the 1 week duration of a retention test. Indices for the detection of a phase transition and discontinuity can be applied to these kinds of data (van Geert et al., 1999). The results of level crossing method analysis (Wagenmakers et al., 2005), performed on the 500 trials of individual performance measures, showed distinct bimodal distributions indicating two stable equilibrium states over the 500 trials.

Zanone and colleagues (Zanone & Kelso, 1992, 1994; Kostrubiec, Zanone, Fuchs, & Kelso, 2012) have shown similar dynamics of change in the learning of a 90-degree phase relation in the two-finger bimanual coordination task. This series of learning studies builds from the landscape of the intrinsic dynamics of bimanual coordination where the in-phase 0-degree and antiphase 180-degree phase relations are intrinsically the two stable states. Practice not only induces a transition in the coordination mode and the evolution of a stable 90-degree relation, but in addition changes the whole finger coordination landscape, including the reduced stability of the 0- and 180-degree relations. Zanone and colleagues have modeled this transition as a nonequilibrium phase transition (Kostrubiec et al., 2012).

The limited study of movement transitions leaves much work to be done to determine the relevance of the dynamical approach to this aspect of the change question in motor learning and development. Given the embedding of the constructs of coordination, control, and skill (Kugler et al., 1980; Newell, 1985), it is the case that the qualitative and quantitative aspects of the dynamics need to be considered as a collective. Nevertheless, the transitions are unique in that they reflect the boundaries of the stability of a coordination mode and the discontinuities in the flow of learning and development over the life span.

Concluding Comments

The studies of learning fixed-point and transition dynamics in movement and action provide support for the proposition that learning dynamics are different in tasks that require the acquisition of a new coordination mode from those that require a rescaling of the task demand within a coordination mode that can be produced by the learner (Newell, 1985; Newell et al., 2001). On the other hand, given sufficient practice and the passage of time, there is likely to always be some change in the properties of the landscape attractor with resultant influence on motor learning and development. The general point, therefore, is that there are qualitative and quantitative changes in motor learning and development, with the relative contribution of the types of change dependent on the continuities and discontinuities of the dynamics that emerge with task demands. The use of the landscape model will be most robust when the timescales of change in the landscape are much longer than those associated with ones given by the eigenvalues of the fixed point (Mayer-Kress et al., 2006).

The framework of the timescales of dynamics provides a coherent set of principles not only to characterize the change in motor learning and development but also to reveal the processes that are contributing to the change. It is our position that there are multiple processes involved in the changes in movement and action over the life span, the timescales of which are continually evolving. Nevertheless, there is growing evidence for the central view presented here: namely, that there is much to be gained from a common theoretical and experimental approach to the changes in motor learning and development, irrespective of age and motor task.

ACKNOWLEDGMENTS

This work was supported in part by Grant No. 0848339 from the National Science Foundation to Karl M. Newell and by "Aim for the Top University Plan" of the National Taiwan Normal University and the Ministry of Education, Taiwan, Republic of China, to Yeou-Teh Liu.

REFERENCES

Adams, J. A. (1952). Warm-up decrement in performance on the pursuit rotor. *American Journal of Psychology, 65*, 404–414.

Adams, J. A. (1961). The second facet of forgetting: A review of warm-up decrement. *Psychological Bulletin, 58*, 257–273.

Adams, J. A. (1971). A closed-loop theory of motor learning. *Journal of Motor Behavior, 3*, 111–150.

Adams, J. A. (1987). Historical review and appraisal of research on learning, retention, and transfer of human motor skills. *Psychological Bulletin, 101*, 41–74.

Anderson, J. R., Fincham, J. M., & Douglass, S. (1999). Practice and retention: A unifying analysis. *Journal of Experimental Psychology: Learning, Memory, and Cognition, 25*, 1120–1136.

Bahrick, H. P., Fitts, P. M., & Briggs, G. E. (1957). Learning curves: Facts or artefacts? *Psychological Bulletin, 54*, 256–268.

Bassingthwaighte, J. B., Liebovitch, L. S., & West, B. J. (1994). *Fractal physiology*. New York: Oxford University Press.

Bernstein, N. A. (1967). *The coordination and regulation of movements*. London: Pergamon Press.

Bernstein, N. A. (1996). On dexterity and its development. In M. L. Latash & M. T. Turvey (Eds.), *Dexterity and its development* (pp. 3–244). Mahwah, NJ: Erlbaum.

Brainerd, C. J. (1978). The stage question in cognitive-development theory. *Brain and Behavioral Sciences, 1*, 173–213.

Bryan, W. L., & Harter, N. (1897). Studies in the physiology and psychology of telegraphic language. *Psychological Review, 4*, 27–53.

Burnham, K. P., & Anderson, D. R. (2002). *Model selection and multimodel inference: A practical information-theoretic approach* (2nd ed.). New York: Springer.

Crossman, E. R. F. W. (1959). A theory of the acquisition of speed–skill. *Ergonomics, 2*, 153–166.

Davids, K., Bennett, S., & Newell, K. M. (Eds.). (2006). *Movement system variability*. Champaign, IL: Human Kinetics.

Edelman, G. M., & Gally, J. (2001). Degeneracy and complexity in biological systems. *Proceedings of the National Academy of Sciences USA, 98*, 13763–13768.

Emde, R. N., & Harmon, R. J. (Eds.). (1984). *Continuities and discontinuities in development*. New York: Plenum Press.

Fitts, P. M. (1964). Perceptual–motor skill learning. In A. W. Melton (Ed.), *Categories of human learning* (pp. 243–285). New York: Academic Press.

Ford, D. H., & Lerner, R. M. (1992). *Developmental systems theory: An integrated approach*. London: Sage.

Fowler, C. A., & Turvey, M. T. (1978). Skill acquisition: An event approach with special reference to searching for the optimum of a function of several variables. In G. E. Stelmach (Ed.), *Information processing in motor control and learning* (pp. 1–40). New York: Academic Press.

Gallistel, C. R., Fairhurst, S., & Balsam, P. (2004). The learning curve: Implications of a quantitative analysis. *Proceedings of the National Academy of Sciences USA, 101*, 13124–13131.

Gesell, A. (1929). Maturation and infant behaviour pattern. *Psychological Review, 36*, 307–319.

Goldfield, E. C. (1995). *Emergent forms: Origins and early development of human action and perception*. Oxford, UK: Oxford University Press.

Gunar, M., & Thelen, E. (Eds.). *Systems in development: The Minnesota Symposium in Child Psychology* (Vol. 22). Hillsdale, NJ: Erlbaum.

Haken, H. (1983). *Synergetics—an introduction: Nonequilibrium phase transitions and self-organization in physics, chemistry, and biology* (3rd ed.). New York: Springer.

Haken, H., Kelso, J. A. S., & Bunz, H. (1985). A theoretical model of phase transitions in human hand movements. *Biological Cybernetics, 51*(5), 347–356.

Heathcote, A., Brown, S., & Mewhort, D. J. K. (2000). The power law repealed: The case for an exponential law of practice. *Psychonomic Bulletin and Review, 7*, 185–207.

Hilgard, E. R., & Bower, G. H. (1975). *Theories of learning*. Englewood Cliffs, NJ: Prentice-Hall.

Jirsa, V. K., & Kelso, J. A. S. (2004). *Coordination dynamics: Issues and trends*. Berlin: Springer.

Kaplan, D., & Glass, L. (1995). *Understanding nonlinear dynamics*. New York: Springer-Verlag.

Kelso, J. A. S. (1995). *Dynamic patterns: The self-organization of brain and behavior*. Cambridge, MA: MIT Press.

Keogh, J., & Sugden, D. (1985). *Movement skill development*. New York: Macmillan.

Kostrubiec, V., Zanone, P. G., Fuchs, A., & Kelso, J. A. (2012). Beyond the blank slate: Routes to learning new coordination patterns depend on the intrinsic dynamics of the learner: Experimental evidence and theoretical model. *Frontiers of Human Neuroscience, 6*, 1–14.

Kugler, P. N., Kelso, J. A. S., & Turvey, M. T. (1980). On the concept of coordinative structures as dissipative structures: I. Theoretical lines of convergence. In G. E. Stelmach & J. Requin (Eds.), *Tutorials in motor behavior* (pp. 1–49). New York: North-Holland.

Kugler, P. N., Kelso, J. A. S., & Turvey, M. T. (1982). On the control and coordination of naturally developing systems. In J. A. S. Kelso & J. E. Clark (Eds.), *The development of movement control and co-ordination* (pp. 5–78). New York: Wiley.

Lane, N. E. (1987). *Skill acquisition rates and patterns: Issues and training implications.* New York: Springer-Verlag.

Latash, M. L. (2012). The bliss (not the problem) of motor abundance (not redundancy). *Experimental Brain Research, 217,* 1–5.

Lewis, C. I. (1946). *An analysis of knowledge and valuation.* La Salle, IL: Open Court.

Lipsitz, L. A., & Goldberger, A. L. (1992). Loss of "complexity" and aging: Potential applications of fractals and chaos theory to senescence. *Journal of American Medical Association, 267,* 1806–1809.

Liu, Y.-T., Mayer-Kress, G., & Newell, K. M. (1999). A piecewise linear map model for the sequential trial strategy of discrete timing tasks. *Acta Psychologica, 103,* 207–228.

Liu, Y.-T., Mayer-Kress, G., & Newell, K. M. (2003). Beyond curve fitting: A dynamical systems account of exponential learning in a discrete positioning task. *Journal of Motor Behavior, 35,* 197–207.

Liu, Y.-T., Mayer-Kress, G., & Newell, K. M. (2006). Qualitative and quantitative change in the dynamics of motor learning. *Journal of Experimental Psychology: Human Perception and Performance, 32,* 380–393.

Liu, Y.-T., Mayer-Kress, G., & Newell, K. M. (2010). Bi-stability of movement coordination as a function of skill level and task difficulty. *Journal of Experimental Psychology: Human Perception and Performance, 36,* 1515–1524.

Mayer-Kress, G., Liu, Y.-T., & Newell, K. M. (2006). Complex systems and human movement. *Complexity, 12,* 40–51.

Mischel, T. (1969). Scientific and philosophical psychology: A historical introduction. In T. Mischel (Ed.), *Human action.* New York: Academic Press.

Molenaar, P. C. M. (1986). Issues with a rule-sampling theory of conservation learning from a structuralist point of view. *Human Development, 29,* 137–144.

Molenaar, P. C. M. (2004). A manifesto on psychology as idiographic science: Bringing the person back into scientific psychology, this time forever. *Measurement, 2,* 201–218.

Molenaar, P. C. M. (2007). On the implications of the classic ergodic theorems: Analysis of developmental processes on an individual level using non-stationary time series modelling. *Developmental Psychobiology, 50,* 60–69.

Molenaar, P. C. M., & Newell, K. M. (2010). *Individual pathways of change: Statistical methods for analyzing learning and development.* Washington, DC: American Psychological Association.

Muchisky, M., Gershkoff-Stowe, L., Cole, E., & Thelen, E. (1996). The epigenetic landscape revisited: A dynamic interpretation. In C. Rovee-Collier (Ed.), *Advances in infancy research* (Vol. 10, pp. 121–159). Norwood, NJ: Ablex.

Nacson, J., & Schmidt, R. A. (1971). Further tests of the activity–set hypothesis for warm-up decrement. *Journal of Experimental Psychology, 90,* 56–64.

Newell, A., & Rosenbloom, P. S. (1981). Mechanisms of skill acquisition and the law of practice. In J. R. Anderson (Ed.), *Cognitive skills and their acquisition* (pp. 1–55). Hillsdale, NJ: Erlbaum.

Newell, K. M. (1985). Coordination, control and skill. In D. Goodman, I. Franks, & R. Wilberg (Eds.), *Differing perspectives in motor learning, memory and control* (pp. 295–317). Amsterdam: North-Holland.

Newell, K. M. (1986). Constraints on the development of coordination. In M. G. Wade & H. T. A. Whiting (Eds.), *Motor skill acquisition in children: Aspects of coordination and control* (pp. 341–360). Amsterdam: Martinies NIJHOS.

Newell, K. M., Kugler, P. N., van Emmerik, R. E. A., & McDonald, P. V. (1989). Search strategies and the acquisition of coordination. In S. A. Wallace (Ed.), *Perspectives on the coordination of movement* (pp. 85–122). Amsterdam: North-Holland.

Newell, K. M., Liu, Y.-T., & Mayer-Kress, G. (2001). Time scales in motor learning and development. *Psychological Review, 108*, 57–82.

Newell, K. M., Liu, Y.-T., & Mayer-Kress, G. (2003). A dynamical systems interpretation of epigenetic landscapes for infant motor development. *Infant Development and Behavior, 26*(4), 449–472.

Newell, K. M., Liu, Y.-T., & Mayer-Kress, G. (2009). Time scales in connectionist and dynamical systems approaches to learning and development. In J. P. Spencer, M. S. C. Thomas, & J. L. McClelland (Eds.), *Toward a unified theory of development?: Connectionism and dynamic systems theory re-considered* (pp. 119–138). New York: Oxford University Press.

Newell, K. M., Mayer-Kress, G., Hong, S. L., & Liu, Y.-T. (2009). Adaptation and learning: Characteristic time scales of performance dynamics. *Human Movement Science, 28*, 655–687.

Newell, K. M., Mayer-Kress, G., Hong, S. L., & Liu, Y.-T. (2010). Decomposing the performance dynamics of learning through time scales. In P. C. M. Molenaar & K. M. Newell (Eds.), *Individual pathways of change in learning and development* (pp. 71–86). Washington, DC: American Psychological Association.

Newell, K. M., Mayer-Kress, G., & Liu, Y.-T. (2006). Human learning: Power laws or multiple characteristic time scales? *Tutorials in Quantitative Methods for Psychology, 2*, 66–76.

Newell, K. M., & McDonald, P. V. (1994). Learning to coordinate redundant biomechanical degrees of freedom. In S. Swinnen, H. Heuer, J. Massion, & P. Casaer (Eds.), *The control and modulation of patterns of inter-limb coordination: A multidisciplinary perspective* (pp. 515–536). New York: Academic Press.

Newell, K. M., & van Emmerik, R. E. A. (1990). Are Gesell's developmental principles general principles for the acquisition of coordination? In J. E. Clark & J. H. Humphrey (Eds.), *Advances in motor development research* (Vol. 3, pp. 143–164). New York: AMS Press.

Piaget, J. (1952). *The origins of intelligence in children*. New York: Norton.

Poulton, E. C. (1957). On prediction in skilled movements. *Psychological Bulletin, 54*, 467–478.

Prechtl, H. F. R. (1986). Prenatal motor development. In M. G. Wade & H. T. A. Whiting (Eds.), *Motor development in children: Aspects of coordination and control* (pp. 53–64). Dordrecht, The Netherlands: Martinus Nijhoff.

Saltzman, E., & Kelso, J. A. S. (1987). Skilled actions: A task dynamic approach. *Psychological Review, 94*, 84–106.

Savelsbergh, G., van der Maas, H., & van Geert, P. (Eds.). (1999). *Non-linear developmental processes*. Amsterdam: Royal Netherlands Academy of Arts and Sciences.

Schmidt, R. A. (1975). A schema theory of discrete motor learning. *Psychological Review, 82*, 225–260.

Schmidt, R. A., & Lee, T. D. (2005). *Motor control and learning: A behavioral emphasis* (5th ed.). Champaign, IL: Human Kinetics.

Schneider, K., Zernicke, R. F., Schmidt, R. A., & Hart, T. J. (1989). Changes in limb dynamics during practice of rapid arm movements. *Journal of Biomechanics, 22*, 805–817.

Schöner, G. (1989). Learning and recall in a dynamic theory of coordination patterns. *Biological Cybernetics, 62*, 39–54.

Schroeder, M. (1991). *Fractals, chaos, power laws*. New York: Freeman.

Shirley, M. M. (1931). *The first two years: A study of twenty-five babies: Vol 1. Postural and locomotor development*. Minneapolis: University of Minnesota Press.

Snoddy, G. S. (1926). Learning and stability: A psychophysiological analysis of a case of motor learning with clinical applications. *Journal of Applied Psychology, 10*, 1–36.

Spencer, J. P., Thomas, M. S. C., & McClelland, J. L. (Eds.). (2009). *Toward a unified theory of development*. Oxford, UK: Oxford University Press.

Spirduso, W. W. (1995). *Physical dimensions of aging*. Champaign, IL: Human Kinetics.

Stratton, S. M., Liu, Y.-T., Hong, S. L., Mayer-Kress, G., & Newell, K. M. (2007). Snoddy (1926) revisited: Time scales of motor learning. *Journal of Motor Behavior, 39*(6), 503–515.

Thelen, E. (1986). Development of coordinated movement: Implications for early human development. In M. G. Wade & H. T. A. Whiting (Eds.), *Motor development in children: Aspects of coordination and control* (pp. 107–124). Dordrecht, The Netherlands: Martinus Hijhoff.

Thelen, E., & Smith, L. B. (1994). *A dynamic systems approach to the development of cognition and action*. Cambridge, MA: MIT Press.

Thurstone, L. L. (1919). The learning curve equation. *Psychological Monographs, 26*(Whole No. 114), 1–51.

Turvey, M. T. (1990). Coordination. *American Psychologist, 45*, 938–953.

Vaillancourt, D. E., & Newell, K. M. (2002). Changing complexity in behavior and physiology through aging and disease. *Neurobiology of Aging: Experimental and Clinical Research, 23*, 1–11.

van der Maas, H. L. J., & Molenaar, P. C. M. (1992). Stagewise cognitive development: An application of catastrophe theory. *Psychological Review, 99*, 395–417.

van Geert, P., Savelsbergh, G., & van der Maas, H. (1999). Transitions and non-linear dynamics in developmental psychology. In G. J. P. Savelsbergh, H. L. J. van der Maas, & P. L. C. van Geert (Eds.), *Non-linear developmental processes*. Amsterdam: Royal Netherlands Academy of Arts and Sciences.

Vereijken, B., & Adolph, K. (1999). Transitions in the development of locomotion. In G. Savelsbergh, H. van der Maas, & P. van Geert (Eds.), *Non-linear developmental processes*. Amsterdam: Royal Netherlands Academy of Arts and Sciences.

Vereijken, B., van Emmerik, R. E. A., Whiting, H. T. A., & Newell, K. M. (1992). Free(z)ing degrees of freedom in skill acquisition. *Journal of Motor Behavior, 24*, 133–142.

Waddington, C. H. (1957). *The strategy of the genes*. London: Unwin & Unwin.

Wagenmakers, E.-J., Molenaar, P. C. M., Grasman, R. P. P. P., Hartelman, P. A. I., & van der Maas, H. L. J. (2005). Transformation invariant stochastic catastrophe theory. *Physica D, 211*, 263–276.

Ward, L. M. (2002). *Dynamical cognitive science*. Cambridge, MA: MIT Press.

Winter, D. A. (2009). *Biomechanics and motor control of human movement* (4th ed.). Hoboken, NJ: Wiley.

Wulf, G., & Shea, C. H. (2002). Principles derived from the study of simple skills do not generalize to complex skill learning. *Psychonomic Bulletin and Review, 9*, 185–211.

Zanone, P. G., & Kelso, J. A. S. (1992). Evolution of behavioral attractors with learning: Nonequilibrium phase transitions. *Journal of Experimental Psychology: Human Perception and Performance, 18*, 403–421.

Zanone, P. G., & Kelso, J. A. S. (1994). The coordination dynamics of learning: Theoretical structure and experimental agenda. In S. P. Swinnen, H. Heuer, J. Massion, & P. Casaer (Eds.), *Interlimb coordination: Neural, dynamical, and cognitive constraints* (pp. 461–490). San Diego, CA: Academic Press.

PART VI

DYNAMICS OF SOCIAL INTERACTION

CHAPTER 13

Differential Equations for Evaluating Theoretical Models of Dyadic Interactions

Emilio Ferrer and Joel Steele

Theoretical perspectives on close relationships emphasize the interreliant nature of close dyads, in which each member of the dyad influences the other's thoughts, feelings, and behaviors (e.g., Thomas & Martin, 1976). For example, interdependence theory (Kelley, 1983; Kelley & Thibaut, 1978) stresses that close relationships are characterized by two chains of causally interconnected events, one for each dyadic actor. Attachment theory (Bowlby, 1969/1982, 1973, 1980), in turn, posits that infants develop relatively stable attachment patterns from their founding interactions with their caregivers. Although these theoretical models offer detailed accounts of processes, they are not accompanied by a parallel development in applicable analytic models that can evaluate the hypotheses empirically. Because of this mismatch, it is not always easy to consider the fundamental fact that intimate pairs form an interdependent, evolving system with mutual influences over time.

In this chapter we illustrate differential equation models (DEMs) as a technique suited to examine the interdependence of dyadic members over time. First, we describe theoretical models developed for dyadic interactions. We then apply differential equations representing such models to daily data on affect from individuals in couples. Third, we evaluate the parameter estimates from each model and compare across the different specifications. Finally, we examine the predictive validity of the models by using their corresponding estimates to predict the couples' relationship quality and status 1 and 2 years later.

Models for Dyadic Interactions

Although multiple theoretical accounts exist to describe interactions of two individuals in a system, not many of these accounts detail the dynamics of such interactions, explicitly describing the interrelationships of the individuals' trajectories, as they unfold over time. We examine here some of these explicit models of dyadic interactions. Whereas most of such models come from empirical work, others are merely theoretical. Yet, all the specifications are written using either differential or difference equations, thus lending themselves to precise predictions of the dyadic behavior over time.

Dyadic Interactions Based on Individuals' Reference Behaviors

Felmlee and Greenberg (1999; Felmlee, 2007) developed a model for dyadic interactions using the premise that members of intimate couples form an interactive system in which the behavior of each person of a couple is influenced by the other person's behavior as well as by goals that each person has for him- or herself. To develop this model, they used differential equations, implying that change in the dyad's behavior occurs in a continuous fashion.

Felmlee and Greenberg's (1999) model can be written as a system of linear differential equations, as

$$\frac{dx}{dt} = a_1 \cdot (x^* - x_t) + a_2 \cdot (y_t - x_t)$$

and

$$\frac{dy}{dt} = b_1 \cdot (y^* - y_t) + b_2 \cdot (x_t - y_t) \qquad \textbf{(13.1)}$$

where x and y represent the behavior—or any other dependent variable—of two actors measured at any time t; a and b are parameters to be estimated from the data; and x^* and y^* refer to ideal types of behavior (e.g., goals) or states (e.g., equilibrium point) to which each individual tends. According to this specification, changes in each person's behavior over time are a function of self (i.e., first term) as well as other person (i.e., second term). In Felmlee and Greenberg's original formulation, the individual factors consider how far each person's current behavior deviates from an ideal behavior, quantified by the coefficients a_1 and b_1 for persons x and y, respectively. The partnered influences consider how far the person's current behavior is from the other person's behavior, and this is quantified by the coefficients a_2 and b_2 for persons x and y, respectively. If all the coefficients are positive, the model represents a dyadic system in which both members are cooperative, as each individual attempts to approach his or her goals as well as the partner's behavior. Under these conditions, the system approaches equilibrium, representing a compromise position between one's goals and the partner's behavior (for a detailed discussion of the analytic equilibrium solution for the model and its derivation, see Felmlee & Greenberg, 1999).

This general specification can accommodate a number of variations. Based on parameter restrictions and valences, this expression can result in different specifications with different predictive behavior of the system over time (Felmlee, 2007; Ferrer & Steele, 2012). Some of these possible long-term patterns of interactions include, for example, smooth convergence of the individuals' trajectories over time, exponential departure of such trajectories, or synchronous cyclical patterns. In previous work, we have applied this model (and its derivations) to empirical data, both self-reports (Ferrer & Steele, 2012; Steele, Ferrer, & Nesselroade, in press) and psychophysiological signals (Ferrer & Helm, 2013).

The Equations of Marriage

One of the most influential researchers of interpersonal relationships and marital interactions is John Gottman (e.g., Gottman, 1994; Gottman & Levenson, 2000; Gottman, Murray, Swanson, Tyson, & Swanson, 2002; Gottman & Notarius, 2000; Gottman, Swanson, & Murray, 1999). Of particular interest here is a nonlinear dynamic system proposed by Gottman and Murray to study the interaction between spouses. This model, labeled "The Equations of Marriage," describes an interacting system in which each individual's behavior at a given time is a function of his or her own dynamics and the interaction with the spouse. This model can be expressed as

$$\begin{aligned} W_{t+1} &= \alpha_{0w} + \alpha_{1w} \cdot W_t + I_W (H_t) \\ H_{t+1} &= \beta_{0h} + \beta_{1h} \cdot H_t + I_H (W_t) \end{aligned} \quad \textbf{(13.2)}$$

This system of equations represents the behavior of the wife, W, and husband, H, at time $t + 1$ as a function of two components. The first two terms in each equation, α_{0w} and α_{1w}, and β_{0h} and β_{1h}, for the wife and husband, respectively, depict an uninfluenced steady state. This is a baseline behavior that is not affected by the partner. The last term in each equation represents the influence of the husband on the wife ($I_W (H_t)$) and the influence of the wife on the husband ($I_H (W_t)$).

In their book *The Mathematics of Marriage*, Gottman and colleagues (2002) examine the predictive behavior of this dynamic system and apply the model to empirical data from couples, investigating the use of various influence functions for studying marital conflict. One important general finding of this work is that conflict-laden marital discussions are, in part, a function of the couples' influence processes. More recently, Hamaker and colleagues (Hamaker, Zhang, & van der Maas, 2009; Madhyastha, Hamaker, & Gottman, 2011) have examined this same system of equations using an alternative estimation procedure based on threshold autoregressive models. Interestingly, their findings indicate that, for many couples, the outcome of conflict discussion is explained by each person's own emotional dynamics and not by the partner's influence.

Dyadic Interactions Based on Appraisals and Imitation

Also derived from theory is the dynamic system formulated by Steenbeek and van Geert (2007) to describe dyadic interactions in children's social development. First, they developed a theory of dyadic interactions based on the concepts of *concern* (i.e., intentions, goals, and interests), *appraisal*, and *contagiousness*. From this theory, they formulated a dynamic system that expresses child–parent interactions. Steenbeek and van Geert's model is expressed in a set of coupled equations, as

$$\Delta I/\Delta_t = a(P_I - I_t) + b(P_I - Y_t) + e(Y_t - I_t)$$

and

$$\Delta Y/\Delta_t = c(P_Y - Y_t) + d(P_Y - I_t) + f(Y_t - I_t) \tag{13.3}$$

where I represents the behavior of a child (child's involvement in the original work) at a given occasion, Y represents the behavior of an interaction partner, and P denotes the preferred level of behavior.

According to these coupled equations, changes in a child's behavior are a function of three components. The first term, quantified by the parameter a (or c for the partner), represents how fast a preferred level of behavior can be achieved. The second term represents the discrepancies between the child's preferred level of behavior and the partner's involvement behavior. This process is quantified by the parameter b (or d, for the partner), which expresses the extent to which the difference between the two interacting partners affects the corresponding child's behavior. The third term is meant to include the possibility of imitation (or contagiousness) of the partner's behavior. In this term, the parameters e and f, for the child and partner, respectively, represent how fast each individual adapts to the other's behavior. For example, the authors contend that in a situation where the partner tends to engage in solitary activities, the child will tend to increase his or her own solitary activity level so as to match that of the partner.

To illustrate the use of this dynamic system, Steenbeek and van Geert (2007) included an empirical application related to dyadic play between children. Using children with varying sociometric status, they generated predictions related to social competence and development. In addition to highlighting the benefits of models for short-term dynamics like this one, they emphasize the need to complement such models with those of long-term dynamics.

Dyadic Interactions Based on Competition

Many models for dyadic interaction were developed in population biology to examine the competition and cooperation among species. Perhaps the most classic model of interacting populations is the Lotka–Volterra predator–prey model (Lotka, 1925; Volterra, 1926). This model captures the interaction between a predator and a prey population; it is written as

$$\frac{dx}{dt} = r_x \cdot x_t - a \cdot (x_t \cdot y_t)$$

and

$$\frac{dy}{dt} = -r_y \cdot y_t + b \cdot (x_t \cdot y_t) \tag{13.4}$$

In the classical predator–prey specification, x and y represent the respective densities of a prey and a predator population, r is the growth (for prey population) or death rate (for predator population), and a and b represent the interspecies interaction parameters. The first term on the right-hand side of the equation for x implies that the prey will grow exponentially in the absence of a predator. The right-hand side of the equation for y, in contrast, implies that the predator population decreases exponentially in the absence of prey. As a consequence of this specification, the interaction between the predator and prey populations is hypothesized to lead to a decrease in the prey population and a proliferation of the predator population, as regulated by the parameters a and b, respectively.

Since its original development, the predator–prey model has been used extensively in population ecology. Its use in the social and behavioral sciences, however, is scarce. In one of the exceptions, Chow, Ferrer, and Nesselroade (2007) applied a predator–prey model to examine the daily emotional interactions of two married people over the course of 6 months. Using a Kalman filter technique to estimate the model parameters, they were able to evaluate hypotheses related to the directionality of the relationship. In particular, a model in which the wife's positive affect was the prey and the husband's negative affect was the predator was more tenable than its opposite (i.e., the husband's positive affect as prey and the wife's negative affect as the predator).

Dyadic Interactions Based on Mutualism

Another important model in population biology to examine the competition and cooperation among species is a model of mutualism (Hirsch, 1982, 1985; Kot, 2001). Of particular interest in human interactions is a model of facultative mutualism (Kot, 2001), wherein the interaction between the dyad members is helpful but not essential. A dyadic interaction model based on this premise can be expressed as

$$\frac{dx}{dt} = \frac{r_1}{K_1} \cdot x \cdot (K_1 - x + a \cdot y)$$

and

$$\frac{dy}{dt} = \frac{r_2}{K_2} \cdot y \cdot (K_2 - y + b \cdot x) \tag{13.5}$$

where r represents the growth rate, K represents the density, and the parameters a and b measure the strength of the influence of individual Y on X (a) and individual X on Y (b). As expressed, Equation 13.5 is a model of facultative mutualism as long as the rate and density parameters are greater than zero (Kot, 2001). That is, under the conditions

$[r_1 > 0, r_2 > 0, K_1 > 0, K_2 > 0]$, the behavior of each person can exist without the partner's influences. Although widely used in mathematical ecology, this model has not been used in social research. Its parameters, however, make it particularly useful for examining dyadic interactions in the social and behavioral sciences, in which the behavior of the dyad members can typically coexist without explicit influences. In this sense, this model appears to be more realistic for dyadic interactions than other models such as the classic predator–prey model, which was developed to examine competition between species.

The models just described are not intended to serve as an exhaustive account of models for dyadic interactions. Rather, they were selected as examples of combining theory and mathematical formulations. Of course, there are many other models that could be used—or have been developed—to study dyadic and social interactions. For example, Boker and Laurenceau's (2006) coupled oscillator model, which examines intimacy and disclosure in married couples, is worth mentioning. Similar cases, such as models developed to study turn-taking in conversations (Buder, 1991; Newtson, 1993), affect coregulation between romantic partners (Butner, Diamond, & Hicks, 2007) and, more broadly, social interactions (Durlauf & Ioannides, 2010).

What the selected models have in common is separate components to account for self-regulatory and coregulatory mechanisms over time. That is, at least one term in each of the models expresses the changes in behavior for each of the dyadic members as a function of the individual's own dynamics. At least one other term of the equation relates such changes to the interaction between both individuals in the dyad (e.g., see Thomas & Martin, 1976).

Using alternative expressions of the previous equations, important similarities among the proposed models are apparent (see Table 13.1). One similarity, in particular, is the expression of the individual dynamics for the first three models. The first term in these models (a_1x^*, α_{0w}, and aP_I, respectively) is a constant that, depending on the theoretical specification, represents ideal or preferred levels of the construct. For all of the models, the individual dynamics are determined, partly or entirely (as is the case for the fourth model), by current levels of the construct. This is expressed as a_1x_t, $\alpha_{1w}W_t$, aI_t, $\pm rx_t$, and rx_t, for the models listed, respectively. This term translates into exponential growth in the observed trajectories once these equations are integrated. In

TABLE 13.1. Mathematical Expression of Proposed Models

	Outcome	Individual dynamics	Partner dynamics
Reference affect	dx/dt	$a_1x^* - a_1x_t$	$b_1y_t - b_1x_t$
Marriage equations	W_{t+1}	$\alpha_{0w} + \alpha_{1w}W_t$	$I_W(H_t)$
Appraisals and imitation	$\Delta I/\Delta t$	$aP_I - aI_t$	$bP_I - bY_t + eY_t - eI_t$
Lotka–Volterra	dx/dt	$\pm rx_t$	$\pm a(x_t y_t)$
Mutualism	dx/dt	$rx_t - rx_t^2/k$	$rx_t ay_t /k$

the fight model, such growth is limited by a scaled quadratic term that determines the point when the peak growth is reached. The fifth model is the only model to include such limiting of the individual dynamics dependent on individual level of the construct. For the other models, the constant can serve as an asymptote of sorts, depending on the sign of the other parameters.

Where the models differ most is in the specification of the partner dynamics. This influence is operationalized as independent from individual levels (i.e., $I_w(H_t)$ for the second model) or as composed of scaled multiplicative effects (i.e., $\pm\, a(x_t\, y_t)$ and $rx_t\, ay_t /k$, for the fourth and fifth models, respectively). The first and third models express the partner influence as a function of discrepancies among individual levels, ideal levels, and levels of the partner. It is with the specification of the partner influences that differences in theoretical orientation toward dyadic interactions emerge most saliently.

To formally evaluate these models, at least two features are necessary. The first is an intense set of measurements that reflects the dyad's fluctuations over time as well as the time dependency of those fluctuations. The second component is an analytic procedure that can accurately and reliably capture such dynamics. In this chapter, we combine these two features to examine the models previously described. We apply each model to empirical time series data of affect from couples and examine its validity in the context of affective processes.

Method

Participants

All the analyses used in this chapter were performed using data from the Dynamics of Dyadic Interactions Project (DDIP), a project focused on the development of models to analyze dyadic interactions (e.g., see Ferrer & Steele, 2102; Ferrer, Steele, & Hsieh, 2012; Ferrer & Widaman, 2008). Participants in the DDIP include couples involved in a romantic relationship. As part of the overall project, all participants were asked to complete a daily questionnaire about their affect for up to 90 consecutive days. In this report we present data from 197 couples who had from 52 days to 108 days of data (M = 70 days, SD = 17). The age of participants ranged from 19.1 to 74.11 years (M = 34.5; SD = 12.6). The time that they had been involved in the relationship ranged from 0.8 to 35.1 years (M = 9.80; SD = 9.31). Of the 197 dyads, four (2.0%) were dating casually, 160 (81%) were living together, six (3%) were engaged, and 28 (14%) were married.

Measures

Relationship-Specific Affect

The daily questionnaire was intended to examine day-to-day fluctuations in affect. As part of this questionnaire, a set of 18 items was intended to tap into the participants' positive and negative emotional experiences specific to their relationship. Participants were asked to complete these items by responding to the instructions "Indicate to what extent you have felt this way about your relationship today." The nine positive items included *emotionally intimate, trusted, committed, physically intimate, free, loved, happy, loving,* and *socially supported*, whereas the nine negative mood items included *sad, blue, trapped, argumentative, discouraged, doubtful, lonely, angry,* and *deceived.* For all items, participants were asked to respond using a 5-point Likert-type scale ranging from 1 (very slightly or not at all) to 5 (extremely). To examine the precision of the measurement of systematic change in persons across days, we computed the reliability of change within person using generalizability analysis (Cranford et al., 2006). The resulting reliability coefficients for positive and negative affect were .85 and .87 (for females) and .82 and .85 (for males).

For all subsequent analyses, we created unit–weight composites for positive and negative affect, for each person, using all the items in the scales. We then constructed a ratio of positive affect in relation to all affect [*positive* / (*positive* + *negative*)]. Our choice was motivated by the balanced state of mind (BSOM; Schwartz, 1997) model, an integrative model of positive and negative affect. Research using this approach has shown that different ratios can describe various states of affect, ranging from psychopathological, to normal, and to optimal states. In particular, ratios of about .70–.80 have been found to indicate normal and optimal functioning in individuals (Schwartz et al., 2002) and couples (Gottman, 1994). Table 13.2 reports the descriptive statistics of the positive and negative affect composites as well as the affect ratios for both females and males.

Relationship Satisfaction and Breakup

Between 1 and 2 years after the initial visit, couples returned to the laboratory for a follow-up assessment. As part of this visit, participants completed a questionnaire that included six items from the Perceived Relationship Quality Component Inventory (PRQC; Fletcher, Simpson, & Thomas, 2000). These items asked couples about the quality of their relationship and required responses using a 7-point Likert scale (1 =

TABLE 13.2. Descriptive Statistics of Positive and Negative Affect

	Mean	*SD*	Minimum	Maximum
Affect scales				
Female positive	3.502	0.902	1.000	5.000
Female negative	1.337	0.548	1.000	5.000
Male positive	3.528	0.846	1.000	5.000
Male negative	1.329	0.515	1.000	5.000
Affect ratios				
Female ratio	0.717	0.115	0.167	0.833
Male ratio	0.722	0.103	0.167	0.833

not at all and 7 = *extremely*). Reliability estimates of the composite scores were .92 for females and .95 for males. Participants also completed information about the status of their relationship. If contacted participants indicated that they were no longer together with their partner, they were classified as broken up.

Data Analysis: DEMs

To examine all the models previously described, we used DEM. DEMs are particularly useful as a technique for examining dyadic interactions, as these models explicitly consider the two members of a dyad as an interdependent system. Moreover, they directly model the changes in the system—or each of its units—as a continuous process.

There are multiple procedures for estimating DEMs. One of the most traditional methods is to pool cross-section and time series data and use weighted generalized least squares. Another approach involves the use of filtering procedures such as the Kalman filter (e.g., Julier, Uhlmann, & Durrant-Whyte, 1995; for examples using this technique, see Chow et al., 2007; Hamaker et al., 2005). There are also approaches for fitting linear and nonlinear dynamical system models directly to the data. One such toolbox is the package ReBEL in Matlab (Recursive Bayesian Estimation Library; van der Merwe, 2003). Other programs such as R and Statistical Analysis Software (SAS) include procedures for ordinary differential equations. For our analyses, we used the procedure MODEL in SAS.

MODEL is a procedure to analyze models in which the relations among the variables comprise a system of one or more nonlinear equations. Some of the principal uses of the procedure include estimation, simulation, and forecasting of simultaneous equation models. This procedure includes various possibilities with regard to methods for parameter estimation, including ordinary least squares, two-stage least squares, seemingly unrelated regression and iterative seemingly unrelated regression, and three-stage least squares. For the analyses reported here, we used full information maximum likelihood (FIML). FIML assumes that the equation errors have a multivariate normal distribution, which is not unreasonable in these data, even if our proposed model is deterministic. Furthermore, FIML helps to deal with the so-called *simultaneous equation bias problem* inherent in systems of equations that are dependent. Cross-dependency in these equations leads to a violation of independence in the errors (Pindyck & Rubinfeld, 1981). FIML is known to overcome, or at least reasonably attenuate, this bias (Amemiya, 1977; Gallant & Holly, 1980). In addition to the estimation features, the MODEL procedure can be used to forecast. When the specified model has variables that enter linearly, forecasting can be done simply by solving the equations. When the variables enter the model in a nonlinear way, an iterative procedure can be used by invoking the subcommand SOLVE in the MODEL procedure.

In our own research, we have utilized this estimation procedure using both daily self-report data (Ferrer & Steele, 2013; Steele et al., in press) and physiological signals (Ferrer & Helm, 2012). Appendix 13.1 includes sample scripts of PROC MODEL for all our models.

Results

Fitting Theoretical Models of Dyadic Interactions

In the first set of analyses, we fitted each of the dynamic systems models to our daily data. For this, we fitted each of the different models to each dyad separately and then compiled the parameter estimates across all dyads to build empirical distributions. In cases when a model did not converge properly, the parameter estimates were not included. In addition, parameter estimates or standard errors of unrealistic values were excluded.

The first model we examined was Felmlee and Greenberg's (1999). As described previously, this model includes values representing "ideal" behavior for both individuals in the system (x^* and y^*, see Equation 13.1). In our analyses, we used .70 as the ideal value of affect. We chose this value to represent meaningful ratios of positive to negative affect, as theoretically determined by Schwartz (1997). Alternative values for these parameters are also possible. For example, they can be specified as the mean of the series or, alternatively, estimated from the data (see Steele et al., in press).

Table 13.3 presents the parameter estimates from this model. The first finding to notice is that the estimates from parameters representing self-regulation (a_1 = 0.780 and b_1 = 0.675) are larger than those representing coregulation (a_2 = 0.223 and b_2 = 0.109), indicating that, on average, daily changes in affect for individuals in these couples are more influenced by their own emotional dynamics (fluctuations toward an emotional set point) than by those of their partner. A second noticeable feature, however, is the large range of values of all estimates. Thus, whereas for some couples the estimates were close to zero, for others they were large and negative, and yet for others, large and positive. Finally, the model converged successfully for 141 of the 197 couples.

The next model that we fitted was Gottman's dynamic system representing the mathematics of marriage. As was the case for the previous analyses, we fitted this model to each dyad's data separately. In addition, we fitted it as a differential equation without the intercept in order to detrend the data and focus on the fluctuations only. The results from these analyses are in the second set of rows in Table 13.3. To facilitate comparisons across models, the parameters shown here differ from the original specification. Parameters a_1 and a_2 in Table 13.3 stand for α_{1w} and I_W in the original notation (see Equation 13.2). Similarly, b_1 and b_2 stand for β_{0h} and I_H. Thus, parameters a_1 and b_1 represent individual dynamics (affect inertia), whereas a_2 and b_2 represent dynamics of the interaction with the partner, for the female and male, respectively.

As was the case for the previous model, all parameter estimates show a large range, indicating great variability in the estimates across dyads. Interestingly, the estimates representing partners' influences (a_2 and b_2) show larger values than those reported in the previous model. This model specification converged successfully for 158 of the 197 available couples.

TABLE 13.3. Descriptive Statistics for Parameter Estimates across Models

	Mean	*SD*	Minimum	Maximum
Reference affect (N = 141)				
a_1	0.780	0.716	−0.680	3.143
a_2	0.223	0.756	−1.702	3.317
b_1	0.675	0.636	−0.410	2.645
b_2	0.109	0.560	−1.596	2.306
Marriage equations (N = 158)				
a_1	−0.741	0.680	−3.988	0.736
a_2	0.709	0.668	−0.748	3.879
b_1	−0.667	0.642	−3.121	0.578
b_2	0.640	0.613	−0.598	3.082
Appraisals and imitation (N = 83)				
a_1	0.515	0.665	−0.065	3.958
a_2	0.460	1.062	−3.031	4.732
b_1	0.525	0.746	−0.356	4.090
b_2	0.515	0.728	−0.555	2.910
L–V: Female is prey (N = 190)				
a_1	−0.029	0.384	−1.623	1.933
a_2	−0.023	0.474	−1.958	2.466
b_1	0.087	0.313	−0.962	2.238
b_2	0.105	0.432	−1.281	3.236
L–V: Male is prey (N = 188)				
a_1	−0.088	0.345	−2.238	0.962
a_2	−0.109	0.473	−3.236	1.281
b_1	0.033	0.385	−1.933	1.623
b_2	0.036	0.484	−2.466	1.957
Mutualism (N = 92)				
a_1	2.031	1.146	−0.350	6.970
a_2	2.389	1.698	−0.380	7.360
a_3	0.531	1.141	−1.070	4.290
b_1	2.020	1.197	−0.990	7.331
b_2	3.096	2.947	−1.080	6.840
b_3	0.696	1.073	−1.640	2.880

For Steenbeek and van Geert's (2007) model, we also made changes in the original specification. In particular, we excluded the second term in the original equation (Equation 13.3), which represented discrepancies between an individual's preferred level of behavior and the partner's behavior. This change was done to simplify the model and facilitate comparison with the other models in the analyses. Moreover, as a marker of preferred behavior (affect in our data), we used daily general affect, as measured by the Positive and Negative Affect Schedule (PANAS; Watson, Clark, & Tellegen, 1988). Thus, the final specification was as follows:

$$\Delta I/\Delta_t = a(P_I - I_t) + e(Y_t - I_t)$$

and

$$\Delta Y/\Delta_t = c(P_Y - Y_t) + f(Y_t - I_t)$$

The results from fitting this model to the data are in the third set of rows in Table 13.3. Here, the parameters a_1 and a_2 stand for a and e, representing the female's rate of approaching her general affect and her rate of interaction with her partner, respectively. Similarly, the parameters b_1 and b_2 stand for c and f, representing the male's rate of approaching his general affect and his rate of interaction with his partner, respectively. Unlike the two previous analyses, there was a large number (114) of dyads for which this model did not converge successfully. The parameter estimates were all positive and of similar values, indicating that, on average, these individuals' daily emotional changes were influenced equally by their own individual dynamics and their partners' influences. As was the case for the previous models, the estimates showed large variability across dyads, revealing that these average estimates do not capture the important differences across all couples and might not represent a single dyad.

The next two sets of rows in Table 13.3 include estimates from a predator–prey model. The first set of estimates is from a model in which the female's affect is the prey; the second set is from a model in which the male's affect is the prey. For the first model, a_1 represents the female's growth rate of affect, a_2 represents the female's rate of "predation" from her male partner, b_1 denotes the male's rate of loss in affect, and b_2 represents the male's predating growth (these parameters are r_x, a, r_y, and b, in the original model; see Equation 13.4). For both specifications, the model converged successfully for most of the dyads; the average estimates were close to zero and did not show apparent differences across the two specifications.

When considering the female as the prey, for example, these estimates indicate a negative growth in her affect in the absence of interactions, and a positive interaction with her partner (negative coefficient multiplied by a negative sign in the model). For him, in turn, the estimates reveal a negative loss in affect (positive coefficient multiplied by a negative sign in the model) as well as positive interaction with ("predation" on) his partner. The estimates for the specification in which the male is the prey can be interpreted similarly.

The last model that we fitted was that of mutualism (Equation 13.5). This model includes six parameters, labeled in Table 13.3 as a_1, a_2, and a_3, for the female, and b_1, b_2, and b_3, for the male. These parameters stand for r_1, k_1, and a_1, as well as r_2, k_2, and a_2, for the female and male, respectively. As described previously, a_1 and b_1 represent the growth rate of affect, a_2 and b_2 represent the "density" of affect, and a_3 and b_3 denote the partnered influences.

The parameter estimates from this model appear in the last set of rows in Table 13.3. Apparent here are the differences in values between the parameters representing individual dynamics and those representing couple dynamics. According to this model, the daily changes in affect for these individuals appear to be more a function of their own dynamics (growth and density) and less related to their partners' influences. The large and positive values of all the parameters relating to individual dynamics suggest that, for a large number of couples in our data, the model meets the criteria for facultative mutualism, wherein all the individual parameters are greater than zero (Kot, 2001). Under these conditions, the affect of each person can exist without the partner's influence. Similar to all previous analyses, these estimates show sizeable standard deviations and ranges, indicating large variability across all couples. Finally, this model only converged successfully for 92 dyads.

Model Evaluation: Utility Value and Predictive Validity

After fitting the different models to the data, one important task was to evaluate the tenability of the model for the dyads. One standard approach is to inspect model fit and make decisions regarding which model is more plausible, given the data. This is typically the case when several models are fitted to data from an entire sample. In our case, we had various models fitted to each of the dyads, and thus it would be necessary to evaluate model fit separately for each model and dyad. One likely result from this approach is that the best-fitting model is different across dyads. In previous work, we have used this procedure to evaluate model fit and establish a classification of dyad prototypes based on different restrictions imposed on a baseline model (Ferrer & Steele, 2012). Another possible approach to consider model fit is to develop a criterion to weight each parameter estimate by the fit from its corresponding model for each dyad. We have employed such an approach using the Akaike information criterion (AIC) for each model to create a final set of parameter estimates that presumably represent each of the different models (Steele et al., in press).

In the current analyses we used model utility as the criterion to determine which model was more representative of the data. In particular, we were interested in the validity of each model to predict measures taken at a follow-up assessment. For this, we compiled the parameter estimates from each model across all dyads and used them as predictors in regression analyses. To take into consideration the precision of the estimates in the analyses (i.e., some large estimates could have a large standard error and thus not differ from zero), we rescaled the estimates by their corresponding standard errors and used these ratios as the predictors. In the first set of regressions, the outcome measure was relationship satisfaction quality at the follow-up. In the second set, the outcome measure was relationship status (i.e., together, broken up). At the follow-up, 148 (75.1%) of the original 197 couples reported being still together, and 49 (24.9%) reported having broken up.

Table 13.4 includes the results from the regression analyses with relationship satisfaction at follow-up as the outcome variable. Each set of estimates/standard errors per model was entered separately. Thus, each block of estimates refers to one model. For example, the first set of estimates represents the results from Felmlee and Greenberg's (1999) model. Here, the coefficient a_2 was predictive of relationship satisfaction (a_2 = 0.112, p = .012), indicating that the female's interaction with her partner was associated with a higher relationship quality at the follow-up. The male's rate of approaching

TABLE 13.4. Parameter Estimates from Regression Analysis Predicting Relationship Satisfaction

	Est.	*SE*	*t*-value	*p*	R^2
Reference affect (N = 95)					
a_1	−0.126	0.067	−1.881	0.062	0.102
a_2	0.112	0.044	2.523	0.012	
b_1	−0.104	0.057	−1.822	0.070	
b_2	−0.028	0.049	−0.564	0.574	
Marriage equations (N = 104)					
a_1	−0.149	0.222	−0.670	0.504	0.033
a_2	−0.280	0.224	−1.250	0.212	
b_1	0.202	0.339	0.601	0.552	
b_2	0.171	0.329	0.524	0.603	
Appraisals and imitation (N = 50)					
a_1	0.075	0.086	0.874	0.387	0.089
a_2	0.058	0.064	0.891	0.373	
b_1	−0.166	0.068	−2.431	0.017	
b_2	0.109	0.177	0.652	0.518	
L–V: Female is prey (N = 123)					
a_1	0.499	0.492	1.018	0.312	0.102
a_2	−0.669	0.525	−1.275	0.203	
b_1	−2.203	0.540	−4.073	< .001	
b_2	2.301	0.556	4.146	< .001	
L–V: Male is prey (N = 122)					
a_1	1.643	0.494	3.323	0.001	0.075
a_2	−1.721	0.517	−3.337	0.001	
b_1	−0.931	0.498	−1.878	0.062	
b_2	1.050	0.532	1.972	0.049	
Mutualism (N = 60)					
a_1	0.217	0.127	1.710	0.091	0.070
a_2	−0.081	0.086	−0.940	0.349	
a_3	−0.060	0.097	−0.621	0.534	
b_1	−0.187	0.110	−1.712	0.090	
b_2	0.076	0.059	1.301	0.198	
b_3	−0.133	0.110	−1.214	0.229	

his ideal, in contrast, was negatively associated with relationship quality ($b_1 = -0.104$, $p = .070$). For the marriage equations model, none of the estimates was predictive of the outcome. With regard to the appraisals and imitation model, the male's rate of approaching his general affect was negatively related to relationship quality ($b_1 = -0.166$, $p = .017$), in line with one of the findings from Felmlee and Greenberg's model.

For the predator–prey models, all the coefficients associated with the males were predictive of the outcome. When the female was the prey, the male's rate of loss in affect was associated with a decrease in relationship quality at follow-up ($b_1 = -2.203$,

$p < .001$). The male's rate of predating, in contrast, was positively associated with relationship quality ($b_2 = 0.231$, $p < .070$). When the male was the prey, the male's growth rate of affect was positively associated and his rate of predation from the female partner was negatively associated with relationship quality ($a_1 = 1.643$ and $a_1 = -1.721$, both $p < .001$). Finally, for the mutualism model, the coefficient representing growth in affect for both genders is close to being different from zero. Whereas the estimate for females is positive ($a_1 = 0.217$, $p = .091$), the one for males is negative ($b_1 = -0.187$, $p = .090$). For all models the R^2 was rather small, indicating that a large amount of variance in the outcome remained unexplained.

The results from the logistic regression predicting relationship status at the follow-up are included in Table 13.5 (0 = *together*, 1 = *break up*). A number of estimates across the various models were predictive of relationship status 1 or 2 years after the daily questionnaire. For example, for Felmlee and Greenberg's (1999) model, the coefficient representing the female's self-regulation (or rate of approaching her ideal affect) was related to staying together (negatively related to breaking up; $a_1 = -0.519$, $p = .007$). For the marriage equations model, both the male's own affective inertia and the influence from his female partner were negatively associated with breaking up ($b_1 = -6.285$ and $b_2 = -6.560$, both $p < .05$). For the appraisals and imitation model, however, none of the coefficients was predictive of the status at follow-up.

As was the case for the regression of relationship satisfaction, all the male coefficients in the predator–prey models were associated with the relationship status. When the female was the prey, the male's rate of loss in affect was associated with breaking up ($b_1 = 4.518$, $p = .003$). The male's rate of predating on his partner's affect, in contrast, was associated with staying together ($b_2 = -4.955$, $p = .001$). When the male was the prey, his rate of growth in affect was related to staying together ($a_1 = -2.637$, $p = .037$), whereas his rate of predation from the female partner was related to breaking up ($a_2 = 2.930$, $p = .028$). Finally, for the mutualism model, the coefficient representing density in affect for females ($a_2 = -0.706$, $p = .005$) and that denoting influence in affect from females on males ($b_3 = -1.334$, $p < .001$) were both related to staying together 1 to 2 years after the couples' daily reports.

Discussion

Summary of Findings

Our goal in this chapter was to examine a number of theoretical models of dyadic interactions. These models were taken from various disciplines and had been developed theoretically or investigated with empirical data. In our analyses, we applied each of these models to daily data on affect from couples. In the first step, we used differential equations to fit each of the models to each dyad separately and then compiled the parameter estimates to build empirical distributions of the estimates for each model. In the second step we evaluated the validity of each model by using the parameter estimates to predict the couples' relationship satisfaction and status at a follow-up.

TABLE 13.5. Parameter Estimates from Logistic Regression Analysis Predicting Relationship Status

	Est.	*SE*	Wald χ^2	*p*	Odds ratio
Reference affect (N = 107)					
a_1	−0.519	0.193	7.181	0.007	0.595
a_2	−0.214	0.131	2.669	0.102	0.807
b_1	−0.144	0.171	0.705	0.400	0.866
b_2	−0.208	0.138	2.249	0.133	0.812
Marriage equations (N = 137)					
a_1	0.127	0.614	0.043	0.835	1.136
a_2	0.311	0.619	0.253	0.615	1.365
b_1	−6.285	3.010	4.119	0.042	0.002
b_2	−6.560	3.093	4.499	0.034	0.001
Appraisals and imitation (N = 57)					
a_1	0.192	0.246	0.611	0.434	1.212
a_2	−0.044	0.150	0.086	0.770	0.957
b_1	0.011	0.168	0.005	0.946	1.011
b_2	−0.084	0.189	0.196	0.658	0.920
L–V: Female is prey (N = 138)					
a_1	0.313	1.360	0.052	0.818	1.368
a_2	−0.113	1.471	0.005	0.938	0.893
b_1	4.518	1.520	8.833	0.003	91.657
b_2	−4.955	1.596	9.636	0.001	0.007
L–V: Male is prey (N = 136)					
a_1	−2.637	1.268	4.320	0.037	0.072
a_2	2.930	1.336	4.810	0.028	18.731
b_1	0.727	1.269	0.328	0.566	2.068
b_2	−0.871	1.373	0.402	0.526	0.419
Mutualism (N = 68)					
a_1	0.152	0.307	0.244	0.621	1.164
a_2	−0.706	0.250	7.970	0.005	0.494
a_3	−0.182	0.294	0.384	0.536	0.834
b_1	−0.689	0.377	3.329	0.068	0.502
b_2	−0.245	0.179	1.869	0.172	0.783
b_3	−1.334	0.367	13.199	< .001	0.264

Note. Probability modeled status = 1 (break up). Estimates are in logit units.

One key finding of the first set of results is the large variability in estimates across dyads. Irrespective of the average values for each of the models, the parameter estimates showed substantial variation across dyads. These estimates were negative for some dyads, zero for others, and positive for some other dyads, and this was true for all models. These results support our previous work in this area (e.g., Ferrer & Widaman, 2008; Ferrer & Steele, 2012) and confirm the idea that modeling dynamics using an entire sample (dyads or individuals) might not be reasonable, as it is likely to obscure important interindividual differences and not represent the dynamics of any particular dyad. Recent work has demonstrated that aggregating data from individuals to describe interindividual changes over time fails to characterize accurately how each person changes over time (Hamaker et al., 2005; Molenaar, 2004; Molenaar & Valsiner, 2005; Nesselroade, 2001, 2002). Our findings in the current chapter add to this argument and indicate the importance of analyzing time series for each unit of analysis separately.

Several findings from the predictive validity analyses are worth mentioning. First, a relatively large number of parameters was predictive of relationship satisfaction 1–2 years after the daily reports of the time series. For some models, these parameters were representative of the person's own dynamics (i.e., self-regulatory mechanisms or affect inertia), whereas for other models, such parameters denoted interpersonal dynamics (i.e., coregulatory mechanisms). And yet for still other models (e.g., predator–prey) both types of parameters were predictive of relationship satisfaction. Second, several parameters were also predictive of relationship status at the follow-up. About half of these were representative of intrapersonal dynamics and the other half were related to interpersonal dynamics, and most of the latter were associated with staying together in the relationship. As was the case in the previous regression analyses, all but one model had parameters associated with the outcome measure. Together, these findings indicate that the selected models are useful to capture patterns of affective exchange between the individuals in the couples. Furthermore, these dynamics seem to carry important features of the couples that help predict their relationship status and quality over time. As evaluated by predictive validity, these models are useful to extract information about the dynamics of affect in dyadic interactions.

Methodological Issues

In our differential equations modeling a number of models failed to converge. These ranged from very few, for the predator–prey models, to many, for the mutualism and appraisals and imitation models. Furthermore, many parameters did not reach significance. Several reasons for this are possible. First, the data consisted of self-reports and might be too noisy to capture the important features of affect that make up a relationship. Moreover, self-reports have been criticized on both measurement and substantial grounds, but have also been defended as a legitimate tool with which to examine people's emotions (Barrett, 2004). Second, the daily recordings might not be an ideal time sampling of affect, as emotions fluctuate daily. Third, and perhaps more salient, the length of the series might not have been sufficient for these differential equation models, which assume that measurement and change occur in a continuous fashion. The individual time series in our data ranged from 50 to 103 data points. Whereas these values are more than reasonable from a psychological standpoint, the shorter time series are on the lower bound for examining any complex models (Song & Ferrer, 2009).

In spite of these shortcomings, the data and models were adequate to identify characteristics about the couples' affective dynamics during 2–3 months, and to link such characteristics to future states of the couple, as measured 1–2 years later. These findings add to previous work with these data in which we have found similar predictions when using similar statistical models (Ferrer & Steele, 2012) as well as exploratory techniques based on individual variability in affect and the overlap in such variability across the two individuals in the couple (Ferrer et al., 2012).

Future Directions

The application of differential equations to empirical data is a relatively novel avenue of research in the social and behavioral sciences. As such, many issues need to be addressed and many paths are yet to be discovered. Some of these issues have to do with the poor measurement properties unique to the social sciences. Unlike other disciplines such as physics, measurement in psychology is typically contaminated with various sources of error. Moreover, it almost never taps directly onto the construct of interest, but indirectly through observed variables. These issues make the application of differential equation modeling to time series data even more challenging. How to address these issues in the context of complex modeling is an important task to advance our understanding of psychological processes.

There are also many specific ways in which our analyses and general methodology can be improved. One has to do with stationarity. Our models assumed that the parameters remained invariant throughout the series. And this might not be the case. It is quite plausible that during the 2–3 months of daily data recording, the couples experienced several distinct cycles of affective dynamics. In addition to the discontinuity in the change, the affective dynamics for some couples could have multiple equilibrium points. Under those situations our models would fail to capture the different underlying processes, and other more complex models would be necessary. It would be important to combine our differential equation models with state–space models (e.g., Ho, Shumway, & Ombao, 2006; Shumway & Stoffer, 2006), functional data analyses (Ramsay & Silverman, 2005), methodology based on regime-switching models (Chow et al., Chapter 15, this volume; Kim & Nelson, 1999), Bayesian change point detection models (Carlin, Gelfand, & Smith, 1992), and individual shifts (Chow, Hamaker, & Allaire, 2009).

Our analytical approach consisted of individual (dyad) modeling and compiling the parameter estimates across all dyads to build empirical distributions. One drawback of this approach is that when the estimates are used in some additional step, as in our regression analyses, their imprecision is lost. In our case, this loss was compensated in part by weighting the parameter estimates by their corresponding standard error. Perhaps a more sophisticated approach would be to fit the model to all dyads simultaneously and estimate random effects in the parameters. This procedure would yield a distribution of parameters that would account for the variability in the sample (e.g., see Chow et al., 2007; Song & Ferrer, 2012).

Finally, all our differential equation models were deterministic. This is an assumption that is both convenient and simplistic. A more realistic representation, however, should include components to account for any stochastic process. In fact, some of the same models used in this chapter have been applied with error measurement. For example, Chow and colleagues (2009) estimated the predator–prey model that included error measurement using an unscented Kalman filter via state–state models, and Hamaker and colleagues estimated the Gottman and Murray models using threshold autoregressive models (Hamaker et al., 2009; Madhyastha et al., 2011).

In sum, in this chapter we applied several models of dyadic interaction to empirical time series data of affect from couples. Our goal was to find ways to formally describe the daily exchange of affect between the two dyad members and to relate such dynamics to future states of the couples. Understanding psychological processes requires the use of methodologies that can capture the complexity of those processes. Differential equation modeling is a powerful tool to identify and characterize those dynamics. We hope our analyses were useful to illustrate some of these possibilities.

ACKNOWLEDGMENTS

This work was supported in part by grants from the National Science Foundation (Nos. BCS-05-27766 and BCS-08-27021) and the National Institutes of Health/National Institute of Neurological Disorders and Stroke (No. R01 NS057146-01) to Emilio Ferrer. We appreciate the help and comments by Fushing Hsieh, Diane Felmlee, David Greenberg, Dave Sbarra, Hairong Song, Jonathan Helm, Laura Castro-Schilo, Michael McAssey, and the members of the DDIP Lab.

REFERENCES

Amemiya, T. (1977). The maximum likelihood estimator and the nonlinear three-stage least squares estimator in the general nonlinear simultaneous equation model. *Econometrica, 45*, 955–968.

Barrett, L. F. (2004). Feelings or words?: Understanding the content in self-report ratings of experienced emotion. *Journal of Personality and Social Psychology, 87*, 266–281.

Boker, S. M., & Laurenceau, J.-P. (2006). Dynamical systems modeling: An application to the regulation of intimacy and disclosure in marriage. In T. A. Walls & J. L. Schafer (Eds.), *Models for intensive longitudinal data* (pp. 195–218). New York: Oxford University Press.

Bowlby, J. (1973). *Attachment and loss: Vol. 2. Separation: Anxiety and anger.* New York: Basic Books.

Bowlby, J. (1980). *Attachment and loss: Vol. 3. Sadness and depression.* New York: Basic Books.

Bowlby, J. (1982). *Attachment and loss: Vol. 1. Attachment* (2nd ed.). New York: Basic Books. (Original work published 1969)

Buder, E. H. (1991). A nonlinear dynamic model of social interaction. *Communication Research, 18*, 174–198.

Butner, J., Diamond, L. M., & Hicks, A. M. (2007). Attachment styles and two forms of affect coregulation between romantic partners. *Personal Relationships, 14*, 431–455.

Carlin, B., Gelfand, A., & Smith, A. (1992). Hierarchical Bayesian analysis of change points problems. *Applied Statistics, 41*, 389–405.

Chow, S.-M., Ferrer, E., & Nesselroade, J. R. (2007). An unscented Kalman filter approach to the estimation of nonlinear dynamical systems models. *Multivariate Behavioral Research, 42*, 283–321.

Chow, S.-M., Hamaker, E., & Allaire, J. C. (2009). Using innovative outliers to detect discrete shifts in dynamics in group-based state–space models. *Multivariate Behavioral Research, 44*, 465–496.

Cranford, J. A., Shrout, P. E., Iida, M., Rafaeli, E., Yip, T., & Bolger, N. (2006). A procedure for evaluating sensitivity to within-person change: Can mood measures in diary studies detect change reliably? *Personality and Social Psychology Bulletin, 32*, 917–929.

Durlauf, S. N., & Ioannides, Y. M. (2010). Social interactions. *Annual Review of Economics, 2*, 451–478.

Felmlee, D. H. (2007). Application of dynamic systems analysis to dyadic interactions. In A. D. Ong & M. H. M. van Dulmen (Eds.), *Oxford handbook of methods in positive psychology* (pp. 409–422). Oxford, UK: Oxford University Press.

Felmlee, D. H., & Greenberg, D. F. (1999). A dynamic systems model of dyadic interaction. *Journal of Mathematical Sociology, 23*, 155–180.

Ferrer, E., & Helm, J. (2013). Dynamical systems modeling of physiological coregulation in dyadic interactions. *International Journal of Psychophysiology, 88*, 296–308.

Ferrer, E., & Steele, J. (2012). Dynamic systems analysis of affective processes in dyadic interactions using differential equations. In G. R. Hancock & J. R. Harring (Eds.), *Advances in longitudinal methods in the social and behavioral sciences* (pp. 111–134). Charlotte, NC: Information Age.

Ferrer, E., Steele, J., & Hsieh, F. (2012). Analyzing dynamics of affective dyadic interactions using patterns of intra- and inter-individual variability. *Multivariate Behavioral Research, 47*, 136–171.

Ferrer, E., & Widaman, K. F. (2008). Dynamic factor analysis of dyadic affective processes with inter-group differences. In N. A. Card, J. P. Selig, & T. D. Little (Eds.), *Modeling dyadic and interdependent data in the developmental and behavioral sciences* (pp. 107–137). Hillsdale, NJ: Psychology Press.

Fletcher, G. J. O., Simpson, J. A., & Thomas, G. (2000). Ideals, perceptions, and evaluations in early relationship development. *Journal of Personality and Social Psychology, 79*, 933–940.

Gallant, A. R., & Holly, A. (1980). Statistical inference in an implicit, nonlinear, simultaneous equation model in the context of maximum likelihood estimation. *Econometrica, 48*, 697–720.

Gottman, J. M. (1994). *What predicts divorce?: The relationship between marital processes and marital outcomes.* Hillsdale, NJ: Erlbaum.

Gottman, J. M., & Levenson, R. W. (2000). The timing of divorce: Predicting when a couple will divorce over a 14–year period. *Journal of Marriage and Family, 62*(3), 737–745.

Gottman, J., Murray, J., Swanson, C., Tyson, R., & Swanson, K. (2002). *The mathematics of marriage: Dynamic nonlinear models*. Cambridge, MA: MIT Press.

Gottman, J. M., & Notarius, C. I. (2000). Decade review: Observing marital interaction. *Journal of Marriage and Family, 62*(4), 927–947.

Gottman, J. M., Swanson, C. C., & Murray, J. D. (1999). The mathematics of marital conflict: Dynamic mathematical nonlinear modeling of newlywed marital interaction. *Journal of Family Psychology, 13*, 3–19.

Hamaker, E. L., Dolan, C. V., & Molenaar, P. C. M. (2005). Statistical modeling of the individual: Rationale and application of multivariate stationary time series analysis. *Multivariate Behavioral Research, 40*, 207–233.

Hamaker, E. L., Zhang, Z., & van der Maas, H. L. J. (2009). Using threshold autoregressive models to study dyadic interactions. *Psychometrika, 74*, 727–745.

Hirsch, M. W. (1982). Systems of differential equations which are competitive or cooperative: I. Limit sets. *SIAM Journal on Mathematical Analysis, 13*, 167–179.

Hirsch, M. W. (1985). Systems of differential equations which are competitive or cooperative: II. Convergence almost everywhere. *SIAM Journal on Mathematical Analysis, 16*, 423–439.

Ho, R. M., Shumway, R., & Ombao, H. (2006). The state–space approach to modeling dynamical processes. In T. A. Walls & J. L. Schafer (Eds.), *Models for intensive longitudinal data* (pp. 148–175). New York: Oxford University Press.

Julier, S. J., Uhlmann, J. K., & Durrant-Whyte, H. F. (1995). A new approach for filtering nonlinear systems. In *Proceedings of the American Control Conference* (pp. 1628–1632). Seattle, WA: IEEE.

Kelley, H. H. (1983). Epilogue: An essential science. In H. H. Kelley, E. Berscheid, A. Christensen, J. J. Harvey, T. L. Huston, G. Levinger, et al. (Eds.), *Close relationships* (pp. 265–314). New York: Freeman.

Kelley, H. H., & Thibaut, J. W. (1978). *Interpersonal relations: A theory of interdependence*. New York: Wiley.

Kim, C.-J., & Nelson, C. R. (1999). *State–space models with regime switching: Classical and Gibbs-sampling approaches with applications*. Cambridge, MA: MIT Press.

Kot, M. (2001). *Elements of mathematical ecology*. Cambridge, UK: Cambridge University Press.

Lotka, A. J. (1925). *Elements of physical biology*. Baltimore: Williams & Wilkins.

Madhyastha, T. M., Hamaker, E. L., & Gottman, J. M. (2011). Investigating spousal influence using moment-to-moment affect data from marital conflict. *Journal of Family Psychology, 25*, 292–300.

Molenaar, P. C. M. (2004). A manifesto on psychology as idiographic science: Bringing the person back into scientific psychology—this time forever. *Measurement, 2*, 201–218.

Molenaar, P. C. M., & Valsiner, J. (2005). How generalization works through the single case: A simple idiographic process analysis of an individual psychotherapy. *International Journal of Idiographic Science*, Article 1. Retrieved October 18, 2005, from *www.valsiner.com/articles/molenvals.htm*.

Nesselroade, J. R. (2001). Intraindividual variability in development within and between individuals. *European Psychologist, 6*, 187–193.

Nesselroade, J. R. (2002). Elaborating the differential in differential psychology. *Multivariate Behavioral Research, 37*, 543–561.

Newtson, D. (1993). The dynamics of action and interaction. In L. B. Smith & E. Thelen (Eds.),

A dynamic systems approach to development: Applications (pp. 241–264). Cambridge, MA: MIT Press.

Pindyck, R. S., & Rubinfeld, D. L. (1981). *Econometric models and economic forecasts* (2nd ed.). New York: McGraw-Hill.

R Development Core Team. (2009). *R: A language and environment for statistical computing* [Computer software manual]. Vienna, Austria: Author. Available from *www.R-project.org.*

Ramsay, J. O., & Silverman, B. W. (2005). *Functional data analysis*. New York: Springer.

Schwartz, R. M. (1997). Consider the simple screw: Cognitive science, quality improvement, and psychotherapy. *Journal of Consulting and Clinical Psychology, 65*, 970–983.

Schwartz, R. M., Reynolds, C. F., III, Thase, M. E., Frank, E., Fasiczka, A. L., & Haaga, D. A. F. (2002). Optimal and normal affect balance in psychotherapy of major depression: Evaluation of the balanced states of mind model. *Behavioural and Cognitive Psychotherapy, 30*, 439–450.

Shumway, R., & Stoffer, D. (2006). *Time series analysis and its applications: With R examples*. New York: Springer.

Song, H., & Ferrer, E. (2009). State–space modeling of dynamic psychological processes via the Kalman smoother algorithm: Rationale, finite sample properties, and applications. *Structural Equation Modeling: A Multidisciplinary Journal, 16*, 338–363.

Song, H., & Ferrer, E. (2012). Bayesian estimation of random coefficient dynamic factor models. *Multivariate Behavioral Research, 47*, 26–60.

Steele, J., Ferrer, E., & Nesselroade, J. R. (in press). An idiographic approach to estimating models of dyadic interactions with differential equations. *Psychometrika*.

Steenbeek, H. W., & van Geert, P. L. C. (2007). A theory and dynamic model of dyadic interaction: Concerns, appraisals, and contagiousness in a developmental context. *Developmental Review, 27*, 1–40.

Thomas, E. A. C., & Martin, J. A. (1976). Analysis of parent–infant interaction. *Psychological Review, 83*, 141–156.

van der Merwe, R. (2003). *ReBEL: Recursive Bayesian Estimation Library* (ReBEL online documentation). Available at *http://choosh.cse.ogi.edu/rebel.*

Volterra, V. (1926). Fluctuations in the abundance of a species considered mathematically. *Nature, 118*, 558–560.

Watson, D., Clark, L. A., & Tellegen, A. (1988). Development and validation of brief measures of positive and negative affect: The PANAS Scales. *Journal of Personality and Social Psychology, 47*, 1063–1070.

APPENDIX 13.1. Sample Script of PROC MODEL Used for the Analyses

```
TITLE 'Model1: Felmlee & Greenberg';
PROC MODEL DATA = ddip ;
      BY id_dyad ;
      PARM a1 a2 b1 b2;
      dert.fs = a1*(.7 - fs) + a2*(ms - fs);
      dert.ms = b1*(.7 - ms) + b2*(fs - ms);
      FIT fs ms
      START=(a1=1 .5 .1 0 -.1 -.5 -1, b1=1 .5 .1 0 -.1 -.5 -1)/
      FIML STARTITER; RUN;


TITLE 'Model2: Gottman & Murray';
PROC MODEL DATA = ddip;
      BY id_dyad ;
      PARM a2 a3 b2 b3;
      dert.fs = a2*fs + a3*ms ;
      dert.ms = b2*ms + b3*fs ;
      FIT fs ms
      START=(a2=1 .5 .1 0 -.1 -.5 -1, b2=1 .5 .1 0 -.1 -.5 -1)/
      FIML STARTITER; RUN;


TITLE 'Model3: Steenbeek & van Geert';
PROC MODEL DATA = ddip ;
      BY id_dyad ;
      PARM a1=.1 a3=.1  b1=.1  b3=.1 ;
      dert.fs = a1*(fg - fs) + a3*(ms - fs) ;
      dert.ms = b1*(mg - ms) + b3*(fs - ms) ;
      FIT fs ms
      START=(a1=1 .5 .1 0 -.1 -.5 -1, b1=1 .5 .1 0 -.1 -.5 -1)/
      FIML STARTITER; RUN;

TITLE 'Model4: Predator-Prey Model (Female = Prey)';
PROC MODEL DATA = ddip ;
      BY id_dyad ;
      PARM a1 a2 b1 b2;
      dert.fs =  a1*(fs) - a2*((fs*ms));
      dert.ms = -b1*(ms) + b2*((fs*ms));
      FIT fs ms
      START=(a1=1 .5 .1 0 -.1 -.5 -1, b1=1 .5 .1 0 -.1 -.5 -1)/
      FIML STARTITER; RUN;

TITLE 'Model5: Predator-Prey Model (Male = Prey)';
PROC MODEL DATA = ddip ;
      BY id_dyad ;
      PARM a1 a2 b1 b2;
```

```
      dert.ms =  a1*(ms) - a2*((ms*fs));
      dert.fs = -b1*(fs) + b2*((fs*ms));
      FIT ms fs
      START=(a1=1 .5 .1 0 -.1 -.5 -1, b1=1 .5 .1 0 -.1 -.5 -1)/
      FIML STARTITER; RUN;

TITLE 'Model6: Protection Mutualism';
PROC MODEL DATA = ddip ;
      BY id_dyad ;
      PARM r1 k1 a1 r2 k2 a2 ;
      dert.fs = ((r1/k1)*fs)*(k1 - fs + a1*ms);
      dert.ms = ((r2/k2)*ms)*(k2 - ms + a2*fs);
      FIT fs ms
      START=(a1=1 .5 .1 0 -.1 -.5 -1, a2=1 .5 .1 0 -.1 -.5 -1)/
      FIML STARTITER; RUN;
```

CHAPTER 14

A Differential Equations Model for the Ovarian Hormone Cycle

Steven M. Boker, Michael C. Neale, and Kelly L. Klump

Recently there has been great interest in understanding developmental, behavioral, and physiological processes from a dynamical systems point of view. One benefit of studying processes as dynamical systems is that data are analyzed as person-specific and only subsequently modeled for sources of individual differences. This is a critical point—as students of human development and behavior, we wish our findings to not only apply to population aggregates, but also to represent the complex patterns and etiologies exhibited by individuals as they develop and behave within their life context.

Underlying dynamical systems methods are statistical models that treat change as the fundamental concept of interest, and data are intensively sampled in time over multiple individuals (Boker, 2012). Statistical models for dynamical systems represent the time dependency of data as the fundamental individual characteristic (Boker, Molenaar, & Nesselroade, 2009). That is, one does not so much make a model of what happened at a particular time as a model of what would have happened given some knowledge about previous states and contexts. In plain terms, dynamical systems models are not so much concerned with *what* happened as they are with *how* it happens.

Dynamical systems models are frequently considered within the theoretical framework of regulation. Thus a central concept is the idea of equilibrium—the time dependency of a system is relative to its equilibrium (Boker, 2013). An equilibrium can be a single point, a cyclic figure such as a circle, or have more complex shapes. The behavior of a system as it regulates itself with respect to its equilibrium is termed the *basin of attraction* of a system. The shape of the attractor for a system will place constraints on the minimum set of model equations that can represent the dynamics of that system. Thus, visualizing the attractor for a system is often a first step to model specification because it can place constraints on model choice.

One technique that is frequently employed for visualizing attractors is called *phase space reconstruction*. Most techniques for phase space reconstruction involve some form of *time delay embedding*. This technique encapsulates the time dependency within short time segments of observed data. An advantage of this method is akin to nonparametric smoothing: One need not make long-term assumptions about the behavior of a system. Instead, one uses short intervals of a few measurements at a time as local measures of the overall manifold of the dynamic. By assuming that the system can be captured locally by a smooth function (e.g., a parabolic basis function), one can recapture the overall geometry of the attractor—the long-term regulatory dynamics defined by the relationships between the derivatives of the attractor manifold (Takens, 1985; Whitney, 1936)—even when the long-term dynamics are not well described by a simple parabolic basis.

In order to make these abstract notions more concrete, sample data from a study of relationships between ovarian hormones and eating behavior are presented. Reconstructed attractors for estradiol and progesterone are plotted and discussed. Next, modeling choices are made to create a biologically plausible differential equations model for estradiol and progesterone. Finally, the model is numerically integrated and compared to the reconstructed attractor from the data for one individual. This is only a first step toward understanding these data, but statistical parameter estimation of a multilevel model for the ovarian hormone cycle with proper treatment of individual differences is beyond the scope of this introductory chapter. Instead, let us focus our attention on how one develops a dynamical systems model suitable for later statistical testing.

All too often, statistical models are chosen because they have been useful in other applications—a case of everything looking like a nail to someone holding a hammer. We strongly recommend that models be developed to serve an explanatory function by tying the dynamics of the model to what is known about the micro- and/or macrostructure of the processes of interest. With this view in mind, we reconstruct empirical attractors for estradiol and progesterone as estimates of the macrostructure of the process and use biological details of the ovarian cycle to inform the microstructure.

The Sample Hormone Data

The data we use as an example are from a small sample (n = 4) from a much larger (N = 238) twin study of eating behavior and ovarian hormones in normally menstruating young (age $\mu = 17.9$, $\sigma = 1.7$) women (Klump et al., 2013). Participants included in the study met the following criteria: (1) menstruation every 22–32 days for the past 6 months; (2) no hormonal, psychotropic, or steroid medications within the past 8 weeks; (3) no pregnancy or lactation within the past 6 months; and (4) no history of genetic or medical conditions known to influence hormone functioning or appetite/weight. The sample of four individuals was selected specifically to exemplify the diversity of patterns of hormones observed in the larger study.

Saliva samples were collected every morning for 45 consecutive days within 30 minutes of waking. Daily estradiol and progesterone levels were measured by enzyme immunoassay (see Klump et al., 2013, for details). Time series plots of normalized hormone levels and self-reported days of bleeding for the four selected participants are shown in Figure 14.1. Two observations are immediately apparent from these plots:

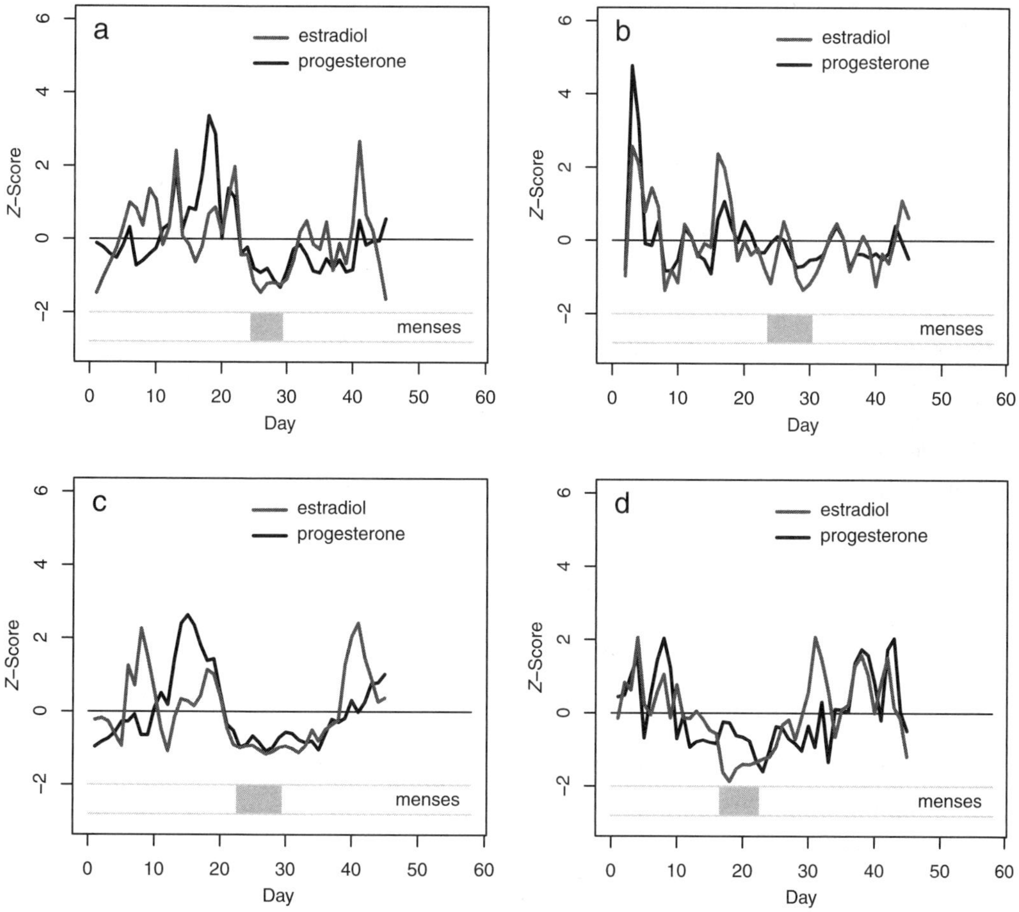

FIGURE 14.1. Time series plots of estradiol and progesterone Z-scores for four selected participants.

There are substantial individual differences in the cycles as well as substantial intraindividual variability within the major cyclic pattern. Of particular interest is the apparent 5-day cycle of estradiol in Figure 14.1. This rapid cycle in estradiol levels was observed in a substantial fraction of women in the study.

Phase Space Reconstruction

In order to plot a phase space reconstruction of an attractor from time series data, the time series must first be converted into a time delay embedded matrix. In the most commonly used methods, two integers are chosen: a delay parameter, τ, and a dimension, D; the time series (e.g., a vector with P elements) is then converted to a $P - D \times D$ matrix where each column of the matrix is lagged by τ elements from the previous column (Sauer, Yorke, & Casdagli, 1991), as shown in Equation 14.1. When using this method, *tau* is chosen so as to spread the attractor out in the D dimensional state–space. In the case of a linear cyclic attractor with a wavelength λ, the optimal value would be $\tau = \tau/4$ (Takens, 1985). Various methods are used to choose τ, the most common of which are the first zero crossing of the autocorrelation and the first minimum of the average mutual information (Abarbanel, Brown, & Kadtke, 1990; Kennel, Brown, & Abarbanel, 1992). Once the data are time delay embedded, they are typically displayed as a scatter plot where, for instance, the first column of the time delay matrix, $x(t)$, is represented by the horizontal axis and the second column, $x(t + \tau)$, by the vertical axis. This creates what is termed a *state–space plot*, since D-dimensional states of the system are directly plotted.

The advantage of a state–space plot is that it preserves the fine-grained structure of the system. However, in data of interest in human dynamics, there is frequently a mixture of time-independent measurement error and time-dependent dynamics. By using a locally parabolic (or higher-order Taylor series) filter, time derivates of a time series can be estimated (Savitzky & Golay, 1964) while creating a local smoothing of the data to separate time-independent noise from the time-dependent dynamics. A recently developed method for performing this task uses generalized orthogonal linear derivative (GOLD) (Deboeck, 2010) weights for generalized local linear approximation (GLLA) (Boker, Deboeck, Edler, & Keel, 2010). This method estimates derivatives for each measurement in a time series, and then these derivatives can be plotted as a *phase space plot*.

In order to use the GLLA method, a time delay embedded matrix must be constructed, and thus the quantities τ and D must still be chosen. However, these choices follow a somewhat different logic than does the state–space reconstruction method. The reasons for this difference are twofold: (1) The choice of the order of the highest derivatives to estimate is decoupled from the choice of D in the time delay embedding; and (2) the choice of τ and D contribute to the degree of smoothing applied to the time series. The details of these two choices are treated more comprehensively in a moment, but first we explicitly define the construction of a time delay embedded matrix and its associated GLLA filter for estimating derivatives.

Time Delay Embedding

Suppose a time series with P elements, $X = \{x(1), x(2), x(3), \ldots, x(P)\}$. Suppose further that we choose a time lag, τ, and dimension, $D = 5$. A time delay embedded matrix, which we designate $\mathbf{X}_{(5)}$, can be constructed as

$$\mathbf{X}^{(5)} = \begin{bmatrix} x_{(1)} & x_{(1+r)} & x_{(1+2r)} & x_{(1+3r)} & x_{(1+4r)} \\ x_{(2)} & x_{(2+r)} & x_{(2+2r)} & x_{(2+3r)} & x_{(2+4r)} \\ x_{(3)} & x_{(3+r)} & x_{(3+2r)} & x_{(3+3r)} & x_{(3+4r)} \\ \vdots & \vdots & \vdots & \vdots & \vdots \\ x_{(P-1-4\tau)} & x_{(P-1-3\tau)} & x_{(P-1-2\tau)} & x_{(P-1-\tau)} & x_{(P-1)} \\ x_{(P-4\tau)} & x_{(P-3\tau)} & x_{(P-2\tau)} & x_{(P-\tau)} & x_{(P)} \end{bmatrix} \quad (14.1)$$

Note that time dependency from the time series X is now represented as multiple short samples in the rows of $\mathbf{X}_{(5)}$. Thus, when we estimate derivatives of the time series, each row can contribute a new estimate of the covariance between these time derivatives. This is an important point: We have *localized* the time derivative estimation over the manifold of the attractor without committing to any particular model for the whole attractor. This is true of any method that uses time delay embedding.

The GLLA method uses a power series approximation to extract the time dependency in each row of the time delay embedded matrix. Suppose we construct a basis function matrix L in a manner similar to the fixed loading matrix for a latent growth curve model such that the first column of L contains the loadings to estimate the latent intercept, the second column of L estimates the latent first derivative (slope), and the third column of L estimates the latent second derivative (curvature). For the $D = 5$ time delay embedding example above, L would be

$$\mathbf{L} = \begin{bmatrix} 1 & -2\Delta t\tau & (-2\Delta t\tau)^2 / 2 \\ 1 & -1\Delta t\tau & (-1\Delta t\tau)^2 / 2 \\ 1 & 0 & 0 \\ 1 & 1\Delta t\tau & (1\Delta t\tau)^2 / 2 \\ 1 & 2\Delta t\tau & (2\Delta t\tau)^2 / 2 \end{bmatrix} \tag{14.2}$$

The first column of L is a basis function for an intercept. The second column of L is a basis function for a unit value linear slope where the elapsed time between successive columns of $\mathbf{X}_{(5)}$ is τ times the elapsed time Δt between successive observations of X. The third column is the indefinite integral of the second column. Now, if the zeroth, first, and second derivatives of X were in the first, second, and third columns of a $P - 4\tau \times 3$ matrix Y, then

$$\mathbf{X}^{(5)} = \mathbf{Y}\mathbf{L}' + \mathbf{E} \tag{14.3}$$

If $\mathbf{L}'\mathbf{L}$ is nonsingular, then the right quasi-inverse $\mathbf{W} = \mathbf{L}(\mathbf{L}'\mathbf{L})^{-1}$ is a least squares solution, such that

$$\mathbf{Y} = \mathbf{X}^{(5)}\mathbf{W} \tag{14.4}$$

Thus, we can obtain least squares estimates of the derivatives of X with a single matrix multiplication if $\mathbf{W}$ is known. An R function to calculate $\mathbf{W}$, given choices for D, τ, and Δt, is available online at *http://people.virginia.edu/~smb3u/GLLAfunctions.R* (Deboeck's GOLD function can be found in Deboeck, 2010).

Choosing τ and D for GLLA Derivative Estimation

When using GLLA to estimate derivatives, the first question is, "What is the highest-order derivative to estimate?" Zeroth-order derivatives will just estimate a polynomial smooth of your time series and store it in a column vector. Zeroth- and first-order derivatives will estimate a polynomially smoothed time series along with a smoothed first derivative time series and store it into a matrix with two columns. Zeroth-, first-, and second-order derivatives add a column for the second derivative to the output matrix, and so on. In practice, it is rarely useful to estimate less than second-order derivatives, even if you only use the first two columns of the resulting matrix. Note that a first-order autoregressive process can be transformed into a first-order differential equation, and so you would need zeroth and first derivatives at a minimum. If the process of interest has fluctuations or oscillations, a second-order differential equation is the minimum order that can represent oscillations.

The next question to ask is, "What is the minimum amount of time that elapses in one complete oscillation or fluctuation?" If the process of interest is a first-order process—for example, growth that approaches but does not pass an equilibrium value—then it is safe to skip to the next paragraph. But if the process could oscillate or fluctuate around an equilibrium, then decide how long the time is, on average, between peaks. An easy way to do this is to use the autocorrelation function (acf) in R. The acf will plot a graph that will show the autocorrelation for a series of lags in a time series. If one finds the first peak to the right of zero, the elapsed time as read on the horizontal axis of the plot will tell the elapsed number of samples, on average, from peak to peak. The lag in samples between the leftmost and rightmost column of the time delay embedded matrix must be less than the elapsed samples in the acf plot. Figure 14.2 plots the autocorrelation function for the four participants whose data were plotted in Figure 14.1. The minimum number of days for the autocorrelation peaks across all four participants is 12 days.

The last question to ask is, "How much smoothing should be applied when calculating the derivatives?" If one wishes to remove as much non-time-dependent information as possible, the total amount of elapsed time between the first and last columns of the time delay embedded matrix should be as long as possible, but of course shorter than the minimum fluctuation wavelength estimated above. If the time series is short, one may wish to modify that decision, since the more time that is covered by the time delay embedded matrix, the fewer the number of rows that result. Notice that the number of rows in the matrix in Equation 14.1 is $P - (D\tau)$. So although the time series originally had P elements, the time delay embedded matrix only has $P - (D\tau)$ rows. As either D or τ become larger, the number of samples of the time dependency become smaller. So, one does not necessarily want to always make D and τ as large as possible.

In practice, it turns out that the total lag from the first column to the last column is what is most important for all of these problems. In general, we recommend setting $\tau = 1$, since this means that there is better noise rejection within each row of the time delay embedded matrix (for a discussion of how time delay embedding improves noise rejection, see Oertzen & Boker, 2010). Thus, the only choice that remains is setting D to be less than the minimum wavelength (in samples) of the time-dependent fluctuations of interest and for derivative estimation of order O, D should be greater than or equal to $O + 2$. This last restriction ensures at least one degree of freedom for smoothing.

In the ovarian hormones example, we wish to estimate a second-order system ($O = 2$). In addition, we are primarily interested in an approximately monthly cycle, but

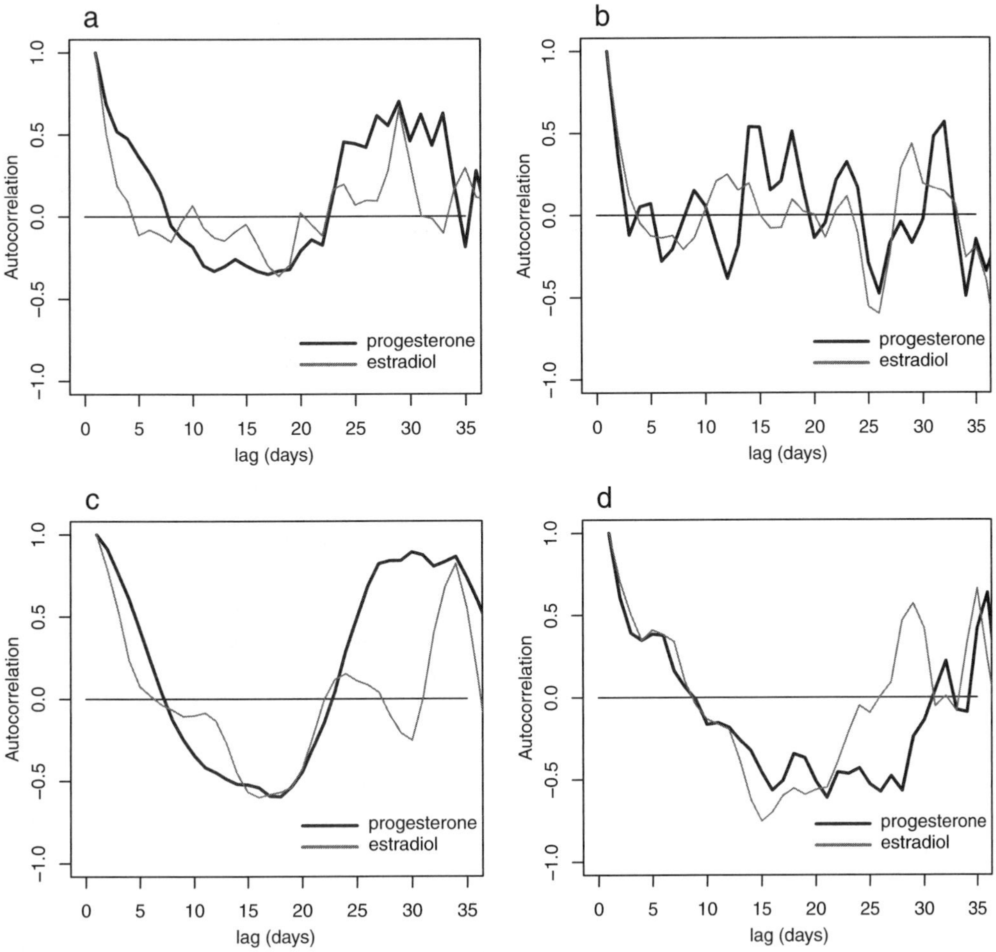

FIGURE 14.2. Autocorrelation plots of estradiol and progesterone for four selected participants. (a) Main oscillation autocorrelation peak for both estradiol and progesterone at 29 days. (b) Noisy but likely peaks for progesterone at 15 and 31 days and for estradiol at 12 and 30 days. (c) Easy-to-read peaks for progesterone at 30 days and for estradiol at 12, 24, and 35 days. (d) Peak for estradiol at 28 days and for progesterone at 35 days.

the shorter cycles may be important too. Thus, we chose $\tau = 1$ and $D = 11$ so that we could still capture the 12-day cycles but had maximum smoothing within the 12-day period. In practice, one may wish to explore other choices for D to see if other shorter- or longer-period features are exposed.

Matrix Scatter Plot of Phase Space

To prepare for phase space visualization, participants' hormone time series were normalized using their individual mean and standard deviation. The data were then time delay embedded using a lag of $\tau = 1$ and number of embedding columns $D = 11$. Zeroth-, first-, and second-order derivatives were estimated by the generalized local linear approximations method (Boker, Deboeck, et al., 2010) using a GOLD (Deboeck, 2010) transformation matrix. The resulting derivatives for estradiol and progesterone for the four participants are shown as matrix scatter plots in Figure 14.3. This method plots each combination of derivatives against one another, including the derivative pairs for the cross-coupling between estradiol and progesterone. Thus, not only is each within-variable reconstructed attractor visualized, but also the attractor formed by the cross-coupled variables. Since this visualization is novel, next we explain and interpret what is being displayed.

Figure 14.3 provides rich information about regulatory processes for estradiol and progesterone. We begin by focusing on Figure 14.3c. First, note that the relationship between almost all of the pairs of derivatives includes some sort of loop—an interior and an exterior. This is very different from a bivariate normal scatter plot! These loops are clear indicators of oscillation. However, if the process were a linear oscillator, we would expect to see ovals, circles, and perhaps lines. Instead, note that, for example, the relationship between estradiol and the first derivative of estradiol looks like it has a spot where there are many dots overlaid as well as a regular loop. The relationship between estradiol and the first derivative of progesterone looks as if it is a figure eight. The relationship between the first and second derivatives of progesterone is heart-shaped, with an area that looks like many overlaid dots, along with a long loop where the dots are spread out. Each of these observations leads to a conclusion that the regulatory system must include some nonlinear process.

If ovarian hormone regulation can be considered as a single system, then the individual's matrix plot in Figure 14.3c displays only one attractor that exists in a higher dimensional space. However, we are looking at its shadow on the plot as if it were lit from different perspectives. That is to say, the geometric figure in the higher dimensional space is projected down onto the two-dimensional surface of the plot. What does this higher dimensional figure look like? Is it possible that a relatively simple structure is being viewed from different angles and casts its shadow in a way that would reproduce all of these varied forms?

In order to answer that question, we consider a concept from dynamical systems called *winding number.* Figure 14.4a displays a system with a winding number of 0: the solid, straight black line plotted onto the top of the tube. This figure is a representation of an attractor for a linear system such as a univariate regression; there is one slope coefficient, and it doesn't change over time. Now consider Figure 14.4c. The solid line is now a circle. We have bent the tube around and joined its two ends to form a torus. This figure represents a sinusoidal oscillating system: The slope is first positive, then negative, then positive again. Imagine mentally rotating the torus so that you see the circle from various perspectives. It would cast the shadow of either a circle, an oval, or a line if you looked at it edge-on.

Next consider Figure 14.4b. The solid line spirals around the tube twice—a winding number of 2. If we perform the same trick of bending the tube and connecting the

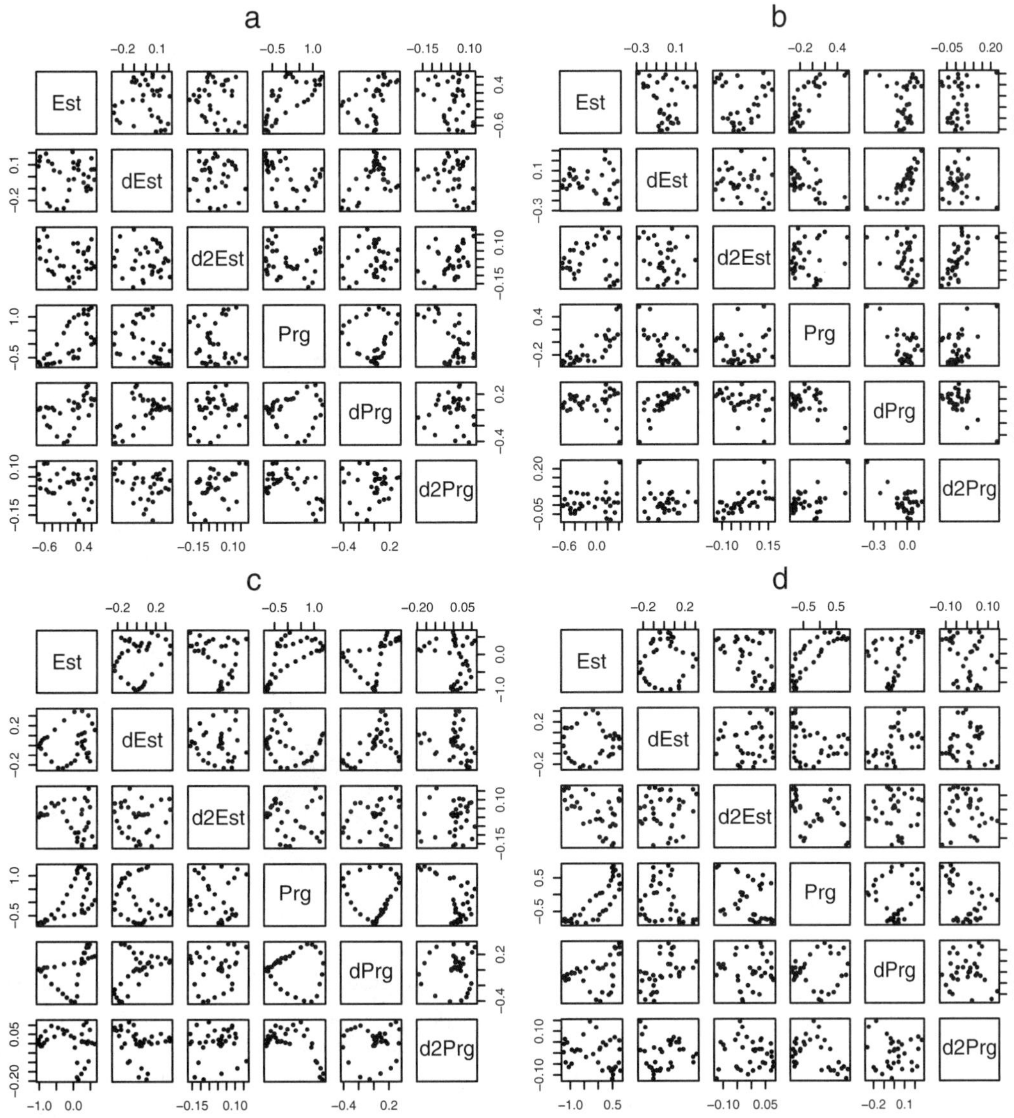

FIGURE 14.3. Scatter plot matrix of zeroth (Est), first (dEst), and second (d2Est) derivatives for estradiol and progesterone from the four participants. A sample R script to produce this type of phase space reconstruction plot from time series is provided in Appendix 14.1.

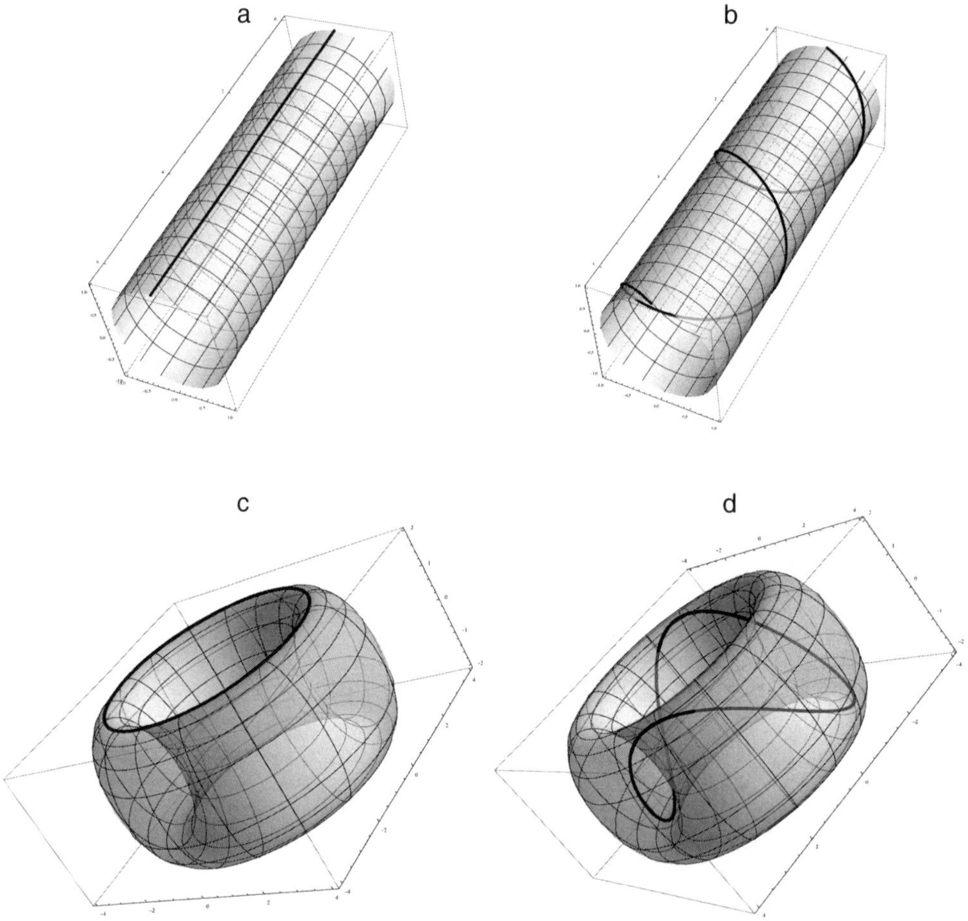

FIGURE 14.4. Visualization of winding numbers 0 (a, c) and 2 (b, d). (a) A system with a winding number of 0 can be plotted as a straight line on a tube. (c) An oscillating system with a winding number 0 can be plotted as a circle around the hole in a torus. (b) A system with a winding number 2 spirals around the tube twice. (d) When the tube is turned into a torus, a system with winding number 2 spirals around the tube of the torus twice for each time it circles the hole in the torus.

ends to form a torus (Figure 14.4d), the black line spirals around the tube of the torus twice for each time it orbits the hole of torus. This figure represents an oscillating system with a winding number of 2.

It is difficult to mentally rotate and project the shadow of the black line spiraling and orbiting on Figure 14.4d, so six projections of that system are displayed in Figure 14.5. The projections of the black line spiraling and orbiting the torus are very similar to those plotted in the empirical phase space reconstructions shown in Figure 14.3. It appears that a system with winding number 2 may be able to model the ovarian hormone data.

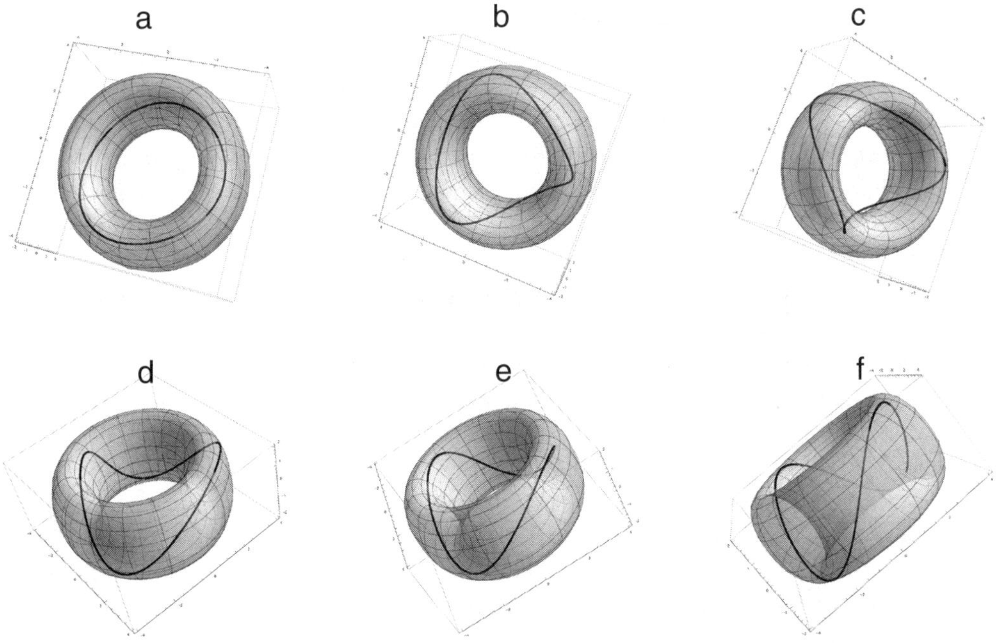

FIGURE 14.5. Six perspective views of the same trajectory: an oscillation with winding number 2. The trajectory makes two spirals around the tube of the torus for each orbit it makes around the hole in the torus. From each perspective, the black line of the trajectory appears to have a different shape.

Constructing a Dynamical Systems Model

If dynamical systems models are to be explanatory rather than merely descriptive, then the construction of models for phenomena must be placed under constraints that are not merely data driven. In other words, constraints from theory and known mechanisms (e.g., from biology, genetics, or physics) must play a major role in constructing alternative models. There is no statistical test that optimally weights the contributions of theory, known mechanisms, and statistical fit to observed data so as to provide an objective measure for model selection. However, it does make sense to direct the model building process at the outset so as to include as many constraints as are well established to restrict the search for a statistically optimal model to the space of plausible explanatory models. Simply put, it may not be useful to fit a model to data simply because it is easy. For instance, fitting a linear regression to time-intensive repeated measures of self-reported hunger would likely result in a model with a significant nonzero intercept and a slope that was not distinguishable from zero—a prediction that everyone was moderately hungry all the time: clearly, a nonsensical result.

So, how does one construct a dynamical systems model? First, it is important to realize that this will be a model for relationships between instantaneous time derivatives of variables and their equilibrium values. To continue the hunger example, one should ask, "How does the current rate of change of hunger relate to the current displacement from equilibrium?" If one is more hungry than one's long-term average (equilibrium) value for hunger, it is likely that a meal or snack will be in one's immediate future, and thus hunger is likely to decrease as a result. Similarly, if one is less hungry than the equilibrium, it is likely that one will not be eating for some time—resulting in an increasing value for hunger. Thus, we predict a negative relationship between the current value of hunger and the current slope for hunger.

In practice, we have used three strategies for constructing dynamical systems models. First, we have used existing theories and intuition to list the plausible relationships between derivatives and built models from the resulting constraints, as was begun in the preceding paragraph. This method can also be empirically driven in that features of the data, such as the phase space reconstructions plotted in Figure 14.3, can restrict the choice of models.

Second, we have looked closely at the language that is used to describe phenomena and tied descriptions of dynamic relationships back to models from physics. For instance, when articles describe a phenomenon such as resiliency as "bouncing back from adversity" (Ong, Bergeman, & Boker, 2009), we have taken these dynamic metaphors seriously and constructed and tested models for elasticity (Boker, Montpetit, Hunter, & Bergeman, 2010). Or when a rapid change in mood is described as a "mood swing," we have looked to physical models of pendulums and momentum (Bisconti, Bergeman, & Boker, 2004; Boker, Leibenluft, Deboeck, Virk, & Postolache, 2008).

The third way we have constructed models is by implementing physiological or biological constraints in order for the model to be biologically plausible and for its parameters to map to physical processes. We use a combination of the first and third methods to construct a differential equations model of the ovarian hormone cycle in this chapter. With these points in mind, the next section describes some of the known biology involved in the production and regulation of ovarian hormones.

Constraints from Biology

The human menstrual cycle has three characteristic hormones (estrogen, luteinizing hormone [LH], and progesterone) that are functionally involved in ovulatory regulation. The menstrual cycle is divided into two phases and sometimes subdivided further. The follicular phase begins at the onset of menstrual bleeding and ends at ovulation. The luteal phase then begins after ovulation and ends at the onset of menstrual bleeding. Ovulation is considered to be a phase in itself. Some refer to the luteal phase as the premenstrual phase, whereas others define a separate premenstrual phase as a shorter period at the end of the luteal phase. Menstrual bleeding lasts an average of 5 days, and 90% of menses are within the range of 3–8 days (Fehring, Schneider, & Raviele, 2006). The mean interval between subsequent onsets of menses (menstrual cycle length) has been estimated to be between 28 (Munster, Schmidt, & Helm, 1992) and 29 days (Fehring et al., 2006), but the variability in this estimate is large. Estimated within-person cycle-to-cycle variability (approximately 40% of women have more than 7 days intracycle variability) is nearly as large as the between-person individual differences (approximately 22–35 days; 95% confidence interval) in cycle length (Fehring et al., 2006; Geirsson, 1991; Munster et al., 1992). This large variability in cycle length provides another hint that substantial nonlinearity is present in this regulatory system.

Figure 14.6 plots mean curves of normalized progesterone, estradiol, and LH levels and their associated 90% confidence intervals from a daily measurement norming assay study (Stricker et al., 2006). Each of these plots is centered around the peak of LH at day 15. Timing variability in these plots is represented by the shaded gray areas. In Figure 14.6b, note the large uncertainty as to the timing of the peak of LH

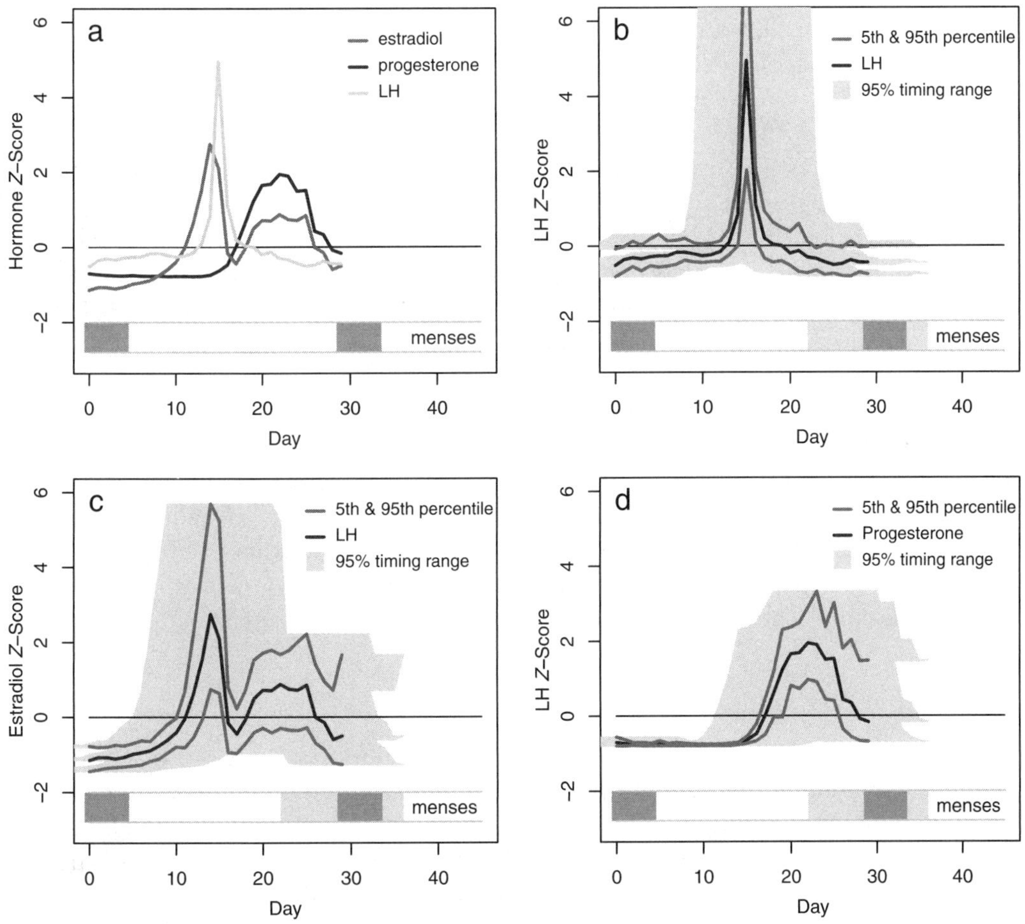

FIGURE 14.6. Visualization of (a) mean luteinizing hormone (LH), estradiol, and progesterone; (b) LH; (c) estradiol; and (d) progesterone cycles. Shaded gray areas represent the 95% confidence intervals for the timing of the mean and percentile curves and onset of subsequent menses as referenced from onset of menses at day 0. All mean curves are referenced to the day of peak LH. Means, 5th, and 95th percentiles are from Stricker et al. (2006), and timing variability estimates are from Fehring et al. (2006).

represented by the large gray area from day 10 to 22. This graphically presents the observed variability in the length of the follicular phase (Fehring et al., 2006). The graphs of estradiol and progesterone also show this large timing uncertainty due to the uncertainty in timing of the LH peak. Timing variability in the individual data curves that were averaged to produce the mean curves results in substantial smoothing and could hide high-frequency within-person fluctuations that may be present in the individual-level data.

After the end of menses, ovarian follicles are stimulated to grow, one of which subsequently matures. As the follicles develop, they release estradiol, and levels of estradiol begin to rise. Once estradiol passes a threshold, it triggers a quick rise in LH, a phenomenon known as the *LH surge*. The mechanism for this in humans has not been well established, but evidence from animal models suggests that two receptors in the hypothalamus, estrogen receptor α and receptor β, are involved (Hu et al., 2008). According to this view, low levels of estrogen activate estrogen receptor α in a negative feedback loop, which in turn suppresses the production of LH. But once estrogen has risen above a threshold, a positive feedback mechanism is triggered by β receptors in the hypothalamus, which produces LH. The positive feedback means that there is rapid exponential growth in LH over an interval of hours—the LH surge. This combination of positive and negative effects of estrogen on the production of LH, along with both a positive and negative feedback network, can result in a nonlinear trigger mechanism with a high degree of variability and nonstationarity. Observed high variability in the length of the follicular phase is consistent with this type of nonlinear effect.

High levels of LH trigger ovulation and the transformation of the dominant mature follicle into the corpus luteus. The corpus luteus begins to produce progesterone, and soon progesterone levels begin to rise. Progesterone in turn suppresses estrogen production, and estrogen levels begin to fall. Once estrogen levels fall sufficiently, receptor β turns off and receptor α quickly suppresses production of LH. The slow surge in progesterone continues either until fertilization results or menstrual bleeding flushes the corpus luteum from the uterus, and the cycle begins again.

Modeling Constraints

At this point, it is useful to list the modeling constraints and integrate them into a plan for building the differential equations model. Two types of constraints are used: what we know from biology that would help make the model biologically plausible and what we know from repeated measures of data—the plots of the daily assay data and the large-scale studies' aggregated mean curves. The differential equations are constructed to satisfy the following eight constraints:

1. Menstruation occurs approximately once every 29 days and lasts approximately 5 days.
2. Estradiol increases after menstruation. It may be that menstruation suppresses estradiol production. It takes about 15 days for estradiol to reach its peak.
3. LH production is suppressed when estradiol is below a threshold.

4. When estradiol is above a threshold (possibly the same threshold), LH then enters a positive feedback production cycle.
5. Progesterone production is triggered by LH.
6. Estradiol production is suppressed by progesterone.
7. From the data, estradiol levels drop more rapidly than could be explained by the slower increase in progesterone. One possible mechanism is that the LH surge itself suppresses estradiol levels.
8. From these data, it appears that there is a fast (approximately 5-day) cycle in the estradiol levels after the LH pulse. This could be a result of a self-regulatory mechanism in estradiol production.

Model Specification

By choice, the modeling constraints do not attempt to explain onset or length of menses. Although this is also an interesting problem, it is outside the scope of this example, for which a model of estradiol and progesterone regulation is of primary interest. However, the data are consistent with menses having a suppressive effect on estradiol. Thus, a discrete daily time series, Z, was constructed and coded such that –1 represented days of menstrual bleeding and 0 for all other days. The variable $z(t)$ was then used as an exogenous variable in the model.

From the data, estradiol appears to have two oscillations: a slow oscillation and a fast oscillation that is only apparent after the LH surge. As a first approximation, the model for estradiol levels was decomposed into two independent, self-regulating, oscillating systems: a slow oscillation with a period of about 30 days and a fast oscillation with a period of about 5 days. The slow estradiol equation was modeled as a damped linear oscillator, with additional damping from menstrual bleeding and progesterone such that

$$\ddot{x} = \eta_x x + \zeta_x \dot{x} + \gamma_{xz} z + \gamma_{xp} \dot{p} \tag{14.5}$$

where x, $\dot{x}$, and $\ddot{x}$ are the level, first, and second derivatives of estradiol, z is –1 during menstrual bleeding and zero otherwise, and $\dot{p}$ is the first derivative of progesterone. The first derivative of progesterone was used since the biological constraint is that progesterone level reduces estradiol production (i.e., the first derivative of estradiol), and so by taking the first derivative of this relationship, we come up with a relationship between the second derivative of estradiol and the first derivative of progesterone.

Similarly, the fast estradiol oscillation was modeled as a damped linear oscillator with a trigger from the LH surge as

$$\ddot{y} = \eta_y y + \zeta_y \dot{y} + \gamma_{yv} v \tag{14.6}$$

where y, $\dot{y}$, and $\ddot{y}$ are the level, first, and second derivatives of the residuals of estradiol after the slow estradiol oscillation is taken into account, and v is the level of LH.

Progesterone was modeled as an overdamped second-order linear system triggered by the LH surge. This means that the progesterone was considered to be self-regulating once its cycle was initiated by the LH surge. Thus, progesterone also takes the form

$$\ddot{p} = \eta_p p + \zeta_p \dot{p} + \gamma_{pv} v \tag{14.7}$$

where p, $\dot{p}$, and $\ddot{p}$ are the level, first, and second derivatives of the residuals of progesterone, and v is the level of LH. Note that although this equation has the same form as Equations 14.5 and 14.6, the parameters ηp and ςp are chosen so that damping occurs within the first cycle and is not considered to be an oscillator, but rather an overdamped system.

Finally, the LH surge was modeled. We did not have access to individual level LH data since the sample study (Klump et al., 2013) did not assay LH levels. Thus, the only data constraint on LH comes from published aggregated average LH levels (Stricker et al., 2006). However, we do have more information about the mechanisms of the LH surge. A reasonable assumption is that neural receptors with thresholds respond according to a sigmoid function. With that in mind, we first developed two sigmoid response functions, one for the estrogen receptor α and one for the estrogen receptor β. Each of these were modeled to respond to the sum of the slow and fast oscillation estradiol equations and were offset from each other so that the estrogen receptor α stopped firing at estrogen levels slightly below the levels where the estrogen receptor β began to fire. The two sigmoids thus were modeled as

$$\alpha = 1/(1 + e^{8(x+y+.5)}) \tag{14.8}$$

$$\beta = 1/(1 + e^{-10(x+y+.5)+3}) \tag{14.9}$$

where x and y are the slow and fast estradiol levels, respectively. Note that the slope of β is a bit steeper than that of α and that β has an offset of 3 from the response point of the α. These effects are depicted graphically in Figure 14.7. The choice of the constants in these equations is somewhat arbitrary, but is constrained by the need for their combined response to approximate the published average LH curves.

The LH surge was then modeled using a first-order nonlinear system as

$$\dot{v} = \zeta_v v + p_\alpha \alpha v + p_\beta \beta v + \gamma_v p v + 0.01 \tag{14.10}$$

where v, $\dot{v}$, and $\ddot{v}$ are the level, first, and second derivatives of LH, α and β are the two estrogen receptor responses, and p is the progesterone level. There are two parts to this equation that are speculative, but were found to be necessary in order to reproduce the published average LH surge. The equation thus makes two predictions for which we were unable to find either positive or negative evidence. First, LH is considered to be self-damped, essentially hypothesizing a depletion model for production of LH in the hypothalmus. The prediction is that a large production of LH produces a refractory period in which LH production is more difficult to stimulate. This is not an implausible

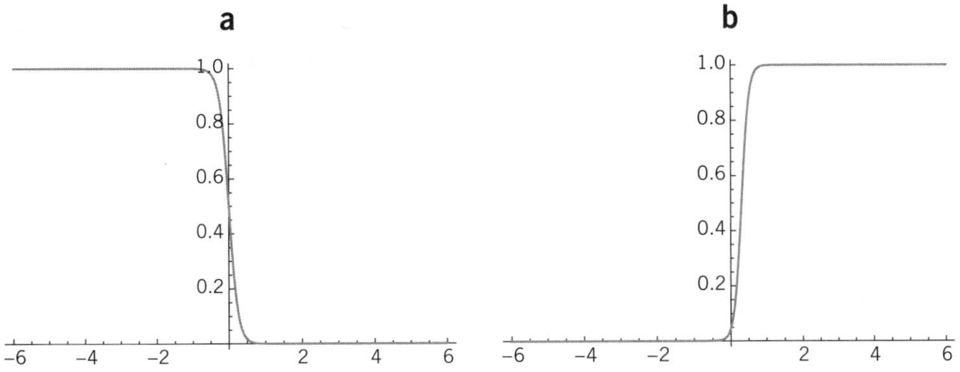

FIGURE 14.7. Sigmoidal response curves for (a) estrogen receptor α and (b) estrogen receptor β.

hypothesis, but we know of no positive evidence to support it. The second hypothesis is that there is a negative interaction between LH and progesterone such that when LH and progesterone are both high, LH production is suppressed. This second hypothesis is on less firm ground, but could be empirically tested.

Substituting the response curves for α and β into Equation 14.10, we now have a system of four simultaneous differential equations

$$\begin{aligned}
\ddot{x} &= \eta_x x + \zeta_x \dot{x} + \gamma_{xz} z + \gamma_{xp} \dot{p} \\
\ddot{y} &= \eta_y y + \zeta_y \dot{y} + \gamma_{yv} v \\
\ddot{p} &= \eta_p p + \zeta_p \dot{p} + \gamma_p v \\
\dot{v} &= \zeta_v v + \rho \alpha v / (1 + e^{8(x+y+.5)}) + \rho \beta v / (1 + e^{-10(x+y+.5)+3}) + \gamma_v p v + 0.01
\end{aligned} \quad (14.11)$$

which can be numerically integrated to predict levels of each hormone over multiple cycles.

Simulated Time Series

We used the data from the individual plotted in Figures 14.1c, 14.2c, and 14.3c as a target and hand-tuned the coefficients of the simultaneous equations to approximate this individual's data as well as the average LH surge shown in Figure 14.6b. This procedure resulted in the constant coefficients substituted into the simultaneous equations as

$$\begin{aligned}
\ddot{x} &= -0.06x - 0.1\dot{x} + 0.2z \\
\ddot{y} &= -1.1y - 0.3\dot{y} + -0.1v \\
\ddot{p} &= -0.06p - .35\dot{p} + 0.1v \\
\dot{v} &= -3.5v - 1.5v / (1 + e^{8(x+y+.5)}) + 4v / (1 + e^{10(-x-y+.5)+3}) - 0.1pv + 0.01
\end{aligned} \quad (14.12)$$

The equations in Equation 14.12 were numerically integrated using Mathematica (*Mathematica 8.0*, 2012) and the resulting trajectories from each equation are plotted in Figure 14.8 for an interval of 90 days. Note that the slow estradiol (Figure 14.8b) and fast estradiol (Figure 14.8c) are added together to form the predicted estradiol levels in Figure 14.8d.

There are minor differences between the first cycle in, for instance, Figures 14.8d and 14.8f, and the subsequent two cycles. That is because there is some carryover in the dynamics from the previous cycle that is not captured in the arbitrary initial conditions used to start the system at time $t = 0$. For this reason, we extracted a 45-day segment of the simulated estradiol and progesterone time series starting at day 38 of the simulation, in order to align the onset of menses in the simulation with the participant's data from Figures 14.1c and 14.2c. The simulated 45-day time series and observed time series for the participant are shown side by side for comparison in Figure 14.9.

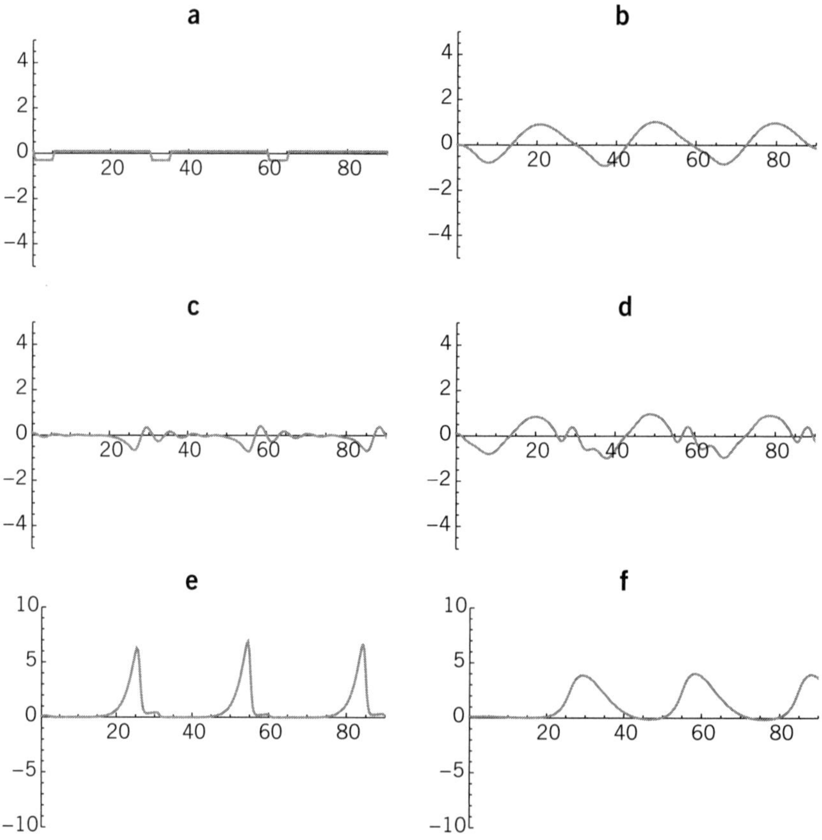

FIGURE 14.8. Simulated 90-day time series for (a) menses discrete time series, (b) slow estradiol cycle levels, (c) rapid estradiol cycle levels, (d) combined estradiol levels, (e) LH levels, and (f) progesterone levels.

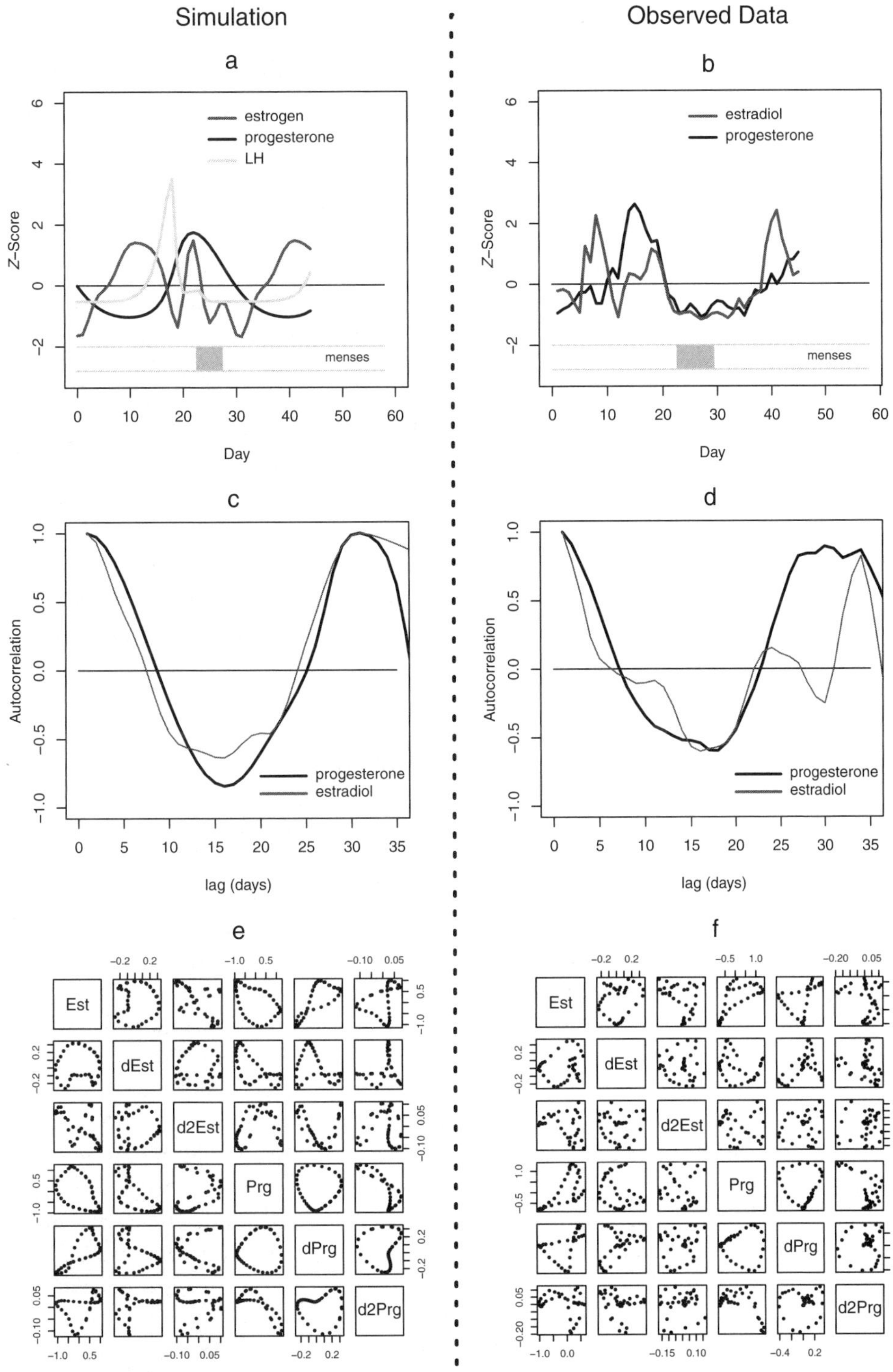

FIGURE 14.9. Comparison of 45 days of simulated data and one study participant's observed data using (a, b) time series, (c, d) autocorrelation functions, and (e, f) phase space reconstruction.

The simulated and observed time series share a number of similar features. However, the simulated time series is smoother and the hormone peaks are more spread out. In addition, the simulation shows more fast oscillation in estradiol than does the observed time series. The spread nature of the hormone peaks in the simulated time series translate into a smoother autocorrelation function plot that deemphasizes the secondary oscillation in estradiol.

Most important are similarities between the phase space plots. These plots allow the relationships between the derivatives of the simulated time series to be compared to the relationships between the derivatives of the observed time series. Most features of the observed phase space plot are captured in the phase space plot of the simulated time series. Note the relationships between the derivatives of progesterone—although the observed time series is a bit noisier, the phase portrait shapes are very well preserved. There is also a striking similarity between the geometric shapes of the phase space of the simulated estradiol derivatives and the observed estradiol derivatives. The least similar are the coupled relationships between the estradiol derivatives and the progesterone derivatives. Although many of the basic geometric features are similar—for example, figure eights remain figure eights and loops remain loops—the details of the shapes are not as well preserved between simulated and observed as they are in the within-hormone dynamics.

Discussion

The aim of this chapter was to show how a complicated dynamical systems model with nonlinear components could be constructed. The process first used graphical display of a few individual time series to identify features that would need to be included in the model. Then, other information that was known about the system, from biology and from aggregate data, was included in order to place constraints on the model design. An integrated list of model constraints was created and then the model was built step by step in order to satisfy these constraints. When the observed time series were not well explained by the biological constraints, new mechanisms were hypothesized and implemented so that the model could reproduce the major features of the data.

The model appears to be able to reproduce at least the major features of the dynamics observed in the individual-level time series—the primary aim of model construction. The next step will be to use model parameter estimation techniques—for example, latent differential equations (Boker, Neale, & Rausch, 2004)—to find maximum likelihood coefficients for the system of differential equations that can account both for within-person dynamics and between-persons individual differences. The results of this modeling effort will then inform models of coupling between daily self-report behavioral variables, such as eating and affect, and the hormone dynamics.

The results presented in this chapter are only a feasibility demonstration. The model includes several hypothesized biological mechanisms for which there is no evidence. There are two possible reasons for this mismatch. First, the model may just be wrong and the data features that we attempted to reproduce in one way perhaps could be reproduced in another way that did not require the novel hypothesized mechanisms. Alternatively, it could be that one or more of these biological mechanisms have not yet been discovered and that the model is pointing to areas for fruitful biological research in ovarian hormone regulation. At the very least, by mathematically specifying a regulatory model using differential equations, a conversation can begin about features in the data that are not currently well explained.

One interesting but unintended consequence of the modeling described here is that the dynamics for the estrogen receptors α and β regulating the LH surge produced a nonlinear instability that is consistent with high variability in the interval between the end of menses and the onset of ovulation but lower variability in the interval between ovulation and onset of subsequent menstrual bleeding. This difference in temporal variability was not part of the model design, but emerged from the dynamics as the biological constraints were implemented. We emphasize this point because when one is building and testing dynamical systems models, one should be alert to recognize emergent consequences of the model—for positive explanatory purposes but also as a way of excluding parts of a model that produce behavior inconsistent with observations.

Conclusions

Differential equations models have the potential to provide explanatory power for understanding the dynamics of individuals' regulation of physiology, behavior, and development. The model building process described in this chapter could be equally well applied to daily diary data, ecological momentary assessment data, or long-term developmental data playing out over periods of decades. Building and fitting these models helps refocus research away from descriptions of what happened on average to a person-centered inquiry about how a system would behave given a particular context and current state. The study of human dynamics should aim for models of *how and why* rather than just models of *what*.

ACKNOWLEDGMENTS

Funding for this work was provided in part by National Institute of Mental Health Grant No. MH082054-01, National Institutes of Health Grant No. 1R21DA024304-01, and the Max Planck Institute for Human Development Lifespan Psychology Group. Any opinions, findings, and conclusions or recommendations expressed in this material are those of the authors and do not necessarily reflect the views of the National Institutes of Health or the Max Planck Society.

REFERENCES

Abarbanel, H., Brown, R., & Kadtke, J. (1990). Prediction in chaotic nonlinear systems: Methods for time series with broadband Fourier spectra. *Physical Review A, 41*(4), 1782–1807.

Bisconti, T. L., Bergeman, C. S., & Boker, S. M. (2004). Emotion regulation in recently bereaved widows: A dynamical systems approach. *Journal of Gerontology: Psychological Sciences, 59*(4), 158–167.

Boker, S. M. (2012). Dynamical systems and differential equation models of change. In H. Cooper, A. Panter, P. Camic, R. Gonzalez, D. Long, & K. Sher (Eds.), *APA handbook of research methods in psychology* (pp. 323–333). Washington, DC: American Psychological Association.

Boker, S. M. (2013). Selection, optimization, compensation, and equilibrium dynamics. *Gero-Psych: The Journal of Gerontopsychology and Geriatric Psychiatry, 26*(1), 61–73.

Boker, S. M., Deboeck, P. R., Edler, C., & Keel, P. K. (2010). Generalized local linear approximation of derivatives from time series. In S.-M. Chow, E. Ferrer, & F. Hsieh (Eds.), *Statistical methods for modeling human dynamics: An interdisciplinary dialogue* (pp. 161–178). Boca Raton, FL: Taylor & Francis.

Boker, S. M., Leibenluft, E., Deboeck, P. R., Virk, G., & Postolache, T. T. (2008). Mood oscillations and coupling between mood and weather in patients with rapid cycling bipolar disorder. *International Journal of Child Health and Human Development, 1*(2), 181–202.

Boker, S. M., Molenaar, P. C. M., & Nesselroade, J. R. (2009). Issues in intraindividual variability: Individual differences in equilibria and dynamics over multiple time scales. *Psychology and Aging, 24*(4), 858–862.

Boker, S. M., Montpetit, M. A., Hunter, M. D., & Bergeman, C. S. (2010). Modeling resilience with differential equations. In P. Molenaar & K. Newell (Eds.), *Learning and development: Individual pathways of change* (pp. 183–206). Washington, DC: American Psychological Association.

Boker, S. M., Neale, M. C., & Rausch, J. (2004). Latent differential equation modeling with multivariate multi-occasion indicators. In K. van Montfort, H. Oud, & A. Satorra (Eds.), *Recent developments on structural equation models: Theory and applications* (pp. 151–174). Dordrecht, The Netherlands: Kluwer Academic.

Deboeck, P. R. (2010). Estimating dynamical systems: Derivative estimation hints from Sir Ronald A. Fisher. *Multivariate Behavioral Research, 45*(4), 725–745.

Fehring, R. J., Schneider, M., & Raviele, K. (2006). Variability in the phases of the menstrual cycle. *Journal of Obstetric, Gynecologic and Neonatal Nursing, 35*, 376–384.

Geirsson, R. T. (1991). Ultrasound instead of last menstrual period as the basis of gestational age assignment. *Ultrasound in Obstetrics and Gynecology, 1*(3), 212–219.

Hu, L., Gustofson, R. L., Feng, H., Leung, P. K., Mores, N., Krsmanovic, L. Z., et al. (2008). Converse regulatory functions of estrogen receptor-α and -β subtypes expressed in hypothalamic gonadotropin-releasing hormone neurons. *Molecular Endocrinology, 22*(10), 2250–2259.

Kennel, M. B., Brown, R., & Abarbanel, H. D. I. (1992). Determining embedding dimension for phase–space reconstruction using a geometrical construction. *Physical Review A, 45*(6), 3403–3411.

Klump, K. L., Keel, P. K., Kashy, D. A., Racine, S., Burt, S. A., Neale, M., et al. (2013). The interactive effects of estrogen and progesterone on changes in binge eating across the menstrual cycle. *Journal of Abnormal Psychology, 122*(1), 131–137.

Mathematica 8.0. (2012). Champaign–Urbana, IL: Wolfram Research.

Munster, K., Schmidt, L., & Helm, P. (1992). Length and variation in the menstrual cycle: A cross-sectional study from a Danish county. *British Journal of Obstetrics and Gynecology, 99*, 422–429.

Oertzen, T. V., & Boker, S. M. (2010). Time delay embedding increases estimation precision of models of intraindividual variability. *Psychometrika, 75*(1), 158–175.

Ong, A. D., Bergeman, C. S., & Boker, S. M. (2009). Resilience in adulthood comes of age: Defining features and dynamic conceptions. *Journal of Personality, 77*(6), 1777–1804.

Sauer, T., Yorke, J., & Casdagli, M. (1991). Embedology. *Journal of Statistical Physics, 65*(3, 4), 95–116.

Savitzky, A., & Golay, M. J. E. (1964). Smoothing and differentiation of data by simplified least squares. *Analytical Chemistry, 36*, 1627–1639.

Stricker, R., Eberhart, R., Chevailler, M.-C., Quinn, F. A., Bischof, P., & Stricker, R. (2006). Establishment of detailed reference values for luteinizing hormone, follicle stimulating hormone, estradiol, and progesterone during different phases of the menstrual cycle on the abbott ARCHITECT analyzer. *Clinical Chemistry and Laboratory Medicine, 44*(7), 883–887.

Takens, F. (1985). Detecting strange attractors in turbulence. In A. Dold & B. Eckman (Eds.), *Lecture notes in mathematics 1125: Dynamical systems and bifurcations* (pp. 99–106). Berlin: Springer-Verlag.

Whitney, H. (1936). Differentiable manifolds. *Annals of Mathematics, 37*, 645–680.

APPENDIX 14.1.

```
# --------------------------------------------------------------------------
# Program: MatrixPlotHormoneDerivatives.R
#  Author: Steven M. Boker
#    Date: Thu Dec 13 11:53:05 EST 2012
#
# This program is an example of how to create a matrix plot of
#   calculated derivatives from time series.
# The two functions that are read in may be obtained from the author's
#   website at
#   http://people.virginia.edu/~smb3u/GLLAfunctions.R
#   http://people.virginia.edu/~smb3u/GOLDestimates.R
# The simulated data can also be obtained from the same website:
#   http://people.virginia.edu/~smb3u/symOvarianSystem120913.csv
# This script can be downloaded at
#   http://people.virginia.edu/~smb3u/MatrixPlotHormoneDerivatives.R
#
# --------------------------------------------------------------------------

# --------------------------------------------------------------------------
# Read libraries and set options.

options(width=130)
source("GLLAfunctions.R")
source("GOLDestimates.R")
library(psych)

# --------------------------------------------------------------------------
# Set constants.

theDeltaT <- 1
embedD <- 11
theTau <- 1

# --------------------------------------------------------------------------
# Read the simulated from the chapter data.

simFrame1 <- read.table("symOvarianSystem120913.csv", header=FALSE,
sep=",",
                col.names=c("day", "bleed", "slowEst", "dSlowEst",
"fastEst",
                            "dFastEst", "LH", "dLH", "progest",
"dProgest"))

describe(simFrame1)

# --------------------------------------------------------------------------
```

```
# Standardize variables and combine the slow and fast estradiol.

ZsimFrame1 <- data.frame(day=simFrame1$day,
                         bleed=simFrame1$bleed,
                         slowEst=scale(simFrame1$slowEst),
                         dSlowEst=scale(simFrame1$dSlowEst),
                         fastEst=scale(simFrame1$fastEst),
                         dFastEst=scale(simFrame1$dFastEst),
                         estradiol=scale(simFrame1$fastEst+.7*simFrame1$s
lowEst),
                         LH=scale(simFrame1$LH),
                         dLH=scale(simFrame1$dLH),
                         progest=scale(simFrame1$progest),
                         dProgest=scale(simFrame1$dProgest))

# ------------------------------------------------------------------------
# Time delay embed the three hormone variables.

embeddedEST <- gllaEmbed(ZsimFrame1$estradiol, embed=embedD, tau=theTau,
                         label="est", idColumn=FALSE)
embeddedPRG <- gllaEmbed(ZsimFrame1$progest, embed=embedD, tau=theTau,
                         label="prg", idColumn=FALSE)
embeddedLH <- gllaEmbed(ZsimFrame1$LH, embed=embedD, tau=theTau,
                         label="lh", idColumn=FALSE)

# ------------------------------------------------------------------------
# Combine first and last time-delay embedded columns into one matrix.

tEmbeddedMatrix <- cbind(embeddedEST[,c(1,embedD)],
embeddedPRG[,c(1,embedD)],
                         embeddedLH[,c(1,embedD)])

# ------------------------------------------------------------------------
# Create a pdf of a 2-D state-space plot for the three variables.

pdf(file=paste("MatrixPlot_StateSpace_Tau_11.pdf", sep=""), height=5,
width=5)
pairs(tEmbeddedMatrix, cex=.25)
dev.off()

# ------------------------------------------------------------------------
# Estimate 0th, 1st, and 2nd derivatives for each variable.

wMatrix <- GOLDW(c(0:(embedD-1)),2)
gllaDataEST <- embeddedEST %*% wMatrix
dimnames(gllaDataEST) <- list(NULL, c("Est", "dEst", "d2Est"))
gllaDataPRG <- embeddedPRG %*% wMatrix
dimnames(gllaDataPRG) <- list(NULL, c("Prg", "dPrg", "d2Prg"))
gllaDataLH <- embeddedLH %*% wMatrix
dimnames(gllaDataLH) <- list(NULL, c("LH", "dLH", "d2LH"))
```

```
# ------------------------------------------------------------------------
# Merge estimated derivatives of all variables into a single data frame

gllaFrame <- data.frame(gllaDataEST, gllaDataPRG, gllaDataLH)
describe(gllaFrame)

# ------------------------------------------------------------------------
# Create a pdf of a phase space plot.

pdf(file=paste("MatrixPlot_PhaseSpace_D_11.pdf", sep=""), height=5,
width=5)
pairs(gllaFrame, cex=.25)
dev.off()
```

PART VII

NONLINEAR DYNAMICAL MODELS OF DEVELOPMENT

CHAPTER 15

A Regime-Switching Longitudinal Model of Alcohol Lapse–Relapse

Sy-Miin Chow, Katie Witkiewitz, Raoul Grasman, R. Shane Hutton, and Stephen A. Maisto

Over the last few decades, some researchers' initial enthusiasm for nonlinear dynamical systems has gone through various periods of waxing and waning. Since the pioneering work of Poincaré (1892) and Lorenz's (1963) computer simulation of a "chaotic" weather system (for an introduction, see Alligood, Sauer, & Yorke, 1996; Gleick, 1987; Hilborn, 1994), some of the properties of nonlinear dynamical systems have continued to intrigue social and behavioral scientists. In particular, the idea that a small difference in initial conditions could lead to drastically discrepant outcomes as time progresses, or that the existence of models, broadly classified as catastrophe models, can capture discontinuous jumps in various systems' change trajectories, have great conceptual appeal to social scientists (e.g., see Vallacher & Nowak, 1994).

The Cusp Catastrophe Model

Catastrophe theory (Thom, 1975, 1993) is the study of the many ways in which continuous changes in a system's parameters can result in discontinuous changes in an outcome variable of interest. Various catastrophe models have been used to represent sudden jumps manifested, for instance, in individuals' driving speed (Cobb, 1981; Poston & Stewart, 1978; Zeeman, 1976), attitude (Wimmers, Savelsbergh, & van der Kamp, 1998), affective states (Allen & Carifo, 1995; Strahan & Conger, 1999), alcohol use (Clair, 1999; Witkiewitz & Marlatt, 2004), paranormal beliefs (Lange & Houran, 2000), and human development in discrete stages (Byrne, Mazanov, & Gregson, 2001; Preece, 1980; Molenaar & Raijmakers, 2000; van der Maas & Molenaar, 1992). Among the seven possible catastrophe models considered by Thom, the cusp catastrophe is one of the simplest and most commonly applied forms in the social sciences. The model posits that changes in the values of an outcome variable (often termed the *behavioral parameter*) depend on two control parameters: a splitting parameter and a normal parameter (Gilmore, 1981; Stewart & Peregoy, 1983). In practice, these parameters are often expressed as linear, fixed functions of one or more covariates (e.g., Cobb & Zacks, 1985; Grasman, van der Maas, & Wagenmakers, 2009; Guastello, 1984; van der Maas & Molenaar, 1992). Thus, they are sometimes referred to as the *splitting* and *normal variables*.

Several approaches have been used to fit various catastrophe models. Cobb (1981; Cobb, Koppstein, & Chen, 1983) formulated the cusp catastrophe model as a stochastic partial differential equation model and used a likelihood approach to estimate the parameters. As an alternative to Cobb's method, Guastello (1982, 1992) used polynomial regression to estimate the parameters in the catastrophe model. This estimation technique has been applied by researchers such as Hufford, Witkiewitz, Shields, Kodva, and Caruso (2003) in modeling the relapse process of alcohol use. Alexander, Herbert, DeShon, and Hanges (1992) pointed out some problems associated with Guastello's approach and suggested the use of Oliva, DeSarbo, Day, and Jedidi's (1987) multivariate confirmatory approach as an alternative. In particular, Guastello (1982), just as Brown (1995) in a different approach, used empirical difference scores as the dependent variable in estimating the parameters for a cusp catastrophe model. Brown used nonlinear least squares together with a Runge–Kutta integration procedure to obtain iterative estimates of the parameters in the cusp catastrophe model, whereas Guastello reformulated the cusp catastrophe model as a polynomial regression model and used ordinary least squares to estimate the corresponding parameters.

For the purpose of this chapter, suffice it to say that the difference score has a long history of controversy in psychology. In some cases use of the difference score is totally justifiable (e.g., Nesselroade & Bartsch, 1977; Nesselroade & Cable, 1974; Hummel-Rossi & Weinberg, 1975; Maxwell & Howard, 1981), but in others, not necessarily so (Bereiter, 1963; Cronbach & Furby, 1970; Harris, 1963; Horn, 1963; Lord, 1956). Methodological advances in the past few decades have enabled researchers to formulate difference score-related approaches within a latent variable framework, which helps strengthen the appeal of such approaches when used with noisy data (e.g., McArdle & Hamagami, 2001; Molenaar & Newell, 2003). This feature has not been incorporated into the modeling approaches summarized above. One other issue pertaining to these approaches is that transformation functions are typically used to reparameterize the cusp catastrophe model into a simpler form (e.g., polynomial regression model) to facilitate model fitting. However, when the transformation functions are not explicitly specified as constraints in the model specification, there is no guarantee that the fitted, simpler model corresponds in any way to the original model (for a discussion of this issue in the context of stochastic differential equation modeling, see Singer, 1992, 1993).

More importantly, although some of the features of catastrophe models have proven useful across a wide range of applications, they impose several assumptions that may be overly restrictive in empirical applications. The goal of this study is to present a mixture structural equation model with regime switching as an alternative approach to account for some of the known data-analytic issues associated with the cusp catastrophe model. Although the new proposed model retains only some of the features of the original model thought to be particularly pertinent to the modeling of alcohol use dynamics, it provides more flexibility than the cusp catastrophe model in a number of ways. Some of these advantages include the following: (1) It is suited for use with longitudinal panel data with multiple subjects; (2) missing data can be readily accommodated; (3) model fitting can be performed at the latent variable level to explicitly distinguish measurement errors from process noises; (4) researchers can test multivariate extensions involving lagged effects; (5) ongoing, over-time fluctuations in drinking behaviors can be represented in more flexible ways; (6) it helps capture heterogeneous timing of relapse within and across subjects; and (7) the presence of a large number of abstainers at any given time point—referred to as the *zero-inflation phenomenon* in statistics—is represented more flexibly using mixture modeling techniques. The proposed model is illustrated using a set of longitudinal drinking data from the COMBINE study (Anton et al., 2006).

Mathematical Background on the Cusp Catastrophe System

Consider a dynamical system whose changes over time are governed by a deterministic potential function, $V(y; \boldsymbol{\theta})$, as

$$\frac{dy}{dt} = \frac{-\partial V(y;\boldsymbol{\theta})}{\partial y} \qquad \textbf{(15.1)}$$

where y is the dependent variable of interest—namely, the behavioral variable in the catastrophe literature—and $\boldsymbol{\theta}$ is a vector of control parameters. One of the simplest forms of catastrophe is the cusp catastrophe, described by the potential function

$$-V(y;\boldsymbol{\theta}) = \frac{1}{2}\beta y^2 + \alpha y - \frac{1}{4}y^4 \qquad \textbf{(15.2)}$$

where $\boldsymbol{\theta} = [\alpha\ \beta]'$ consists of the two control parameters in the system. For reasons that will become clear in a moment, the parameter α is referred to as the *normal variable* or *asymmetry variable*, whereas β is often referred to as the *splitting* or *bifurcation variable* (Gilmore, 1981; Grasman et al., 2009; Poston & Stewart, 1978). The equilibrium points of the cusp catastrophe systems are values of y for which $\frac{-\partial V(y;\boldsymbol{\theta})}{\partial y} = 0$. That is, they are the y values the system eventually settles into over time. These points are functions of the control parameters α and β, and they can be obtained by differentiating Equation 15.2 with respect to y and setting it to zero as

$$\beta y + \alpha - y^3 = 0 \qquad \textbf{(15.3)}$$

to obtain the roots or equilibrium points that arise under different values of α and β. It is well known that (e.g., see Poston & Stewart, 1978) the number and nature of the equilibrium points that arise under different values of α and β can be deduced from the Cardan's discriminant function, expressed as

$$D = 4\beta^3 + 27\alpha^2 \tag{15.4}$$

A stable equilibrium is defined as a point toward which a system gravitates and settles into in the long run when small perturbations are imposed to move the system away from the equilibrium point. The value of D helps classify the equilibrium points of the system under different α–β configurations into four broad categories or sets, which are plotted in Figure 15.1 (i.e., A, B, C_1, and C_2). Specifically, the shaded region labeled "A" corresponds to values of D that are < 0, and there are three equilibrium points when α and β assume values in this region. The four quadrants labeled "B" outside of the shaded region all correspond to $D > 0$, and each quadrant is characterized by only one stable equilibrium point. The exact value of the equilibrium point differs by quadrant, as is indicated approximately by the modes of the first three conditional densities plotted in the right panel of Figure 15.1. The third set simply consists of the origin at $\alpha = 0$ and $\beta = 0$. At this point, there is only one stable equilibrium point at $y = 0$.

The two branches or borders of the shaded region, marked as C_1 and C_2 in Figure 15.1, constitute the last set of values for which $D = 0$. On each of these branches, the system is characterized by one unstable equilibrium, namely, an inflection point, and one stable equilibrium (indicated by one of the two peaks of the last conditional density in Figure 15.1). The location of the stable equilibrium differs depending on whether the system shows changes in α and β values that cross the branches (C_1 or C_2) from left to right, or from right to left. When either one of the two branches is crossed, the original stable minimum is annihilated and a new stable minimum arises. The asymmetry in the values of the equilibria when the branches are crossed from left to right versus from right to left constitutes one of the key catastrophe flags, known as *hysterisis*, which we elaborate further below.

It is often more intuitive to understand the behavioral dynamics of a catastrophe system geometrically in the catastrophe manifold,[1] which shows all the possible equilibrium points at different combinations of α and β values. Witkiewitz and colleagues (Witkiewitz & Marlatt, 2004, 2007; Witkiewitz, van der Maas, Hufford, & Marlatt, 2007) have introduced the cusp catastrophe model as a useful representation of alcohol lapse–relapse dynamics. Their specific example provides a helpful introduction to the various concepts and key constituent variables of the cusp catastrophe model, and we adopt the terminologies in their application to illustrate the features of the model.

Catastrophe in Alcohol Use Dynamics

According to the framework presented by Witkiewitz and Marlatt (2004), distal and proximal processes serve as the splitting and normal parameters in the cusp catastrophe model, respectively, to affect drinking outcomes (i.e., the behavioral parameter). Distal risks consist of background variables (e.g., alcohol dependence) and relatively

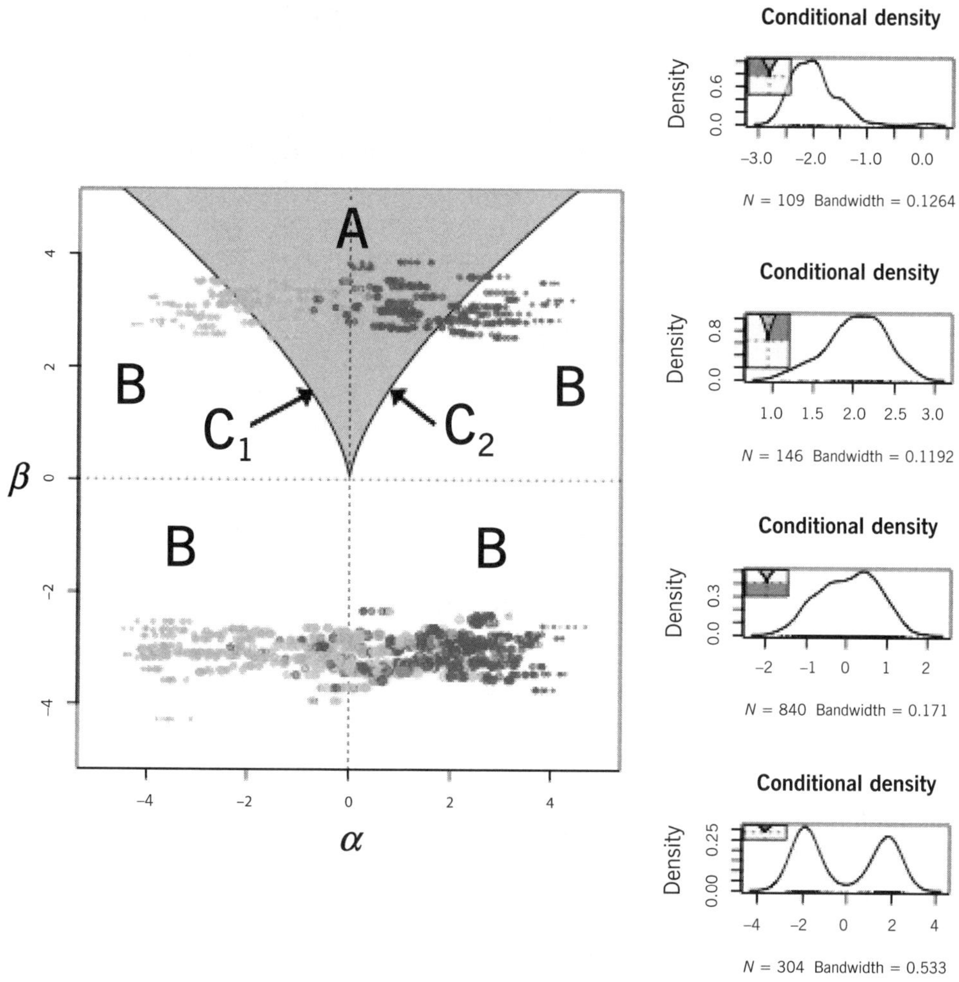

FIGURE 15.1. A plot summarizing the broad categories of behaviors in the cusp catastrophe system as a function of different combinations of the normal (α) and splitting (β) parameters. The plot was created using the *cusp* package in R written by Grasman et al. (2009). The points located on the α–β plane are simulated data generated using the cusp catastrophe model. The conditional densities located in the right panel show the densities of the data points located in different quadrants of the α–β plane.

stable pretreatment characteristics (e.g., alcohol expectancies). Proximal risks, in contrast, include transient precipitants (e.g., stressful situations, negative affect) that change an individual's tendency to drink on a moment-to-moment basis. The resultant catastrophe manifold depicting the equilibrium values of drinking (represented as the vertical axis) at different values of distal risks (the splitting parameter, β) and proximal risks (the normal parameter, α) is portrayed in Figure 15.2. The bottom surface of the model, called the *control plane*, is defined by the range of possible values for the two control parameters. The top surface, called the *behavior plane*, is defined as the range of predicted values of the outcome variable (i.e., drinking) at different values of the two control parameters. The folded part of the behavior plane represents the discontinuity that is characteristic of the model, whereby the same values of the control parameters can result in two widely differing values of the behavior variable. While in this region, intermediate outcome values are unlikely to occur or are not sustainable (i.e., they do not constitute a stable equilibrium). This property renders these intermediate outcome values to be called the *inaccessible region* (Gilmore, 1981; Zeeman, 1978).

Putting the cusp catastrophe model in the context of alcohol use, Witkiewitz and colleagues (Witkiewitz & Marlatt, 2004, 2007; Witkiewitz et al., 2007) stipulated that for individuals with low distal risk, drinking changes linearly with changes in proximal risk (Path A). However, for those with high distal risk, a small increase in proximal risk (e.g., an argument with a significant other) may push an individual over the edge, leading to an episode of lapse (Path B). A prolapse (the return from heavy drinking to abstinence; Path C) then requires a substantially larger reduction in proximal risk to help the individual regain abstinence. These unique characteristics of the cusp catastrophe model are just some of the flags associated with it. The seven

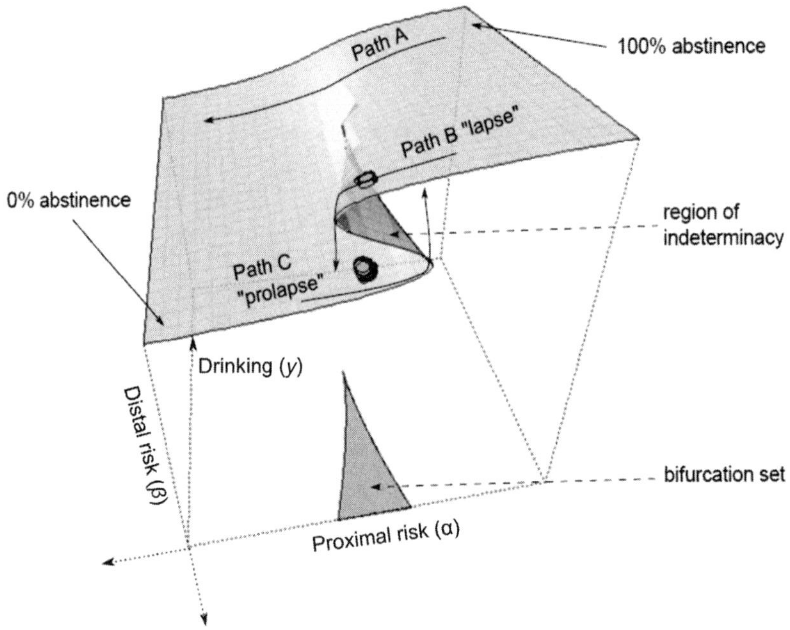

FIGURE 15.2. The cusp catastrophe model adopted by Witkiewitz and colleagues to represent the multiple pathways of AUD clinical course. The circles represent possible predictions of this model for individuals with high distal risk, which put an individual into one of two possible modes of behavior. The plot was created using the *cusp* package in R written by Grasman et al. (2009).

catastrophe flags associated with this model include multimodality, sudden jumps, inaccessibility, hysteresis, anomalous variance, divergence of linear response, and critical slowing down. Each of these flags is described briefly here.

Multimodality is indicated by the presence of more than one mode of outcome variable based on the same values of the control parameters (or in the context of alcohol use, distal and proximal risks). The multiple modes arise because of the presence of multiple equilibria points in the system. Thus, continuous changes in the distal risk (β in Figures 15.1 and 15.2) can yield a sudden, qualitative shift from a single mode of outcome (when an individual is located near Path A of the cusp landscape) to the coexistence of two modes of outcome (when an individual is located near the fold). In this case, bifurcation is said to have occurred. Thus, distal risk (or the splitting parameter) is also termed the *bifurcation variable*. The range of distal risk values at each specific value of proximal risk, α, under which bifurcation into two modes of outcome is observed constitutes the bifurcation set.

Individuals may show *sudden jumps* or abrupt transitions from one mode of behavior to another mode of behavior (i.e., from one equilibrium point to another). Such sudden jumps between the two behavioral modes render any intermediate value between the two modes unlikely to occur, or in other words, *inaccessible*. The combination of parameter values that gives rise to such behavior constitutes a *region of indeterminacy*. That is, even if values of the control parameters are known, it is impossible to determine which of the two possible modes the outcome variable would manifest unless previous history of the outcome variable is known.

Hysteresis refers to the asymmetry in the values of the normal parameter (i.e., proximal risk) that trigger a sudden jump in the direction of a prolapse (i.e., a transition from heavy drinking to abstinence, as indicated by Path C) as opposed to a lapse (i.e., a transition from abstinence to heavy drinking, denoted as Path B in Figure 15.2). That is, the points at which the system shows a sudden transition in behavioral mode depend on the previous values of the system and the direction of change. Thus, although a particular value of proximal risk (e.g., stress) is enough to cause an individual to "fall off the wagon" and transition from abstinence to drinking, a much lower level of proximal risk is needed before the same individual can return from drinking to abstinence (i.e., to be back "on the wagon").

Divergence is present when greater bimodality is evident in the outcome variable with an increase in the splitting parameter (i.e., distal risk). In the context of alcohol use, this means that individuals who have higher levels of distal risk also tend to manifest more extreme transitions between heavy drinking and abstinence, as opposed to the gradual changes in drinking level characterizing those with lower distal risk (e.g., see Path A in Figure 15.2). Closely associated with this flag is the *anomalous variance* flag—namely, the increase in variance in the outcome variable near the points of sudden transition.

Divergence of linear response and *critical slowing down* are two catastrophe flags that emerge when a system is perturbed, or, in other words, when slight changes are induced in the control parameters, either mathematically or experimentally. Divergence of linear response occurs when perturbations of a system lead to larger fluctuations near the transition points, whereas critical slowing down corresponds to the feature that longer time is needed for the system to return to equilibrium if the system is perturbed near the transition points. Such manipulations are difficult to implement in applications in the social and behavioral sciences, and these flags have thus not been discussed as often as the other flags (e.g., see Zeeman, 1978).

Mixture Structural Equation Model with Regime Switching

Structural equation modeling (SEM; Jöreskog, 1973) is a statistical technique for representing multivariate relationships among observed and latent variables. Composed of a measurement model that relates the observed indicators to a set of underlying latent variables, and a structural model that specifies the interrelationships among the latent variables, SEMs and related latent variable models have been used extensively in the social and behavioral sciences. When systematic heterogeneities exist across subpopulations but the grouping variable that distinguishes the subpopulations is unobserved (i.e., latent), mixture SEMs (Arminger & Muthén, 1998; Dolan & van der Maas, 1998; Jedidi, Jagpal, & DeSarbo, 1997; Vermunt & Magidson, 2005; Yung, 1997) can be employed.

Mixture structural equation modeling (MSEM) is an extension of SEM that allows for heterogeneities in the means and covariance structures of an SEM model conditional on a series of latent, unobserved groups, Ci ($i = 1, \ldots, n$), typically referred to as latent *classes*. More recently, SEM statistical packages such as Mplus also allow users to fit longitudinal extensions of MSEMs and estimate how individuals transition among different latent classes over time (Kaplan, 2008; Muthén & Asparouhov, 2011; Nylund-Gibson, Muthén, Nishina, Bellmore, & Graham, 2013). The use of different latent classes over time to represent the different phases of a process is consistent with the notion of regime switching in the time series and econometric literature (Hamilton, 1994; Kim & Nelson, 1999). Here we use the terms *latent class* and *regime* interchangeably. The resultant model is referred to as a mixture structural equation model with regime switching (MSEM-RS).

In an MSEM-RS, the measurement model for an individual i in latent class k, namely, $Ci = k$, can be expressed as

$$\begin{aligned} \mathbf{y}_{i|C_i=k} &= \boldsymbol{\nu}_k + \boldsymbol{\Lambda}_k \boldsymbol{\eta}_i + \boldsymbol{\varepsilon}_i \\ (\boldsymbol{\varepsilon}_i | C_i = k) &\sim \mathcal{N}(0, \boldsymbol{\Omega}_k) \end{aligned} \quad \textbf{(15.5)}$$

where $\mathbf{y}_i$ is a $p \times 1$ vector of continuous observed variables for individual i, $\boldsymbol{\nu}_k$ is a $p \times 1$ vector of intercepts, $\boldsymbol{\Lambda}_k$ is a $p \times w$ matrix of factor loadings, $\boldsymbol{\eta}_i$ is a $w \times 1$ vector of latent variables, and $\boldsymbol{\varepsilon}_i$ is a $p \times 1$ vector of measurement errors. The subscript k highlights the class-specific nature of the parameters. In cases involving ordinal observed variables, the corresponding elements in $\mathbf{y}_i$ are unobserved. The lth unobserved continuous variable, y_{il}, is then linked to the corresponding manifest ordinal response, y^*_{il} (Jöreskog & Moustaki, 2001; Muthén, 1984) as

$$y_{il}^{*} = s \Leftrightarrow \tau_{l,s-1} < y_{il} \leq \tau_{l,s}, s = 1, \ldots, S \tag{15.6}$$

where $\tau_{l,h}$, is a set of threshold values for variable l that is held invariant across individuals.

Conditional on class k, a structural model serves to specify the relationships among the latent variables in $\boldsymbol{\eta}_i$ as

$$\begin{aligned} \boldsymbol{\eta}_{i|C_i=k} &= \boldsymbol{\alpha}_k + \boldsymbol{\beta}_k \boldsymbol{\eta}_i + \boldsymbol{\zeta}_i \\ (\boldsymbol{\zeta}_i | C_i = k) &\sim \mathcal{N}(0, \boldsymbol{\Psi}_k) \end{aligned} \tag{15.7}$$

where $\boldsymbol{\alpha}_k$ is a vector of regime-dependent intercepts,[2] $\boldsymbol{\beta}_k$ is a $w \times w$ matrix of regime-dependent regression effects among the latent variables, and $\boldsymbol{\zeta}_i$ is a $w \times 1$ vector of disturbances. When multiple time points of the same variables are present, y_i and $\boldsymbol{\eta}_i$ are expanded to include $\mathrm{y}_i = [\mathrm{y}'_{i1} \ldots \mathrm{y}_{iT'}]'$ and $\boldsymbol{\eta}_i = [\boldsymbol{\eta}'_{i1} \ldots \boldsymbol{\eta}_{iT'}]'$. Other corresponding modeling components are also expanded accordingly to accommodate the multiple time points.

Multinomial logistic regression models are typically used to represent the initial class (or regime) probabilities as

$$\Pr(C_{i1} = h | \boldsymbol{x}_{i1}) \triangleq \pi_{h,i1} = \frac{\exp(a_{h,1} + c'_{h,1} x_{i1})}{\sum_{s=1}^{K_1} \exp(a_{s,1} + c'_{s,1} x_{i1})} \tag{15.8}$$

where $a_{h,1}$ is the logit intercept for class h at Time 1; $\boldsymbol{x}_{i1}$ is a vector of covariates used to predict initial class membership, c_{h1} is the associated vector of logit slopes, and K_1 is the number of latent classes at Time 1. For identification purposes, one of the classes has to be designated to be the reference class, and the logit intercept and slopes for this class, $a_{K_1,1}$ and $c_{K_1,1}$ have to be set to zeroes.

A multinomial logistic regression model is, again, used to describe each individual i's transition in class membership from Time t-1 to Time t. In the general MSEM-RS framework, the transition in class membership can depend on all previous class membership information (Asparouhov & Muthén, 2011). We adopt a simpler specification and only allow this transition to depend on the class membership at Time t-1 and the initial class membership at Time 1. The corresponding multinomial logistic regression model is written as

$$\Pr(C_{it} = k | C_{i,t-1} = j, C_{i1} = h, \boldsymbol{x}_{it}) \triangleq \pi_{jk,h,it} = \frac{\exp(a_{k,h,t} + \boldsymbol{b}'_{jk,h,t} \boldsymbol{x}_{it})}{\sum_{s=1}^{K_t} \exp(a_{s,h,t} + \boldsymbol{b}'_{js,h,t} \boldsymbol{x}_{it})} \tag{15.9}$$

where $\pi_{jk,h,it}$ is individual i's transition probability of moving from class j at Time t-1 to class k at Time t, conditional on initial class membership in class h; Kt denotes the number of classes at Time t, highlighting the possibility that latent classes may emerge or diminish over time. $a_{k,h,t}$ is the logit intercept for the kth class at Time t, given that individual i is in initial class h at Time 1, $\boldsymbol{x}_{it}$ is a vector of fixed covariates for predicting the transition probabilities, with an associated vector of logit slopes, $\boldsymbol{b}_{jk,h,t}$.

Included in x_{it} is a binary constant reflecting the deviation in log odds of transitioning from regime j at Time t-1 to regime k at Time t relative to transitioning from regime j to regime K_t, given initial class membership in class h. Note that no binary constants are included in x_{i1} in Equation 15.8 because at Time 1, there are no transition probabilities to be predicted, only class probabilities.

Similar to the need to select a reference class at Time 1 for identification purposes, a reference class also has to be selected for each time point and all the logit-related parameters for the reference class. This has the effect of setting each row of person i's $K_{t-1} \times K_t$ conditional transition probability matrix at Time t for each initial latent class to sum to 1. For instance, if two regimes exist at Time t-1 as well as at Time t,

$$\pi_{.,h,it} = \begin{bmatrix} \pi_{11,h,it} & \pi_{12,h,it} \\ \pi_{21,h,it} & \pi_{22,h,it} \end{bmatrix} \tag{15.10}$$

where the element $\pi_{jk,h,it}$, as obtained using Equation 15.9, represents the probability of transitioning from regime j at Time t-1 to regime k at Time t for person i, conditional on the initial class membership and the time-varying covariates at Time t. This ensures that once an individual is assumed to be in one of the K_{t-1} regimes at Time t-1, the individual can only switch to one of the K_t hypothesized regimes at Time t.

Maximum likelihood estimates of the parameters in the MSEM-RS can be obtained via the expectation–maximization (EM; Dempster, Laird, & Rubin, 1977; Everitt & Hand, 1981; Titterington, Smith, & Markov, 1985) algorithm. The estimation details have been documented elsewhere (e.g., see Asparouhov & Muthén, 2011; Muthén & Shedden, 1999) and are not reiterated here.

A Cusp Catastrophe-Inspired MSEM-RS Model for Representing Alcohol Use Dynamics

An MSEM-RS is proposed as an alternative to the cusp catastrophe model. Two unobserved initial regimes at Time 1 are first obtained by "slicing" the catastrophe landscape along the direction of the β-axis (i.e., the distal risk axis) to yield two initial regimes: one corresponding to high distal risk and another to low distal risk. That is, the initial regime variable partitions the α–β plane in Figure 15.1 into two segments, one with $\beta > 0$ and the other with $\beta < 0$. Then, conditional on whether individuals are in one of these initial regimes, we further describe the catastrophe landscape in terms of a "high drinking" regime and a "low drinking" regime. In contrast to the cusp catastrophe model, we explicitly incorporate some added dynamic functions to represent the within-person, over-time variability in the levels of the behavioral variable while individuals are in these two regimes.

Initial Class Membership

At Time 1, the only latent variable of interest is an individual's latent distal risk based on baseline individual difference measures. We assume that two regimes or classes exist at Time 1, with the first class representing a high distal risk class (denoted as C_{i1} = high distal), and the second class representing a low distal risk class (denoted as C_{i1} = low distal). The same measurement model is assumed to be at work for both distal risk classes, with a series of distal risk measures serving as indicators of an individual's latent distal risk in the form of Equations 15.5 and 15.6, except that there is no class dependency on the parameters in the measurement model.

A class-specific structural model is then used to define the distribution of individuals' latent distal risk. For those in the high distal risk class, the initial distribution of latent distal risk is specified to be

$$(Distal_{it} | C_{i1} = \text{high distal}) \sim N(m_{high}, \psi_{high})$$

whereas for those in the low distal risk class, the corresponding initial distribution is specified to be

$$(Distal_{it} | C_{i1} = \text{low distal}) \sim N(m_{low}, \psi_{low}) \quad \textbf{(15.11)}$$

In other words, we allow the two classes that exist at Time 1 to differ in their average levels of distal risk (with means m_{high} and m_{low}, respectively), as well as variances in distal risk levels (i.e., given by ψ_{high} and ψ_{low}, respectively). To ensure that the first class is a high distal risk class, one may impose the constraint that $m_{high} > m_{low}$.

Time 2 and Beyond

For $t = 2$ and beyond, a measurement model in the form of Equations 15.5 and 15.6 is again used to describe the individuals in all latent classes/regimes, with no class dependency in measurement parameters. In this case, the latent variable vector η_{it} reduces simply to one latent variable, $Drink_{it}$, representing individual i's latent drinking tendency at Time t. y_{it} consists of continuous (observed and unobserved) measures of the individual's underlying drinking tendency.

At the structural level, individuals' levels of drinking tendency are hypothesized to show distinct changes depending on the regime the person is in at Time t, expressed as

Regime	Model
R1: High drinking	$(Drink_{it} - Drink^{*}_{high}) = \alpha_{high}(Drink_{i,t-1} - Drink^{*}_{high}) + \zeta_{drink,it}$
R2: Low drinking	$(Drink_{it} - Drink^{*}_{low}) = \alpha_{low}(Drink_{i,t-1} - Drink^{*}_{low}) + \zeta_{drink,it}$

(15.12)

where the two regimes are distinguished by within-person, over-time fluctuations around two different baseline drinking levels, $Drink^{*}_{high}$ and $Drink^{*}_{low}$, with autoregressive parameters, α_{high} and α_{low}, respectively. That is, while in each of these two regimes, deviations in drinking tendency are hypothesized to fluctuate over time following an autoregressive model of order 1 (i.e., an AR(1) model), as driven by process noise, $\zeta_{drink,it}$, at Time t and a lag-one autoregression parameter (α_{high} or α_{low}). To identify the two regimes, the high drinking regime is constrained to have a higher baseline, $Drink^*_{\text{high}}$, than the low drinking regime baseline of $Drink^*_{low}$. The process noise variable, $\zeta_{drink,it}$, is often referred to as "random shock" because it encompasses all remaining uncertainties that affect the latent variables in ways that are not modeled explicitly (Browne & Nesselroade, 2005). The process noise variable is assumed to be normally distributed with zero mean and variance, ψ_{high} and ψ_{low}, respectively, for the high drinking regime and low drinking regime.

The hypothesized over-time dynamics of individuals' drinking levels can be better understood by taking a closer examination of the value of the AR(1) parameter. If the absolute value of the AR(1) parameter is less than 1, any deviations in drinking tendency from baseline are projected to diminish exponentially over time in the absence of new shocks imposed by $\zeta_{drink,it}$. The speed with which past influences diminish over time is determined by the AR(1) parameter. That is, the closer it is to 0, the faster the rate of return to the regime-specific baseline drinking level. In the special case where a substantial portion of the individuals in a sample is able to maintain abstinence over long periods of time (e.g., in an especially successful treatment program), the structural equation for the low drinking regime can be readily modified by setting $Drink^*_{low}$ to zero. In addition, the process noise variance in the regime can be freed to yield a "nearly abstinent" regime during which only some fluctuations in drinking level are observed, but always around a baseline of zero. Alternatively, both α_{low} and the process noise variance in this regime can be set to zero to yield a complete abstinence regime, namely, one with no variability in drinking level while individuals are in this regime. Higher-order AR processes can also be included by allowing drinking tendency from more distant time points (e.g., from Time t-2, t-3, and so on) to influence the current drinking tendency at Time t.

One of the key appeals of the cusp catastrophe model lies in its ability to allow for the presence of multiple modes of behavioral outcome (i.e., multimodality), as well as the possibility that individuals may show sudden jumps from one mode to another depending on their distal risk levels. In the context of alcohol use, this means that for individuals with low distal risk, the tendency for alcohol use increases proportionately with increase in proximal risk. In contrast, for those with high distal risk, sudden jumps in such tendency may be observed. The proposed regime-switching model captures these properties by allowing the effect of proximal risk on an individual's transition probabilities to depend on the individual's distal risk class membership at Time 1, C_{i1}. Specifically, a multinomial logistic regression model is specified for the transition probability as

$$\pi_{jk,h,it} = \frac{\exp[a_{k,t} + b_{1,jk,t} + (b_{2,jk,t} + b_{C_{i1}=high\ distal,jk,t})proximal_{it}]}{\sum_{s=1}^{K_t} \exp[a_{s,t} + b_{1,js,t} + (b_{2,js,t} + b_{C_{i1}=high\ distal,js,t})proximal_{it}]} \quad (15.13)$$

Compared to Equation 15.9, we can see that the logit intercept, $a_{k,h,t} = a_{k,t}$, is constrained to be invariant across both distal risk classes, $\boldsymbol{b}'_{jk,h,t} = [b_{1,jk,t} \; b_{2,jk,t} + b_{C_{i1}=high\ distal,jk,t}]$, with $b_{C_{i1}=high\ distal,jk,t}$ freely estimated for those with C_{i1} = high distal and $b_{C_{i1}=high\ distal,jk,t}$ = zero otherwise; $\boldsymbol{x}_{it} = [1 \; proximal_{it}]'$, where the covariate, $proximal_{it}$, is a composite score indicating individual *i*'s proximal risk at a particular time point. The term $b_{2,jk,t}$ indicates the deviation in log odds for those in the low distal risk class at Time 1 of switching to regime *k* at Time *t* from regime *j* relative to switching to regime *k* from the reference regime with each unit of increase in proximal risk. $b_{C_{i1}=high\ distal,jk,t}$ is the corresponding deviation in the logit slope associated with proximal risk for individuals who are in the high distal risk class at Time 1 compared to those who are in the low distal risk class.

Thus, for someone in the high distal risk class at Time 1, the effect of proximal risk (as encapsulated in $b_{2,jk,t} + b_{C_{i1}=high\ distal,jk,t}$ in Equation 15.13) is hypothesized to be more pronounced compared to those in the low distal risk class. That is, we expect $b_{C_{i1}=high\ distal,jk,t}$ to be greater than zero in two specific transition probability parameters. First, for those in the high distal risk class, the same amount of increase in proximal risk may lead to a disproportionate increase in the probability of "falling off the wagon"—namely, to switch from complete abstinence or low level of drinking to heavy drinking. In a similar vein, we also hypothesize that for those in the high distal risk class, proximal risk would have a disproportionately large effect of increasing the probability for this class of individuals to stay within the heavy drinking regime compared to those who are in the low distal risk class. Thus, once individuals with high distal risk have transitioned into the heavy drinking regime, a prolapse is hypothesized to occur with much lower probability than those in the low distal risk class, even with the same amount of reduction in proximal risk.

Another catastrophe flag of interest is *hysteresis*, the asymmetry in the values of proximal risk that trigger a lapse versus a prolapse. This property is already inherent to the proposed regime-switching model due to the general inequality between the probability of transitioning from the low to the high drinking regime, and transition in the reverse direction.

Empirical Illustration: The COMBINE Study

The data for this study are from the COMBINE study (Combined Pharmacotherapies and Behavioral Interventions for Alcohol Dependence; Anton et al., 2006), a multisite randomized trial. A total of 1,383 subjects across 11 research sites were randomized into nine treatment groups. Treatment was provided for 16 weeks, and participants were followed for 1 year posttreatment.

In the COMBINE study, the participants were randomly assigned to one of nine treatment groups.[3] The sample was recruited from inpatient and outpatient referrals at the study sites and throughout the community. Prior to baseline, 4,965 volunteers were screened by telephone to determine whether the individuals met eligibility criteria. Participants were excluded if they were dependent on another drug besides alcohol, nicotine, or cannabis, recently used opioids, had a serious mental illness, had any other medical condition that could disrupt study participation, had taken one of the study medications 30 days prior to baseline, or took medication that could raise the potential risks of the study. To be included in the study, subjects needed to have a minimum of 14 drinks (females) or 21 drinks (males) average per week over 30 days in the 90-day period prior to beginning abstinence. In addition, participants needed to have 2 or more days of heavy drinking in the 90-day period, with the last drink being within 21 days of enrollment. A heavy drinking day was defined as four drinks for females and five drinks for males. Following meeting eligibility criteria, subjects were required to produce a breath alcohol level of zero before completing consent and baseline assessments.

Measures

Outcome Measure

In the COMBINE study, drinking measures were collected at baseline prior to the beginning of treatment (denoted herein as Week 0), during treatment (at Weeks 1, 2, 4, 6, 8, 10, 12, and 16) and posttreatment (at Weeks 26, 36, and 52). Consistent with common practice in the alcohol use disorder (AUD) modeling literature, we used each participant's proportion of heavy drinking days (PHD) since the last assessment occasion (or up to 30 days prior to the next occasion, whichever one was shorter) as our key dependent variable of interest. Using information deduced from the Form–90 interview (Miller & Del Boca, 1994), heavy drinking was defined as four or more drinks per day for women and five or more drinks per day for men. For model fitting purposes, we used data spanning every 2 weeks from Week 0 to Week 16, yielding a total of nine equally spaced time points, one of which was missing. To evaluate whether information from different points of the treatment process served to indicate longer-term drinking outcomes, the participants' drinking status at Week 52 was used as the target outcome we sought to predict using information available from model fitting.

Time series plot and histogram of the PHD scores of 300 randomly selected participants over the course of the treatment program are shown in Figures 15.3A and 15.3B. The presence of a large number of abstainers at any single time point is evident in both of the plots. Two other prominent characteristics of the data are (1) the highly positive skewed nature of alcohol data, and (2) the considerable variability in drinking levels among the nonabstainers. To account for the positive skew of the data, we performed model fitting using the log transformed PHD scores (plotted in Figure 15.3C). PHD scores that were exactly zero (corresponding to complete abstinence) were fixed to a small nonzero constant (i.e., .01) to enable us to take log transformation. It can be seen from the plot that the log-transformed PHD scores still deviated from normality due to the presence of a large number of abstainers and right-censored responses. As noted earlier, the first issue is commonly referred to as the *zero-inflation phenomenon*. The latter arose because the participants could not have a PHD score that exceeded

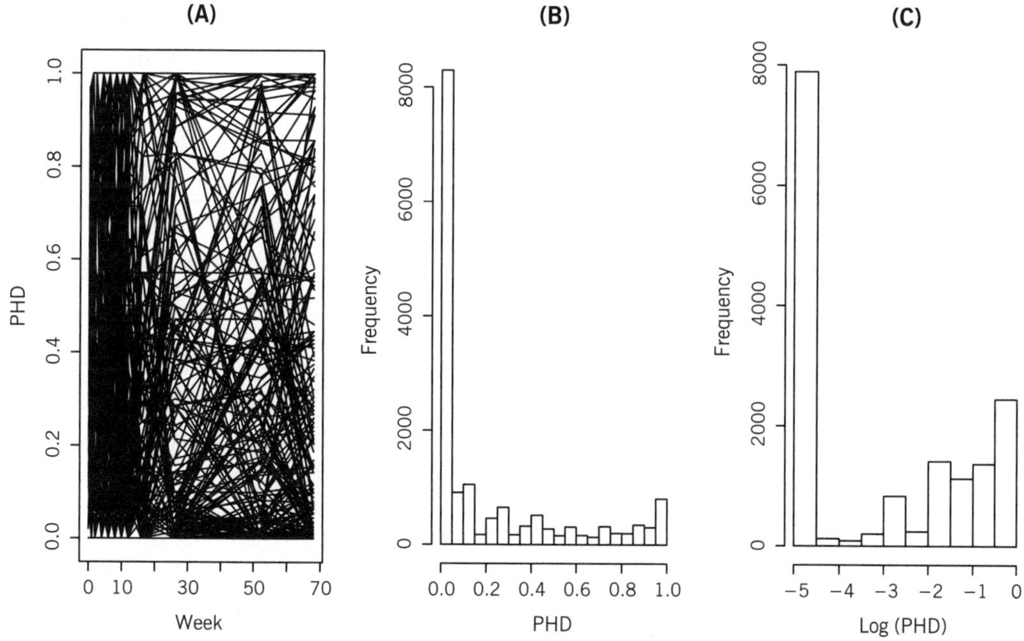

FIGURE 15.3. Data from 300 randomly selected participants from the COMBINE dataset, with (A) a plot of the biweekly PHD scores over the course of the treatment program, (B) a histogram showing the frequencies of the PHD scores, and (C) a histogram of the log-transformed PHD scores. Scores corresponding to 100% abstinence (i.e., zeroes) were converted into a small, nonzero constant (i.e., .01) prior to being subjected to log transformation.

1.0 (i.e., with greater than 100% of heavy drinking days for a particular assessment period). Thus, the scores were "censored" at the upper value of 1.0. Ways of handling these deviations from normality are described in the context of our modeling results.

Measures of Proximal Risk

Negative emotion, perceived stress, and craving were incorporated as measures to derive a composite indictor of proximal risk. The Profile of Mood States—Brief (POMS; McNair, Lorr, & Droppleman, 1971) was administered as a measure of negative emotions. Ratings on 30 Likert-type items were combined to yield a composite score of negative emotions. The Perceived Stress Scale (PSS; Cohen, 1983) was used to measure the participants' levels of perceived stress, and the Obsessive Compulsive Drinking Scale (OCDS; Anton, Moak, & Latham, 1995) was used as a measure of craving.

For model-fitting purposes, we computed a composite proximal risk score for each individual and each time point using the measures described earlier. Because the different measures of proximal risk all conformed to very diverse scales, we first standardized each proximal risk indicator using each individual's within-person, over-time mean and standard deviation on that indicator prior to computing a composite score using all the variables. This standardization procedure was appropriate in the current context because we were primarily interested in using proximal risk as a time-varying covariate. That is, we were not interested in the between-person differences in overall proximal risk level; rather, we sought to investigate how within-person, over-time variations in proximal risk would influence individuals' probabilities of manifesting lapses and prolapses on an occasion-by-occasion basis.

Measures of Distal Risk

Alcohol dependence measures based on self-reported symptoms and a structured clinical interview were used as potential indicators of distal risk at baseline. The measures included self-reported alcohol dependence severity, the number of alcohol dependence symptoms manifested by each participant, and a clinician's rating of the participant's severity of alcohol dependence. Self-reported alcohol dependence severity was based on scores from the Alcohol Dependence Scale (ADS; Skinner & Horn, 1984), a 25-item measure of alcohol withdrawal symptoms, impaired control over drinking, awareness of compulsions to drinking, tolerance, and drink-seeking behavior. The number of alcohol dependence symptoms was determined according to the *Diagnostic and Statistical Manual of Mental Disorders—Fourth Edition* (DSM-IV; American Psychiatric Association, 1994) criteria, using Module E of the Structured Clinical Interview for DSM-IV (SCID; First, Spitzer, Gibbon, & Williams, 1997). This measure is a count variable and is denoted herein as *ALCDEP*.[4] A final measure was an ordinal item from the SCID interview (item 22 from the alcohol evaluation module of the SCID) indicating a clinician's rating on the participant's severity of alcohol dependence during "the worst week of the past month" using one of three response categories: *mild, moderate*, and *severe*. This ordinal item is referred to herein as *ORD-DEPSEV*. In addition, we were not interested in the scaling differences between these measures. Thus, *ALCDEP* and *ADS* were both standardized using the means and standard deviations of all participants at Week 0 prior to model fitting.

Model-Fitting Results

The MSEM-RS model summarized in Equations 15.11–15.13 was fitted to the COMBINE data. Based on the plots of the data in Figure 15.3C, there was a substantial portion of abstainers in the sample. To accommodate this data characteristic, commonly encountered in empirical AUD data, we modified the structural model for regime 2 (the low drinking regime) to define a complete abstinence regime by setting α_{low} and the process noise variance in this regime (i.e., ψ_{low}) to be zero, and $Drink^*_{low}$ to log(.01). In addition, $t = 2$ corresponded to data from the second week immediately following the initiation of treatment. Thus, the effects of treatment were particularly salient and drinking levels were generally low. We thus allowed the intercept and initial variance of the high drinking regime at $t = 2$ to be different from the rest of the study span, and these parameters are denoted as $m_{high,1}$ and $\psi_{high,1}$, respectively. In a similar vein, the logit parameters at $t = 2$ were allowed to differ from the corresponding parameters at the remaining time points (i.e., for $t = 3, \ldots, T$). From $t = 3$ and beyond, we constrained these parameters to be invariant over time. This is because at $t = 2$, the transition probabilities captured the probabilities of transitioning from different distal risk classes into the two drinking regimes (i.e., complete abstinence vs. drinking). These transition probabilities were thus conceptually distinct from the transition probabilities for the remaining time points, where they reflected the probabilities of transitioning between the two drinking regimes. Another caveat noted earlier was that the log-transformed PHD scores were right-censored at the value of log(1). Here we relied on the use of the sandwich-type standard error estimator in Mplus to circumvent the non-normality of the log-transformed PHD scores while individuals were in the drinking regime.

Statistically significant parameters from fitting the proposed model are summarized in Table 15.1. Based on the value of the logit intercept for the high distal risk class (i.e., $a1,1$), 87% of the participants were classified to be in the high distal risk class, whereas the remaining 13% were classified to be in the low distal risk class. To help shed light on the between-regime differences in transition patterns for the remaining time points, the logit intercept and slope parameters in Table 15.1 were used to compute the transition probabilities associated with different periods of the study. The parameter $b1,11,2$ was not significantly different from zero, indicating that on the days with average proximal risk, there was no significant difference in the log odds of the two distal risk classes to switch into the drinking regime at $t = 2$. Thus, from pretreatment to the first assessment occasion during the treatment phase ($t = 1$–2, corresponding to 2 weeks after the initialization of treatment), the transition probability matrix on the days with an average level of proximal risk was governed by:

$$\begin{array}{r} \\ \text{High distal risk}_{t=1} \\ \text{Low distal risk}_{t=1} \end{array} \begin{array}{c} \begin{array}{cc} \text{Drink}_{t=2} & \text{Abs}_{t=2} \end{array} \\ \begin{bmatrix} .27(0.02) & .73(0.02) \\ .27(0.02) & .73(0.02) \end{bmatrix} \end{array}$$

The patterns of transition probabilities indicate that within the first 2 weeks of the treatment program, a large proportion of the participants was able to maintain abstinence on the days with average proximal risk. There was no significant difference in the two distal risk classes in their probabilities for a lapse. In contrast, on the days with high proximal risk (i.e., one standard deviation above the mean), the transition probability matrix was given by:

$$\begin{array}{r} \\ \text{High distal risk}_{t=1} \\ \text{Low distal risk}_{t=1} \end{array} \begin{array}{c} \begin{array}{cc} \text{Drink}_{t=2} & \text{Abs}_{t=2} \end{array} \\ \begin{bmatrix} .64(0.04) & .36(0.04) \\ .27(0.02) & .73(0.02) \end{bmatrix} \end{array} \quad (15.14)$$

TABLE 15.1. Parameter Estimates Obtained from Fitting the Regime-Switching Alcohol Use Model

Parameters	Meanings	Estimates (*SE*)
$m_{highdistal}$	Average distal risk of high distal risk class	−0.01 (0.03)
$m_{lowdistal}$	Average distal risk of high distal risk class	−0.21 (0.09)
$\psi_{highdistal}$	Variance in distal risk of high distal risk class	0.37 (0.03)
$\psi_{lowdistal}$	Variance in distal risk of low distal risk class	0.31 (0.05)
$a_{1,1}$	Logit intercept at Time 1	1.90 (0.22)
$m_{high,1}$	Initial average alcohol use tendency at $t = 2$	−1.24 (0.05)
$\psi_{high,1}$	Initial variance in alcohol use tendency at $t = 2$	0.47 (0.03)
$Drink^{*}_{high}$	Baseline drinking level in R1	−1.32 (0.05)
α_{high}	AR(1) parameter in R1 starting from $t = 3$	0.35 (0.02)
ψ_{high}	Random shock variance in R1 starting from $t = 3$	0.56 (0.02)
$a_{1,2}$	Logit intercept at $t = 2$	−1.00 (0.09)
$b_{1,11,2}$	Deviation in log odds of being in R1 at $t = 2$	= 0
$b_{2,11,2}$	Effect of proximal risk on the log odds of transitioning from C_{i1} = high distal class to C_{i2} = drinking (R1)	1.57 (0.17)
$b_{2,21,2}$	Effect of proximal risk on the log odds of transitioning from C_{i1} = low distal class to C_{i2} = drinking (R1)	= 0
$a_{1,t}$, $t = 3, \ldots, T$	Logit intercept for other t	−1.71 (0.08)
$b_{1,11,t}$, $t = 3, \ldots, T$	Deviation in log odds of staying in R1 for other t	3.09 (0.13)
$b_{2,11,t}$, $t = 3, \ldots, T$	Effect of proximal risk on the log odds of staying in the drinking regime from t-1 to t given C_{i1} = low distal class	1.03 (0.16)
$b_{2,21,t}$, $t = 3, \ldots, T$	Effect of proximal risk on the log odds of transitioning from abstinence at t-1 to drinking at t given C_{i1} = low distal class	−2.12 (0.39)
$b_{C_{i1}=high\ distal,11,t}$, $t = 3, \ldots, T$	Deviation in the effect of proximal risk on $b_{2,11}$ given C_{i1} = high distal risk class	= 0
$b_{C_{i1}=high\ distal,21,t}$, $t = 3, \ldots, T$	Deviation in the effect of proximal risk on $b_{2,21}$ given C_{i1} = high distal risk class	3.97 (0.37)

Note. R1, drinking regime; R2, complete abstinence regime.

That is, on the days with high proximal risk, individuals with high distal risk showed a substantially higher probability for a lapse compared to those who were low in the distal risk class. For those in the low distal risk class, the parameter $b_{2,21,2}$ was not significantly different from zero, suggesting that proximal risk did not have a significant effect on the probability for the low distal risk class to manifest a lapse during the first 2 weeks of the treatment program. On the contrary, the parameter $b_{2,11,2}$ was positive and significantly different from zero. Thus, for those in the high distal risk class, one unit of increase in proximal risk led to an abrupt increase in the probability of a lapse from .27 to .64.

The transition probabilities characterizing the rest of the study span helped indicate the relative stability of the drinking and abstinence regimes once individuals transitioned into these regimes. While individuals were in the drinking regime, the AR(1) parameter of 0.35 indicated that when individuals deviated from the baseline drinking level of $Drink^*_{high} = -1.32$, they tended to return to it relatively quickly. The closer the AR(1) parameter was to 0 in absolute value, the quicker the return.

On the days with average proximal risk, the same transition probability matrix was assumed to govern individuals' drinking patterns regardless of initial distal risk status. The corresponding transition probability matrix was observed to be

Average proximal risk, all distal risk classes:

$$\begin{array}{c} \\ \text{Drink}_{t-1} \\ \text{Abs}_{t-1} \end{array} \begin{array}{c} \begin{matrix} \text{Drink}_t & \text{Abs}_t \end{matrix} \\ \begin{bmatrix} .80(0.01) & .20(0.01) \\ .15(0.01) & .85(0.01) \end{bmatrix} \end{array} \qquad (15.15)$$

That is, on the days with average proximal risk, the probability for a lapse was relatively low [Pr(Drinking at Time t | Abstinence at Time t-1) = .15], but the probability for a prolapse was also very low [Pr(Abstinence at Time t | Drinking at Time t-1) = .20]. On the days with high proximal risk (i.e., 1 *SD* above the mean), some differences in transition probability patterns were observed. Consistent with our hypothesis, those with high distal risk were much more likely to show a lapse when proximal risk was high. Consequently, for individuals in the low distal risk class, their transition probability matrix was given by:

High (+1 SD) proximal risk, low distal risk:

$$\begin{array}{c} \\ \text{Drink}_{t-1} \\ \text{Abs}_{t-1} \end{array} \begin{array}{c} \begin{matrix} \text{Drink}_t & \text{Abs}_t \end{matrix} \\ \begin{bmatrix} .92(0.02) & .08(0.02) \\ .02(0.01) & .98(0.01) \end{bmatrix} \end{array} \qquad (15.16)$$

whereas for those in the high distal risk class, the corresponding transition probability matrix was given by:

High (+1 SD) proximal risk, high distal risk:

$$\begin{array}{c} \\ \text{Drink}_{t-1} \\ \text{Abs}_{t-1} \end{array} \begin{array}{c} \begin{array}{cc} \text{Drink}_t & \text{Abs}(t) \end{array} \\ \begin{bmatrix} .92(0.02) & .08(0.02) \\ .54(0.05) & .47(0.05) \end{bmatrix} \end{array} \tag{15.17}$$

Specifically, individuals with high distal risk were found to show a significantly greater increase in the probability for a lapse when proximal risk was high (i.e., from .15 to .54), as indicated by the statistical significance of the parameter $b_{C_{i1}=high\ distal,21,t}$. In contrast, for those with low distal risk, the probability for a lapse actually decreased slightly on the days with high proximal risk. This may be due in part to the modeling constraint that the two distal risk classes were specified to show the same transition probability patterns on the days with average proximal risk. That is, based on postulates of the cusp catastrophe model, we did not include the main effects of distal risk on the transition probabilities in Equation 15.13. Instead, distal risk was posited to play only a moderating role on the effects of proximal risk on the transition probabilities. Regardless, the detrimental effects of high proximal risk in triggering a lapse among those with high distal risk is clear throughout the treatment phase.

Contrary to our initial hypothesis, we did not find a significant difference between the two distal risk classes in their probability of bouncing back from a lapse (i.e., to manifest a prolapse). That is, there was no significant difference in the log odds for the two distal risk classes to manifest a prolapse on the days with high proximal risk (i.e., $b_{C_{i1}=high\ distal,11,t}$ was not significantly different from zero; see Table 15.1). Whereas the estimated prolapse rate was generally low (.20) for both distal risk classes even on the days with average proximal risk, the probability of a prolapse reduced by more than half (i.e., the probability decreased from .20 to .08) when proximal risk was high.

Overall, some of our key results indicated that individuals with high distal risk were especially susceptible for a lapse when proximal risk was high within the first 2 weeks of the treatment program. Specifically, individuals who were high in distal risk were more likely to transition into the drinking regime when faced with a heightened proximal risk level within the first 2 weeks of treatment (P = .64) than individuals with low levels of distal risk (P = .27). In fact, the probability for the low distal risk class to manifest a lapse during the initial treatment phase appeared unrelated to the fluctuations in proximal risk reported by the participants. It is thus critical to monitor the progress of the high distal risk group closely during the early stages of a treatment program.

Conclusions

The MSEM-RS model summarized in Equations 15.11–15.13 was proposed as an alternative to the cusp catastrophe model in the context of alcohol use. Although it may not be as mathematically elegant as the cusp catastrophe model (e.g., the cusp catastrophe model is capable of producing the rich array of dynamics described earlier with only one nonlinear potential function), it relaxes some of the constraints of the cusp catastrophe model that may be overly restrictive in empirical analytic situations. First, the proposed model can be fit using existing SEM programs such as Mplus (Muthén & Muthén, 2001). The corresponding estimation procedures implemented in these programs can readily handle longitudinal panel data with multiple subjects as well as the presence of missing data hypothesized to be missing at random (Little & Rubin, 2002). Because the broader modeling framework assumes that an SEM underlies each latent regime, the proposed model can fully capitalize on the flexibility of the SEM framework in handling (linear) structural relations among latent variables. Furthermore, the proposed model incorporates the lagged effects of previous drinking on current drinking, and it can be expanded, as needed, to test other multivariate extensions involving other variables of interest. In addition, by using person- and time-specific covariates to predict the probabilities of transitioning between regimes, the new approach also allows for heterogeneous timing of lapse and prolapse within and across subjects. Finally, the zero-inflation phenomenon often observed in alcohol use data is now explicitly modeled by specifying a complete abstinence regime for the abstainers.

Despite some of the promises of the proposed regime-switching dynamics models, some methodological challenges remain. For instance, the increase in computational costs was substantial when more than five time points were included in the model. The increase in computational time was even more substantial when we attempted to fit variations of the proposed model with three drinking regimes. This imposes added burden in fitting mixture models because users are often advised to fit mixture models using multiple starting values to check the sensitivity of the estimation results to different starting values, which is a computationally intensive pursuit in itself. To this end, some of the alternative formulations in the time series and state–space literature as well as related estimation procedures can be adopted (Hamilton, 1994; Kim & Nelson, 1999; Yang & Chow, 2010).

Although the proposed model was inspired by the cusp catastrophe model and was able to retain some of its features, by no means should it be regarded as a replacement of the cusp catastrophe or other related catastrophe models. For instance, through the use of one nonlinear potential function, the cusp catastrophe model is capable of generating a wide array of interesting dynamic properties such as hysterisis, sudden jumps, and the coexistence of multiple equilibria. Our use of two different sets of transition probability patterns to characterize the differing pathways of change manifested by individuals with different degrees of distal risk also did not fully capture the distinction between the linear and the nonlinear pathways of change posited in the cusp catastrophe model. In addition, as opposed to treating distal risk as a continuous variable (e.g., the way the splitting parameter is utilized in the cusp catastrophe model), we only distinguished between the trajectories of those in the low versus high distal risk class. In other words, distal risk was treated in our proposed model as a categorical, as opposed to a continuous latent variable. Thus, as emphasized earlier, the proposed model is intended to serve as an alternative to the cusp catastrophe in certain empirical applications. Whereas it helps to circumvent several data-analytic issues commonly encountered in the analysis of AUD data, we still acknowledge the general appeal of the cusp and other related catastrophe models as a representation of change.

AKNOWLEDGMENTS

Funding for this study was provided by a grant from the National Science Foundation (No. BCS-0826844). We are grateful to Peter Molenaar and Han van der Maas for their comments on earlier versions of the proposed model, and for their work on catastrophe systems, which has inspired us to undertake the modeling work in the present chapter. We are also grateful to the comments of R. Shane Hutton in the early stages of model building and to the help of Zhaowei Hua and Ruixin Guo in organizing data from the COMBINE study.

NOTES

1. A manifold is simply a higher-dimensional analogue of a smooth curve or surface (Poston & Stewart, 1978).

2. Typically, only one set of intercept parameters—either ν_k in Equation 15.5 or α_k in Equation 15.7—is estimated, unless other constraints are imposed.

3. The Medical Management groups ($n = 607$) included naltrexone, acamprosate, naltrexone + acamprosate, and placebo. The Combined Behavioral Intervention (CBI) groups ($n = 776$) consisted of naltrexone + CBI, acamprosate + CBI, naltrexone + acamprosate + CBI, placebo + CBI, and CBI only (COMBINE Study Research Group, 2003). Subjects received treatment for a total of 16 weeks; participants receiving study medication were offered 9 Medical Management visits, and those who received CBI had a maximum of 20 sessions.

4. We treated *ALCDEP* as a continuous observed indicator in all model fitting. Compared to specifying *ALCDEP* as a continuous indicator, we found that specifying it as a count variable with a Poisson link function led to slower computational speed and estimation difficulties in fitting some of the more complex models, while yielding only negligible differences in the modeling results.

REFERENCES

Alexander, R. A., Herbert, G. R., DeShon, R. P., & Hanges, P. J. (1992). An examination of least-squares regression modeling of catastrophe theory. *Psychological Bulletin, 111*(2), 366–374.

Allen, B. D., & Carifo, J. (1995). Nonlinear analysis: Catastrophe theory modeling and Cobb's cusp surface analysis program. *Evaluation Review, 19*(1), 64–83.

Alligood, K. T., Sauer, T. D., & Yorke, J. A. (1996). *Chaos: An introduction to dynamical systems.* New York: Springer-Verlag.

American Psychiatric Association. (1994). *Diagnostic and statistical manual of mental disorders* (4th ed.). Washington, DC: Author.

Anton, R. F., Moak, D. H., & Latham, P. (1995). The Obsessive Compulsive Drinking Scale: A self-rated instrument for the quantification of thoughts about alcohol and drinking behavior. *Alcoholism: Clinical and Experimental Research, 19,* 92–99.

Anton, R. F., O'Malley, S. S., Ciraulo, D. A., Cisler, R. A., Couper, D. M., Donovan, D. M., et al. (2006). Combined pharmacotherapies and behavioral interventions for alcohol dependence. The COMBINE study: A randomized controlled trial. *Journal of the American Medical Association, 295*(17), 2003–2017.

Arminger, G., & Muthén, B. (1998). A Bayesian approach to nonlinear latent variable models using the Gibbs sampler and the Metropolis-Hastings algorithm. *Psychometrika, 63,* 271–300.

Asparouhov, T., & Muthén, B. O. (2011). *C on C and X* (Mplus technical appendix) [Computer software manual]. Unpublished manual.

Bereiter, C. (1963). Some persisting dilemmas in the measurement of change. In C. W. Harris (Ed.), *Problems in measuring change* (pp. 3–20). Madison: University of Wisconsin Press.

Brown, C. (1995). *Chaos and catastrophe theories.* Thousand Oaks, CA: Sage.

Browne, M. W., & Nesselroade, J. R. (2005). Representing psychological processes with dynamic factor models: Some promising uses and extensions of autoregressive moving average time series models. In A. Maydeu-Olivares & J. J. McArdle (Eds.), *Contemporary psychometrics: A Festschrift for Roderick P. McDonald* (pp. 415–452). Mahwah, NJ: Erlbaum.

Byrne, D. G., Mazanov, J., & Gregson, R. A. M. (2001). A cusp catastrophe analysis of changes to adolescent smoking behavior in response to smoking prevention programs. *Nonlinear Dynamics, Psychology, and Life Sciences, 5*(2), 115–137.

Clair, S. (1999). A cusp catastrophe model for adolescent alcohol use: An empirical test. *Nonlinear Dynamics, Psychology, and Life Sciences, 2*(3), 217–241.

Cobb, L. (1981). Parameter estimation for the cusp catastrophe model. *Behavioral Science, 26* (1), 75–78.

Cobb, L., Koppstein, P., & Chen, N. H. (1983). Estimation and moment recursion relations for multimodal distributions of the exponential family. *Journal of the American Statistical Association, 78,* 124–130.

Cobb, L., & Zacks, S. (1985). Applications of catastrophe theory for statistical modelling in the biosciences. *Journal of the American Statistical Association, 80*(392), 793–802.

Cohen, J. (1983). The cost of dichotomization. *Applied Psychological Measurement, 7*(3), 249–253.

Cronbach, L. J., & Furby, L. (1970). How should we measure "change"—or should we? *Psychological Bulletin, 74*(1), 68–80.

Dempster, A. P., Laird, N. M., & Rubin, D. B. (1977). Maximum likelihood from incomplete data via the EM algorithm. *Journal of the Royal Statistical Society, Series B, 39*(1), 1–38.

Dolan, C. V., & van der Maas, H. L. J. (1998). Fitting multivariate normal finite mixtures subject to structural equation modeling. *Psychometrika, 63*(3), 227–253.

Everitt, B. S., & Hand, D. J. (1981). *Finite mixture distributions.* London: Chapman & Hall.

First, M., Spitzer, R. L., Gibbon, M., & Williams, J. B. W. (1997). *Structured Clinical Interview for DSM-IV Axis I Disorders—Patient Edition.* New York: New York State Psychiatric Institute.

Gilmore, R. (Ed.). (1981). *Catastrophe theory for scientists and engineers.* New York: Dover.

Gleick, J. (1987). *Chaos: Making a new science.* New York: Viking.

Grasman, R. P. P. P., van der Maas, H. L. J., & Wagenmakers, E.-J. (2009). Fitting the cusp catastrophe in R: A cusp package primer. *Journal of Statistical Software, 32*(8), 1–27.

Guastello, S. J. (1982). Color matching and shift work: An industrial application of the cusp-difference equation. *Behavioral Science, 27,* 131–139.

Guastello, S. J. (1984). A catastrophe theory evaluation of a policy to control job absence. *Behavioral Science, 29,* 263–269.

Guastello, S. J. (1992). Clash of the paradigms: A critique of an examination of the polynomial regression technique for evaluating catastrophe theory hypothesis. *Psychological Bulletin, 111*(2), 375–379.

Hamilton, J. D. (1994). *Time series analysis.* Princeton, NJ: Princeton University Press.

Harris, C. W. (Ed.). (1963). *Problems in measuring change.* Madison: University of Wisconsin Press.

Hilborn, R. C. (1994). *Chaos and nonlinear dynamics.* New York: Oxford University Press.

Horn, J. L. (1963). The discovery of personality traits. *Journal of Educational Research, 56,* 460–465.

Hufford, M. R., Witkiewitz, K. A., Shields, A. L., Kodva, S., & Caruso, J. C. (2003). Relapse as a nonlinear dynamic system: Applications to patients with alcohol use disorders. *Journal of Abnormal Psychology, 112*(2), 219–227.

Hummel-Rossi, B., & Weinberg, S. L. (1975). Practical guidelines in applying current theories to the measurement of change. Part I: Problems in measuring change and recommended procedures. Part II: Numerical illustrations of research designs concerned with measuring change. Abstracted in the *JSAS Catalog of Selected Documents in Psychology, 5,* 226.

Jedidi, K., Jagpal, H. S., & DeSarbo, W. S. (1997, 21). Stemm: A general finite mixture structural equation model. *Journal of Classification, 14* (1), 23–50. Available at *http://dx.doi.org/10.1007/s003579900002.*

Jöreskog, K. G. (1973). A general method for estimating a linear structural equation system. In A. S. Goldberger & O. D. Duncan (Eds.), *Structural equation models in the social sciences* (pp. 85–112). New York: Seminar Press.

Jöreskog, K. G., & Moustaki, I. (2001). Factor analysis of ordinal variables: A comparison of three approaches. *Multivariate Behavioral Research, 36*(3), 347–387.

Kaplan, D. (2008). An overview of Markov Chain methods for the study of stage-sequential developmental processes. *Developmental Psychology, 44*(2), 457–467.

Kim, C.-J., & Nelson, C. R. (1999). *State-space models with regime switching: Classical and Gibbs-sampling approaches with applications.* Cambridge, MA: MIT Press.

Lange, R., & Houran, J. (2000). Modeling Maher's attribution theory of delusions as a cusp catastrophe. *Nonlinear Dynamics, Psychology, and Life Sciences, 4*(3), 235–254.

Little, R. J. A., & Rubin, D. B. (2002). *Statistical analysis with missing data.* New York: Wiley.

Lord, F. M. (1956). The measurement of growth. *Educational and Psychological Measurement, 16,* 421–437.

Lorenz, E. N. (1963). Deterministic nonperiodic flow. *Journal of Atmospheric Science, 20,* 130–141.

Maxwell, S. E., & Howard, G. S. (1981). Change scores—necessary anathema? *Educational and Psychological Measurement, 41,* 747–756.

McArdle, J. J., & Hamagami, F. (2001). Latent difference score structural models for linear dynamic analysis with incomplete longitudinal data. In L. Collins & A. Sayer (Eds.), *New methods for the analysis of change* (pp. 139–175). Washington, DC: American Psychological Association.

McNair, D. M., Lorr, M., & Droppleman, L. F. (1971). *Manual for the Profile of Mood States.* San Diego, CA: Educational and Industrial Testing Services.

Miller, W. R., & Del Boca, F. K. (1994). Measurement of drinking behavior using the Form 90 family of instruments. *Journal of Studies on Alcohol, 12*(Suppl.), 112–118.

Molenaar, P. C. M., & Newell, K. M. (2003). Direct fit of a theoretical model of phase transition in oscillatory finger motions. *British Journal of Mathematical and Statistical Psychology, 56*, 199–214.

Molenaar, P. C. M., & Raijmakers, M. E. J. (2000). A causal interpretation of Piaget's theory of cognitive development: Reflections on the relationship between epigenesis and nonlinear dynamics. *New Ideas in Psychology, 18*(1), 41–55.

Muthén, B. O. (1984). A general structural equation model with dichotomous, ordered categorical, and continuous latent variables indicators. *Psychometrika, 49*, 115–132.

Muthén, B. O., & Asparouhov, T. (2011, July). *LTA in Mplus: Transition probabilities influenced by covariates* [Mplus Web Notes: No. 13]. Available at *www.statmodel.com/examples/LTAwebnote.pdf.*

Muthén, B. O., & Shedden, K. (1999). Finite mixture modeling with mixture outcomes using the EM algorithm. *Biometrics, 55*, 463–469.

Muthén, L. K., & Muthén, B. O. (2001). *Mplus: The comprehensive modeling program for applied researchers: User's guide.* Los Angeles, CA: Author.

Nesselroade, J. R., & Bartsch, T. W. (1977). Multivariate perspectives on the construct validity of the trait-state distinction. In R. B. Cattell & R. M. Dreger (Eds.), *Handbook of modern personality theory* (pp. 221–238). Baton Rouge, LA: Hemisphere.

Nesselroade, J. R., & Cable, D. G. (1974). "Sometimes, it's okay to factor difference scores"—The separation of state and trait anxiety. *Multivariate Behavioral Research, 9*, 273–282.

Nylund-Gibson, K., Muthén, B. O., Nishina, A., Bellmore, A., & Graham, S. (2013). *Stability and instability of peer victimization during middle school: Using latent transition analysis with covariates, distal outcomes, and modeling extensions.* Manuscript submitted for publication.

Oliva, T. A., DeSarbo, W. S., Day, D. L., & Jedidi, K. (1987). GEMCAT: A general multivariate methodology for estimating catastrophe models. *Behavioral Science, 32*, 121–137.

Poincaré, H. (1892). *Les méthods nouvelles de la mécanique céleste.* Paris: Gauthier-Villars.

Poston, T., & Stewart, I. N. (1978). *Catastrophe theory and its applications.* London: Pitman.

Preece, P. F. W. (1980). A geometrical model of Piagetian conservation. *Psychological Reports, 46*, 143–148.

Singer, H. (1992). The aliasing-phenomenon in visual terms. *Journal of Mathematical Sociology, 14*(1), 39–49.

Singer, H. (1993). Continuous-time dynamical systems with sampled data, errors of measurement, and unobserved components. *Journal of Time Series Analysis, 14*, 527–545.

Skinner, H. A., & Horn, J. L. (1984). *Alcohol dependence scale: Users guide.* Toronto: Addiction Research Foundation.

Stewart, I. N., & Peregoy, P. L. (1983). Catastrophe theory modeling in psychology. *Psychological Bulletin, 94*(2), 336–362.

Strahan, E. Y., & Conger, A. J. (1999). Social anxiety and social performance: Why don't we see more catastrophes. *Journal of Anxiety Disorders, 13*(4), 399–416.

Thom, R. (1975). *Structural stability and morphogenesis* (D. H. Fowler, Trans.). Reading, MA: Benjamin.

Thom, R. (1993). *Structural stability and morphogenesis: An outline of a general theory of models*. Reading, MA: Addison-Wesley.

Titterington, D. M., Smith, A. F. M., & Markov, U. E. (1985). *The statistical analysis of finite mixture distributions*. New York: Wiley.

Vallacher, R. R., & Nowak, A. (Eds.). (1994). *Dynamical systems in social psychology*. San Diego, CA: Academic Press.

van der Maas, H. L. J., & Molenaar, P. C. M. (1992). Stagewise cognitive development: An application of catastrophe theory. *Psychological Review, 99*(3), 395–417.

Vermunt, J. K., & Magidson, J. (2005). Structural equation models: Mixture models. In B. Everitt & D. Howell (Eds.), *Encyclopedia of statistics in behavioral science* (pp. 1922–1927). Chichester, UK: Wiley.

Wimmers, R. H., Savelsbergh, G. J., & van der Kamp, J. (1998). A developmental transition in prehension modeled as a cusp catastrophe. *Developmental Psychobiology, 32*, 23–35.

Witkiewitz, K. A., & Marlatt, G. A. (2004). Relapse prevention for alcohol and drug problems: That was zen, this is tao. *American Psychologist, 59*(4), 224–235.

Witkiewitz, K. A., & Marlatt, G. (2007). Modeling the complexity of warning signs following treatment for addictive and non-addictive behavior: It's a rocky road to relapse. *Clinical Psychology Review, 27*, 724–738.

Witkiewitz, K. A., van der Maas, H. L. J., Hufford, M., & Marlatt, G. A. (2007). Nonnormality and divergence in post-treatment alcohol use: Reexamining the project MATCH data "another way." *Journal of Abnormal Psychology, 116*, 378–394.

Yang, M., & Chow, S.-M. (2010). Using state–space model with regime switching to represent the dynamics of facial electromyography (EMG) data. *Psychometrika: Application and Case Studies, 74*(4), 744–771.

Yung, Y. F. (1997). Finite mixtures in confirmatory factor-analysis models. *Psychometrika, 62*, 297–330.

Zeeman, E. C. (1976). Catastrophe theory. *Scientific American, 234*(4), 65–83.

Zeeman, E. C. (1978). *Catastrophe theory: Selected papers, 1972–1977*. Reading, MA: Addison-Wesley.

PART VIII

NONERGODIC DEVELOPMENTAL SYSTEMS

CHAPTER 16

Idiographic Applications

Issues of Ergodicity and Generalizability

Wayne F. Velicer, Steven F. Babbin, and Richard Palumbo

Idiographic methods are procedures that focus on the time-dependent variation within a single individual or unit (intrasubject variability), in contrast to nomothetic methods that focus on group-level relationships (intersubject variability). This important class of intensive measurement designs typically involves single subjects (or other research units) that are measured repeatedly at regular intervals to include a large number of observations. Time series analysis is a statistical methodology appropriate for this type of data (Glass, Willson, & Gottman, 2008; Velicer & Molenaar, 2013). A time series analysis can help us understand underlying naturalistic processes, assess the pattern of change over time, or evaluate the effects of either a planned or unplanned intervention. Advances in information systems technology are making idiographic designs and time series analysis an increasingly feasible method for studying important behavioral science phenomena.

Idiographic methods have several advantages. First, idiographic methods can address different research questions and may provide more insight into the data. Second, idiographic methods can be used in applied settings, such as businesses, schools, clinics, and hospitals. More traditional between-subject research designs may not be appropriate, or, in some instances, impossible to implement in such settings. Third, idiographic methods have important advantages for investigating patterns of change across time. Some questions that can be investigated in this context include whether intervention effects are temporary or permanent, and whether the intervention produces the same pattern of change for each person or unit. Fourth, idiographic methods can be used to determine the generating function for the behavior of interest. Fifth, idiographic methods can address the relationship between variables over time, including the presence of an intervention.

Idiographic methods involve two important challenges. First, an interindividual analysis may yield different results than an intraindividual analysis. Equivalent results will occur only if the two conditions specified by the ergodic theorems are met: (1) Each individual trajectory has to obey the same dynamic laws, and (2) each individual trajectory must have equal mean levels and serial dependencies. These theorems appear unlikely to be met in practice, but often data are not adequate to test them. Available examples do not provide support for these conditions. Second, the widely accepted methods of generalizability are difficult to apply when using idiographic methods. Alternatives such as generalizability to homogeneous subgroups, generalizabilty across different contexts, and generalizability across occasions represent important alternative approaches.

This chapter reviews some of the main issues involved in idiographic analysis, including its historical role in behavioral science, the importance of the ergodic theorems, generalization issues, the temporal dimension, and an overview of the idiographic–nomothetic continuum. These sections are followed by examples meant to highlight some of the unique questions that can be addressed through these techniques.

Idiographic Analysis Rediscovered

Historically, idiographic approaches such as narrative case studies formed the basis for the behavioral sciences. However, they were often viewed as inferior due to the lack of an objective, quantitative basis. Even in present-day research, with readily available statistical techniques such as time series analysis, many traditional data collection procedures are incapable of obtaining the high-quality psychological data appropriate for idiographic analysis. Many classic psychological assessments, such as IQ and personality tests, are simply not designed to be administered repeatedly to the same person. These factors contributed to the modest attention given to idiographic techniques over the last 50 years by the behavioral sciences, instead emphasizing the nomothetic approach. However, during that time, idiographic methods have been extensively developed and employed in many other branches of science.

In early idiographic designs, especially as used within psychology, there was a heavy emphasis on graphical analysis to describe and interpret results (Kazdin, 2011). Although graphical methods are useful and still provide important ancillary information to the understanding of a time series process, developments in techniques that bring an advanced statistical methodology to bear on this class of data have revolutionized the area of single-subject research. Modern time series analysis and related research methods represent a sophisticated leap in the ability to analyze longitudinal data gathered on single subjects or units. This has led to a renewed interest in idiographic methods in the behavioral sciences. Additionally, the development of telemetric measures (Goodwin, Velicer, & Intille, 2008) and a shift in interest from laboratory to real-world research have combined to fuel the resurgence.

The Temporal Dimension

A key shift with idiographic methods is their focus on the temporal dimension, whereas most current data-analytic methods address the variable dimension. The distinction can be visualized through Cattell's data cube (see Figure 16.1), in which each direction represents a different dimension of the data: the person dimension (top to bottom), variable dimension (left to right), and time dimension (front to back). The variable dimension focuses on group-level relationships between variables (intersubject variability), utilizing well-known methods including analysis of variance (ANOVA), multiple regression, and structural equation modeling. Alternatively, idiographic methods utilize the time dimension, focusing on the pattern of change in an individual over time (intrasubject variability). Methods include autoregressive integrated moving average (ARIMA) models, dynamic factor analysis, and state–space modeling.

In many cases, group-level analyses have the potential to give misleading answers. A key assumption with nomothetic methods is the additive constant, which assumes that each participant in a sample is equally influenced by an intervention. This assumption can obscure the problem rather than assisting in the solution. Because typically, no one is average, the derived conclusion may not describe the processes of any individual in the population. This is an important consideration, as the questions being asked by a researcher should determine the methodology being used. For example, a clinical psychology professor may be interested in nomothetic results on the effectiveness of treatments for anxiety. Population tendencies would be of interest,

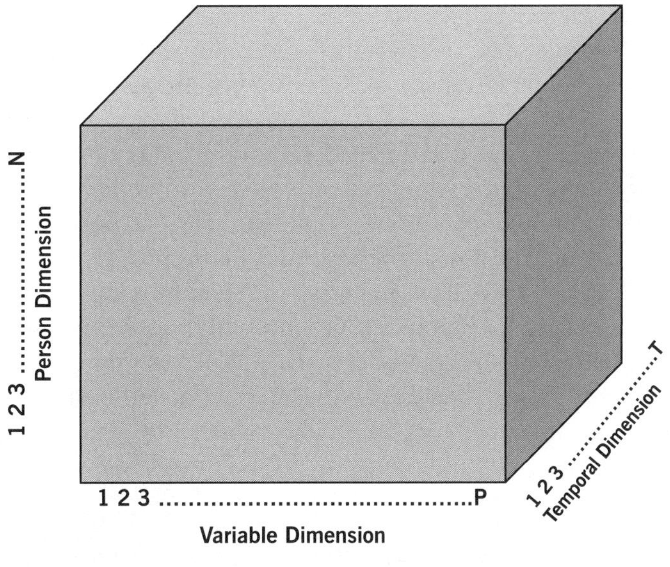

FIGURE 16.1. Cattell's data cube represents three dimensions of data: the person dimension (*top to bottom*), variable dimension (*left to right*), and time dimension (*front to back*).

so that clinical students can be informed of interventions found to be effective for most people. Nomothetic conclusions will give a generalized understanding of whether the group, as a whole, is influenced and in what way. Though the nomothetic results can be informative, they cannot predict how any individual will be impacted, nor can they describe the process of change in any person over time. Alternatively, a practicing clinician may be more interested in the trajectory of an individual patient. By regularly collecting data on the anxiety levels of a client and his or her daily activities, a clinician would be able to assess the effectiveness of the treatment as it is applied to that particular person. This detailed, individualized data could inform treatment, help track progress, and determine whether alternative interventions are necessary—or if recovery has been achieved. In most cases, this type of process can be understood only through longitudinal data and idiographic analyses (see Molenaar, 1987, for an early application).

Ergodic Theorems

The discrepancies between results from cross-sectional nomothetic data and those from longitudinal idiographic data can be understood through the ergodic theorems (Choe, 2005; Molenaar, 2008). The ergodic theorems pertain to arbitrary measurable processes, whereas the two conditions described below to render Gaussian processes ergodic derive from a special corollary of these theorems. The special corollary states that equivalent results will occur only if the two conditions specified by the ergodic theorems are met: (1) Each individual trajectory has to obey the same dynamic laws, and (2) each individual trajectory must have equal mean levels and serial dependencies. If these conditions are not met, then results from nomothetic analyses will not capture the processes of the individuals that make up a sample. Inappropriately inferring from a group to an individual is known as an *ecological fallacy*, and is a common issue with nomothetic methods.The ergodic theorems are based on a general theory about the relationships between *inter*system and *intra*system variability (where a system can be any unit: person, family, school, etc.). Ergodic theory addresses the relationships between individual units and groups of those units in the most general setting possible: namely, for all measurable processes and using different metrics. A very special case is the set of Gaussian processes. Ergodicity for Gaussian processes is associated with the two conditions specified by the ergodic theorems. Whereas these two criteria are sufficient for Gaussian processes, they are necessary (but not sufficient) for non-Gaussian processes. These fundamental issues are rarely evaluated by researchers, often due to too few data points (2–3 per unit is common and 8–10 in single-subject research).

Generalizability Issues

A major issue for idiographic research is generalizability. How can the results from a single individual or unit be generalized to a larger population? In the classic approach, researchers generalize from an individual or group to a general population. However, with idiographic methods, alternative goals are more appropriate.

Since each person is viewed as unique, idiographic methods do not use statistical inference as a method of generalization. In the past, logical inference was used, but as it relies solely on subjective judgment, it is no longer appropriate. One potential solution is systematically replicating an idiographic study with different participants in each trial. In time series designs, Barlow, Nock, and Hersen (2009) recommend such replication across five or more units. This approach, sometimes called a *boxscore* approach, relies on a frequency count across studies with no qualitative rating of input. Alternatively, meta-analysis for time series designs combines data using effect size estimates from multiple studies. The approach relies on published summary statistics and can be applied to time series if there is no slope present with an alternative effect size estimate. Another technique that has been attempted is the pooled time series, in which all units are combined into a single design. This requires access to all the data being evaluated and employs a super vector of $Z_1, Z_2, \ldots, Z_k$ to compare k units (Velicer & McDonald, 1991). A design matrix is then used to form comparisons. In practice, this is difficult to implement and quickly becomes prohibitively complex (Hoeppner, Goodwin, Velicer, & Heltshe, 2007). Despite the options, these approaches ignore the individual, as each replication is done with a different participant.

Due to the complexities involved in generalizing idiographic research to populations of interest, it is more appropriate to consider levels of generalization. Typically, there are three distinct levels of generalization: individual, in which a unique individual is the focus; subgroups, in which there are distinct homogeneous subgroups; and population, the more classic approach.

At the idiographic level, there are multiple approaches to exploring generalizability. The first distinct idiographic generalization is within an individual, across contexts. If a treatment is established as effective for a unit in one context, the generalization of interest would be to other contexts. For example, if an intervention is successful in one classroom, will it also be maintained in another classroom? This is sometimes referred to as the *generalization gradient.* A second approach is to generalize across occasions. As the focus of idiographic methods is on variability within the unit across occasions, generalization should also be across occasions. For example, to evaluate the effect of changes in speed limits on car accidents in the United States, the focus would be on the pattern of change over time, not on generalizing to other countries. A third approach is to generalize to homogeneous subgroups. If the ergodic theorems are not supported, generalizing to a heterogeneous population is not logical. However, there are often homogeneous subgroups that meet the ergodic theorems, and generalizing to the subgroups is appropriate. This inductive approach applies a bottom-up rather than a top-down philosophy. By gaining a more detailed and complex understanding of individual process, we can not only better understand individuals, but also gain a clearer understanding of the ways in which people group together. When homogeneous subgroups are identified through idiographic data, nomothetic analyses and generalizations can be applied to those groups. Procedures for identifying homogeneous subgroups are explored in some of the examples discussed later.

The Idiographic–Nomothetic Continuum

A strict adherence to either an idiographic or nomothetic approach may impair overall understanding of a psychological phenomenon. The two perspectives may not be separable, in that a unique individual is still a member of some population and is therefore a manifestation of that population. Though an individual may represent a unique individualized set of processes, the possibility of population-level generalization is not necessarily invalidated. The idiographic–nomothetic split may reside more in current constraints in study design and data analyses than in divergent or incompatible inference from one perspective or the other. This may not always be the case, as methodological and statistical designs may advance to combine both ends of the continuum.

Another way of looking at the continuum of idiographic–nomothetic conceptions holds that any observation can be considered in idiographic (unique) or nomothetic (general) terms. That is, no observation needs to be framed as inherently one or the other. Accordingly, Silverstein (1988) suggested that events can be abstracted into universal or particular descriptions to regard specific (idiographic) or general (nomothetic) processes. Similarly, Windelband (1998) argued that the particular is subordinate to the general and the general must ever accommodate the particular. Allport (1962) also supported this view by suggesting that a morphogenic position starts with either the idiographic or nomothetic description and incorporates the other in assigning treatment.

These views suggest that the vantage points along the idiographic–nomothetic continuum are simultaneously important. Theoretically, descriptions of all individual (idiographic) processes comprise the general (nomothetic) definition of the processes. Therefore, nomothetic descriptions are meant to include all cases of the particulars they describe, and to be a general description of what is always common to all (Lamiell, 1998). The general law is meant to cover any instance of that phenomenon (Lamiell, 1998). This is in line with the original definition of nomothetic by Windelband (1998), which holds that statements that are true in general include all instances of the particular they cover.

In this view, the term *nomothetic* refers to the utilization of comprehensive models to test and explain process, assuming each individual's uniqueness within the generality (Silverstein, 1988). Unfortunately, there is typically a mismatch between the philosophical intent of research designs (looking for general laws applicable to all individuals) and the statistical procedures applied to the data. As most nomothetic analyses are based on central tendencies, the results provide an aggregated trajectory from the unique paths of all individuals. Unless the ergodic theorems are met, however, the aggregate will not represent an individual path, only the center of all individuals combined. For example, if three individuals are given a medication, group-level analysis could show that prior to the medication all individuals were sick, after the medication there was some overall improvement, and at two follow-up tests, all individuals had recovered. However, individually, person 1 may have gotten better immediately following the intervention, whereas persons 2 and 3 got worse. The overall average between the three would still show some overall improvement. At later follow-ups, person 2 may have shown steady improvement and finally recovery, whereas person 3 showed slight decline. Once aggregated, the data will not match the trajectory of any individual, yet will suggest significant improvement. Though this may be important information, it is often misinterpreted as generalizable to the individual, though the results do not map to anyone specifically.

In the purely idiographic perspective, each case is unique and so cannot be defined by a general nomothetic description. Work by Molenaar (2004) adheres to this view, as he argues that intraindividual variation is not equal to interindividual variation. Molenaar suggests that the complete set of life histories of a population of human participants can be represented as a collection of trajectories in the same behavior space. This suggests that people are too different to allow us to gain accurate information through generalities, so they cannot be defined by a sample-based nomothetic description (Münsterberg, 1899). What may be most important, then, is a clear understanding of the advantages and limitations of idiographic and nomothetic approaches when designing research studies.

In the following sections, we give more detailed examples to help demonstrate some idiographic approaches to research and generalizability. In Example 1, we illustrate generalizability across contexts; in Examples 2 and 3, we illustrate generalizability to homogeneous subgroups. Our intent is to illustrate some of the unique research questions that emerge from idiographic research.

Example 1: Telemetric Assessment of Stereotypy in Children with Autism

Stereotypical behaviors are a core diagnostic feature of autism spectrum disorders (ASDs). They have been found to be one of the most disruptive classes of behaviors occurring in children with an ASD. This type of behavior can lead to social stigmatization and may complicate social interactions. If stereotypy becomes a dominant behavior, it can interfere with the acquisition of new skills and the performance of established skills. In severe cases it may lead to self-injurious behavior. There is therefore a need for accurate and timely measures to facilitate the development of interventions.

The goal of this project (Goodwin, Intille, Albinali, & Velicer, 2011) was to evaluate the use of wireless accelerometers and pattern recognition software to automatically detect stereotypical movements in real time. The sensors being developed—wireless three-axis accelerometers—can be unobtrusively placed on multiple limb locations to detect the two most common stereotypy: body rocking and hand flapping. Figure 16.2 presents an example of such a device. The study followed six participants during regular school hours over the course of 12 months. Each participant completed one 15- to 30-minute trial per week. Three phases of trials were implemented: (1) training the algorithm, (2) testing the algorithm in the laboratory, and (3) testing the algorithm in the natural environment (e.g., classroom). All trials were videotaped and double-coded

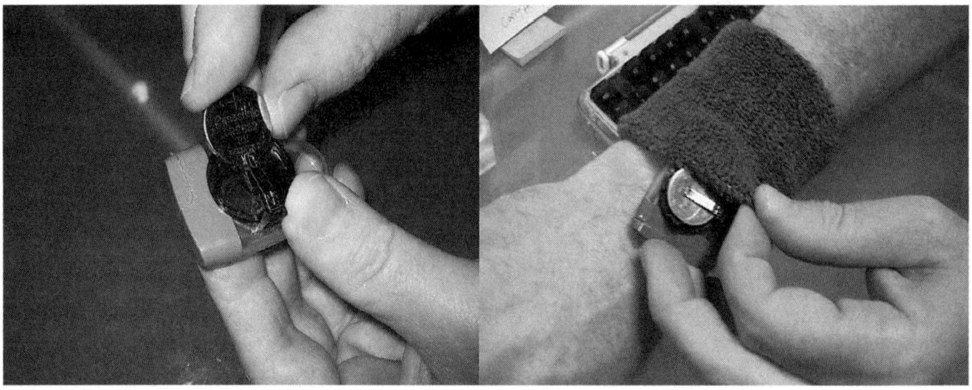

FIGURE 16.2. A wireless three-axis accelerometer can be used to track body movements.

by observers for stereotypy. Data were collected on the occurrence of a stereotypy, the type of movement (i.e., topography), the frequency of movement, the duration of the bout, and the setting in which the event occurred. Classification algorithms used acceleration data to differentiate between activities.

Table 16.1 provides the results for the six subjects in the two different contexts. Results indicated that accelerometer data were sufficient for the detection of stereotypy by the algorithm. In comparison to the observer coding, the algorithm accurately detected an average of 91.1% of stereotypy in the laboratory and 90.4% accuracy in the classroom.

There are numerous applications for these devices. For example, they can be employed in future collection of research data, further developed as intervention tools for therapists and caregivers, and utilized as outcome measures for behavioral and pharmacological interventions.

It is important to note that the application of group-level analysis would make little sense with this study design, as the algorithms are trained to the distinct patterns

TABLE 16.1. Percent Correct Identification of Stereotypic Behaviors

Student[a]	Study 1: Laboratory	Study 2: Classroom
JK	82.5%	85.9%
EH	92.9%	88.9%
CF	91.0%	90.6%
TM	93.7%	93.7%
CS	90.0%	90.2%
DB	96.4%	93.0%
Mean	91.1	90.4

[a]Not real initials.

of each individual. Therefore, data must be collected and analyzed on an individual basis. In this case, generalization across contexts was very high, showing that once the algorithm was trained to an individual, it was able to accurately detect stereotypy in that person across multiple settings.

Example 2: Longitudinal Patterns from a Nicotine Harm Reduction Study

Hoeppner, Goodwin, Velicer, Mooney, and Hatsukami (2008) examined individual patterns in daily smoking rates following the end of an intensive smoking reduction program. The primary study evaluated the relationship of tobacco biomarkers with reduced levels of smoking (Hatsukami et al., 2005). The sample analyzed involved a subset of 151 active smokers recruited for the study and consisted of three phases: baseline, reduction, and maintenance. The Hoeppner and colleagues follow-up study focused on the 6-week-long maintenance phase and utilized data from 57 participants who completed daily records for the first 40 days of the maintenance phase.

The study involved two phases. In the first phase, the timeline of cigarettes smoked per day was plotted for each participant, and the series was analyzed using time series analysis. The level (intercept), autocorrelation (serial dependence), variance, and linear trend parameters (i.e., slope) were estimated for each individual. On average, the 57 participants started the maintenance phase with a daily smoking rate of 7.85 cigarettes per day. Approximately half of the participants (50.9%) had significant first-order autoregressive (AR[1]) parameters, and 17.5% had significant second-order autoregressive (AR[2]) parameters. AR(1) parameter estimates averaged 0.56, which suggests a moderate level of dependence on the previous day's smoking levels for most subjects. For others, smoking levels in the 2 days before impact the current day; some of these AR(2) models may be indicative of stochastic oscillatory behavior (see Box & Jenkins, 1976, p. 59). The results suggest that different generating functions may underlie the individual smoking patterns (Velicer, Redding, Richmond, Greeley, & Swift, 1992). Slopes were significant for more than half of the sample (59.6%), indicating that there were significant changes in smoking rates over time. Of the statistically significant slope parameters, the majority was positive (58.8%), indicating an increase in the number of cigarettes smoked per day after the conclusion of the cigarette reduction program. However, a substantial portion of the statistically significant slope parameters was negative (41.2%), indicating a continuing decline over the maintenance phase. This finding is important to note, as strictly nomothetic analysis would mask these different trajectories.

The second phase applied a dynamic typology analysis to the results of the time series analysis to determine if homogeneous subgroups, which met the conditions of the ergodic theorems, existed. A dynamic typology is a cluster analysis applied to longitudinal data. Cluster analysis classifies units based on the level, scatter, and shape of the variables used for clustering. The analysis was performed using the squared Euclidean distance metric and Ward's minimum variance algorithm. Several methods were used to determine the number of clusters, and a three-cluster solution was determined to provide the best fit. The clusters represent three distinct types of maintenance patterns: (1) increasing (47.4%), (2) constant (12.3%), and (3) decreasing (40.4%). These descriptive labels correspond to the slope parameters estimated in the individual time series analyses. Figure 16.3 illustrates the three dynatype clusters and the overall average trajectory across the 57 subjects.

The results of this study demonstrate that an idiographic approach yields different findings from those in a group-level analysis. Group-level analyses would conclude that subjects who achieved a 75% decrease in the number of cigarettes smoked were able to maintain that level. In this study, idiographic analysis revealed that that conclusion is only accurate for a small minority of smokers (12.3%). The two larger groups were the increasing group (47.4%), which was increasing in number of cigarettes smoked per day, and the decreasing group (40.4%), which was tending toward smoking cessation. The data suggest that an intervention focused on the increasing group is needed. Generalization should focus on the subgroup level.

From the perspective of the ergodic theorems, it is clear that neither theorem was satisfied. The autocorrelations varied widely—a clear violation of the first theorem—and the trajectories clearly did not have the same statistical characteristics—a clear violation of the second theorem. For this example, the idiographic approach provides

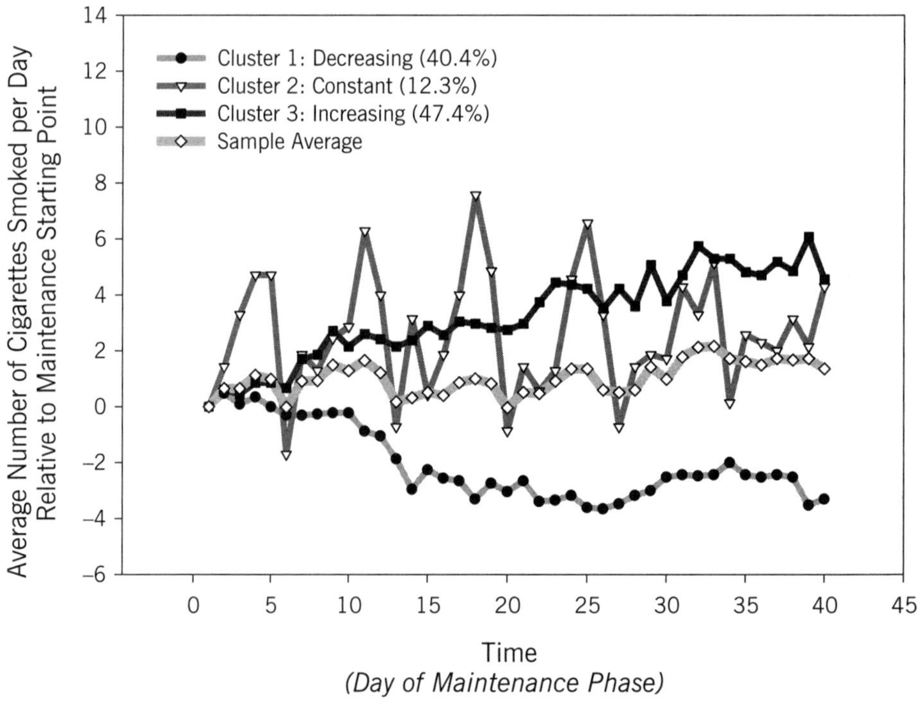

FIGURE 16.3. These three clusters from Hoeppner et al. (2008) represent three distinct types of maintenance patterns: increasing (47.4%), constant (12.3%), and decreasing (40.4%).

different results from the group-level analysis. The important next step in this research is to find reliable predictors of subgroup membership, and to develop interventions for the increasing group members to prevent their regression.

Example 3: Patterns of Adherence in Sleep Apnea

A study by Aloia and colleagues (2008) provides another example of both utilizing time series methodology to assess individual patterns and determining whether homogeneous subgroups exist. The study focused on the patterns of adherence to the most common treatment for obstructive sleep apnea (OSA), positive airway pressure (PAP). The disorder is characterized by fragmented sleep resulting from repeated cessations of breathing during the night. The treatment requires the nightly use of a PAP device, which is a machine that delivers positive airway pressure through a face mask. Adherence is suboptimal in this area (DiMatteo, 2004), making examination of adherence data an important scientific concern. In addition to delivering therapeutic pressure, PAP devices also monitor usage. This feature provides an extremely useful amount of adherence data.

This study involved two phases. First, time series analysis was used to examine adherence patterns objectively. Seventy-one participants with moderate to severe OSA, each with 365 days of treatment data, were studied. A separate time series analysis was conducted for each participant. Time series parameters assessed for hours of PAP use per night included the mean, level, variance, slope, and autocorrelation. Second, a team of raters evaluated the individual time series graphs to visually classify the participants into groups of similar adherence patterns. Seven homogeneous subgroups were found: (1) good users, (2) slow improvers, (3) slow decliners, (4) variable users, (5) occasional attempters, (6) early dropouts, and (7) nonusers. Good users (24%) were characterized by high levels of use, low variability, and a near-zero slope. See Figure 16.4 for examples from each of the subgroups. Slow improvers (13%) had positive slopes and variability that decreased over time. Slow decliners (14%) had negative slopes and variability that increased over time. Variable users (17%) were characterized by high variability and a near-zero slope. Occasional attempters (8%) had fewer days of PAP use and a lower mean. Early dropouts (13%) did not continue PAP use past the first few months, and nonusers (11%) did not continue PAP use past the first few weeks. The dramatic differences among these subgroups emphasize the importance of evaluating individual differences, as group-level analyses would simply combine very different users into groups and mask the significant heterogeneity. The analysis provides a much more nuanced view of adherence and can serve as the basis for future interventions.

A follow-up study by Babbin, Velicer, Aloia, and Kushida (2011) assessed patterns of adherence to PAP in a larger sample, with some methodological differences. This study involved an international sample of 161 participants, each with approximately 180 nights of data. The study involved three phases. First, as in Aloia and colleagues (2008), a separate time series analysis was performed for each individual. Second, similar to Hoeppner and colleagues (2008), a dynamic typology was performed to identify

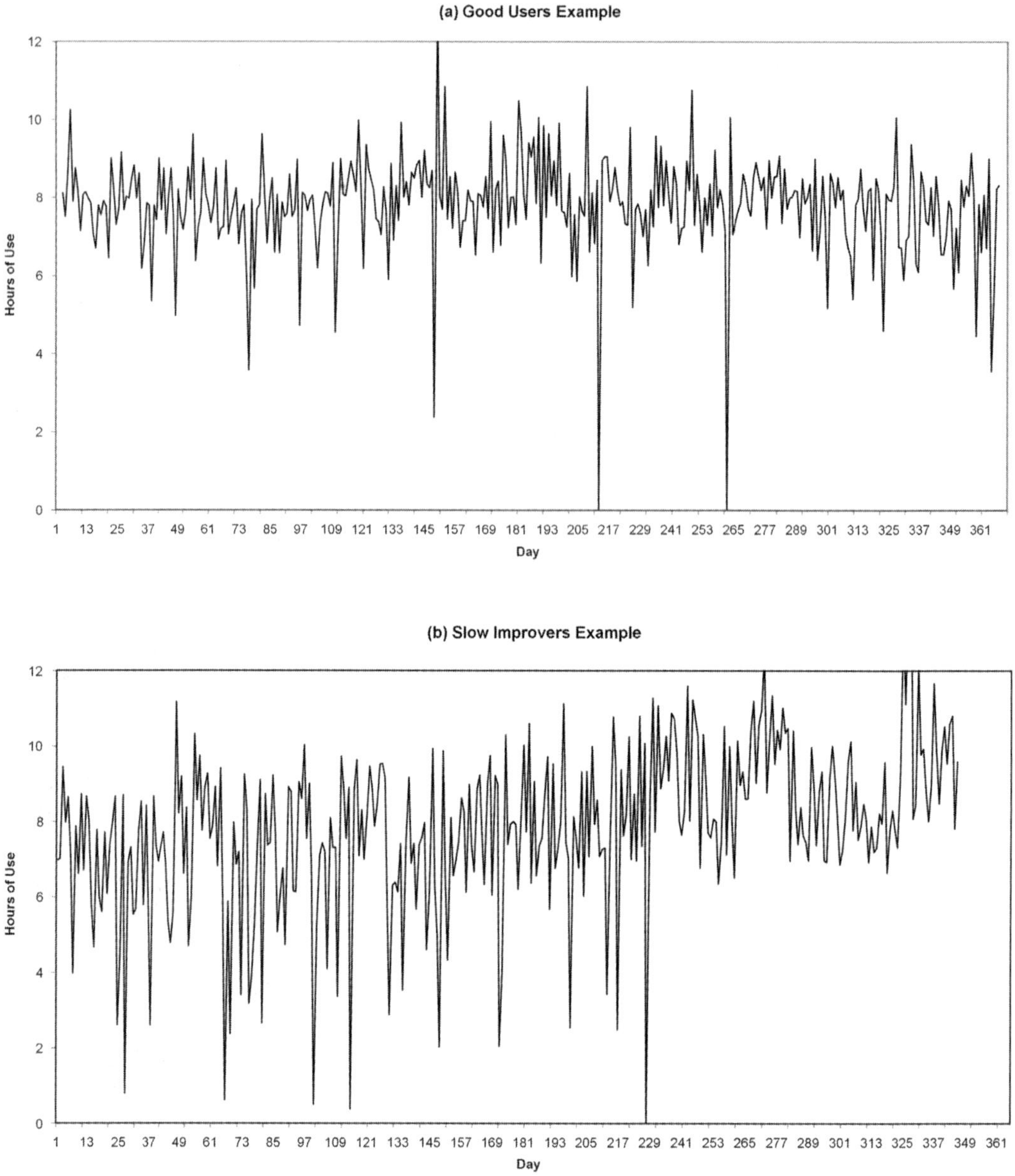

FIGURE 16.4. Examples of homogeneous subgroups of PAP users from Aloia et al. (2008): (a) good users, (b) slow improvers, (c) slow decliners, and (d) occasional attempters.

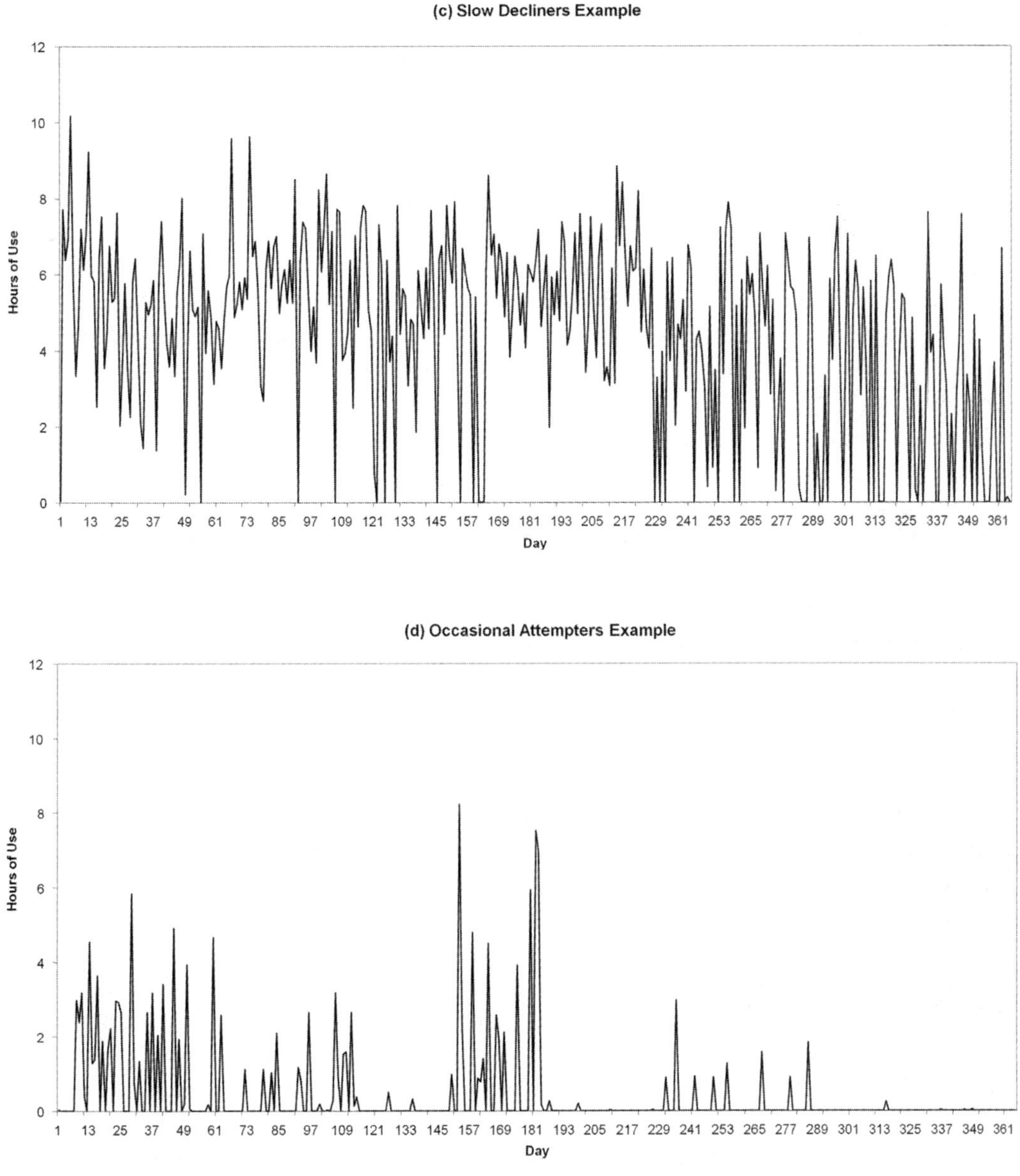

FIGURE 16.4. *(continued)*

homogeneous subgroups of individuals with similar adherence patterns. A four-cluster solution was found, and the subgroups were labeled based on their time series parameters: great users (17.2%; high mean and level, no slope), good users (32.8%; moderate mean and level, no slope), poor users (22.7%; low mean and level, negative slope), and slow decliners (moderate mean and level, negative slope, high variance). Third, participants in the identified subgroups were compared on a number of variables that were not involved in the dynamic typology to establish external validity and to identify variables that could be useful for intervention purposes. Some of the most notable findings included these: Great users reported the most self-efficacy (confidence to use PAP); poor users reported the most sleepiness; and great users reported the highest quality of sleep.

Results of this study offer an important replication of the study by Aloia and colleagues (2008). Participants in both studies demonstrated individual differences in series mean, level, slope, and variance. Additionally, most autocorrelations were nonsignificant in both studies. In Aloia and colleagues, 14 of 71 (19.7%) participants had a significant autocorrelation, and 4 of these 14 (28.6%) had a large ($r_1 > .30$) autocorrelation. In Babbin and colleagues (2011), 63 of 161 (39.1%) had a significant autocorrelation, and 17 of these 63 (27%) had a large autocorrelation. Thus, only a minority of adult PAP users appears to be influenced by the previous night's usage. Subgroups were also replicated. Whereas Aloia and colleagues identified seven subgroups compared to the four subgroups identified in Babbin and colleagues, there are two clear reasons for these differences. First, the subgroups were identified in different ways (qualitative, visual inspection vs. quantitative, dynamic typology). Second, the length of the study was different. In the Babbin and colleagues study, some potential subgroups simply did not form within the 180 days of data. For both studies, an important focus was on the necessity of assessing individual differences for improving adherence. Psychological variables relevant to adherence patterns, such as self-efficacy, could be the focus of interventions to increase PAP usage. The timing and content of interventions should be tailored to adherence patterns in order to maximize adherence.

In both Aloia and colleagues (2008) and Babbin and colleagues (2011), the ergodic theorems were not satisfied. The first condition appears to be violated. Although the inequality of autocorrelations was not formally tested, they appear to vary among participants, with participants demonstrating significant and positive autocorrelations, significant and negative autocorrelations, and nonsignificant autocorrelations. The second condition was also violated: Participants clearly do not have the same statistical characteristics. However, the identified subgroups have similar trajectories with similar statistical characteristics. Thus, the subgroups arguably satisfy the second condition. This improvement in ergodicity is important to generalizability; inferences about subgroup-level findings can be more validly applied to individuals.

Research Issues in Idiographic Research

There are major challenges for researchers employing idiographic methods. One issue involves missing data. When a single unit is studied for a long series of observations, missing data are almost inevitable. A partial solution is to automate data collection as much as possible. PAP devices, which automatically record data when they are active, are a good example of this kind of data collection. Another solution is to estimate missing data using procedures that have been developed over the last three decades, such as maximum likelihood and multiple imputation (Schafer & Graham, 2002). Recent simulation studies suggest that these methods can produce very accurate results for time series even when up to 40% of the data is missing and assumptions have been violated (Velicer & Colby, 2005a, 2005b).

A second issue involves identifying the generating functions. Idiographic data permit us to study the generating function that produces the types of data that we are observing. This is an emerging research area with few good examples at present. Identification of the function can be based on the autocorrelation structure, as Velicer and colleagues (1992) did, in an attempt to determine the most appropriate nicotine regulation model for smoking behavior. The nature of the function is important for intervention development.

A third issue involves the development of appropriate measures. Many widely used measures in the behavioral sciences are simply not appropriate for regular assessment at short intervals. Among the problems are recall bias, measure reactivity, and measure pollution. Advances in information systems technology have permitted the development of telemetrics (Goodwin, Velicer, & Intille, 2008), which represents a promising new approach to developing appropriate measures.

Conclusions

Idiographic designs have tremendous potential in the behavioral sciences to address research questions that could not be addressed, or only addressed indirectly, by nomothetic methods. Time series analysis is one of the large number of computational procedures that has been developed during the last 40 years specifically for the analysis of longitudinal data. All these recently developed procedures share the common characteristic of requiring high-speed computers, which are becoming increasingly ubiquitous, to perform analyses. Advances in computer technology are also producing new and more time-intensive sources of data. The combination of computational advances and new sources of data has increased the range of potential applications for idiographic methods. Some of the early drawbacks to idiographic designs have been largely overcome in the last two decades. The development of statistical methods to address the issue of missing data has provided good methods of dealing with this inevitable problem. The problem of finding data appropriate for intensive longitudinal designs is being addressed by the development of new measures such as those provided by telemetrics.

Issues of generalizability are being addressed by identifying homogeneous subgroups that better meet the ergodic theorems, and recognizing that sometimes the solution is to generalize intraindividually in the time dimension. Advances in other areas are still in the early stages. The issue of identifying the generating function for a time series and the utilization of this type of information is still in the formative stages. The area of multivariate time series has been conceptually developed, but there are few examples of the applications of these methods currently available. Perhaps the issue that deserves the most attention is the ergodic theorems. Accumulating evidence indicates that they are unlikely to be met in practice. At a minimum, this means that the results provided by idiographic designs are likely to be different from those provided by nomothetic methods. The issue also has unexplored consequences for the types of conclusions that can be drawn from nomothetic methods.

Idiographic designs have the potential to address research questions that cannot be addressed by nomothetic methods. Matching methodology involved in data collection and analysis to the research questions being asked is essential. In the above examples, patterns of behavior over time were examined. Individual patterns were identified, and sometimes homogeneous subgroups were found. Studying variations in these patterns of complex behaviors with idiographic methods is crucial to developing individually tailored interventions. Since most nomothetic research seems to violate the ergodic theorems, idiographic approaches are necessary to both understanding changes in behavior over time and to generalizing about the changes.

REFERENCES

Allport, G. W. (1962). The general and the unique in psychological science. *Journal of Personality, 30*, 405–422.

Aloia, M. S., Goodwin M. S., Velicer, W. F., Arnedt, J. T., Zimmerman, M., Skrekas, J., et al. (2008). Time series analysis of treatment adherence patterns in individuals with obstructive sleep apnea. *Annals of Behavioral Medicine, 36*, 44–53.

Babbin, S. F., Velicer, W. F., Aloia, M. S., & Kushida, C. A. (2011). Abstract: Identifying longitudinal patterns of adherence to treatment for obstructive sleep apnea. *Multivariate Behavioral Research, 46*(6), 1005–1006.

Barlow, D. H., Nock, M. K., & Hersen, M. (2009). *Single case experimental designs* (3rd ed.). Boston: Pearson & AB.

Box, G. E., & Jenkins, G. M. (1976). *Time series analysis: Forecasting and control* (rev. ed.). San Francisco: Holden Day.

Choe, G. H. (2005). *Computational ergodic theory*. Berlin: Springer.

DiMatteo, M. R. (2004). Variation in patients' adherence to medical recommendations: A quantitative review of 50 years of research. *Medical Care, 42*, 200–209.

Glass, G. V., Willson, V. L., & Gottman, J. M. (2008). *Design and analysis of time series experiments*. Charlotte, NC: Information Age.

Goodwin, M. S., Intille, S. S., Albinali, F., & Velicer, W. F. (2011). Automated detection of stereotypical motor movements. *Journal of Autism and Developmental Disorders, 41*, 770–782.

Goodwin, M. S., Velicer, W. F., & Intille, S. S. (2008). Telemetric monitoring in the behavior sciences. *Behavior Research Methods, 40*, 328–341.

Hatsukami, D. K., Kotlyar, M., Allen, S., Jensen, J., Li, S., Le, C., et al. (2005). Effects of cigarette reduction on cardiovascular risk factors and subjective measures. *Chest, 128*(4), 2528–2537.

Hoeppner, B. B., Goodwin, M. S., Velicer, W. F., & Heltshe, J. (2007). An applied example of pooled time series analysis: Cardiovascular reactivity to stressors in children with autism. *Multivariate Behavioral Research, 42*, 707–727.

Hoeppner, B. B., Goodwin, M. S., Velicer, W. F., Mooney, M. E., & Hatsukami, D. K. (2008). Detecting longitudinal patterns of daily smoking following drastic cigarette reduction. *Addictive Behaviors, 33*, 623–639.

Kazdin, A. E. (2011). *Single-case research design* (2nd ed.). New York: Oxford University Press.

Lamiell, J. T. (1998). "Nomothetic" and "idiographic": Contrasting Windelband's understanding with contemporary usage. *Theory and Psychology, 8*, 23–38.

Molenaar, P. C. M. (1987). Dynamic assessment and adaptive optimization of the therapeutic process. *Behavioral Assessment*, 9, 389–416.

Molenaar, P. C. M. (2004). A manifesto on psychology as idiographic science: Bringing the person back into scientific psychology, this time forever. *Measurement, 2*(4), 201–218.

Molenaar, P. C. M. (2008). Consequences of the ergodic theorems for classical test theory, factor analysis, and the analysis of developmental processes. In S. M. Hofer & D. F. Alwin (Eds.), *Handbook of cognitive aging* (pp. 90–104). Thousand Oaks, CA: Sage.

Münsterberg, H. (1899). Psychology and history. *Psychological Reviews, 6*, 1–31.

Schafer, J. L., & Graham, J. W. (2002). Missing data: Our view of the state of the art. *Psychological Methods, 7*, 147–177.

Silverstein, A. (1988). An Aristotelian resolution of the idiographic versus nomothetic tension. *American Psychologist, 43*(6), 425–430.

Velicer, W. F., & Colby, S.M. (2005a). A comparison of missing data procedures for ARIMA time series analysis. *Educational and Psychological Measurement, 65*, 596–615.

Velicer, W. F., & Colby, S. M. (2005b). Missing data and the general transformation approach to time series analysis. In A. Maydeu-Olivares & J. J. McArdle (Eds.), *Contemporary psychometrics. A Festschrift to Roderick P. McDonald* (pp. 509–535). Hillsdale, NJ: Erlbaum.

Velicer, W. F., & McDonald, R. P. (1991). Crosssectional time series designs: A general transformation approach. *Multivariate Behavioral Research, 26*, 247–254.

Velicer, W. F., & Molenaar, P. (2013). Time series analysis. In J. Schinka & W. F. Velicer (Eds.), *Research methods in psychology: Vol. 2. Handbook of psychology* (pp. 628–660). New York: Wiley.

Velicer, W. F., Redding, C. A., Richmond, R., Greeley, J., & Swift, W. (1992). A time series investigation of three nicotine regulation models. *Addictive Behaviors, 17*, 325–345.

Windelband, W. (1998). History and natural science. *Theory and Psychology, 8*, 5–22.

CHAPTER 17

New Trends in the Inductive Use of Relational Developmental Systems Theory

Ergodicity, Nonstationarity, and Heterogeneity

Peter C. M. Molenaar and John R. Nesselroade

The Romantic era with its focus on organicism and systems thinking (Beiser, 2003) was an important formative time for developmental science, as witnessed by the work of psychologists such as Carus and Tetens (cf. Baltes, 1979). Several traces of this early worldview recur in relational developmental systems theory (RDS), the dominant theoretical paradigm in current developmental science. It therefore is no exaggeration to state that the systems perspective has been an essential ingredient throughout the history of developmental science. In this chapter we consider foundational methodological–statistical aspects that arise when RDS is used in an *inductive mode*; that is, *when dynamic systems models inspired by RDS are fitted to empirical data.* For this purpose we only need to consider selected aspects of RDS; the reader is referred to the chapters of Overton (Chapter 2) and Geldhof and colleagues (Chapter 3) in this volume for more encompassing discussions of RDS.

The aspects of RDS that are important for our purposes are the following. According to RDS, development is the result of multiple interacting influences that are context sensitive and contingent. This idea implies that development is inherently subject-specific and stochastic because the contexts within which a subject develops have contingent, subject-specific effects that continuously build up within the developing system due to ongoing interactions. According to RDS, development also is a constructive process in which nonlinear epigenetic influences play important roles. As is explained below, such nonlinear epigenetic influences create substantial subject-specific variations that reinforce the subject-specific effects due to contingent contextual influences. Last but not least, RDS focuses on the potential for change evolving at multiple timescales and at multiple levels. This idea implies that dynamic systems models inspired by RDS will include time-varying parameters located at different levels and changing with different rates. In sum, our discussion of dynamic systems models used in inductive RDS focuses on subject specificity, stochasticity, and pervasive change.

Methodological implications of some aspects of development emphasized by RDS have been considered previously. For instance, Sidman (1960) presents an overview of early methodological work addressing the (lack of) relationship between individual and average learning curves, starting with Merrill (1931). Wohlwill (1971) criticizes the use of interindividual variation in studying developmental processes and in its stead recommends a focus on individual developmental functions. In what follows we considerably strengthen and generalize these early approaches. That is, first we consider a general mathematical theory, ergodic theory, which specifies necessary conditions for a valid inductive RDS based on fitting subject-specific stochastic dynamic systems models with time-varying parameters to appropriate empirical data. Although ergodic theory was founded more than a century ago, its surprisingly direct relevance to inductive RDS was made explicit only recently (Molenaar, 2004). Next, important implications of ergodic theory for inductive RDS are elaborated; in particular, how to deal with pervasive subject specificity and change in fitting appropriate dynamic systems models to the data. In this context we consider a new approach inspired by RDS to estimate subject-specific heritabilities.

Before proceeding, the following preliminary remarks are in order. We already indicated that inductive RDS is to be understood as the fit of dynamic systems models inspired by RDS to appropriate empirical data. The class of dynamic systems models under consideration is specified below, but its role in inductive RDS is similar to the role of, for instance, factor models in psychometrics. That is, each dynamic systems model contains free parameters that are estimated in the direct fit to empirical data (for several examples, see Molenaar & Newell, 2010; Newell & Molenaar, 1998). This point is all that we mean by the denotation *inductive*. In using this denotation we do not intend to take a particular stance in the philosophy of science. The main contrast of inductive RDS in the indicated sense is with applications of dynamic systems theory that do not involve the actual fit of parametric systems models to empirical data but instead involve more indirect comparisons of implications of systems models with empirical evidence. For instance, one such indirect approach would involve the simulation of systems models and then comparing in qualitative ways the outcomes thus obtained with selected aspects of actual developmental processes. An example of the latter approach is McClelland and Jenkins's (1991) connectionist simulation of stagewise development (see Raijmakers, van Kooten, & Molenaar, 1996, for important criticism).

Classical Ergodic Theory

The standard approach to statistical analysis in psychology is to draw a random sample of subjects from a presumably homogeneous population of subjects, analyze the structure of interindividual variation in this sample, and then generalize the results thus obtained to the population. Such analyses of interindividual variation underlie all known statistical techniques, including analysis of variance, regression analysis, factor analysis, multilevel (latent growth curve) modeling, mixture modeling, and so on. Consequently the standard approach to psychological data analysis aims to describe the state of affairs at the population level, not at the level of individual subjects. Accordingly, the individuality of each of the persons in the sample and population is deemed immaterial: The subjects are considered to be replications devoid of individuality. This idea is expressed by the assumption that subjects are homogeneous in all respects relevant to the analysis. This essential assumption of homogeneity allows for the averaging (pooling) of the scores of the sampled subjects in the estimation of statistics (means, variances, correlations, etc.) to be generalized to the population. Pooling across subjects is the hallmark of analyses of interindividual variation.

However, this standard approach based on analysis of interindividual variation can be shown to be incorrect if a psychological process under investigation does not obey stringent conditions (Molenaar, 2004). The proof is based on classical ergodic theorems—theorems of extreme generality that apply to all measurable processes irrespective of their content (see Choe, 2005, for a modern proof of the first ergodic theorem of Birkhoff, 1931). To appreciate the implications of the classical ergodic theorems, it is helpful to first characterize the elementary methodological situation in psychological measurement. Instead of postulating an abstract population of subjects, consider an ensemble of actually existing human subjects whose measurable psychological processes are functions of time and place (Elder, 1998; Elder, Modell, & Parke, 1993). To simplify the following discussion, without affecting its generality, the focus is on time as the basic dimension along which psychological processes are evolving. The ensuing basic scientific representation of a human subject in psychology therefore is a high-dimensional dynamic system generating a set of time-dependent processes. The system includes important functional subsystems, such as the perceptual, emotional, cognitive, and physiological systems, as well as their dynamic interrelationships. The complete set of measurable time-dependent variables characterizing the system's behavior can be represented as the coordinates of a high-dimensional space, referred to here as the *behavior space*. The behavior space contains all the scientifically relevant information about a person (cf. De Groot, 1954).

Within the behavior space, interindividual variation is defined as follows: (1) Select a fixed subset of variables; (2) select one or a few fixed time points as measurement occasions; and (3) determine the variation of the scores on the selected variables at the selected time points by pooling across subjects. Analysis of interindividual variation thus defined is called the *R-technique* by Cattell (1952). In contrast, intraindividual variation is defined as follows: (1) Select a fixed subset of variables; (2) select a fixed subject; and (3) determine the variation of the scores of the single subject on the selected variables by pooling across a sampled time interval. Analysis of intraindividual variation thus defined is called *P-technique* by Cattell.

With these preliminary specifications in place, the following heuristic description of the content of the classical ergodic theorems can be given. These theorems detail the conditions that must be met in order to generalize from analyses of interindividual variation to analyses of intraindividual variation, and vice versa. In contrast, a process is "nonergodic" if the results of the analyses of interindividual variation do not generalize to the level of intraindividual change over time, and vice versa. In what follows we consider only Gaussian processes. The conditions that Gaussian processes have to meet in order to be ergodic are twofold (cf. Hannan, 1970).

1. The process has to be homogeneous in time, having constant mean levels, no cycles, and sequential dependencies that depend only on relative time differences (lags). Such a process is called *weakly stationary*.
2. The process has to be homogeneous across different subjects in the population. That is, the process obeys exactly the same dynamic model for different subjects in the population (ensemble).

If a Gaussian process is either nonstationary (violating condition 1) or heterogeneous across subjects (violating condition 2), or both, then this process is nonergodic. This means that there is no lawful relationship between its structure of interindividual variation at the population level and the structures of intraindividual variation at the level of individual subjects belonging to the population. Put another way, if the conditions of ergodicity are violated, no a priori relationship exists between results obtained in an analysis of interindividual variation (R-technique) and results obtained in an analogous analysis of intraindividual variation (P-technique).

The consequences of the classical ergodic theorems affect all psychological statistical methodology (Borsboom, 2005; Molenaar, Huizenga, & Nesselroade, 2003). Because *development* generally implies that some kind of growth or decline occurs, developmental processes are almost always nonstationary (violating condition 1) and are therefore nonergodic. Generally, developmental scientists consider change that occurs in average or mean levels of a process. However, change may also occur in variances or sequential dependencies over time. The nonergodicity of developmental processes implies that their analysis has to be based on intraindividual variation in order to obtain valid information at the level of individual development. It is argued below that starting with analyses of intraindividual variation does not preclude valid generalization across subjects. In fact, a successful new method is presented here to accomplish just this, even if the subjects concerned are also heterogeneous (violating condition 2).

At the close of this section we present the following, more general remarks. Attention was confined to Gaussian processes, for which two conditions were specified that are necessary and sufficient for ergodicity. For measurable processes in general (including all non-Gaussian processes), these two conditions are necessary but not sufficient. It turns out to be exceedingly difficult to prove whether some general measurable process is ergodic. This proof has been accomplished for only a few such processes. But apart from the mathematical difficulties in providing complete proofs that a given measurable process is (non)ergodic, what can be said about the prevalence of ergodic processes in reality? As to this latter question, a result by Emch and Liu (2002) is quite revealing. These authors consider homogeneous Hamiltonian systems, which are the prime candidates in physics for being ergodic. They show that ergodicity is not a generic property of homogeneous Hamiltonian systems, which means that almost all such systems are nonergodic. This important result makes clear that ergodicity is a very rare property that never should be a default assumption.

Nonergodicity Due to Nonstationarity

In the remainder of this chapter we consider ways to deal with violations of the two conditions for ergodicity of Gaussian processes. First the statistical analysis of nonstationary developmental processes is addressed. This focus requires that we first outline the analysis of weakly stationary processes based on dynamic factor models, after which we generalize these models to linear state–space models with arbitrarily time-varying parameters.

Dynamic Factor Analysis of Stationary Processes

Dynamic factor analysis is factor analysis of a single-subject multivariate time series. It constitutes a generalization of Cattell's P-technique (Cattell, 1952) in that it takes account of lead–lag patterns in the dynamic relationships between latent factor series and observed series. In contrast, P-technique involves a straightforward application of standard factor analysis to multivariate time series without accommodation of lead–lag sequential dependencies; the reader is referred to Molenaar and Nesselroade (2009) for further discussion of the domain of application of the P-technique.

In what follows, time series are assumed to be normally distributed (i.e., Gaussian series) and observed at equidistant discrete time points. Generalizations of these restrictive assumptions are discussed in the concluding section of this chapter. In this section, it is additionally assumed that the series are stationary (defined shortly). Boldface lowercase letters denote column vectors; boldface uppercase letters denote matrices; an apostrophe attached to vectors or matrices denotes transposition. Let $\mathbf{y}(t) = [y_1(t), y_2(t), \ldots, y_p(t)]'$ be a p-variate time series, $p \geq 1$, observed at equidistant time points $t = 1, 2, \ldots, T$. The mean of $\mathbf{y}(t)$ at each time point t is: $E[\mathbf{y}(t)] = \boldsymbol{\mu}(t)$. Considered as a function of t, $\boldsymbol{\mu}(t)$ denotes the p-variate mean function (trend) of $\mathbf{y}(t)$. If $\boldsymbol{\mu}(t) = \boldsymbol{\mu}$—that is, if the mean function is constant in time—then $\mathbf{y}(t)$ has a stationary mean function. The sequential covariance of $\mathbf{y}(t)$ between a given pair of time points t_1 and t_2 is defined as $\boldsymbol{\Sigma}(t_1, t_2) = \text{cov}[\mathbf{y}(t_1), \mathbf{y}(t_2)']$. Considered as a function of two-dimensional time, $\boldsymbol{\Sigma}(t_1, t_2)$ denotes the (p,p)-variate covariance function of $\mathbf{y}(t)$. If $\boldsymbol{\Sigma}(t_1,$

t_2) only depends on the relative time difference $t_1 - t_2 = u$, that is, $\mathbf{\Sigma}(t_1, t_2) = \mathbf{\Sigma}(t_1 - t_2) = \mathbf{\Sigma}(u)$, $u = 0, \pm 1, \ldots, \pm T - 1$, then $\mathbf{y}(t)$ has a stationary covariance function depending only on lag u. If both the mean function and covariance function of $\mathbf{y}(t)$ are stationary, then $\mathbf{y}(t)$ is called a weakly stationary p-variate time series. Because Gaussian series are completely characterized by the first two-moment functions, weakly stationary Gaussian series are also (strongly) stationary in the distributional sense.

In the first publication on the use of dynamic factor analysis in psychology, the following model for stationary multivariate Gaussian series was considered (Molenaar, 1985):

$$\mathbf{y}(t) = \boldsymbol{\mu} + \mathbf{\Lambda}(0)\boldsymbol{\eta}(t) + \mathbf{\Lambda}(1)\boldsymbol{\eta}(t-1) + \ldots + \mathbf{\Lambda}(s)\boldsymbol{\eta}(t-s) + \boldsymbol{\varepsilon}(t) \qquad \textbf{(17.1a)}$$

where $\mathbf{y}(t)$ is an observed p-variate time series, $\boldsymbol{\eta}(t)$ is a latent q-variate factor series, and $\boldsymbol{\varepsilon}(t)$ is a p-variate measurement error series. Because our main interest is not in the constant mean function $\boldsymbol{\mu}$, it is conveniently assumed that $\boldsymbol{\mu} = 0$. Then $\mathbf{y}(t)$, $\boldsymbol{\eta}(t)$, and $\boldsymbol{\varepsilon}(t)$ are zero mean stationary series.

The $\mathbf{\Lambda}(u)$, $u = 0,1, \ldots, s$, are (p,q)-dimensional matrices of lagged factor loadings, where $s \geq 0$ is the maximum lag. These lagged factor loadings allow for the possibility that the realization of the latent factor series $\boldsymbol{\eta}(t)$ at each time t not only has an instantaneous effect on $\mathbf{y}(t)$, but also may have delayed effects at later time points $t + 1, \ldots, t + s$. The linear combination $\mathbf{\Lambda}(0)\boldsymbol{\eta}(t) + \mathbf{\Lambda}(1)\boldsymbol{\eta}(t-1) + \ldots + \mathbf{\Lambda}(s)\boldsymbol{\eta}(t-s)$ is called a *convolution*.

For later reference the limiting case of Equation 17.1a is considered in which $s = 0$, that is, the case in which there are no lagged factor loadings:

$$\mathbf{y}(t) = \mathbf{\Lambda}(0)\boldsymbol{\eta}(t) + \boldsymbol{\varepsilon}(t) \qquad \textbf{(17.1b)}$$

Equation 17.1b is a special instance of Equation 17.1a. It has been assigned several labels, including *state–space model* (e.g., Molenaar, 1985) and *process factor model* (e.g., Browne & Nesselroade, 2005). In what follows, Equation 17.1b is referred to as a *state–space model*. As is explained shortly, Equation 17.1b has a special property not shared by Equation 17.1a.

The dynamic factor model in Equation 17.1a was inspired by Brillinger's (1975) principal component analysis of multivariate weakly stationary time series. It differs from Brillinger's approach in a number of respects, perhaps the most important of which is that in Equation 17.1a the convolution of lagged factor loadings and latent factor series is finite and only depends at each time point t upon earlier realizations $\boldsymbol{\eta}(t), \ldots, \boldsymbol{\eta}(t-s)$ of the latent factor series, whereas in Brillinger's model this convolution is infinite and also may depend on future realizations $\boldsymbol{\eta}(t+1)$, $\boldsymbol{\eta}(t+2), \ldots$. Another important difference relates to the statistical characteristics of the measurement error series in each of the two models.

To complete the definition of the dynamic factor model under consideration, the covariance functions of the time series occurring in Equation 17.1a and Equation 17.1b have to be specified. Let diag-$\mathbf{A}$ denote a square diagonal matrix $\mathbf{A}$ (all off-diagonal elements being zero). Then the covariance functions associated with the right-hand side of Equation 17.1a are

$$\mathrm{cov}[\boldsymbol{\varepsilon}(t), \boldsymbol{\varepsilon}(t-u)'] = \text{diag-}\boldsymbol{\Theta}(u); u = 0, \pm 1, \ldots \quad \textbf{(17.1c)}$$

$$\mathrm{cov}[\boldsymbol{\eta}(t), \boldsymbol{\eta}(t-u)'] = \boldsymbol{\Psi}(u), u = 0, \pm 1, \ldots$$

The first equation of Equation 17.1c defines the covariance function of the measurement error process. The univariate measurement error process $\varepsilon_k(t)$ associated with the kth observed univariate series $y_k(t)$, $k \in \{1, \ldots, p\}$, is allowed to have nonzero sequential covariance: $\mathrm{cov}[\varepsilon_k(t), \varepsilon_k(t-u)] \neq 0$ for $\forall u$. However, measurement error processes $\varepsilon_k(t)$ and $\varepsilon_m(t)$ associated with different observed univariate series $y_k(t)$ and $y_m(t)$, $k \neq m \in \{1, \ldots, p\}$, are assumed to be uncorrelated at all lags u: $\mathrm{cov}[\varepsilon_k(t), e_m(t-u)] = 0$ for $\forall u$. The second equation in Equation 17.1c defines the covariance function of the latent factor series.

It was proven in Molenaar (1985) that under certain conditions the covariance function of the latent factor series defined in Equation 17.1c is not identifiable. That means that under certain conditions, the variances and sequential covariances in $\boldsymbol{\Psi}(u)$, $u = 0, \pm 1, \ldots$, cannot be estimated, but have to be fixed a priori. The conditions concerned are twofold. First, the maximum lag s of the matrices of factor loadings $\boldsymbol{\Lambda}(u)$, $u = 0, 1, \ldots, s$, has to be larger than zero: $s > 0$. Second, all factor loadings in $\boldsymbol{\Lambda}(u)$, $u = 0, 1, \ldots, s$, should be free parameters (apart from minimal constraints to guarantee the distinctness of different component factor series in $\boldsymbol{\eta}(t)$). That is, the dynamic factor model should be exploratory, having no a priori restrictive pattern of fixed factor loadings.

If both these conditions obtain, then the covariance function of the latent factor series is not identifiable and therefore has to be fixed a priori. $\boldsymbol{\Psi}(u)$, $u = 0, \pm 1, \ldots$, then can be fixed at any stationary covariance function without affecting the goodness of fit of the model. In Molenaar (1985) the simplest covariance function for $\boldsymbol{\eta}(t)$ was chosen: $\mathrm{cov}[\boldsymbol{\eta}(t), \boldsymbol{\eta}(t-u)] = \delta(u)\mathbf{I}_q$, where $\delta(u)$ is the Kronecker delta ($\delta(u) = 1$ if $u = 0$; $\delta(u) = 0$ if $u \neq 0$) and $\mathbf{I}_q$ is the (q,q)-dimensional identity matrix. This particular choice implies that the latent factor series lacks instantaneous as well as sequential dependencies. Accordingly, $\boldsymbol{\eta}(t)$ can be conceived of as a sequence of random shocks, often referred to in the engineering literature as a *white noise sequence*. But quite other choices can fruitfully be considered (cf. Molenaar & Nesselroade, 2001).

If the dynamic factor model is confirmatory—that is, if an a priori restrictive pattern of fixed factor loadings has been specified in $\boldsymbol{\Lambda}(u)$, $u = 0, 1, \ldots, s$—then the covariance function of the latent factor series is identifiable. Also if $s = 0$—that is, if the state–space model in Equation 17.1b applies, then the covariance function of the latent factor series is identifiable. In these cases $\boldsymbol{\Psi}(u)$, $u = 0, \pm 1, \ldots$, can be freely estimated or, alternatively, a parametric stationary time series model for $\boldsymbol{\eta}(t)$ (and hence for its covariance function) can be considered. For instance, the latent factor series $\boldsymbol{\eta}(t)$ can be represented by a vector autoregression: $\boldsymbol{\eta}(t) = \mathbf{B}\boldsymbol{\eta}(t-1) + \zeta(t)$, where $\mathbf{B}$ is a (q,q)-dimensional matrix of regression coefficients and $\zeta(t)$ is a q-variate white noise sequence.

Given that for the state–space model in Equation 17.1b, the covariance function (or a parametric time series model) of the latent factor series always is identifiable, it would seem rational to restrict attention to this type of model. At least for exploratory dynamic factor analyses, this restriction would preclude the need to have to arbitrarily fix $\mathbf{\Psi}(u)$, $u = 0, \pm 1, \ldots$, which is necessary in such applications of Equation 17.1a. However, it can be shown that for certain types of psychological time series the state–space model is too restrictive. In particular when the effect of $\boldsymbol{\eta}(t)$ on $\mathbf{y}(t)$ is delayed, the state–space model is inappropriate. For these time series one needs the general dynamic factor model in Equation 17.1a, in which $s > 0$. Such delays occur, for instance, in multilead electroencephalograph (EEG) registrations of electrocortical brain fields caused by a finite set of underlying neural sources (e.g., Molenaar, 1987; Nunez, 1981). The activity of the neural sources is represented by the latent factor series $\boldsymbol{\eta}(t)$ and the EEG registrations by the manifest series $\mathbf{y}(t)$. The effects of each neural source travel with finite speed along long-range axons to their target regions. Because target regions are located at different distances from a given neural source and the effects of a volley of action potentials of a neural source on the ongoing activity of target areas take time to dissipate, the relationships between $\boldsymbol{\eta}(t)$ and $\mathbf{y}(t)$ will show complex patterns of delays.

Analysis of stationary multivariate time series based on the dynamic factor model (17.1a, 17.1b, 17.1c) can proceed in various ways: An overview of different statistical methods to fit (17.1a, 17.1b, 17.1c), together with references to freely obtainable software, is given in Molenaar and Lo (2012). Excellent recent discussions and illustrations of dynamic factor analysis of (replicated) stationary psychological time series can be found in, for instance, Browne and Nesselroade (2005), Browne and Zhang (2007), Ferrer and Nesselroade (2003), Ferrer (2006), Hamaker, Dolan, and Molenaar (2005), Kim, Zhu, Chang, Bentler, and Ernst (2007), Mumma (2004), Nesselroade, McArdle, Aggen, and Meyers (2002), Sbarra and Ferrer (2006), Shifren, Hooker, Wood, and Nesselroade (1997), and Wood and Brown (1994). Dynamic factor modeling is increasingly prominent in econometrics (e.g., see Forni, Hallin, Lippi, & Reichlin, 2005; Stock & Watson, 2005).

Dynamic Factor Analysis of Nonstationary Processes

A new approach is presented to test for stationarity and model nonstationary processes. The new approach is based on a state–space model with arbitrarily time-varying (nonstationary) parameters. The model concerned is

$$\begin{aligned}\mathbf{y}(t) &= \mathbf{\Lambda}[\boldsymbol{\theta}(t)]\boldsymbol{\eta}(t) + \mathbf{v}(t)\\ \boldsymbol{\eta}(t+1) &= \mathbf{B}[\boldsymbol{\theta}(t)]\boldsymbol{\eta}(t) + \boldsymbol{\zeta}(t+1)\\ \boldsymbol{\theta}(t+1) &= \boldsymbol{\theta}(t) + \boldsymbol{\xi}(t+1)\end{aligned} \tag{17.2a}$$

In Equation 17.2a $\mathbf{y}(t)$ denotes the observed p-variate time series; $\boldsymbol{\eta}(t)$ the q-variate latent factor series (state process). The first equation of Equation 17.2a shows that factor loadings in $\mathbf{\Lambda}[\boldsymbol{\theta}(t)]$ depend on a time-varying parameter vector $\boldsymbol{\theta}(t)$. The second equation describes the time evolution of the state process $\boldsymbol{\eta}(t)$; the autoregressive weights in $\mathbf{B}[\boldsymbol{\theta}(t)]$ depend on $\boldsymbol{\theta}(t)$ and therefore can be arbitrarily time-varying. The third equation in Equation 17.2a describes the time-dependent variation of the unknown parameters. The r-variate parameter vector process $\boldsymbol{\theta}(t)$ obeys a random walk with Gaussian white noise innovations $\xi(t)$. The covariance functions associated with Equation 17.2a are given in Equation 17.2b:

$$\begin{aligned} \text{cov}[\mathbf{v}(t), \mathbf{v}(t-u)'] &= \delta(u)\text{diag} - \boldsymbol{\Xi} \\ \text{cov}[\zeta(t), \zeta(t-u)'] &= \delta(u)\text{diag} - \boldsymbol{\Psi} \\ \text{cov}[\xi(t), \xi(t-u)'] &= \delta(u)\text{diag} - \boldsymbol{\Phi} \end{aligned} \tag{17.2b}$$

To fit the state–space model with time-varying parameters to an observed multivariate time series, use is made of a combination of the expectation–maximization algorithm and the extended Kalman filter/smoother. First, an extended state process is defined. The extended state process consists of the original latent factor series and the time-varying parameter process: $\mathbf{x}(t)' = [\boldsymbol{\eta}(t)', \boldsymbol{\theta}(t)']$. Then, using the extended state process $\mathbf{x}(t)$, Equation 17.2a is rewritten as the following nonlinear state–space model:

$$\begin{aligned} \mathbf{y}(t) &= \mathbf{h}[\mathbf{x}(t),\text{t}] + \mathbf{v}(t) \\ \mathbf{x}(t+1) &= \mathbf{f}[\mathbf{x}(t),t] + \mathbf{w}(t) \end{aligned} \tag{17.3}$$

The vector-valued nonlinear functions $\mathbf{h}[\mathbf{x}(t),t]$ and $\mathbf{f}[\mathbf{x}(t),t]$ consist of products of the entries of $\mathbf{x}(t)$. The $(q + r)$-dimensional innovations process $\mathbf{w}(t)$ is defined as the composition of the innovation processes $\zeta(t)$ and $\xi(t)$: $\mathbf{w}(t)' = [\zeta(t)', \xi(t)']$.

The model in Equation 17.3 can be fitted by means of the raw data likelihood method using the second-order extended Kalman filter for estimation of the latent state process $\mathbf{x}(t)$ (Bar-Shalom, Li, & Kirubarajan, 2001). Because the nonlinearities in Equation 17.3 are confined to products of the entries of $\mathbf{x}(t)$, the second-order extended Kalman filter is locally exact. If an estimated variance along the diagonal of $\text{cov}[\xi(t), \xi(t-u)'] = \delta(u)\text{diag-}\boldsymbol{\Phi}$ is close to zero then the associated parameter is about constant in time. In contrast, if an estimated variance of diag-$\boldsymbol{\Phi}$ is relatively large, then the associated parameter varies substantially in time. The raw data likelihood method yields estimated trajectories of each parameter across the whole observation interval. Because the estimation algorithm has been developed only recently, a limited number of applications of Equation 17.3 has been reported until now. The first application was to father–stepson interactions and was presented in Molenaar and colleagues (2009). Zhou and colleagues (2010) reported the results of application of Equation 17.3 to the adaptive optimal control of type 1 diabetes in real time under normal living conditions. Additional applications to brain imaging data are in preparation (Molenaar, Gates, & Wilson, 2013).

Nonergodicity Due to Heterogeneity

The two criteria that a Gaussian process simultaneously has to meet in order to be ergodic are independent of each other. That is, whether or not a process is stationary is independent from the question whether or not the subjects in a population obey the same dynamics. In this section we consider innovative ways to deal with violations of the latter homogeneity criterion. But first some general remarks are in order.

The assumption that a population is homogeneous (or in mixture modeling, that each of a finite set of subpopulations is homogeneous) is standard in analysis of interindividual variation. For instance, in confirmatory factor analysis of interindividual variation it is assumed that each subject in the population obeys the same factor model in which the number of factors and the values of the factor loadings are invariant across subjects. The derivation of estimators, such as the maximum likelihood estimator, is heavily dependent upon this homogeneity assumption. One therefore would expect that deviations from this homogeneity assumption would lead to severe problems in fitting models to interindividual variation. Yet, surprisingly, this situation is not the case. If data are simulated according to a scenario in which each subject has its own factor model (subject-specific number of factors, factor loadings, measurement error variances, etc.) and this maximally heterogeneous dataset is subjected to standard factor analysis of interindividual variation, then a nicely fitting factor solution is obtained, which, in reality, does not describe any of the subject-specific models used to generate the data. An overview of the results obtained in such simulation experiments, including the use of longitudinal and quantitative genetic factor models, is given in Molenaar (2007). The conclusion is that analysis of interindividual variation is almost blind to large-scale deviations from the homogeneity assumption underlying the derivation of its estimators. This conclusion is corroborated in a formal analysis by Kelderman and Molenaar (2007). The best way to detect heterogeneity is to carry out dedicated analyses of intraindividual variation based on replicated time series data.

While the possible existence of large-scale heterogeneity (each subject in the population obeying its own model) is practically invisible in analyses of interindividual variation, it might be suggested that such heterogeneity rarely occurs in reality anyway. Strong arguments can be provided, however, that large-scale heterogeneity is the rule in human populations. These arguments are based on the importance of self-organizing epigenetic influences during ontogenesis, creating substantial structural variation in neural networks underlying human information processing. The reader is referred to Molenaar (2007) for further details and references to the relevant literature.

The Dangers of Pooling

To illustrate the confounding effects of treating heterogeneous data as homogeneous, a simple simulation experiment is carried out. Suppose there are three families, each consisting of four persons. The degree of rivalry by each person within a family during an ongoing social interaction is measured at $T = 100$ equidistant time points. The data $\mathbf{y}(t)$ for each family are generated according to the following so-called unified structural equation modeling (SEM; Gates et al., 2010), which can be represented in state–space format akin to Equations 17.2 and 17.3:

$$\mathbf{y}(t) = \boldsymbol{\eta}(t)$$
$$\boldsymbol{\eta}(t) = \mathbf{A}\boldsymbol{\eta}(t) + \boldsymbol{\Phi}\boldsymbol{\eta}(t-1) + \zeta(t) \tag{17.4}$$

The first equation of Equation 17.4 is an identity, indicating that the four-dimensional state process is observed (not latent as in Equations 17.2 and 17.3). The (4,4)-dimensional matrix **A** in the second equation describes contemporaneous relationships among the four-component series of $\boldsymbol{\eta}(t)$. These contemporaneous relationships can be understood as manifesting themselves within the span of the sampling interval. Of course, the diagonal of **A** only has zero entries. The (4,4)-dimensional matrix $\boldsymbol{\Phi}$ describes lagged relationships among the component series of $\boldsymbol{\eta}(t)$. The diagonal of $\boldsymbol{\Phi}$ contains the autoregressive relationships (regression of each component series of $\boldsymbol{\eta}(t)$ on itself one time step earlier). The four-dimensional white noise series $\zeta(t)$ has a covariance function of $\text{cov}[\zeta(t), \zeta(t-u)'] = \delta(u)\text{diag-}\boldsymbol{\Psi}$.

The simulation models for Family 1 and Family 2 are identical: $a_{13} = .8$, $a_{23} = .7$, $a_{43} = .6$, $\phi_{11} = \phi_{22} = \phi_{33} = \phi_{44} = .8$, $\phi_{41} = .6$. The simulation model for Family 3 is identical to the model for Families 1 and 2, save for the cross-lagged regression between Persons 4 and 1, which in Family 3 has the reverse direction: $\phi_{14} = .6$. All other elements of **A** and $\boldsymbol{\Phi}$ are zero. This set of simulation models can be characterized as almost homogeneous: Only the third family differs from the other two in a single parameter. The top two quadrants and the lower left quadrant in Figure 17.1 present the results of separate model fits to each of the families. Continuous arrows depict contemporaneous relationships and dashed arrows depict lagged relationships. Dashed arrows that start and end at the same person depict lagged autoregressive relationships. The thickness of each arrow is proportional to the absolute value of its regression weight. The estimation algorithm is completely data driven and automatically selects, without a priori information, the best fitting model based on likelihood criteria (see Gates et al., 2010, for all details). It appears that the true (simulation) model for each family is faithfully recovered, showing that Person 3 is a kind of organizing center as far as contemporaneous relationships are concerned. In particular, the correct direction of the cross-lagged relationship between Persons 1 and 4 is recovered in each family.

The lower right quadrant shows the model fit to the average of the covariance functions across the three families. It appears that two so-called *phantom paths* appear in the fitted model that do not pertain to the true model for any of the three families: a contemporaneous relationship $a_{42} = .22$ and a cross-lagged relationship $\phi_{42} = .31$. In addition, the cross-lagged relationships between Persons 1 and 4 are not recovered. Instead negative bidirectional contemporaneous relationships between Persons 1 and 4 appear in the model fit. This result clearly shows the dangers of averaging even almost homogeneous data.

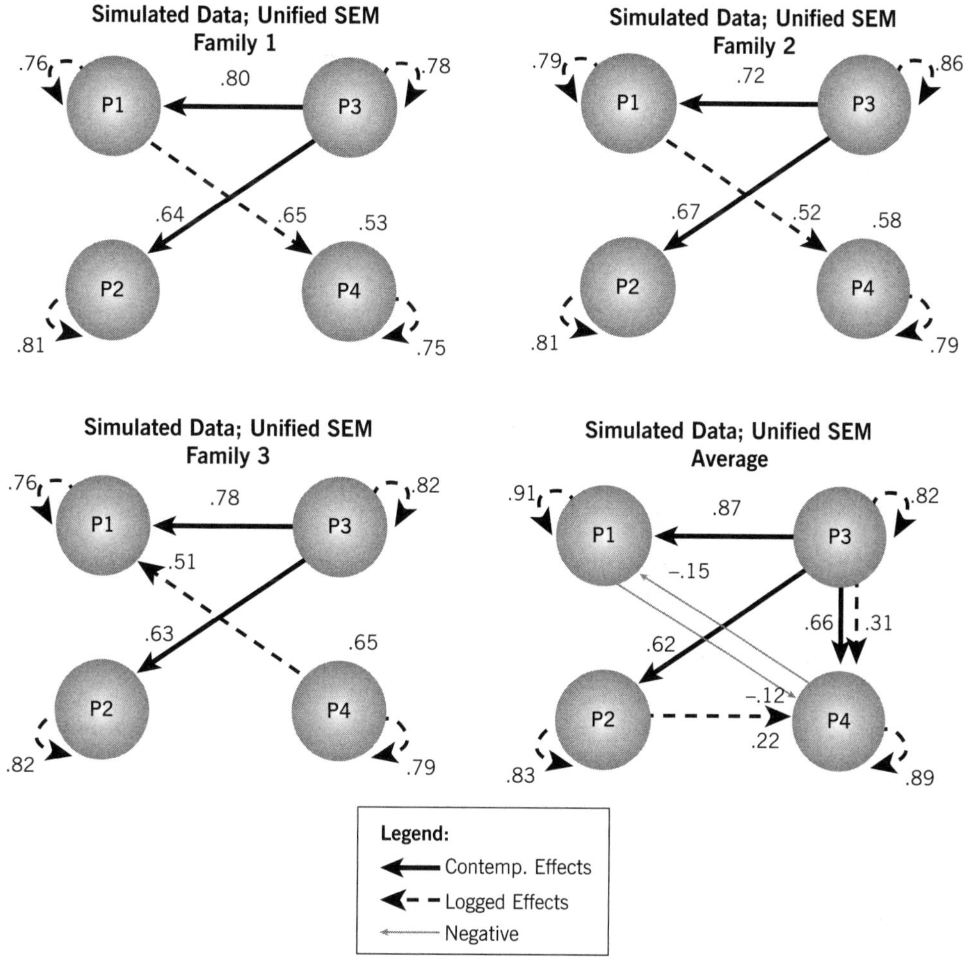

FIGURE 17.1. Fit of unified SEM to simulated four-variate time series for three families, each consisting of four persons. The two upper quadrants and the lower left quadrant are model fits to individual families. The lower right quadrant is the model fit to the average sequential covariance structure.

An Alternative Way to Deal with Heterogeneity: IF and iFACE

For about a century measurement equivalence has been operationalized in terms of invariance of factor loadings in multigroup or longitudinal factor models of interindividual variation. Nesselroade, Gerstorf, Hardy, and Ram (2007) were the first to break with this longstanding tradition and introduce an alternative operationalization of measurement equivalence in terms of intraindividual variation. The key to their alternative definition of measurement equivalence was the recognition that the way in which latent factors manifest themselves in observed variables can be subject-specific.

For instance, stress can manifest itself mainly by gastrointestinal problems in one subject and by tremor and perspiration in another subject. Another important example is conditioning, which manifests by a wide variety of behavioral changes in different organisms. Nesselroade and colleagues therefore proposed the concept of the idiographic filter (IF), which allows for subject-specific factor loadings while defining measurement equivalence by means of invariances at the latent level. The most appropriate way to specify the IF is in terms of the following set of linear state–space models for multivariate weakly stationary Gaussian time series replicated with N subjects:

$$\mathbf{y}^k(t) = \mathbf{\Lambda}^k\eta^k(t) + \varepsilon^k(t); \qquad \eta^k(t+1) = \mathbf{B}\eta^k(t) + \zeta^k(t+1);$$
$$\text{cov}[\zeta^k(t), \zeta^k(t-u)'] = \delta(u)\text{diag-}\mathbf{\Psi}; \; k = 1, 2, \ldots, N \qquad \textbf{(17.5)}$$

It is apparent from Equation 17.5 that the factor loadings $\mathbf{\Lambda}^k$ are subject-specific, whereas the covariance function of the latent state process $\eta^k(t)$ is invariant across subjects. It is shown in Molenaar and Nesselroade (2012) that the IF is testable; that is, that rotation of the state process cannot induce invariance of the covariance function of the latent state process across subjects if there is none. Hence the IF constitutes a viable approach to accommodate subject-specific relationships between latent and observed developmental processes.

The IF has been generalized to the analysis of multivariate time series obtained with a pair of genetically related subjects. This so-called *iFACE* (idiographic Filter for Additive genetic, Common environmental, and specific Environmental) factor model (Molenaar, Smit, Boomsma, & Nesselroade, 2012) allows for subject-specific heritabilities and environmental effects. Moreover, it estimates the actual additive genetic correlation among the genetically related pair of subjects. This is an important feature because there exists abundant information based on epigenetic and developmental biological studies indicating that the additive genetic correlation among monozygotic twins is not 1.0, but smaller. How much smaller may be subject-specific because of stochastic epigenetic effects (cf. Charney, 2012, for an extensive overview). The most general version of the iFACE, which incorporates sequentially dependent measurement error and higher-order vector autoregressive models for the latent state process, was shown to be identifiable by means of alphanumerical evaluation of the first derivative of the model with respect to the parameters (cf. Bekker, Merckens, & Wansbeek, 1994). A first application of the iFACE to multilead EEG obtained in an oddball task showed clear evidence for considerable subject specificity of heritable and environmental effects. Consequently, the assumption that the population is homogeneous regarding brain responses triggered by the oddball task does not appear to hold.

Another Alternative Way to Deal with Heterogeneity: GIMME

Whereas the IF can accommodate heterogeneity due to subject-specific factor loadings, the alternative approach described in this section can accommodate subject specificity of both factor loadings and any other parameters in state–space models. It is called group iterative multiple model estimation (GIMME; Gates & Molenaar, 2012) and is based on the following rationale. Suppose one has p-variate time series obtained with N possibly heterogeneous replications (subjects) and, to ease the presentation, it can be assumed that each subject obeys a model given by Equation 17.4, namely a unified SEM in which the vector autoregressive part is of order 1 (i.e., the model contains only lagged regressions one time step back). Then GIMME first determines an instance of Equation 17.4 that has a common group structure across all N subjects. That is, the common instance of Equation 17.4 has the same free parameters in $\mathbf{A}$ and $\mathbf{\Phi}$ across all N replications, although the actual values of these parameters are allowed to differ arbitrarily between subjects. In Figure 17.1 these free parameters have been depicted by a network in which the vertices are linked by directed arrows. Hence, in the first phase GIMME determines a common network structure for the complete group of N replications, while allowing that the weights associated with each link are subject specific.

GIMME determines the common group model in an automatic data-driven way in which new parameters (directed links) in $\mathbf{A}$ and $\mathbf{\Phi}$ are freed sequentially, one by one, starting from a model that in default mode contains no free parameters. It is an option to start the sequential search with a model containing already free parameters in $\mathbf{A}$ and/or $\mathbf{\Phi}$, where these parameters have been selected based on (theoretical) a priori knowledge. It also is an option to forbid the freeing of a subset of selected parameters in $\mathbf{A}$ and/or $\mathbf{\Phi}$ during the sequential search, where this subset again is determined based on a priori knowledge. At each step in the sequential model search it is determined which one of the subset of eligible parameters that have not yet been freed up will maximally increase the likelihood ratio across the N subjects. If this increase in the likelihood ratio is significant for at least a fixed proportion P of the N subjects, then this parameter (directed link) is added to the network structure, the subject-specific values of the parameter are estimated, and the sequential search moves to the next step. If not, the next phase of GIMME, described shortly, starts. Based on extensive simulation studies, the proportion P of the N subjects for which the increase in the likelihood ratio should be significant has been fixed at 75%.

In the second phase of GIMME the common group network structure determined in the first phase constitutes the starting model in a sequential model search for each individual subject separately. In this search for each subject $i \in \{1,2, \ldots, N\}$ subject-specific directed links are added one by one until the likelihood ratio is acceptable or until no additional subject-specific directed links can be found that significantly improve the likelihood ratio for this subject. A flow diagram of GIMME is given in Figure 17.2.

Although GIMME has been introduced by referring to Equation 17.4 above, the present implementation is based on a generalization of this equation, which is given by Gates and colleagues (2011):

$$\mathrm{y}(t) = \boldsymbol{\eta}(t)$$
$$\boldsymbol{\eta}(t) = \mathbf{A}\boldsymbol{\eta}(t) + \mathbf{\Phi}\boldsymbol{\eta}(t-1) + \mathbf{\Gamma}\mathrm{u}(t) + \mathbf{\Pi}\boldsymbol{\eta}(t-1)u(t-1) + \zeta(t) \tag{17.6}$$

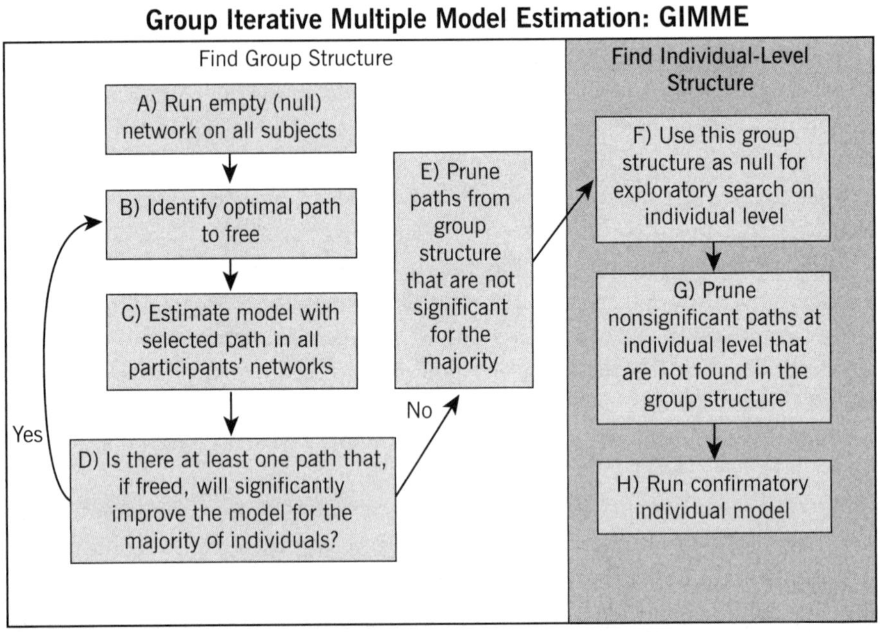

- Group-level analysis
- Individual-level analysis
- Group Iterative Multiple Model Estimation (GIMME)

FIGURE 17.2. Flow diagram of group iterative multiple model estimation (GIMME).

As in Equation 17.4 the (p,p)-dimensional matrix $\mathbf{A}$ in the second equation describes contemporaneous relationships among the p component series of $\boldsymbol{\eta}(t)$, and the (p,p)-dimensional matrix $\boldsymbol{\Phi}$ describes lagged relationships among the component series of $\boldsymbol{\eta}(t)$. The p-dimensional white noise series $\zeta(t)$ has a covariance function of $\text{cov}[\zeta(t), \zeta(t-u)'] = \text{d}(u)\text{diag-}\boldsymbol{\Psi}$. The univariate process $u(t)$ denotes an observed input process, which, in brain imaging applications, is associated with external stimulation, and in medical applications with dose of medication. $\boldsymbol{\Gamma}$ is a p-dimensional vector of regression coefficients quantifying the effects of the time-varying input on $\boldsymbol{\eta}(t)$. $\boldsymbol{\Pi}$ is a (p,p)-dimensional matrix of regression coefficients. The bilinear term $\boldsymbol{\Pi\eta}(t-1)u(t)$ can be rewritten as the product of $\boldsymbol{\Pi}u(t-1)$ and $\boldsymbol{\eta}(t-1)$, where $\boldsymbol{\Pi}u(t-1)$ can be interpreted as a subnetwork of directed lagged links, the strengths of which are modulated by the time-varying input. This interpretation is why this bilinear term is referred to in the brain imaging literature as denoting modulating effects. The inclusion of the bilinear term $\boldsymbol{\Pi\eta}(t-1)u(t-1)$ introduces a mild form of nonstationarity in Equation 17.6.

As alluded to earlier, GIMME has been validated in extensive simulation studies. These studies include application of GIMME to the large and multifaceted simulated dataset considered in Smith and colleagues (2011), who tested 38 methods to estimate connectivity networks based on multivariate functional magnetic resonance imaging (fMRI) time series. The methods included Bayesian net approaches and vector autoregressive modeling. None of these methods did well in terms of recovering the presence of directed links among component series. This surprising result, obtained with realistic simulation of blood-oxygen-level-dependent (BOLD) activities of brain regions of interest under a wide variety of conditions (varying signal-to-noise, varying dimensionality of the observed time series, varying number and strength of links, etc.), created quite a stir in the brain imaging community. When GIMME was applied to these data, all directed links and the numerical values of their weights were recovered with close to, or exactly, 100% fidelity (see Gates & Molenaar, 2012, for a complete report of these results). At present GIMME has been applied only to brain imaging data, but applications to behavioral processes are in progress. GIMME can be freely accessed at *www.personal.psu.edu/kmg311/GIMME*.

Discussion and Conclusions

The future of the inductive use of RDS is challenging because of the necessity to focus on the structure on intraindividual variation in the study of nonergodic psychological processes. This necessity follows directly from the classical ergodic theorems. In case subjects are heterogeneous, that is, in case person-specific dynamics describe the intraindividual variation of particular processes, one can only obtain valid information about such nonergodic processes by means of a replicated time series analysis. In a similar vein, in case psychological processes are nonstationary, such as developmental and learning processes, but also many clinical and biomedical processes, one also can only obtain valid information about such nonergodic processes by means of a nonstationary time series analysis.

The future of the inductive use of RDS also is exciting because a wealth of statistical dynamic modeling tools has become available, enabling the direct fit of RDS-inspired dynamic models to appropriate empirical time series data. This work opens up the possibility to focus investigation on themes that are especially emphasized by RDS, such as the context-dependent unfolding of multiple interacting influences on development as well as the self-organization driven by nonlinear epigenetic processes. As far as context-dependent unfolding of multiple interacting influences on development is concerned, these influences can be expected to occur in a hierarchy of timescales. Some are fast-acting whereas others take much more time to materialize. The state–space model with time-varying parameters in Equation 17.2 is well equipped to accommodate different timescales. Moreover, the model can be straightforwardly extended with the inclusion of external time-varying input. The observed process $\mathbf{y}(t)$ and latent state process $\boldsymbol{\eta}(t)$ in Equation 17.2 evolve at a timescale that is much faster than the parameter process $\boldsymbol{\theta}(t)$. Hence here we have a hierarchy of two levels, and the higher level is distinguished from the lower level by a slower timescale at which the dynamics evolve. This hierarchy can be directly extended with the inclusion of a third and higher levels characterized by increasingly slower timescales. Of course, empirical applications of such extensions of Equation 17.2 will require the availability of high-dimensional processes $\mathbf{y}(t)$, which have been intensively measured at high sampling rates (i.e., short sampling intervals) over long periods. Chapter 16 in this volume discusses innovative measurement techniques that will make this application possible.

As far as self-organization driven by nonlinear epigenetic processes is concerned, it should be emphasized that self-organization emerges only from nonlinear dynamics. Linear dynamics cannot undergo the bifurcations or phase transitions that cause self-organization. Chapters 5, 6, 7, 10, and 15 in this volume all attest to this important observation and present several approaches inspired by RDS to investigate self-organization. Within the context of the present chapter, we would like to emphasize that the linear Gaussian dynamical models considered above can be generalized rather straightforwardly to nonlinear equivalents in several ways. For instance, the linear lagged dependencies in GIMME can be replaced by nonlinear functional relationships. Following the arguments based on the Pearson system of distributions given by Ozaki (2012), it would seem that the Gaussian assumption often is defensible, even in nonlinear dynamic models. But nonlinearity of dynamics is essential for self-organization.

It was shown how one can deal with violations of the two jointly necessary and sufficient conditions for Gaussian ergodicity; namely, nonstationarity and subject specificity (heterogeneity). A general state–space model with arbitrarily time-varying parameters was proposed to accommodate violations of stationarity. Viable alternative ways exist to accommodate (local) nonstationarity—for instance, the regime-switching dynamic models considered in Chow and colleagues (Chapter 15, this volume). In addition, the current implementation of GIMME can handle violations of stationarity as a function of external input. The GIMME model itself can be extended in ways that increase its capacity to accommodate nonstationarity as a function of external input by increasing the bilinear component with additional lagged bilinear and higher-order (so-called *Volterra expansions*) terms and/or increasing the dimension of the input process.

Special emphasis has been given in this chapter to alternative ways to accommodate subject specificity (heterogeneity). The success of GIMME in handling heterogeneous data, as determined in large-scale simulation experiments, appears to be impressive. Yet GIMME can be extended in several ways in addition to the extensions mentioned above. In its present form GIMME identifies a single common group model in the first phase of the data-driven model search. We are currently developing an extension of GIMME in which during the first phase a finite set of common group models is identified, similar to mixture modeling.

The IF not only constitutes a viable approach to accommodate subject specificity of the way in which latent processes manifest themselves in observed time series, but also it involves an alternative definition of measurement equivalence. It is our conviction that developmental science requires such alternative definitions of measurement equivalence because the currently standard operationalization in terms of invariant factor loadings is too static and based on a conception of psychological constructs as trait-like. In addition to this standard definition, which has its obvious merits, developmental science needs definitions aimed at dynamic measurement equivalence. The latter alternative definitions should address the question as to whether it is the same dynamic process that is measured across different occasions. In the IF this dynamic measurement equivalence is operationalized in terms of the constraint that the latent state process has to be invariant across subjects (whereas the factor loadings are allowed to be subject-specific). Reasoning further along these lines, it is perhaps feasible to consider definitions of dynamic measurement equivalence in terms of sufficiently smooth time-dependent variations of all parameters in a state–space model. Whereas the standard definition of measurement equivalence is associated with linear transformations (factor rotations) described by group theory (Ahmavaara & Tourco, 1958), dynamic measurement equivalence would be associated with differential transformations described by Lie group theory (e.g., Steeb, 1996). Dynamic measurement equivalence thus conceived would capture smooth time-dependent changes in factor loadings and state process model parameters and therefore would be of potential interest for detecting invariances and symmetries in developmental science.

The implications of the classical ergodic theorems have a very broad impact, not only in psychometrics and psychology but also in the biological and medical sciences. We are presently involved in a series of pilot studies using stochastic person-specific control of day-to-day intraindividual variation in disease processes such as asthma, diabetes, and daily stress. In these projects dynamic factor analysis of nonstationary time series is used to track momentary changes in a disease process as a function of medication dose and environmental, emotional, and contingent stressors. Then, using predictive control methods based on the fitted time-varying state–space patient models, optimal medication dose is determined at each point in time in patient-specific ways. This combination of real-time process modeling and optimal control also will enable optimal guiding of developmental processes in general.

In conclusion, the future of the inductive use of RDS is challenging and exciting for the reasons mentioned in this section. The integration of advanced theoretical models, new intensive measurement techniques, and sophisticated statistical process modeling will further establish RDS as the dominant paradigm in developmental science.

REFERENCES

Ahmavaara, Y., & Tourco, M. (1958). *The unified factor model.* Helsinki: Finnish Foundation for Alcohol Studies.

Baltes, P. B. (1979). Life-span developmental psychology: Some converging observations on history and theory. In P. B. Baltes & O. G. Brim, Jr. (Eds.), *Life-span development and behavior* (Vol. 2, pp. 255–279). New York: Academic Press.

Bar-Shalom, Y., Li, X. R., & Kirubarajan, T. (2001). *Estimation with applications to tracking and navigation.* New York: Wiley.

Beiser, F. C. (2003). *The Romantic imperative: The concept of early German romanticism.* Cambridge, MA: Harvard University Press.

Bekker, P. A., Merckens, A., & Wansbeek, T. J. (1994). *Identification, equivalent models, and computer algebra.* Boston: Academic Press.

Birkhoff, G. (1931). Proof of the ergodic theorem. *Proceedings of the National Academy of Sciences USA, 17,* 656–660.

Borsboom, D. (2005). *Measuring the mind: Conceptual issues in contemporary psychometrics.* Cambridge, UK: Cambridge University Press.

Brillinger, D. R. (1975). *Time series: Data analysis and theory.* New York: Holt, Rinehart & Winston.

Browne, M. W., & Nesselroade, J. R. (2005). Representing psychological processes with dynamic factor models: Some promising uses and extensions of ARMA time series models. In A. Maydeu-Olivares & J. J. McArdle (Eds.), *Psychometrics: A Festschrift to Roderick P. McDonald* (pp. 415–452). Mahwah, NJ: Erlbaum.

Browne, M. W., & Zhang, G. (2007). Developments in the factor analysis of individual time series. In R. C. MacCallum & R. Cudeck (Eds.), *Factor analysis at 100: Historical developments and future directions* (pp. 265–291). Mahwah, NJ: Erlbaum.

Cattell, R. B. (1952). The three basic factor-analytic designs: Their interrelations and derivatives. *Psychological Bulletin, 49,* 499–520.

Charney, E. (2012). Behavior genetics and postgenomics. *Behavioral and Brain Sciences, 35,* 331–410.

Choe, G. H. (2005). *Computational ergodic theory.* Berlin: Springer-Verlag.

De Groot, A. D. (1954). Scientific personality diagnosis. *Acta Psychologica, 10,* 220–241.

Elder, G. H., Jr. (1998). The life course and human development. In W. Damon & R. M. Lerner (Eds.), *Handbook of child psychology: Vol. 1. Theoretical models of human development* (pp. 939–991). New York: Wiley.

Elder, G. H., Jr., Modell, J., & Parke, R. D. (Eds.). (1993). *Children in time and place: Developmental and historical insights.* New York: Cambridge University Press.

Emch, G., & Liu, C. (2002). *The logic of thermo-statistical physics.* Berlin: Springer.

Ferrer, E. (2006). Application of dynamic factor analysis to affective processes in dyads. In A. Ong & M. van Dulmen (Eds.), *Handbook of methods in positive psychology* (pp. 41–58). Oxford, UK: Oxford University Press.

Ferrer, E., & Nesselroade, J. R. (2003). Modeling affective processes in dyadic relations via dynamic factor analysis. *Emotion, 3,* 344–360.

Forni, M., Hallin, M., Lippi, F., & Reichlin, L. (2005). The generalized dynamic factor model: One-sided estimation and forecasting. *Journal of the American Statistical Association, 100,* 830–840.

Gates, K. M., & Molenaar, P. C. M. (2012). Group search algorithm recovers effective connectivity maps for individuals in homogeneous and heterogeneous samples. *NeuroImage, 63,* 310–319.

Gates, K. M., Molenaar, P. C. M., Hillary, F. G., Ram, N., & Rovine, M. J. (2010). Automatic search for fMRI connectivity mapping: An alternative to Granger causality testing using formal equivalences among SEM path modeling, VAR and unified SEM. *NeuroImage, 50,* 1118–1125.

Gates, K. M., Molenaar, P. C. M., Hillary, F. G., & Slobounov, S. (2011). Extended unified SEM approach for modeling event-related fMRI data. *NeuroImage, 54,* 1151–1158.

Hamaker, E. L., Dolan, C. V., & Molenaar, P. C. M. (2005). Statistical modeling of the individual: Rationale and application of multivariate stationary time series analysis. *Multivariate Behavioral Research, 40,* 207–233.

Hannan, E. J. (1970). *Multiple time series.* New York: Wiley.

Kelderman, H., & Molenaar, P. C. M. (2007). The effect of individual differences in factor loadings on the standard factor model. *Multivariate Behavioral Research, 42,* 435–456.

Kim, J., Zhu, W., Chang, L., Bentler, P. M., & Ernst, T. (2007). Unified structural equation modeling approach for the analysis of multisubject, multivariate functional MRI data. *Human Brain Mapping, 28*, 85–93.

McClelland, J. L., & Jenkins, E. (1991). Nature, nurture, and connections: Implications of connectionist models for cognitive development. In K. van Lehn (Ed.), *Architectures for intelligence: The twenty-second Carnegie Mellon symposium on cognition* (pp. 41–73). Hillsdale, NJ: Erlbaum.

Merrill, M. (1931). The relationship of individual growth to average growth. *Human Biology, 3*, 37–70.

Molenaar, P. C. M. (1985). A dynamic factor model for the analysis of multivariate time series. *Psychometrika, 50*, 181–202.

Molenaar, P. C. M. (1987). Dynamic factor analysis in the frequency domain: Causal modeling of multivariate psychophysiological time series. *Multivariate Behavioral Research, 22*, 329–353.

Molenaar, P. C. M. (2004). A manifesto on psychology as idiographic science: Bringing the person back into scientific psychology, this time forever. *Measurement: Interdisciplinary Research and Perspectives, 2*, 201–218.

Molenaar, P. C. M. (2007). On the implications of the classical ergodic theorems: Analysis of developmental processes has to focus on intra-individual variation. *Developmental Psychobiology, 50*, 60–69.

Molenaar, P. C. M., Gates, K. M., & Wilson, S. (2013). *State space modeling of time-varying contemporaneous and lagged brain network relationships.* Manuscript in preparation.

Molenaar, P. C. M., Huizenga, H. M., & Nesselroade, J. R. (2003). The relationship between the structure of interindividual and intraindividual variability: A theoretical and empirical vindication of developmental systems theory. In U. M. Staudinger & U. Lindenberger (Eds.), *Understanding human development: Dialogues with lifespan psychology* (pp. 339–360). Dordrecht, The Netherlands: Kluwer Academic.

Molenaar, P. C. M., & Lo, L. (2012). Dynamic factor analysis and control of developmental processes. In B. Laursen, T. D. Little, & N. A. Card (Eds.), *Handbook of developmental research methods* (pp. 333–349). New York: Guilford Press.

Molenaar, P. C. M., & Nesselroade, J. R. (2001). Rotation in the dynamic factor modeling of multivariate stationary time series. *Psychometrika, 66*, 99–107.

Molenaar, P. C. M., & Nesselroade, J. R. (2009). The recoverability of P-technique factor analysis. *Multivariate Behavioral Research, 44*, 130–141.

Molenaar, P. C. M., & Nesselroade, J. R. (2012). Merging the idiographic filter with dynamic factor analysis to model process. *Applied Developmental Science, 16*, 210–219.

Molenaar, P. C. M., & Newell, K. M. (2010). *Individual pathways of change: Statistical models for analyzing learning and development.* Washington, DC: American Psychological Association.

Molenaar, P. C. M., Sinclair, K. O., Rovine, M. J., Ram, N., & Corneal, S. E. (2009). Analyzing developmental proceses on an individual level using non-stationary time series modeling. *Developmental Psychology, 45*, 260–271.

Molenaar, P. C. M., Smit, D. J. A., Boomsma, D. I., & Nesselroade, J. R. (2012). Estimation of subject-specific heritabilities from intra-individual variation: iFACE. *Twin Research and Human Genetics, 15*, 393–400.

Mumma, G. H. (2004). Validation of idiosyncratic cognitive schema in cognitive case formulations: An intraindividual idiographic approach. *Psychological Assessment, 16*, 211–230.

Nesselroade, J. R., Gerstorf, D., Hardy, S. A., & Ram, N. (2007). Idiographic filters for psychological constructs. *Measurement: Interdisciplinary Research and Perspectives, 5*, 217–235.

Nesselroade, J. R., McArdle, J. J., Aggen, S. H., & Meyers, J. M. (2002). Dynamic factor analysis models for representing process in multivariate time series. In D. S. Moskowitz & S. L. Hershberger (Eds.), *Modeling intraindividual variability with repeated measures data: Methods and applications* (pp. 235–265). Mahwah, NJ: Erlbaum.

Newell, K. M., & Molenaar, P. C. M. (1998). *Applications of nonlinear dynamics to developmental process modeling.* Florence, KY: Taylor & Francis.

Nunez, P. L. (1981). *Electric fields of the brain: The neurophysics of EEG.* Oxford, UK: Oxford University Press.

Ozaki, T. (2012). *Time series modeling of neuroscientific data.* Boca Raton, FL: CRC Press.

Raijmakers, M. E. J., van Kooten, S., & Molenaar, P. C. M. (1996). On the validity of simulating stagewise development by means of PDP networks: Application of catastrophe analysis and an experimental test of rule-like network performance. *Cognitive Science, 20*, 101–136.

Sbarra, D. A., & Ferrer, E. (2006). The structure and process of emotional experience following nonmarital relationship dissolution: Dynamic factor analysis of love, anger, and sadness. *Emotion, 6*, 224–238.

Shifren, K., Hooker, K., Wood, P., & Nesselroade, J. R. (1997). Structure and variation of mood in individuals with Parkinson's disease: A dynamic factor analysis. *Psychology and Aging, 12*, 328–339.

Sidman, M. (1960). *Tactics of scientific research: Evaluating experimental data in psychology.* New York: Basic Books.

Smith, S. M., Miller, K. L., Salimi-Korshidi, G., Webster, M., Beckmann, C. F., Nichols, T. E., et al. (2011). Network modeling methods for fMRI. *NeuroImage, 54*, 875–891.

Steeb, W. H. (1996). *Continuous symmetries, Lie algebras, differential equations and computer algebra.* Singapore: World Scientific.

Stock, J. H., & Watson, M. W. (2005). *Implications of dynamic factor models for VAR analysis.* Unpublished manuscript. Princeton University, Princeton, NJ.

Wohlwill, J. F. (1971). *The study of behavioral development.* New York: Academic Press.

Wood, P., & Brown, D. (1994). The study of intra-individual differences by means of dynamic factor models: Rationale, implementation, and interpretation. *Psychological Bulletin, 116*, 166–186.

Zhou, J., Wang, Q., Molenaar, P. C. M., Ulbrecht, J., Gold, C., & Rovine, M. (2010). Receding horizon control of type 1 diabetes based on data-driven linear time-varying state-space model. In *Proceedings of the 2010 American Control Conference* (pp. 2033–2038). New York: IEEE.

PART IX

COMPLEX SYSTEMS MODELS IN HUMAN DEVELOPMENT

Reevaluation and Future Directions

CHAPTER 18

The Landscape of Inductive Developmental Systems

Phillip K. Wood

I believe it is safe to say that anyone interested enough by the title of this book to pick it up and read the words here will find much to captivate and challenge them in the chapters that follow. Each chapter is a creative development or application of inductive developmental systems (IDS), and it's quite easy to get excited by the possibilities that careful study of these chapters would repay.

Although I don't usually enjoy chapters that begin with anecdote and digression, I think, in this case, a conversation with a colleague communicates how this book makes for exciting, challenging reading but in a qualitatively different way from the usual volume of new research programs. Specifically, when I told a colleague about the chapters in this volume and how exciting I found them, she interrupted: "That's really interesting, but what kind of scientist is this book pitched to?" I told her I thought that inductive developmental systems, although a "mouthful," had broad application to behavior but maybe more especially to the study of development. She then refined her response: "This all sounds pretty deep. I just wanted to know what kind of science it's geared toward, or is this the kind of heavy-lifting cross-disciplinary stuff that only senior scientists can afford to get into? Some of it almost sounds like philosophy instead of science."

Science and/or Philosophy

That reaction gave me pause. First, it is almost as if the term *scientist* carries implicit assumptions about the distinct and disjointed nature of inquiry in which some questions can only be asked and answered by individuals within the appropriate subdiscipline. I admit I found myself somewhat agreeing with the motivation for the question, however, given that sometimes creative scholars have difficulties in academic promotion or securing funding because of difficulties in getting novel research published or in "placing" research in a single academic home. I thought that we need to be specific about what we mean by *science* and whether we lose sight of some larger picture by our use of the term *scientist.*

Recall that even Galileo, Newton, Descartes, and Leibniz did not consider themselves *scientists*, for example, because the term had not yet been coined. The term *scientist* is an Industrial Age term proposed by Whewell, who noted in 1834, half jokingly, that if "artists" produce "art" then "scientists" must produce "science." It was not an automatically popular term and was unfavored on both sides of the Atlantic for many years, with some English, mistaken in the origins of the term, describing it as "an American barbarous trisyllable" (Ellis, 1874). Even the premiere science journal *Nature* refused to use the term until 1924, when the journal's editor begrudgingly determined that authors would be permitted use of the term, but not staff (Gregory, 1924).[1]

The favored term (and that used by "scientists" before this time) was "natural philosopher," a term that admits that inquiry can also involve critical examination of one's philosophical assumptions. As such, *science* is neither opposite nor distinct from *philosophy*. As discussed below, many of the creative ideas in this volume are "good science" precisely because the questions examined are really good "natural philosophy." Overton (Chapter 2, this volume) explicitly addresses philosophical questions concerning the boundaries of systems approaches or the different types of elements and relations that constitute the posits of IDS. By contrast, new analytic approaches or applications in this volume are developed by first making some novel basic assumptions about the nature of the relationship between constructs over time or the processes of growth and change. They thereby involve both discussion of philosophy and empirical inquiry.

The contributions of this volume also operationalize or cast their analytic approach in a way that challenges the dominant "business as usual" of social science statistical analysis. Much of this formalizing involves some rather sophisticated mathematics, which advances earlier mathematical systems approaches proposed by individuals such as von Bertalanffy (1969) and Thom (1989). Mathematical treatments, however, are also not disjoint from philosophical questions or scientific inquiry: Recall that the humanist Age of Enlightenment promoted the use of mathematical models as the best way to most precisely communicate scientific principles.

Science as IDS

This is to say that even though we often treat any individual scientific question as if it were a well-defined, autonomous area of inquiry, the practice of science itself is actually an IDS that competes with earlier approaches to the study of behavior. Although such a thesis could be a chapter of its own, I use this observation to guide discussion of general themes across chapters. Although this idea is commonplace to many already interested in systems approaches, a few sentences of explanation are worthwhile to some scientists unfamiliar with ideas from the philosophy of science.

Specifically, scientific models constitute competing "webs of belief" in which the terms of explanation are internal to theory and in which some terms are so central to a given web that they are ordinarily (but incorrectly) considered unrevisable (Quine, 1951). (This view replaces the logical positivist view of science as summarizing observable elements into a theory.) Scientific models, in order to be useful, need to be falsifiable because in this way science proceeds by the successive rejection of falsified theories (Popper, 1959). Although one may defend theories by addition of ad hoc hypotheses, at some point the scientist makes a decision to abandon an old theory in favor of a new one. These changes are not randomly incremental, nor do they occur solely on the basis of scientific inquiry. As Kuhn (1996) noted, competing scientific models are more similar to the punctuated equilibria of evolution in that they appear to undergo periods of rapid adjustments and change, during which radically new paradigms are adopted—a process described by Kuhn as "scientific revolutions." Although some changes may result on the basis of social or even psychological characteristics, new technologies may also be the engine for such revolutions. This is not to say, as Lakatos (1970) noted, that such revolutions are merely relative, capricious, or "all-or-none." At any given time, multiple competing theories may exist that shield their core assertions from falsification via a protective belt of auxiliary hypotheses. The successful evolution of a new scientific theory depends on whether research based on the theory is "progressive" or "degenerating." Progressive research programs are characterized by growth and novel or more precise predictions. In contrast, degenerative programs' lack of growth and their protective belt of auxiliary hypotheses do not yield novel, confirmed predictions.

IDSs themselves constitute a novel, progressive theory of human behavior, which, in many ways, provides a qualitative alternative to standard social science research. The relatively rapid increase in developmental systems research has arisen in part due to rapid technological advances in data collection, storage, and analysis, which enable analysis of large datasets to a degree of precision and time resolution not previously feasible. Although it is not possible to design what Francis Bacon and later Newton referred to as an "*experimentum crucis*" or "crossroads experiment" in which the outcome must clearly favor either the IDS or status quo approaches, increasingly the technological advances just mentioned and the increasingly specific IDS theory and application generate testable hypotheses about behavior that challenge some of the "protective belt" of traditional social science research.

In the following discussion, I do not attempt an interlinear concordance across chapters or identification of the "best ideas" contained in each chapter. Each chapter must "do its own work" systematically and with its own priorities. Of course, the best ideas contained here depend also critically on the priorities of the reader. Some common themes, however, are repeated (with variation) across the chapters and contrast markedly with what often constitutes the dominant paradigm approach in these areas.

It's often popular to summarize one's message in an "elevator speech," a roughly 30-second summary of the most important issues you wish to communicate. I believe that the average social scientist curious about dynamic IDS would be most interested to hear the following points (elaborated in more detail later):

1. *IDS are models of qualitative differences.* IDS models explain *qualitative* differences between individuals or even within a single individual over time. By the term *qualitative*, however, something fundamentally different is meant than the polynomial (or other mathematical function) relationships used as alternatives to the linear effects often found in statistical models. IDS models explain, for example, how a child's pattern of growth over time and subsequent adult functioning may be quite different, depending on whether the necessary resources are available at certain critical time points (e.g., Chow, Witkiewitz, Grasman, Hutton, & Maisto, Chapter 15, this volume; Cunha & Heckman, Chapter 9, this volume). Qualitative differences can be identified in any system in addition to the individual, such as a family unit, a social network, or even an entire society or nation. Although IDS can be mathematically complex, they can be applied to a variety of real-world problems.

2. *Qualitative differences are often expressed as conditional relationships.* Operationally, qualitative differences are manifest as different patterns of conditional relationships between variables of interest. Often, these conditional relationships are defined as differential predictions of future behavior based on current behavior or other contemporaneous variables in the system. For example, an individual's craving for alcohol today may be a function of his or her mood, but this prediction may be different depending on what yesterday's mood or alcohol consumption was. State–space models are frequently used to assess such relationships. Such time-bound relationships to explain level of functioning are not the only possibility, however, as several have noted that (a) individual IDS may be exquisitely sensitive to initial conditions (Overton, Chapter 2, this volume; Raijmakers, Kan, Ploeger, & van der Maas, Chapter 7, this volume); (b) conditional relationships may depend on the *change* in one variable as a conditional cause for another variable (e.g., Boker, Neale, & Klump, Chapter 14, this volume); or (c) even absent data gathered over time, predictions of one component of a system may be made differential based on some other component (Ho, Chapter 5, this volume). These dynamics may be quite different across different individuals and systems.

3. *Qualitative differences as a function of critical regions.* Some IDS posit the existence of crucial junctures of time and/or variables during which enrichment, deprivation, or other causes of variation will result in dramatic and lasting subsequent behavior. These models often have their origins in (cusp) catastrophe models and are often operationalized using the analogy of a terrain, as shown in Waddington's (1957) epigenetic landscape (discussed below).

4. *Multisystem IDS.* Because IDS focus on qualitative differences that imply different patterns of behavior across individuals or systems, some may have reservations about whether IDS alternatives are parsimonious relative to traditional theories and statistical models. IDS models need not, however, be tedious serial descriptions of how each individual or system functions as an idiosyncratic "law unto itself." By positing a relatively small number of explanatory mechanisms, IDS can explain a variety of observed patterns that traditional approaches would be hard-pressed to integrate. As such, IDS represent an alternative to traditional variance decomposition approaches

in the identification of systematic interindividual differences in intraindividual change (e.g., Nesselroade, 1991; Nesselroade & Ram, 2004).

This parsimony is achieved in a way different than is customary in statistical models, however. In traditional statistical models, linear parameters are added (or redacted) to achieve a model at the appropriate level of complexity. In IDS, however, different patterns of behavior may result from a common system "type" or "template." Varying one or few parameters of the system can result in summaries across many systems. For example, as is discussed below, in some IDS, performance is explained as analogous to a journey through a contoured landscape with barriers or junctures at different points in development (as discussed in the example of Waddington's epigenetic landscape below). Measured outcomes may appear to form subgroups as a result of canalized longitudinal behavior patterns. A relatively simple landscape can, therefore, explain a wide variety of data patterns within and across many measurement occasions in a way that would involve multiple parameters under a traditional classical testing approach. Although not detailed in the chapters here, even more complicated patterns of systems could emerge by including a relatively small number of additional parameters, such as differences in the degree of tilt in a landscape (as discussed below). Other relatively simple mathematical models of self-organizing systems, such as Turing, or reaction–diffusion models, can be used to explain qualitatively different patterns of a behavior (e.g., differential patterning of seashells or skin/hide pattern variation) (e.g., Kondo & Miura, 2010; Schöner, Chapter 8, this volume; Watanabe & Kondo, 2012, as cited in Ho, Chapter 5, this volume).

Multisystem IDS need not be a simple aggregation of individual IDS or "additive" systems (Overton, Chapter 2, this volume). In many cases, higher-order systems exist in a state of interdependence or as "holistic" systems (Overton, Chapter 2, this volume). As holistic systems, what is a part or element at one level of a system may merge with other elements to form a single unit within a system at another level. For example, one IDS may explain how parents and children function relative to each other, but another system may involve explanations of how a family may be related to a larger context (e.g., a community or school system) (cf. Overton, Chapter 2, this volume; for examples, see Ram, Gerstorf, Lindenberger, & Smith, 2011). The very boundaries of the objects, interfaces, and dependencies at one level of system analysis may change as well (Overton, Chapter 2, this volume). The very "stuff" of an IDS may be the relationship itself, without being explicit about what the parts are or lacking a full understanding of the mechanism. Although this account may seem incomplete, it is helpful to remember that science sometimes begins at this relationships stage; recall that Newton's law of universal gravitation was a theory of a relationship between two bodies absent a mechanism for this relationship, and that Faraday's field theory of electromagnetism similarly describes lines of force as a fundamental reality of relationship between objects.

5. *Testing IDS is possible, but requires the appropriate data and models.* It is simply not possible to model system functioning unless the appropriate data are available for analysis via appropriate statistical or mathematical models. The main point

here is that appropriate systems models will not arise inductively by even careful analysis of (often cross-sectional) data using traditional techniques. This observation is reminiscent of Cattell's (1988) observation that multivariate theories cannot arise from the inductive aggregation of several bivariate experiments, but require the analysis of multivariate data. The chapters in the present volume illustrate how IDS can be proposed and applied to identify the elements and the form of their interrelationships within a system. Methodologically, they involve consideration of mathematical models such as state–space (Cunha & Heckman, Chapter 9, this volume; Molenaar & Nesselroade, Chapter 17, this volume), cusp (Chow et al., Chapter 15; Saunders, Chapter 6; van der Maas, Kan, Hofman, & Raijmakers, Chapter 10, all this volume) and Bayesian models (Frühwirth-Schnatter, 2004). The relationships may be models of time-bound dependencies, feedback loops, differential forecasts based on prior information, or estimation of components that may exist for some systems under study but are not present for other systems. The data required may be based on the serial analysis of longitudinally intensive data (e.g., Velicer, Babbin, & Palumbo, Chapter 16, this volume) or may involve strategically timed periods of interspersed intensive "microburst" assessments (e.g., Nesselroade, 1991).

Even if appropriate "systems-level" output are considered for analysis, use of standard statistical analyses will not produce the "correct" statistical model even over the long term. Aggregate data across multiple, qualitatively different relationships can produce a portrait of an "average" individual, which is far from typical of any actual individual. Even when the appropriate data are considered and the IDS that generated the data produce departures from the model assumptions of standard statistical techniques, common statistical practice is often to transform or adjust to the data so that the data appear to meet standard statistical model assumptions.

When expressed in such a brief form, these five points suggest only the promise of the conceptual and mathematical details outlined in the chapters in this volume. They do help, however, identify relationships and similar themes between and across chapters. One way to illustrate these points is to return to a figure that served as an analogy across many presentations: Waddington's (1957) epigenetic landscape of cell development.

The Epigenetic Landscape

Across many chapters, Waddington's epigenetic landscape figure serves, with many creative modifications, as a central analogy for IDS models. I would also like to reprise a simple form of it here because it permits useful comparisons and contrasts across the chapters. As shown in Figure 18.1, Waddington's landscape views the developmental course as analogous to a marble rolling down a terrain to a point of lowest elevation. Although usage and translation of the terrain are different across contributions to this volume, here I would like to characterize this manifold (M) as having a horizontal axis that represents a score or level of performance of interest (S), with the marble representing some unit of interest (U) (e.g., an individual, organization, cell [Ho, Chapter

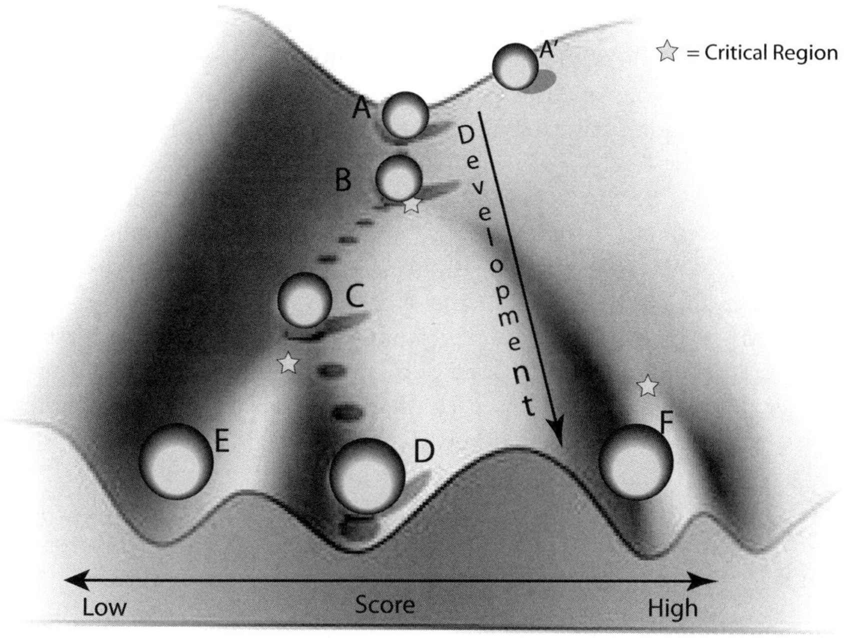

FIGURE 18.1. Epigenetic landscape for possible trajectories of development.

5, this volume], family [Urban, Osgood, Okamoto, Mabry, & Lich, Chapter 4, this volume], or dyad [Ferrer & Steele, Chapter 13, this volume]) at a given time (*T*). The front-to-back axis of the manifold is taken to be a model of development (D). This manifold is assumed to generally slope from back to the front, as shown in the figure, and it is useful to think of this axis as representing the notion of "developmental level," "organismic time" (Salthe, 1985; Slife, 1993), or measurement "occasions" that are operationalized as changes in environmental conditions (Cattell, 1988, p. 77). This choice is made in order to distinguish measured chronological time from the level of development of the individual. Although in many situations chronological and "organismic" time may be highly or even perfectly correlated, in other situations there may be reason to explicitly adjust the analysis to account for this discrepancy as might occur, for example, in the study of systematic differences in stature across children.

Performance

Scores observed across various units will then assume a probability distribution determined by the shape of the manifold at any given level of development. Variation in observed scores at any given time relative to prior score information is jointly determined by the terrain of the manifold, other variables that act on the unit to perturb the unit on the manifold over time, or even factors such as the velocity of the unit. Scores can often assume the form of random stochastic variation that causes the unit to deviate with respect to the *S*-axis but can also take the form of any external variables thought to have a short-term influence on the score. The effect of some variables may be localized to areas of development, such as Cunha and Heckman's (Chapter 9, this volume) observation that parental inputs and interventions for verbal ability have more effect at earlier levels of development than at later levels of development. These "shocks" may take the form of day-to-day stresses (Jahng, Wood, & Trull, 2008), short-term interventions designed to promote performance (Cunha & Heckman, Chapter 9, this volume; Diamond, 2010, in press), or micro- or macro-environmental effects (Ram et al., 2011). These random shocks may be simply inherent to the unit, such as constructs of affective instabiliy (Jahng et al., 2008; Trull et al., 2008), developmental assets (Geldhof et al., Chapter 3, this volume) that have a localized effect on the individual, or simply day-to-day performance variation that may be more marked for some units or individuals than others (Jahng et al., 2008).

Advantages of the Approach

The primary attraction of this epigenetic landscape metaphor is that it begins by explicitly assuming the individual as the proper unit of analysis. The individual's score is *not* assumed to be a joint function of the two implicit unmeasurables of "true score" and "error" assumed by classical test theory (which are almost always estimated as aggregate abstractions based on cross-sectional observations of other individuals). The landscape also identifies the level of development as the appropriate focus of analysis, which may be quite distinct from the observed level of functioning of the individual at any particular measurement occasion. Levels of performance of the individual at points A or B, for example, are roughly the same; however, the developmental prediction associated with interventions administered at point B are dramatically different from those at point A because of the long-term implications for the individual's likely subsequent location on the manifold. Alternatively, the score of the unit located at A′ can be reasonably expected to decrease at the next measurement occasion, but this situation does not imply that the developmental model is incorrect or that this decrease necessarily has hazardous long-term implications for the individual. Instead the expected negative change in the individual's performance is merely a localized "regression to the mean" effect or the kind of change in scores expected under classical test theory (Lord & Novick, 1968).

The epigenetic landscape for IDS is also a concrete operationalization of how patterns of increased differentiation and complexity occur as a function of degree of development. Early in development, variability in performance is often observed to be relatively homogeneous, and this would seem reasonable in the model given the expected distribution of scores at point A, for example. As development progresses, increasing variability and discontinuities in performance are observed as a function of development. These are most dramatically shown if one considers the observations likely to occur at the terrain shown at the front of Figure 18.1, in which the observed distribution of scores shows both a highly variable and multimodal form. In some contexts these distributions may appear to be associated with relatively discrete emergent "groups," "types," or levels (e.g., zero-inflated distributions [Chow et al., Chapter 15, this volume]) or clusters of co-occurring dyadic trajectories (Ferrer & Steele, Chapter 13, this volume). This is in marked contrast to simpler stochastic models consisting of a relatively simple terrain involving only a single straight "valley" during the course of development, or categorical statistical models that assume that group membership is static across developmental levels. Note that a developmental model expressed via such a landscape does not entail the assumption that all individuals will eventually attain the highest possible level of development. For example, some models, such as the reflective judgment model (Fischer & van Geert, Chapter 11, this volume) recognize that the ability to reason at higher levels of abstractions may not be attained by large numbers of individuals.

Comments on the Form of Development Suggested by the Landscape Metaphor

Although Figure 18.1 shows a fairly general case, it is interesting to note that other landscape possibilities may correspond more closely to particular developmental models. It is possible, for example, to use the landscape analogy as a way of specifying a nondynamic alternative in which behavior progresses via a single valley for all individuals. Such a model need not necessarily involve the ancillary assumptions of normality present in standard statistical models. In doing so, however, researchers can construct mathematical models to represent alternative conceptions that a reasonable skeptic might propose for the data. In this fashion, a probabilistic approach to model comparison can be used to document the extent that IDS are indeed better explanations of the phenomena of interest, and it can focus the discussion as a choice between IDS as a superior model and the need to reexamine other assumptions of the model, such as the underlying measurement theory of the model.

Common End State, Multiple Critical Points

As another alternative topography, it is possible to consider the same sawtooth-shaped valley of Figure 18.1, but to reverse the tilt of the figure. Such a model is suggestive of developmental theories in which all individuals tend toward a common developmental level. The presence of multiple critical points, as noted in Figure 18.1 by the stars, occurs when one or more ridges are added at a later point in development. The earliest point at which this ridge is present in the landscape is referred to as a *critical point* (point B in the diagram) and constitutes a time during which random variation in performance (or targeted interventions) has dramatic implications for the subsequent course of development. The presence of multiple critical regions, however, implies that interventions to improve performance at later development may have to overcome multiple hurdles. This rather pessimistic scenario for intervention at later levels is raised in some of the presentations (e.g., Cunha and Heckman's [Chapter 9, this volume] analysis of the stability of verbal ability across development and the research cited by Geldhof et al. [Chapter 3, this volume]). This is not to say that such individuals are unimprovable; deficits in one ability may be countered in other areas. Examples of this include the developmental assets mentioned by Geldhof and colleagues (Chapter 3, this volume) or the promotion of noncognitive skills identified by Cunha and Heckman (Chapter 9, this volume). Note, however, in Figure 18.1, that an individual at point E in the diagram has a larger probability of moving into the "valley" occupied by the individual at point D than the individual in point D has of joining the individual at point F because of the different slopes associated with the intervening valleys. If, as in this discussion, terrain depth represents "development" rather than chronological time, it is also possible to include the notion of a developmental regression as an additional possibility for the individual before subsequent increases in performance are seen. Optimal strategies for changing the individual make a distinction between "nondevelopmental change," in which attempts are made to scale a hill to accomplish a new level of functioning, and developmental regressions that permit the individual to revisit or revert to an earlier stage of development that would make interventions less difficult. Fans of philosophy may note an affinity between these notions of "developmental change" and "stochastic change" as being conceptually similar to Aristotle's notions of "natural" motion, in which change moves toward its end, and "violent" motion, which is temporary and requires the activity of external, nonintrinsic forces.

Interindividual Differences in a Manifold Model of Development

Although this general discussion of the epigenetic landscape is heuristic, it is hopefully specific enough to illustrate that additional models can be envisioned as mathematically tractable and that the examples discussed in the chapters in this book are not exhaustive of the possibilities. The IDS discussed in the chapters differ in the degree to which they assume that measurement assumptions and a common form of qualitative differences are generalizable across different individuals. Velicer and colleagues' (Chapter 16, this volume) survey of the implications of ergodic theory for the aggregation of IDS data is particularly relevant to this point. To use the landscape analogy again, if it can be assumed that the manifold is the same for all individuals and that all individuals are first observed at a common starting point (such as point A in Figure 18.1), then statistical analysis of multiple individuals over time can be used to determine critical junctures in development (e.g., point B in the diagram). These would be identifiable as those points after which dramatically different developmental outcomes are observed across individuals. If developmental theory contains predictions of the specific timing of these bifurcation points, experiments could be designed to specifically test for the presence of abrupt change at specific junctures. If, however, there is evidence that a common developmental model does not explain the data, some features of the landscape must be different across individuals. Researchers must then either explicitly categorize individuals into subgroups or use some statistical model to accomplish this differentiation. In some settings, obvious groupings may be evident, such as Velicer and colleagues' data on smoking or continuous positive airway pressure (CPAP) use, whereas in other contexts, statistical models may be employed, such as the zero-inflated models discussed by Chow and colleagues

(Chapter 15, this volume), or the multilevel models considered by Nisenbaum, Links, Eynan, and Heisel (2010).

Patterns of Unit Change as a Research Focus

Extensions of the landscape metaphor model can also be used to explain characteristics of a variety of other families of developmental systems. For example, it may not be the location of the individual on the manifold that is important, but the rate of change, as illustrated in Boker and colleagues' analysis of nonlinear cycling of ovarian hormones in women, or as specified in an analysis of generalized local linear derivatives (Boker, Deboeck, Edler, & Keel, 2010; Boker et al., Chapter 14, this volume). Study of such patterned emergent relationships over time as a research focus is similar to Faraday's (1839, 1844, 1855) study of "field theories" of force mentioned earlier.

Additional Modeling Possibilities

Other IDS models are possible by varying relatively simple characteristics of the manifold across individuals. For example, it may be that in any given data analysis, individuals actually start at different points on the developmental manifold, despite their similar chronological ages, or they may begin at different points due to initial benefits or liabilities external to the developmental system. Other single characteristics of the manifold could be individual differences as well, such as the general tilt of the manifold (which might affect transition probabilities between the valleys of the manifold if the figure in this chapter were tilted to the right or left, for example). Changes in the relative size of the "marble" used to represent the individual could also be used to represent the idea of differences in the "inertia" of change across individuals. Assuming that the available data and measurement model have been articulated, it is also possible to specify models that seek to identify subgroups of individuals with their own distinct topographies, or topographies that vary in some single salient fashion, such as in the steepness of their "valleys," reflecting different degrees of canalization of behavior. When considering larger IDS (e.g., multisystem or multiconstruct theories), it is also possible to envision the joint behavior of two individuals or components of a system via multiple landscapes operating with reciprocal effects.

Data Analysis Using the Manifold

Use of a developmental manifold to identify critical regions or distinct or common trajectories under this approach requires a level of intensive data collection and measurement sophistication not often found in psychological research. IDS, however, provide the researcher with a justification for the design of research that would enable more precise assessments—and which, in turn, would enable tests of the types of qualitative change that they posit. To begin, for some regions of a developmental manifold, it may not be necessary to gather as much data as at other points. For example, assessment of the individual at early points in development may be continuous in nature, as in point A in Figure 18.1. If particular points are identified by the model (e.g., as can be found in Fischer and van Geert's skill theory predictions about the chronological timing of the intellectual development in Chapter 11, this volume), a researcher can design a study in which intensive assessments are targeted to the critical junctures associated with the kind of bifurcated change predicted by the model. Alternatively, if the researcher desires to model how short-term variability and patterns of variation inform the larger pattern of change, he or she might consider the micro-burst longitudinal assessments proposed by Nesselroade (1991), in which longitudinally intensive assessments are interspersed between longer periods of time.

Many of the chapters outline the use of cusp models as the mathematical basis of the identification of developmental discontinuities. In their simplest form, cusp models define discontinuities as a pattern of at least two abrupt changes in opposite directions as a function of the conditional probability associated with a behavior relative to another variable. The simplest form of such an abrupt transition is the S-curve shown in Figure 15.2 and discussed in more detail by Chow and colleagues (Chapter 15, this volume). S-shaped curves of behavior are all around us, but it seems that, as researchers we are often prone to assume smooth functions. As any parent of an infant can tell you, for example, if a caregiver begins to walk away from an infant, the distance at which the infant will begin to cry is much farther away than the short distance necessary to get him or her to stop crying. In that example, the y-axis criterion could be the child's crying behavior, and the x-axis would be a parent's distance. The conditional probabilities modeled of interest in the example would then be the probability of crying given distance from parent and the fact that the child is already crying, and the probability of crying given distance from the parent and the fact that the child is not already crying. This type of differential patterns of variability as a function of marked conditional change is known in systems theory as *hysteresis*, and is a more localized exploration of systems behavior than the full landscape model (e.g., Gregson, 1983).

Just as the IDS paradigm focuses on measurement issues in delineating carefully modeled changes as a function of developmental level, it also forces the researcher to consider how assessment of development relies critically on the appropriate assessment conditions for development, environmental effects, and the identification of which subject populations are appropriate exemplars of development. As an example of the former case, Fischer and van Geert (Chapter 11, this volume) explain how developmental levels are in evidence only under conditions of support for optimal, as opposed to normative, performance. The average performance scores of individuals on tasks may well show only a smooth pattern of slow increases, whereas the best performance of which one is capable tracks the systems-level changes posited by skill theory. Alternatively, these support conditions for development may be global and environmental in nature; the higher-level cross-systems reasoning posited at the most advanced stages of skill theory may arise only in those environments that require this type of reasoning. Individuals required to do this type of reasoning may be found only in environments that require strategic planning (e.g., government, business, or intelligence communities) or research environments.

Finally, not all populations may demonstrate the nonlinear dependencies predicted by the epigenetic landscape. For example, given that constructs such as intelligence are genetically multidetermined, study of genetically homogeneous populations, such as laboratory rat strains, may not show the range of response (heteresis) in behaviors that more genetically complex wild strains will show (Henderson, 1975). As another example, some have also speculated that healthy heart functioning appears to follow a chaotic, systems-level form, whereas less healthy heart functioning appears more stochastic in form (Louridas & Louridas, 2012).

IDS as Generating Data and Theories That Challenge More Traditional Models

As mentioned at the beginning, technological advances permit researchers to gather multivariate data often finely resolved in time across many individuals. IDS appear especially appropriate to the analysis of such data and such analysis can be used to challenge some of the core assumptions of inference from standard statistical models.

One of the fundamental areas of challenge concerns the development and improvement of our measurement instruments of behavior. Although our measurement instruments are the tools used to generate the (hopefully noncontroversial) data to defend and develop theory, it is important to remember that our research assumptions determine the construction of our research instruments as well as how we maintain, interpret, and adjust those instruments. One expected result from IDS, for example, is that patterns of subgroupings of scores that arise in the data should be conceptually important for the modeling of the developmental landscape. The simple presence of evident subgroups in a population or even observed patterns of differential change across individuals is not, by itself, sufficient grounds for abandoning standard statistical practice because these anomalies are instead used as justifications for further statistical adjustments to the data (e.g., latent response variable reexpressions of the data [Millsap & Yun-Tien, 2004] or other psychometric scaling techniques).

Alternately, even if the existence of subgroups is considered under a nonsystems approach, such subgroups are assumed to conform to the assumptions of traditional statistical analysis. A systems approach to the analysis of longitudinal data, however, can point to reasons for believing that critical regions exist during the process of development; that systematic qualitative differences exist between individuals that cannot be arbitrarily "adjusted away" in a more sophisticated analysis. Cunha and Heckman (Chapter 9, this volume) provide a good example of such an approach, in which the metric of the later adult earned income of the child is used as less arbitrary anchor for the metric of analysis than the psychometrically scaled versions of academic aptitude traditionally used to assess educational achievement. In similar fashion, skill theory as conceptualized by Fischer and van Geert (Chapter 11, this volume) describes the periods of rapid gains in performance associated with assessments under conditions supporting "optimal performance," which constitute differential predictions by level of development. The pragmatic implication of both of these examples is that they could be used to inform educators and other policymakers about how best to target educational resources to most efficiently promote children's development.

Summary

Given that quite a bit of ground has been covered, I would like to revisit the idea presented at the beginning of this chapter, namely that IDS theories are part of a larger ecology of scientific inquiry. IDS as scientific theories of behavior are alternate to, and at times competitor "webs of belief" with, more traditional scientific models. IDS are complex systems models, which include nonlinear dynamics, evolution, game theory, and network analysis, and may well prove to constitute a new "scientific revolution," given the need for better models to analyze the multisystemic, multivariate, and often intensely longitudinal data now possible in the current era of "big data" made possible by the latest information technology. The tools of state–space modeling and identification of phase transitions and other cusp topographic models are the analytic tools needed for identifying such system-specific relatedness.

Said less mathematically, because IDS focus on qualitative in addition to quantitative change, they frequently operationalize qualitative relationships in terms of conditional relationships. When longitudinally intensive data are available, IDS can be used to describe individual developmental trajectories and invoke, as explanatory constructs, notions of critical region, initial conditions for development, and developmental canalization. More or less parsimonious IDS models are specifiable for any given situation, and the adequacy of a model can be adjudicated by model comparison or by appeal to the actuarial efficacy of a model. Extremely parsimonious systems expressions for such data can correspond, analytically, to the more standard stochastic statistical models. When data are less than adequate for specification of a systems approach that explains relationships within and across the systems under observation, strategic periods of longitudinal assessment (e.g., the "burst" designs described by Nesselroade & Ram, 2004) or reconsideration of more gross longitudinal assessments via cusp or phase transition models (e.g., Chow et al., Chapter 15, this volume) may be the experimental design of choice. In some situations, however, the role of IDS may not be to articulate a better systems approach to the analysis of a given set of data, but instead to point out that the posits of traditional statistical analysis may not be warranted and to specify the types of data that ought to be collected instead.

It seems possible that in many situations IDS models are now at the stage of sophistication where they not only offer alternative conceptions of behavior to traditional models, but can now present some reasons as to why traditional approaches to the data may be inadequate (as seen in the discussions of ergodic assumptions of many traditional models; see Velicer et al.'s presentation and evaluation of the ergodicity of behavioral assessments in Chapter 16, this volume). In some settings, it now seems possible to enunciate IDS approaches with actuarially better policy recommendations, given the predictions of differential efficacy of intervention that are possible under the approach. In some arenas, it may be possible to successfully adjudicate the superior qualities of the IDS approach. It is perhaps, though, triumphalist to view the IDS and traditional theories as locked in some form of combat, when other interactions are also possible. For example, the "progressive nature" of developments in the fields of network analysis, Bayesian inference, agent-based modeling, or "swarm behavior" models constitute only a few of the new models of behavior that may be usefully recombined with existing IDS. (See van der Maas et al., Chapter 10, this volume, and Urban et al., Chapter 4, this volume, for discussions of these approaches.)

Some Observations and Comments on the Promotion of IDS in the Ecology of Scientific Theories

I would like to end my comments here with some ideas in service of the promotion of systems researchers and systems research. During the conference that produced these chapters, I took some time to chat informally with as many individuals as possible about their research programs and what they saw as the major benefits and challenges associated with using systems approaches in those programs. I organize these comments here along with some general advice for readers who are motivated to use systems research in their own area.

The Problem of Modularity in Systems Research

As noted above, IDS, as systems approaches, often end up being multidisciplinary inquiries, even despite a researcher's initial intentions. Part of the reason for this "natural" multidisciplinary approach is related to the systems theory notion of "modularity." Briefly, any system is, itself, composed of subsystems or "modules" (e.g., the hand or endocrine system are subsystems of the body). Any system is also, of course, a subsystem or module of a larger system (e.g., a family, a community, an ecological system). To be productive, a researcher must determine the points of delineation that address the major goals of inquiry. This step is necessary to avoid, on one hand, an unproductive overload of ever-larger systems embedding (we are all, after all, embedded in our communities, our civilizations, our ecologies, and the planet), while also avoiding an infinite hall of mirrors of ever smaller subsystems of organization. The task is made more difficult given that systems or modules do not neatly nest within each other like Russian dolls; the circulatory system is co-extant with the nervous sytem within an individual, and the economic environment of some families imposes limits on the social interactions between parents and children. Our chosen systems research may not fit tidily into a particular established discipline or even with other systems approaches to clearly related questions, but such prioritizing will make it clear what services we expect our research to provide.

The modularity if IDS and the cross-disciplinary approach it often entails is, of course, a good thing: All other things being equal, the inherent correctives of a multidisciplinary approach can serve to produce a closer approximation to the phenomenon and also produce a greater potential for impact of the work across larger audiences. That said, one theme often mentioned at the conference was that researchers (especially those just starting out an academic career) should carefully consider the intellectual and often interdisciplinary investment that IDS typically involves. Some important questions include:

- "Can you afford to make the investment of time and effort to master these other schools of inquiry or the mathematical/statistical skills needed to do the analysis?"
- "Are the time and labor constraints for a systems approach to your question feasible given your professional goals regarding tenure and promotion?"

Reading of some of the more technical mathematical contributions or the wide varieties of disciplines engaged in systems work should convince the reader that there is much material to master. The timeline for mastery of this material may be long and must be balanced against your research productivity needs. Developing small, lower-cost systems research using a relatively proscribed set of constructs may be preferable to assuming the risk inherent in blocking out the large amount of time needed to become literate in the disciplines most relevant to your research questions.

Sometimes new approaches to data informatics or mathematical/statistical modeling inspire some (particularly new faculty) to embark on a long program of study to gain these latest and most exciting skill sets. Although such investments are profitable in the long term, it is important to balance this activity against the need for ongoing research productivity.

- "Will you need to collaborate with others in order to execute your research? Can your contribution to shared scholarship be clearly communicated?"

Some IDS projects require a great variety and amount of data, which often necessitates a team approach to research. The intellectual stimulation from this effort was mentioned by several conference participants as a major advantage of IDS approaches. Some of the projects described in this volume are beyond the resources of a single researcher, and it is interesting to see how several individual research programs can be generated from a larger project with multiple disciplinary "homes." That said, the logistics of collaboration often require all involved to take special care in the clear communication of responsibilities and research contributions.

- "Is your research project clearly within your department's disciplinary home?"
- "Is it possible for your research to find a home in leading journals in the field?"

Enthusiasm for your research from outside your own academic home is evidence of the utility of the research. Recall, though, that many of the promotion and tenure considerations for faculty (especially in the United States) require evidence of scholarly production in your field of inquiry as understood by your academic home department. This does not mean that faculty should forfeit their research interests in IDS, but it does mean that it is important that you be able to clearly communicate how your research fits within your academic home. This question points to the need for good departmental and institutional communication skills for new faculty in which the narrative motivation for the systems approach is articulated.

- "If you plan to seek external funding for your research, will funding agencies clearly understand how this research conforms to their funding priorities, and will those reviewing the application understand, conceptually, the merits of this approach relative to more familiar approaches?"

Some of the more intriguing conversations and anecdotes about using systems models to understand behavior occurred in discussions of the challenges and hurdles associated with securing funding for research projects. In the "feud chain" of the research review process, it is sometimes difficult for external reviewers to understand, let alone endorse, novel and at times relatively expensive research designs. This cross-disciplinary critique is not a new experience for many of us. For example, I have learned to provide additional justification for what (to me) was the obvious need for multiple measures of a construct in anticipation of epidemiologically minded reviewers who often assume a construct well measured via a single Likert-format item. It appears that successful strategies in proposing a systems approach to a problem entail carefully recruiting the interdisciplinary team and providing the agency with concrete descriptions of the separate responsibilities of each group on the project. Given the technical nature of the mathematical and statistical models involved, it may be wise to anticipate reservations in this area by including the needed expertise in the form of co-investigators, referencing successful substantive applications of the approach or (when possible) suggesting the appropriate (and scientifically disinterested) experts in the field who can review the technical merits of a proposal.

Geldhof and colleagues (Chapter 3, this volume) discuss, for example, how systems approaches can be cogently organized in the example of positive youth development. On an optimistic note, it appears again that technology is a friend to systems approaches to behavior: Urban and colleagues (Chapter 4, this volume) outline the increased attention that social network, agent-based modeling, and system dynamics approaches have received in systems-level research projects and the increased funding opportunities for such research in federal agencies.

Priorities and Issues for Systems Research

Other conference participants reflected on how best to promote systems approaches more generally. Some of these comments centered around the fact that systems research has no single "home" as a research tradition. Some individuals remarked that if they had known about the required expertise, they would have trained in another area first in order to then apply these models to their chosen field. This seemed particularly true for those most keen to apply the newer mathematical or statistical models. Others wondered about recruitment and training—specifically, how can one identify graduate students well suited for this kind of work? Alternatively, how can we best train the graduate students we have, given the demands for immediate research productivity by these graduate students, and the competing required skill sets in experimental design, interdisciplinary expertise, and the standard canon of statistical methods? Finally, others expressed the desire for more research outlets in the field of systems research, given that successful applications of systems approaches, although often published in well-respected journals, are scattered across a wide variety of disciplines. Some of this frustration was especially noted for cross-disciplinary applications because, rather than rejecting a manuscript on technical grounds, editors tended to shunt it across disciplinary outlets they felt more appropriate (alternate areas mentioned included, e.g., psychology, sociology, biology, family studies, statistics/methodology, or policy analysis). Other concerns centered on the novel analytic methods of a systems approach, with editors or reviewers expecting a thorough defense of why, actuarially, the systems approach taken performed better than more traditional methodologies. Although this requirement is a stringent one for a novel approach, it points to a need to better communicate to researchers how these techniques lead to the production of more useful theories of behavior, either empirically or in terms of future potential research.

I would like to close, as I began, with an anecdote. When I asked one of the presenters about the challenges of using systems theory, she smiled and asked "Where do I begin?" Thinking she simply had many comments, I answered "Begin anywhere you'd like" and got ready to take notes. She surprised me by saying "No, that's really the problem I have: *Where do I begin?* With a policy question? With a statistical model? With a dataset? Where do you start and which way do you go to build these systems theories we talk about?" I'm taken by the fact that the chapters in this volume present several answers to that question. Some present mathematical or statistical models, others begin with data, and others introduce policy or developmental theory as starting points. As Overton (Chapter 2, this volume) notes, the "relationship" examined under some systems models may even be the "unit of analysis" in another systems approach.

Although I share some of this frustration and occasionally wonder how we can know if we are making "scientific progress" using systems approaches, I take some comfort in knowing that this uncertainty is inherent in the scientific process. As Geldhof and colleagues (Chapter 3, this volume) note, the process of proposing and testing scientific inquiry consists of equal parts of inductive observation, speculation, theory, and deductive tests, which form part of what Cattell (1988) termed the "inductive–hypothetico-deductive spiral." Under this view of the process of science, rudimentary and even capricious observation leads to an inductive hypothesis, which, as the rudiments of a theory, then becomes tested. If this theory appears validated, it then is used to guide further observations, which yield further refinements in theory. I believe that this "system of inquiry," if you will, would serve readers well as they work through the contributions of this volume.

In line with ideas of Kuhn and Cattell, begin your journey by finding what surprises or interests you most in terms of research questions, philosophy, mathematics, or even the figures and diagrams in this book. You can then build connections to the remaining material as applications or better understanding of the foundations of the approach. Hopefully, the applications to your own work will result in your own "progressive" program of research. As a final observation, it is interesting to note that the term *hypothetico–deductive spiral* did not originate with Cattell, nor is it confined to psychological inquiry. The author of the term is the same William Whewell (1837, 1840) who also gave us the word *scientist*. Can such an approach be anything less than a "philosophically natural" way to proceed?

NOTE

1. It is interesting to note that the term Whewell thought most appropriate in form was the German word *Naturforscher*, which he translated as "nature poker" or "nature peeper" and that this term, like "natural philosopher," died a similar usage death in the years after World War II in favor of the term *Wissenschaftler*.

REFERENCES

Boker, S. M., Deboeck, P. R., Edler, C., & Keel, P. K. (2010). Generalized local linear approximation of derivatives from time series. In S.-M. Chow, E. Ferrar, & F. Hsieh (Eds.), *Statistical methods for modeling human dynamics: An interdisciplinary dialogue* (pp. 161–178). Boca Raton, FL: Taylor & Francis.

Cattell, R. B. (1988). The principles of experimental design and analysis in relation to theory building. In J. R. Nesselroade & R. B. Cattell (Eds.), *Handbook of multivariate experimental psychology* (2nd. ed., pp. 21–68). New York: Plenum Press.

Diamond, A. (2010). The evidence base for improving school outcomes by addressing the whole child and by addressing skills and attitudes, not just content. *Early Education and Development, 21*, 780–793.

Diamond, A. (in press). Why improving and assessing executive functions early in life is critical. In L. Cardle, L. Freund, & J. A. Griffin (Eds.), *Executive function in preschool age children: Integrating measurement, neurodevelopment and translational research*. Washington, DC: American Psychological Association.

Ellis, A. J. (1874). "Scientist, etc." *Academy, 124*, 321–322.

Faraday, M. (1839). *Experimental researches in electricity* (Vol. 1). London: Richard and John Edward Taylor.

Faraday, M. (1844). *Experimental researches in electricity* (Vol. 2). London: Richard and John Edward Taylor.

Faraday, M. (1855). *Experimental researches in electricity* (Vol. 3). London: Richard Taylor and William Francis.

Frühwirth-Schnatter, S. (2004). Estimating marginal likelihoods for mixture and Markov switching models using bridge sampling techniques. *Econometrics Journal, 7*, 143–167.

Gregory, R. (1924). The word "scientist" or its substitute. *Nature, 114*, 788.

Gregson, R. A. M. (1983). *Time series in psychology*. Hillsdale, NJ: Erlbaum.

Henderson, N. D. (1975). Short exposures to enriched environments can increase genetic variability of behaviour in mice. *Developmental Psychobiology, 9*, 549–553.

Jahng, S., Wood, P. K., & Trull, T. J. (2008). Analysis of affective instability in ecological momentary assessment: Indices using successive difference and group comparison via multilevel modeling. *Psychological Methods, 13*, 354–375.

Kondo, S., & Miura, T. (2010). Reaction–diffusion model as a framework for understanding biological pattern formation. *Science, 329*, 1616–1620.

Kuhn, T. S. (1996). *The structure of scientific revolutions* (3rd ed.). Chicago: University of Chicago Press.

Lakatos, I. (1970). Falsification and the methodology of scientific research programs. In I. Lakatos & A. Musgrave (Eds.), *Criticism and the growth of knowledge* (pp. 91–138). Cambridge, UK: Cambridge University Press.

Lord, F. M., & Novick, M. R. (1968). *Statistical theories of mental test scores.* Reading, MA: Addison-Wesley.

Louridas, G. E., & Louridas, A. G. (2012). Impact of chaos in the progression of heart failure. *International Journal of Applied Science and Technology, 2*, 24–30.

Millsap, R. E., & Yun-Tein, J. (2004). Assessing factorial invariance in ordered-categorical measures. *Multivariate Behavioral Research, 39*, 479–515.

Nesselroade, J. R. (1991). Interindividual differences in intraindividual changes. In L. M. Collins & J. L. Horn (Eds.), *Best methods for the analysis of change: Recent advances, unanswered questions, future directions* (pp. 92–105). Washington, DC: American Psychological Association.

Nesselroade, J. R., & Ram, N. (2004). Studying intraindividual variability: What we have learned that will help us understand lives in context? *Research in Human Development, 1*, 9–29.

Nisenbaum, R., Links, P. S., Eynan, R., & Heisel, M. J. (2010). Variability and predictors of negative mood intensity in patients with borderline personality disorder and recurrent suicidal behavior: Multilevel analyses applied to experience sampling methodology. *Journal of Abnormal Psychology, 119*, 433–439.

Popper, K. (1959). *The logic of scientific discovery.* New York: Basic Books.

Quine, W. V. (1951). Two dogmas of empiricism. *Philosophical Review, 60*, 20–43.

Ram, N., Gerstorf, D., Lindenberger, U., & Smith, J. (2011). Developmental change and intraindividual variability: Relating cognitive aging to cognitive plasticity, cardiovascular lability, and emotional diversity. *Psychology and Aging, 26*(2), 363–371.

Salthe, S. (1985). *Evolving hierarchical systems: Their structure and representation.* New York: Columbia University Press.

Slife, B. D. (1993). *Time and psychological explanation.* Albany: State University of New York Press.

Thom, R. (1989). *Structural stability and morphogenesis: An outline of a general theory of models.* Reading, MA: Addison-Wesley.

Trull, T. J., Solhan, M. B., Tragesser, S. L., Jahng, S., Wood, P. K., Piasecki, T. M., et al. (2008). Affective instability: Measuring a core feature of borderline personality disorder with ecological momentary assessment. *Journal of Abnormal Psychology, 117*, 647–661.

von Bertalanffy, L. (1969). *General system theory.* New York: George Braziller.

Waddington, C. H. (1957). *The strategy of the genes.* London: Allen & Unwin.

Watanabe, M., & Kondo, S. (2012). Changing clothes easily: Connexin41.8 regulates skin pattern variation. *Pigment Cell and Melanoma Research, 25*, 326–330.

Whewell, W. (1837). *History of the inductive sciences from the earliest to the present time.* New York: Appleton.

Whewell, W. (1840). *Philosophy of the inductive sciences founded upon their history.* London: J. W. Parker.

Author Index

C

D

G

H

I

J

K

L

M

N

O

P

Q

R

S

T

U

V

W

Y

Z

Subject Index

Page numbers followed by *f* indicate figure, *t* indicate table

D

N

Q

R

S

T

About the Editors

Peter C. M. Molenaar, PhD, is Distinguished Professor of Human Development and Psychology at The Pennsylvania State University. His research focuses on the development, testing, and application of person-centered analysis techniques; psychophysiological research; and psychological process modeling. The recipient of the 2013 Sells Award for Distinguished Multivariate Research from the Society of Multivariate Experimental Psychology, Dr. Molenaar has also received the Pauline Schmitt Russell Distinguished Research Career Award from Penn State and the Aston–Gottesman Award from the University of Virginia.

Richard M. Lerner, PhD, is the Bergstrom Chair in Applied Developmental Science and Director of the Institute for Applied Research in Youth Development at Tufts University. Dr. Lerner's research focuses on the relations between life-span human development and social change, and the relations between adolescents and their peers, families, schools, and communities. He is a Fellow of the American Association for the Advancement of Science, the American Psychological Association (APA), and the Association for Psychological Science, and is a recipient of the Urie Bronfenbrenner Award for Lifetime Contribution to Developmental Psychology in the Service of Science and Society from Division 7 of the APA.

Karl M. Newell, PhD, is the Marie Underhill Noll Chair of Human Performance and Professor of Kinesiology and Biobehavioral Health at Penn State. His research focuses on the coordination, control, and skill of normal and abnormal human movement across the life span; developmental disabilities and motor skills; and the influence of drugs and exercise on movement control. Dr. Newell was named an Alliance Scholar by the American Alliance for Health, Physical Education, Recreation, and Dance, and is a recipient of the Distinguished Scholar Award from the North American Society for the Psychology of Sport and Physical Activity.

Contributors

Steven F. Babbin, MA, Department of Psychology, University of Rhode Island, Kingston, Rhode Island

Steven M. Boker, PhD, Department of Psychology, University of Virginia, Charlottesville, Virginia

Edmond P. Bowers, PhD, Institute for Applied Research in Youth Development, Tufts University, Medford, Massachusetts

Sy-Miin Chow, PhD, Department of Human Development and Family Studies, The Pennsylvania State University, University Park, Pennsylvania

Flavio Cunha, PhD, Department of Economics, University of Pennsylvania, Philadelphia, Pennsylvania

Emilio Ferrer, PhD, Department of Psychology, University of California, Davis, Davis, California

Kurt W. Fischer, PhD, Harvard Graduate School of Education, Cambridge, Massachusetts

G. John Geldhof, PhD, School of Social and Behavioral Health Sciences, Oregon State University, Corvallis, Oregon

Raoul Grasman, PhD, Department of Psychological Methods, University of Amsterdam, Amsterdam, The Netherlands

James Heckman, PhD, Department of Economics, University of Chicago, Chicago, Illinois

Rachel Hershberg, PhD, Institute for Applied Research in Youth Development, Tufts University, Medford, Massachusetts

Lacey Hilliard, PhD, Institute for Applied Research in Youth Development, Tufts University, Medford, Massachusetts

Mae-Wan Ho, PhD, Institute of Science in Society, London, United Kingdom

Abe Hofman, MSc, Department of Psychology, University of Amsterdam, Amsterdam, The Netherlands

R. Shane Hutton, MA, Department of Psychology, University of North Carolina at Chapel Hill, Chapel Hill, North Carolina

Sara K. Johnson, PhD, Institute for Applied Research in Youth Development, Tufts University, Medford, Massachusetts

Kees Jan Kan, PhD, Department of Biological Psychology, VU University, Amsterdam, The Netherlands

Kelly L. Klump, PhD, Department of Psychology, Michigan State University, East Lansing, Michigan

Jacqueline V. Lerner, PhD, Department of Counseling, Developmental, and Educational Psychology, Boston College, Boston, Massachusetts

Richard M. Lerner, PhD, Institute for Applied Research in Youth Development, Tufts University, Medford, Massachusetts

Kristen Hassmiller Lich, PhD, Department of Health Policy and Management, University of North Carolina at Chapel Hill, Chapel Hill, North Carolina

Yeou-Teh Liu, PhD, Department of Athletic Performance, National Taiwan Normal University, Taipei, Taiwan

Patricia Mabry, PhD, Office of Behavioral and Social Sciences Research, National Institutes of Health, Bethesda, Maryland

Stephen A. Maisto, PhD, Department of Psychology, Syracuse University, Syracuse, New York

Peter C. M. Molenaar, PhD, Department of Human Development and Family Studies, The Pennsylvania State University, University Park, Pennsylvania

Michael C. Neale, PhD, Departments of Psychiatry and Human Genetics, Virginia Commonwealth University, Richmond, Virginia

John R. Nesselroade, PhD, Department of Psychology, University of Virginia, Charlottesville, Virginia

Karl M. Newell, PhD, Department of Kinesiology, The Pennsylvania State University, University Park, Pennsylvania

Janet Okamoto, PhD, Mayo Clinic, Scottsdale, Arizona

Nathaniel Osgood, PhD, Departments of Computer Science and Community Health and Epidemiology, University of Saskatchewan, Saskatoon, Saskatchewan, Canada

Willis F. Overton, PhD, Department of Psychology, Temple University, Philadelphia, Pennsylvania

Richard Palumbo, MA, Department of Psychology, University of Rhode Island, Kingston, Rhode Island

Annemie Ploeger, PhD, Department of Psychology, University of Amsterdam, Amsterdam, The Netherlands

Maartje E. J. Raijmakers, PhD, Department of Psychology, University of Amsterdam, Amsterdam, The Netherlands

Peter T. Saunders, PhD, Department of Mathematics, King's College London, and Institute of Science in Society, London, United Kingdom

Gregor Schöner, PhD, Institut für Neuroinformatik, Ruhr-Universität Bochum, Bochum, Germany

Joel Steele, PhD, Department of Psychology, Portland State University, Portland, Oregon

Jennifer Brown Urban, PhD, Department of Family and Child Studies, Montclair State University, Montclair, New Jersey

Han L. J. van der Maas, PhD, Department of Psychology, University of Amsterdam, Amsterdam, The Netherlands

Paul van Geert, PhD, Department of Developmental Psychology and Heijmans Institute, University of Groningen, Groningen, The Netherlands

Wayne F. Velicer, PhD, Cancer Prevention Research Center, University of Rhode Island, Kingston, Rhode Island

Katie Witkiewitz, PhD, Department of Psychology, University of New Mexico, Albuquerque, New Mexico

Phillip K. Wood, PhD, Department of Psychological Sciences, University of Missouri, Columbia, Missouri